MEASUREMENT CONVERSIONS

English to Metric

English Measure	Multiply by	Metric Equivalent
Length		
Inches (in)	2.54	Centimeters (cm)
Feet (ft)	0.30	Meters (m)
Yards (yd)	0.91	Meters (m)
Miles (mi)	1.61	Kilometers (km)
Statute mile	1.15	Nautical mile
Area		
Square inches (in^2)	6.45	Square centimeters (cm^2)
Acres (a)	0.41	Hectare (ha)
Square miles (mi^2)	2.59	Square kilometers (km^2)
Volume		
Cubic inches (in^3)	16.39	Cubic centimeters (cm^3)
Cubic feet (ft^3)	0.028	Cubic meters (m^3)
Cubic miles (mi^3)	4.17	Cubic kilometers (km^3)
Quarts (qt), U.S.	0.95	Liters (l)
Quarts (qt), imperial	1.14	Liters (l)
Gallons (gal), U.S.	3.8	Liters (l)
Gallons (gal), imperial	4.55	Liters (l)
Water measurements		
Acre-feet	325,872	Gallons (gal)
Mass		
Ounces (oz)	28.35	Grams (gm)
Pounds (lb)	0.45	Kilograms (kg)
Short ton (tn), U.S.	0.91	Metric ton (tonne) (t)
Velocity		
Miles/hour (mph)	0.448	Meters/second (mps)
Miles/hour (mph)	1.61	Kilometers/hour (kmph)
Miles/hour (mph)	0.86	Knots (kn) (nautical mph)
Temperature		
Degrees Fahrenheit (°F)	0.556 (after subtracting 32)	Degrees Celsius(°C)
Fahrenheit degree (F°)	0.556	Celsius degree (C°)

GEOSYSTEMS
An Introduction
to Physical Geography

GEOSYSTEMS
An Introduction
to Physical Geography

Robert W. Christopherson

Macmillan Publishing Company
New York

Maxwell Macmillan Canada
Toronto

Maxwell Macmillan International
New York   Oxford   Singapore   Sydney

*To all the students of Earth, our home planet,
and a sustainable future.*

Front cover photo and frontispiece © Galen Rowell 1990. Back
cover photo courtesy of NASA.

Editor: Paul Corey
Production Editor: Sharon Rudd
Art Coordinator: Raydelle Clement
Photo Researcher: Barbara Scott
Text Designer: Cynthia M. Brunk
Cover Designer: Cathleen Norz
Production Buyer: Pamela D. Bennett

This book was set in Garamond by The Clarinda Company and
was printed and bound by R. R. Donnelley & Sons Company.
The Cover was printed by Lehigh Press, Inc.

Macmillan Publishing Company
866 Third Avenue
New York, New York 10022

Macmillan Publishing Company is part of the
Maxwell Communication Group of Companies.

Maxwell Macmillan Canada, Inc.
1200 Eglinton Avenue East, Suite 200
Don Mills, Ontario M3C 3N1

Library of Congress Cataloging-in-Publication Data

Christopherson Robert W.
    Geosystems : an introduction to physical geography / Robert W.
    Christopherson.
        p.   cm.
    Includes bibliographical references and index.
    ISBN 0-02-322443-6
    1. Physical geography.   I. Title.
    GB54.5.C48   1992
    910'.02--dc20   91-31429
                                                    CIP

Printing: 2 3 4 5 6 7 8 9   Year: 2 3 4 5

Preface

Geosystems explores the dynamic discipline of physical geography. The goal of physical geography is to explain the spatial dimension of Earth's natural systems—its energy, air, water, weather, climates, landforms, soils, plants, and animals. Earth is a place of great physical and cultural diversity, yet people generally know little of it. Recent headlines have warned: "We Are Geographically Ignorant," or "Environmental Concerns Get Lost on the Map!" The 1990s will be a time of critical questions and developments concerning human-environment themes, for many significant changes already are under way. Geography, as a spatial human-Earth science, is in a unique position among the sciences to synthesize and integrate the great physical and cultural diversity facing us. As consumers, voters, and inhabitants of this planet, it is vital to our future that we investigate the life-sustaining environment that surrounds and infuses our lives.

To make this text both informative and enjoyable, particular attention was given to creating a clear writing style and to fully integrating the illustrations with the text. *Geosystems* is carefully organized to help you learn through a logical order and flow of topics. Coverage of essential core materials and inclusion of nonquantitative analyses of Earth's systems results in a text appropriate for both nonmajor and major science students, regardless of your science background.

Geosystems is structured into four parts, each containing related chapters according to the flow of individual systems, or consistent with time and the flow of events. Chapter 1 presents the essentials of physical geography as a foundation, including a discussion of geography, systems analysis, latitude, longitude, time, and the science of mapmaking (cartography). With these essentials learned, each of the four parts then can be covered, either in their presented order or in any sequence.

Part One exemplifies the systems organization of the text, beginning with the origin of the Solar System and the Sun. Solar energy passes across space to Earth's atmosphere (Chapter 2), and through the atmosphere to Earth's surface (Chapter 3). From Earth's surface, seasonal variations and surface energy balances (Chapter 4) generate patterns of world temperature (Chapter 5) and general and local atmospheric circulations (Chapter 6). Thus, the disciplines of astronomy, geodesy, physics, and other atmospheric sciences are dealt with in Chapters 2 through 6.

Part Two presents aspects of hydrology, meteorology and weather, oceanography, and climate, discussed in a flowing sequence in Chapters 7 through 10. Geography is linked to Earth-system sciences—an influence seen in Chapters 11 through 17 of Part Three—for there we discuss the physical planet and related processes. Earth's surface is a place of an enormous ongoing struggle between the processes that build the landscape and those that tear it down.

Finally, Part Four brings the content of the first three parts together in a section covering biogeography, including soils, plants, animals, and Earth's major terrestrial biomes (Chapters 18 through 20). The text culminates with Chapter 21, "The Human Denominator," a unique capstone chapter that overviews human-environment interactions pertinent to physical geography. This chapter is sure to stimulate further thought and discussion, dealing as it does with the Gaia hypothesis and perhaps the most profound issue of our time, Earth's stewardship.

The text contains many features to assist you as a student of physical geography:

- Each chapter includes a heading outline, review questions, key terms list, and suggested readings.
- Key chapters (climate, soils, biomes) present large, integrative tables to help you synthesize content.
- The glossary provides basic definitions of the key terms and concepts that are printed in **boldface** where they are defined within the text; entries include chapter location. Other important terms are in *italics* for emphasis.
- A student study guide is available to provide additional learning tools, examples, and self-tests.
- The text and all figures use SI—metric/English—measurement equivalencies appropriate to this transition period in the United States and for science courses in general. A complete set of measurement conversions is presented in an easy-to-use arrangement inside the front cover.
- Appendix A contains weather, water balance,

and other data for many cities worldwide, arranged by climate. Appendix B provides address listings for important geographic and environmental organizations, agencies, and general reference works.

- *Geosystems* features over 200 color photos and orbital images, 110 full color maps, 125 creative drawings, and many other illustrations throughout, all carefully selected and prepared to enhance your learning experience.
- Twenty Focus Study essays provide additional explanation and background information on key topics as diverse as the scientific method, the Mount Saint Helens eruption, the status of the Colorado River, fire ecology, and biosphere reserves.
- Although political boundaries come and go, the maps in this book are as up-to-date as possible. They reflect the recent unions in Germany and Yemen. At this writing, the sovereignty status of Western Sahara is unresolved and elections are pending. The situation is also evolving with the republics of the Soviet Union.

Acknowledgments

I begin by thanking my parents for the books and encyclopedias they provided that stimulated my interest in geography. To Mom and Dad, to my sister and brothers Lynne, Randy, and Marty, and to our children Keri, Matt, Reneé, and Steve, and the next generation—Chavon and Bryce—go my love and appreciation.

This text is a teaching text, so past mentors, among many, deserve special thanks: Dick Vermeil, Gordon Ferrin, and Duke Campangna of early years, and several master teachers of geography who made the world exciting, including Dr. Robert Layton, Dr. Peter J. Kakela, Dr. John L. Thompson, Dr. Bruce E. Bechtol, and Dr. Margaret E. Trussell. Thanks also to the authors and scientists who published the research, articles, and textbooks that I used in refining this text. And of course, my gratitude to all the students over these past 22 years at American River College for defining the importance of Earth's future, for their questions and enthusiasm—*to them this text is dedicated*.

I owe many thanks to the following individuals who consulted with Macmillan Publishing and reviewed portions of the manuscript in several developmental stages. These most-helpful individuals and their affiliations are:

Ted J. Alsop
Utah State University

Ward Barrett
University of Minnesota

David R. Butler
University of Georgia

Mario P. Delisio
Boise State University

Don W. Duckson, Jr.
Frostburg State University

Christopher H. Exline
University of Nevada—Reno

Michael M. Folsom
Eastern Washington University

David E. Greenland
University of Colorado

John Hall
Louisiana State University—Shreveport

David A. Howarth
University of Louisville

Patricia G. Humbertson
Youngstown State University

David W. Icenogle
Auburn University

Philip L. Jackson
Oregon State University

J. Peter Johnson, Jr.
Carleton University

Guy King
California State University at Chico

Peter W. Knightes
Central Texas College

Richard Kurzhals
Grand Rapids Junior College

Robert D. Larson
Southwest Texas State University

Lawrence C. Nkemdirim
The University of Calgary

John E. Oliver
Indiana State University

Robin J. Rapai
University of North Dakota

William C. Rense
Shippensburg University

Wolf Roder
University of Cincinnati

Daniel A. Selwa
U.S.C. Coastal Carolina College

Armando M. da Silva
Towson State University

Daniel J. Smith
University of Saskatchewan

Stephen Stadler
Oklahoma State University

Susanna T.Y. Tong
University of Cincinnati

David Weide
University of Nevada—Las Vegas

Brenton M. Yarnal
Pennsylvania State University

I thank the staff of Macmillan Publishing for their support throughout the publishing process. Their skill and attention to detail is in evidence throughout this text, as is their dedication to the craft of modern textbook publishing. My appreciation to science editor Paul Corey, Sharon Rudd, Cindy Brunk, Raydelle Clement, and copy editor Fred Schroyer; and to Barbara Duffy who made the initial encouraging contact.

Finally, to my special partner and collaborator, Bobbé Christopherson, I assign extraordinary gratitude for her tireless efforts in the production of this text: researching photographs, preparing tables, obtaining permissions, and reading through many drafts of the manuscript, for I know my word processing and writing benefited. It is her unique sensitivity to Earth and love for the smallest of living plants and animals that provides such an inspiration.

Physical geography teaches us the intricate supporting web that is Earth's environment and the dramatic changes occurring in many human-Earth relationships. All things considered, this is a good time to be enrolled in a relevant geography course! The best to you in your studies.

Robert W. Christopherson
Citrus Heights, California

Brief Contents

Contents

1
Essentials of Geography 1

PART ONE

The Energy-Atmosphere System 32

2

Solar Energy to Earth 34

3
Earth's Modern Atmosphere 55

4
The Seasons and Energy Balances 83

5

World Temperatures 111

6

Atmospheric and Oceanic Circulation 137

PART TWO

The Water, Weather, and Climate System 164

7

Water and Atmospheric Moisture 167

PART THREE

The Earth-Atmosphere Interface 290

11

The Dynamic Planet 293

12

Tectonics and Volcanism 331

15

Eolian Processes and Arid Landscapes 425

16

Coastal Processes and Landforms 451

17
Glacial Processes and Landforms 479

PART FOUR
Soils, Ecosystems, and Biomes 510

18
The Geography of Soils 513

19
Ecosystem Essentials 547

20
Terrestrial Biomes 577

21
The Human Denominator 607

GEOSYSTEMS

An Introduction
to Physical Geography

Courtyard of the Old Royal Observatory at
Greenwich. Photo from the National Maritime
Museum, Greenwich, England.

1

Essentials of Geography

Our study of geosystems begins with the essentials of physical geography. Physical geography is an important discipline for understanding Earth's physical systems and our increasingly complex relationship to them and to the environmental challenges we face. You need a knowledge of physical geography to answer the where, the why, and the how questions about the environment and its related physical processes. Thus, we begin with a look at the discipline of physical geography itself and its sometimes elusive definition.

Because Earth operates as a vast interplay of systems, physical geographers employ systems anal-

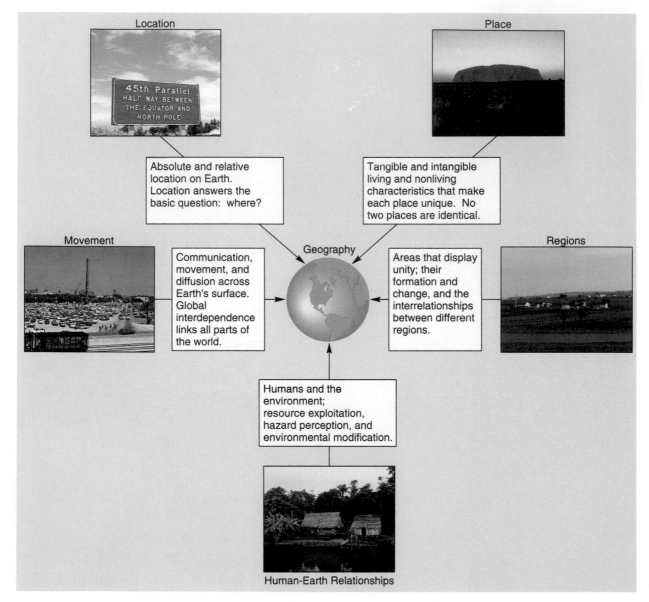

FIGURE 1–1
Fundamental themes in geography. Human-Earth relationships photo by Gael Summer-Hebdon; Place photo by Michael Dunning/TSW; other photos by author.

ysis to study the environment. Thus, we introduce system methodologies and Earth's interrelated systems. We then consider location, a key theme of geographic inquiry—the latitude, longitude, and time coordinates that delimit Earth's surface. Maps are critical tools employed to portray physical and cultural data and analyses, and they are used throughout this text.

THE DISCIPLINE OF GEOGRAPHY

Human beings are spatial actors, both affecting and being affected by Earth. The term *spatial* refers to the nature and character of physical space. Geography is the science that studies the interdependence among geographic areas, natural systems, processes, society, and cultural activities over space. We influence vast areas because of our technology, mobility, and numbers. For example, think of your own route to the classroom or library today, your knowledge of street patterns, traffic trouble spots, parking spaces, or bike rack locations used to minimize walking distances.

As a synthesizing discipline, geography integrates spatial elements to form a coherent whole. Geography is governed by a method—a spatial approach— rather than by a specific content or body of knowledge, like chemistry, biology, anthropology, or English. Various words denote the geographic context of **spatial analysis**: space, territory, zone, pattern, distribution, place, location, region, sphere, province, and distance. Because few physical phenomena are purely spatial, **process** is central to geographic

synthesis—that is, analyzing a set of actions or mechanisms that operate in some special order. Spatial analysis, then, is a process-oriented integrative approach central to geographic inquiry. Physical geography, therefore, centers on the spatial analysis of all the physical elements and processes that comprise the environment: energy, air, water, weather, climate, landforms, soils, animals, plants, and Earth itself.

The Association of American Geographers (AAG) and the National Council for Geographic Education (NCGE) have categorized the discipline into five key themes for modern geographic education: **location**, **place**, **human-Earth relationships**, **movement**, and **region**. Each of these subdivisions of geographic thought is portrayed in Figure 1–1. From a physical geography perspective, this text will draw on traditions found within each theme.

The Geographic Continuum

To further illustrate the discipline of modern geography, and physical geography in particular, Figure 1–2 shows a continuous distribution—a continuum—along which the content of geography is arranged. Disciplines in the physical and life sciences are listed on the continuum's left; those in the cultural/human sciences are listed on the right. Specific subdisciplines within geography that draw from these many academic disciplines are listed within the appropriate spheres at either end of the continuum. Geography by its very nature is eclectic, integrating a wide range of subject matter.

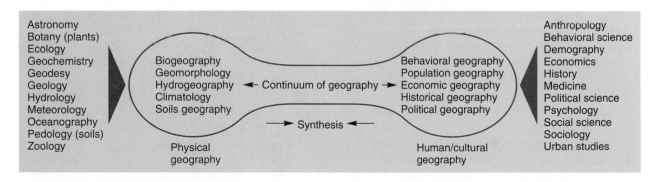

FIGURE 1–2
Distribution of geographic content along a continuum. Synthesis of Earth topics and human topics is suggested by movement toward the middle of the continuum.

FIGURE 1–3

An open system and a closed system. In either type of system, inputs are transformed and stored as the system operates.

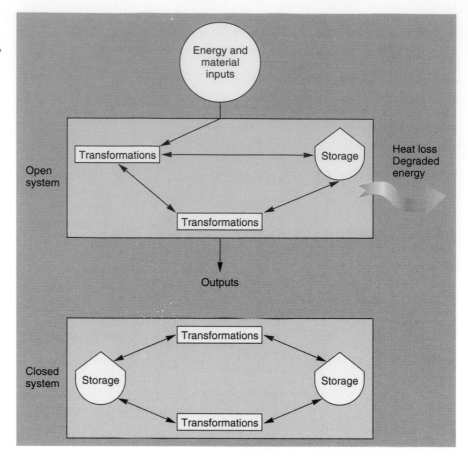

This continuum reflects a basic duality within geography, and this duality is paralleled in society by the tendency of those in developed countries to separate themselves from their supportive life-sustaining environment and to think of themselves as artificially removed from Earth. In contrast, however, many people in Third World countries live closer to nature and are acutely aware of its importance in their daily lives. Regardless of our philosophies toward Earth, we depend on functioning Earth systems to provide oxygen, water, nutrients, and other resources. Thus, we all possess within us the physical identity of this planet and are in a natural sense its representatives.

Our modern world requires that we move our study of geographic processes, and perhaps our philosophical vantage point, more toward the center of the continuum to attain a more holistic, or complete, approach. Modern environmental problems call for academic disciplines to examine rela-

tionships among Earth, society, and cultures throughout space and time. Examples of critical environmental concerns that are integral to physical geography are:

1. Global ozone depletion in the upper atmosphere.
2. Possible human-induced climatic warming through increases in carbon dioxide in the atmosphere.
3. Worsening air pollution hazards, particularly in metropolitan areas.
4. Identification of natural hazards such as earthquakes and floods that pose risk to society.
5. The deliberate destruction of Earth's equatorial and tropical rain forests.
6. The increasing losses of plant and animal diversity.

Furthermore, geography is in a unique position to synthesize the spatial environmental and human aspects of catastrophic events. Examples are the 1986 Chernobyl nuclear power station accident in the Ukraine, or the 1989 crash of the *Exxon Valdez* oil tanker in Prince William Sound, Alaska, or the 1980 eruption of Mount Saint Helens in Washington State, or the multitude of environmental impacts and disruptions from the 1991 Persian Gulf War.

Certainly, the spatial patterns of change we face today are unique, for we are taxing Earth's systems in new ways. Some past civilizations successfully adapted to crises, whereas others failed. Perhaps this ability to adapt is the key. If so, geographic analysis and understanding of our relationship to Earth's physical geography is of great importance to human survival, for the innumerable physical processes in the environment operate as interacting life-support systems.

EARTH SYSTEMS CONCEPTS

You no doubt are familiar with the word *system:* "check the cooling system"; "how does the grading system work"; "there is a weather system approaching." Emanating from engineering disciplines during World War II, systems analysis has moved to the forefront as a method for understanding operational behavior in many disciplines. Geographers use systems methodology as an analytical tool at varying levels of complexity.

Systems Theory

Simply stated, a **system** is any ordered, interrelated set of objects, things, components, parts, and their attributes, as distinct from their surrounding environment. The elements within a system can be arranged in a series or interwoven with one another. A system can comprise any number of subsystems. In Earth's systems, transformation of energy and storage/retrieval of energy and matter occur within a system. They also form links between subsystems.

A natural system generally is not self-contained: inputs of energy and matter flow into the system, and outputs flow from the system. Such a system is called an **open system**. Earth is an open system *in terms of energy,* as are most natural systems. A system that is shut off from the surrounding environment so that it is self-contained is a **closed system**. Earth itself is essentially a closed system *in terms of physical matter and resources.* Figure 1–3 illustrates simple open and closed systems.

Figure 1–4 illustrates a simple open-flow system, using plant photosynthesis as an example. Photosyn-

FIGURE 1–4
Simple flow system of a plant leaf. Plants are capable of taking sunlight, carbon dioxide (CO_2), water (H_2O), and nutrients as system inputs to produce outputs of oxygen (O_2) and carbohydrates (sugars) through the photosynthesis process.

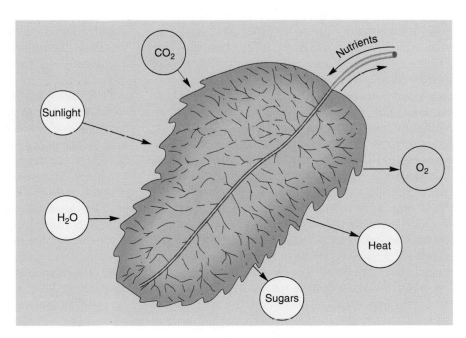

FIGURE 1-5
Steady-state equilibrium in two
systems.

thesis is the growth process by which plants convert solar energy, water, nutrients, and carbon dioxide into stored chemical energy (carbohydrates), releasing oxygen as a by-product. Plants derive energy for their operations from a reverse of photosynthesis, a process called respiration. In respiration, the plant consumes chemical energy (carbohydrates) and releases carbon dioxide, water, and heat into the environment. Thus, a plant acts as an open system.

As a system operates, information that can influence continuing system operations is generated in the system output. This information is returned to various points in the system via pathways called **feedback loops**. Feedback can cause changes that in turn guide further system operations. If the information amplifies or encourages response in the system, it is called **positive feedback**. If the information slows or discourages response in the system, it is called

negative feedback. Such negative feedback provides the basis for self-regulation in a natural system, maintaining the system within its tolerable performance range.

In Earth's systems, negative feedback is far more common than positive feedback. Unchecked positive feedback in a system can create a runaway condition until a critical limit is reached, leading to instability, disruption, or death. In the leaf-flow system (Figure 1-4), any increase or decrease in daylength (photoperiod), carbon dioxide, or water availability will produce feedback information that affects specific responses in plant operations.

Most systems maintain their general structure and character over time. An energy or material system that remains balanced over a given time frame is at **equilibrium**. When the rates of inputs and outputs in the system are equal and the amounts of energy and matter in storage within the system are constant (or more realistically, as they fluctuate around a stable average), the system is in **steady-state equilibrium**. A system fluctuating around an average value may demonstrate a trend over time and represent a changing equilibrium state through time. Such is the status of long-term climatic changes and the present global warming. Figure 1–5 illustrates these two states of system equilibria.

A **model** is a simplified version of a system and represents an idealized part of the real world. Models are designed with varying degrees of abstraction. Physical geographers construct simple system models to demonstrate complex associations occurring in the environment. The inherent simplicity of models makes the nature of systems easier to comprehend and allows predictions based on them to be made, but such predictions are only as good as the assumptions built into the model. Imposing a model too rigidly on a natural setting can lead to misinterpretation, so it is best to view a model for what it is—a simplification. We discuss many system models in this text, including the hydrologic cycle, water balance, surface energy budgets, Earth's internal structure, glacier mass budgets, soil profiles, and various ecosystems.

Earth as a System

Because it receives energy from an outside source—the Sun—Earth operates as an open system. In terms of matter, however, Earth is a closed system.

Earth's Energy Equilibrium: An Open System.
Most Earth systems are dynamic because of the tremendous infusion of radiant energy from thermonuclear reactions deep within the Sun. This energy penetrates Earth's atmosphere and cascades through the terrestrial systems, transforming along the way into various other forms of energy. Eventually, Earth radiates this energy back to the cold vacuum of space in an amount essentially equal to that which entered the system.

Researchers are examining the dynamics of Earth's energy equilibrium in more detail than ever before to distinguish natural changes from those produced by human intervention. Tremendous breakthroughs in understanding and prediction should occur as general circulation models of Earth's energy-atmosphere-water system become even more accurate. By the mid-to-late 1990s, at least four polar-orbiting satellites are planned—part of an Earth Observation System, or EOS—to introduce a new era in the monitoring of Earth's open energy system.

Earth's Physical Matter: A Closed System.
In terms of physical matter—air, water, and material resources—Earth is nearly a closed system. Since the initial formation of the planet, no significant new quantities of matter have entered the system. Just as importantly, no significant quantities have left the system, either. This is it! Thus, Earth's physical materials are finite; no matter how numerous and daring the technological reorganizations of matter become, the physical base is fixed.

The sooner this fact is understood by the population at large, the sooner a serious effort at recycling and efficiency within the system can be initiated. Society simply loses track of its resources in what is essentially a once-through resource stream from virgin materials to the proverbial "urban ore mine" landfill (Figure 1–6). Recycling costs for some of these materials may be prohibitive, but glass, some plastics, newspaper, aluminum, and some other metals—which constitute about half the content of an average landfill—can be extracted economically. The fact that Earth is a closed material system makes such considerations inevitable.

Earth's Four Spheres

Earth's surface is where four immense open systems interface or interact. Figure 1–7 shows three **abiotic** (nonliving) systems overlapping to form the realm of the **biotic** (living) system. The abiotic spheres are the **atmosphere**, **hydrosphere**, and **lithosphere**. The biotic sphere is called the **biosphere**. Because these four models are not independent units in nature, their boundaries must be understood as transition zones rather than sharp delimitations.

Atmosphere. The atmosphere is a thin, gaseous veil surrounding Earth, held to the planet by the force of gravity. Formed by gases arising from within Earth's crust and interior, and the exhalations of all

FIGURE 1–6
Typical urban landfill. Photo by author.

life over time, the lower atmosphere is unique in the Solar System. It is a combination of nitrogen, oxygen, argon, carbon dioxide, and small amounts of trace gases.

Hydrosphere. Earth's waters exist in the atmosphere, on the surface, and in the crust near the surface, in liquid, solid, and gaseous forms. Water occurs in two forms, fresh and saline (salty), and exhibits important heat properties as well as its extraordinary role as a solvent. Among the planets in the Solar System, only Earth possesses water in any quantity.

Lithosphere. Earth's crust and a portion of the upper mantle directly below the crust form the lithosphere. The crust is quite brittle compared to the layers beneath it, which are in motion in response to an uneven distribution of heat and pressure. In a broader sense, the lithosphere is sometimes thought of as the entire solid planet.

Biosphere. The intricate web that connects all organisms with their physical environment is the biosphere. Sometimes called the **ecosphere**, the biosphere is the area in which physical and chemical factors form the context of life. The biosphere exists in the overlap among the abiotic spheres, extending from the seafloor to about 8 km (5 mi) into the atmosphere, the limits between which life is sustainable. In turn, life processes also have powerfully shaped the other spheres through various interactive processes. The biosphere has evolved, reorganized itself at times, faced extinction, gained new vitality, and managed to flourish overall. Earth's biosphere is the only one known in the Solar System; thus, life as we know it is unique to Earth.

A SPHERICAL PLANET

Information about Earth's shape and size is basic to physical geography. The roundness or sphericity of Earth is not as modern a concept as many believe. For instance, the Greek mathematician and philosopher Pythagoras (ca. 580–500 B.C.) determined through observation that Earth is round. We don't actually know what Pythagoras observed to conclude that Earth is a sphere, but we can speculate. He might have noticed ships sailing beyond the horizon and apparently sinking below the surface, only to arrive back at port with dry decks. He probably noticed Earth's curved shadow cast on the lunar surface during an eclipse of the Moon. He might have deduced that the Sun and Moon are not just disks but are spherical, and that Earth must be a sphere as well. Earth's sphericity was generally accepted by the educated populace as early as the first century A.D.

FIGURE 1–7
Earth's four spheres.

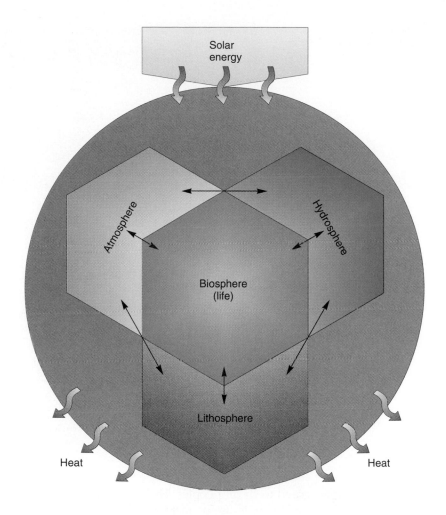

In 1687 Sir Isaac Newton postulated that the round Earth, along with the other planets, could not be perfectly spherical. Until that time the spherical-perfection model was a basic assumption of **geodesy**, the science that attempts to determine Earth's shape and size by surveys and mathematical calculations. But Newton reasoned that the more rapid rotational speed at the equator—the equator being farthest from the central axis of the planet and therefore moving faster—would produce an equatorial bulge in response to a greater centrifugal force pulling Earth's surface outward. He was convinced that Earth is slightly misshapen into what he termed an oblate (flattened) spheroid, with that oblateness occurring at the poles. Earth's equatorial bulge and its oblateness are universally accepted and readily confirmed by modern satellites. The modern era of Earth

measurement is one of tremendous precision and is called the "geoidal epoch" because Earth is considered a **geoid**, which means literally that "the shape of Earth is Earth-shaped."

Figure 1–8 gives measurements for Earth's circumference and its polar and equatorial diameters. Earth's circumference was first measured over 2200 years ago by the Greek geographer, astronomer, and librarian Eratosthenes. The ingenious reasoning by which he arrived at his calculation is presented in Focus Study 1–1.

LOCATION AND TIME ON EARTH

To know specifically where something is located on Earth's surface, a coordinated grid system is needed,

FIGURE 1–8
Earth's diameter and circumference.

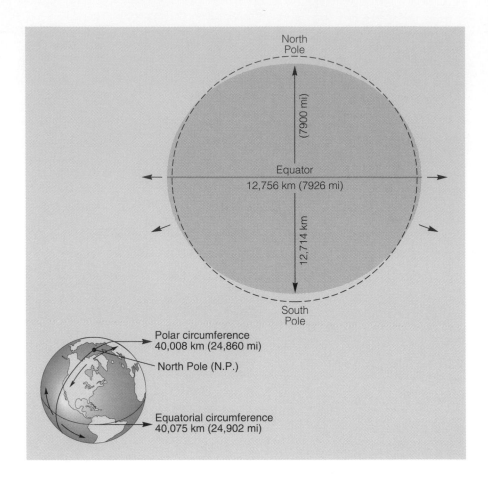

North Pole

(7900 mi)

Equator
12,756 km (7926 mi)

12,714 km

South Pole

Polar circumference
40,008 km (24,860 mi)

North Pole (N.P.)

Equatorial circumference
40,075 km (24,902 mi)

preferably one that is agreed to by all peoples. The terms latitude and longitude were in use on maps as early as the first century A.D., with the concepts themselves dating back to Eratosthenes (ca. 276–195 B.C.) and others.

Ptolemy (ca. A.D. 90–168) contributed greatly to modern maps, and many of his terms and configurations still are used today. On his map of the known world (Figure 1–9), Ptolemy described 8000 places according to their north-south and east-west coordinates. He also placed north at the top and east at the right, orienting maps in a manner familiar to us.

Ptolemy divided the circle into 360 degrees (360°), with each degree comprising 60 minutes (60'), and each minute including 60 seconds (60"), in a manner adapted from the Babylonians. He located places using these degrees, minutes, and seconds, although the precise linear value of a degree of latitude and a degree of longitude on Earth's surface remained unresolved for the next 17

centuries. We'll now examine each of these grid coordinate elements.

Latitude

On a map or globe, lines denoting angles of latitude run east and west, parallel to Earth's equator. **Latitude** is the angular distance north or south of the equator, measured from the center of Earth—Figure 1–10(a). Because Earth's equator divides the distance between the North Pole and the South Pole exactly in half, it is assigned the value of 0° latitude. The North Pole is 90° north latitude, and the South Pole is 90° south latitude.

A line connecting all points along the same latitudinal angle is called a **parallel**. Thus, in Figure 1–10(b), latitude is the name of the angle (49° north latitude), parallel names the line (49th parallel), and both indicate distance north or south of the equator. In the figure, an angle of 49° north latitude is

FOCUS STUDY 1–1

Measuring Earth in 247 B.C.

E ratosthenes, foremost among early geographers, served as the librarian of Alexandria in Egypt. His was a position of scientific leadership, for Alexandria's library was the greatest in the ancient world. Among his achievements was calculation of Earth's circumference to a high level of accuracy, quite a feat for 247 B.C.

Travelers told Eratosthenes that on June 21 they had seen the Sun's rays shine directly to the bottom of a well at Syene, the location of present-day Aswan, Egypt (Figure 1). Eratosthenes knew from his own observations that the Sun's rays never were directly overhead at Alexandria, even at noon on June 21, the longest day of the year and the day on which the Sun is at its most northward position in the sky. Instead, objects always cast a daytime shadow in Alexandria. Using the considerable geometric knowledge of the era, Eratosthenes conducted an experiment.

In Alexandria at noon on June 21, Eratosthenes carefully measured the angle of a shadow cast by an obelisk, a perpendicular column used for telling time by the Sun. Knowing the height of the obelisk and measuring the length of the shadow from its base, he solved the triangle for the angle of the Sun's rays, which he determined to be 7.2°. However, at Syene on the same day, the angle of the Sun's rays was 0° from a perpendicular—that is, the Sun was directly overhead. Geometric principles told Eratosthenes that the distance on the ground between Alexandria and Syene formed an arc of Earth's circumference equal to the angle of the Sun's rays at Alexandria. Since 7.2° is roughly 1/50 of the 360° in Earth's total circumference, the distance between Alexandria and Syene must represent approximately 1/50 of Earth's total circumference. So, he determined the surface distance between the two locations and multiplied that amount by 50. Eratosthenes's calculations convert to roughly 46,250 km (28,738 mi), which is remarkably close to the correct answer of 40,075 km (24,902 mi) for Earth's equatorial circumference.

Eratosthenes's work teaches the value of observing carefully and integrating all observations with previous learning. Measuring Earth's circumference required application of his knowledge of Earth-Sun relationships, geometry, and geography to his keen observations to prove his suspicions about Earth's size. *(Focus Study 1–1 continues on page 12.)*

measured and, by connecting all points at this latitude, the 49th parallel is designated. The 49th parallel is a significant one in the Western Hemisphere, for it is the location of the western and central portions of the Canadian–United States border from Minnesota to the Pacific.

Latitude is readily determined by using celestial bodies as points of reference, a method dating to ancient times. Because Polaris, the North Star, is almost directly overhead at the North Pole, persons anywhere in the Northern Hemisphere can determine their actual latitude simply by sighting Polaris

FIGURE 1
Eratosthenes's determinations.

FIGURE 1–9

Ptolemy's map of the world shown in an A.D. 1522 Renaissance version. Ptolemy published an early version of this map in A.D. 140. Courtesy of the Geography and Map Division of the Library of Congress.

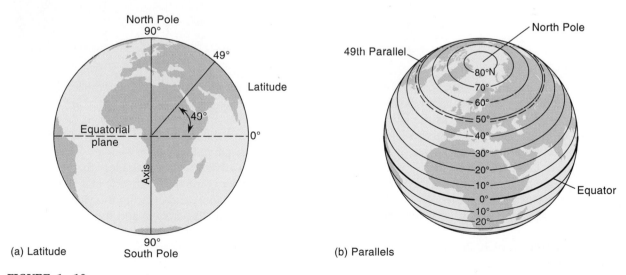

(a) Latitude

(b) Parallels

FIGURE 1–10

Angles of latitude measured north or south of the equator (a) determine parallels (b).

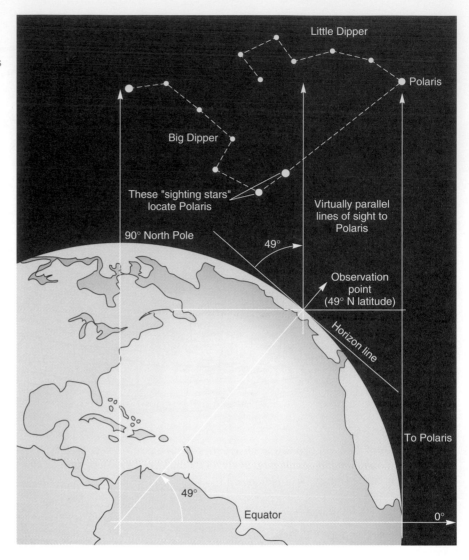

FIGURE 1–11
Determining latitude by using Polaris (the North Star). Polaris is so far from Earth that all lines of sight appear parallel when viewed from different points in the Northern Hemisphere.

and measuring its angular distance above the local horizon (Figure 1–11). The angle of elevation of Polaris above the horizon equals the latitude of the observation point. In the Southern Hemisphere, such measurements are accomplished by sighting on the Southern Cross (Crux Australis) constellation. During daylight hours the angle of the Sun above the horizon likewise indicates the observer's latitude, when season and time of day are taken into account.

Latitudinal Geographic Zones. Figure 1–12 portrays general latitudinal geographic zones, their locations, and their names: **equatorial, tropical, subtropical, midlatitude, subarctic** or **subantarctic,** and **arctic** or **antarctic.** These generalized latitudinal zones are convenient concepts, but are not rigid delineations. The Tropic of Cancer (23.5° north parallel) and the Tropic of Capricorn (23.5° south parallel), discussed further in Chapter 4, are the most extreme north and south parallels that experience perpendicular (directly overhead) rays of the Sun at local noon at some time during the year. The Arctic Circle (66.5° north parallel) and the Antarctic Circle (66.5° south parallel) are the parallels farthest from the poles that experience 24 consecutive hours of night, or day, at some time during the year. "Lower latitudes" are those nearer the equator, whereas "higher latitudes" refer to those nearer the poles.

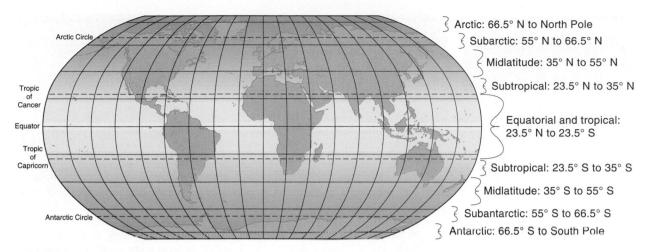

FIGURE 1–12
Latitudinal geographic zones. These generalized zones phase into one another over broad transitional areas.

Longitude

On a map or globe, lines designating angles of longitude run north and south at right angles (90°) to the equator and to all parallels. **Longitude** is an angular distance east or west of a point on Earth's surface, measured from the center of Earth— Figure 1–13(a). A line connecting all points along the same longitude is called a **meridian**. Thus, longitude is the name of the angle, meridian names the line, and both indicate distance east or west of an arbitrary **prime meridian**—Figure 1–13(b). Earth's prime meridian was not generally agreed to by most nations until 1884.

Because meridians of longitude converge toward the poles, the actual distance on the ground covered by a degree of longitude decreases away from the equator (where meridians separate to their widest distance apart) to the poles (where they converge). The distance covered by a degree of latitude,

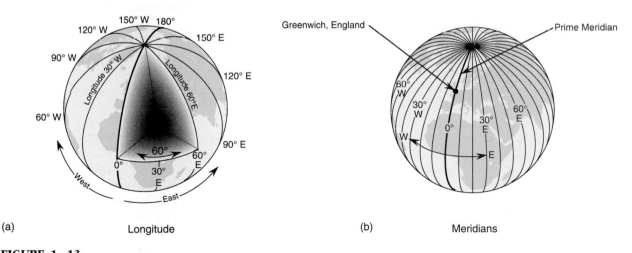

(a) Longitude

(b) Meridians

FIGURE 1–13
(a) Longitude is measured in degrees east or west of a 0° longitude line. (b) Angles of longitude measured from Earth's prime meridian determine other meridians.

The Search for Longitude

Unlike latitude, longitude cannot readily be determined from celestial bodies. Even with precise observations of the Moon and a sophisticated use of mathematics, results are significantly inaccurate. This inability to know longitude at sea persisted far into the eighteenth century. Amerigo Vespucci made this note in his diary, as quoted by Daniel J. Boorstin in *The Discoverers*:

> 5 September 1499—As to longitude, I declare that I found so much difficulty in determining it that I was put to great pains to ascertain the east-west distance I had covered. (p. 247)

In his historical novel *Shogun,* author James Clavell expresses a similar perspective through his pilot Blackthorn:

> Find how to fix longitude and you're the richest man in the world. . . . the Queen, God bless her, 'll give you ten thousand pound and dukedom for the answer to the riddle. . . . Out of sight of land you're always lost, lad. (p. 10)*

In the early 1600s Galileo explained that longitude could be measured by using two clocks. Any point on Earth takes 24 hours to travel around the full 360° of one rotation (one day). If you divide 360° by 24 hours, you find that any point on Earth travels through 15° of longitude every hour. Thus, if there were a way to measure time accurately at sea, a comparison of two clocks could give a value for longitude. One clock could be reset at local noon each day, as determined by the highest Sun position in the sky (solar zenith). The other clock would not be reset; it would indicate the time back at home port (Figure 1).

If the shipboard clock reads local noon and the clock set for home port reads 3:00 P.M., ship time is three hours earlier than home time. Therefore, calculating three hours at 15° per hour puts the ship at 45° west longitude from home port. Unfortunately, the pendulum clocks of Galileo's day did not work on the rolling deck of a ship at sea!

In 1707 the British lost four ships and 2000 men in a sea tragedy that was blamed specifically on the longitude problem. Parliament passed an act—"Publik Reward . . . to Discover the Longitude at Sea"—and authorized a prize of £20,000 sterling (equal to over $1 million today) to the first successful inventor. The Board of Longitude was established to judge any devices submitted. John Harrison, a self-taught country clockmaker, began work on the problem in 1728 and finally produced his chronometer, known as Number 4, in 1760. The clock was tested during a

*From *Shogun* by James Clavell, copyright © 1975 by James Clavell, reprinted by permission of Delacorte Press, a division of Bantam, Doubleday, Dell Publishing Group, Inc.

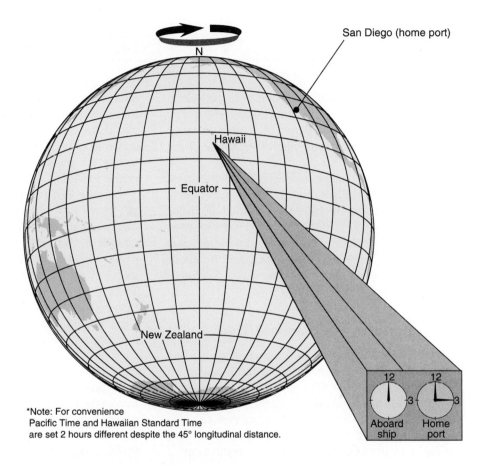

FIGURE 1
Using two clocks to determine longitude.

San Diego (home port)

N

Hawaii

Equator

New Zealand

*Note: For convenience
Pacific Time and Hawaiian Standard Time
are set 2 hours different despite the 45° longitudinal distance.

12 3
Aboard
ship

12 3
Home
port

nine-week trip to Jamaica in November 1761. When taken ashore to test it against Jamaica's land-based longitude, Harrison's Number 4 was only five seconds slow, an error that translates to only 1.25′, or 2.3 km (1.4 mi), well within Parliament's standard. Following many delays, Harrison finally received most of the prize money in his last years of life. From that time it was possible to determine longitude accurately on land and sea, as long as everyone agreed upon a prime meridian to use as the reference for time comparisons.

In this modern era of atomic clocks and satellites in mathematically precise orbits, we have far greater accuracy available for the determination of longitude on Earth's surface. Measurements of Earth's surface from satellites are accurate within millimeters, and surface laser surveys are precise within micrometers.

TABLE 1–1

Physical distances represented by degrees of latitude and longitude. Note that latitude degrees vary little in length, but longitude degrees vary greatly.

Latitudinal Location	Latitude Degree Length		Longitude Degree Length	
	km	(mi)	km	(mi)
90°	111.70	(69.41)	0	(0)
60°	111.42	(69.23)	55.80	(34.67)
50°	111.23	(69.12)	71.70	(44.55)
40°	111.04	(69.00)	85.40	(53.07)
30°	110.86	(68.89)	96.49	(59.96)
0°	110.58	(68.71)	111.32	(69.17)

however, varies only slightly, owing to deviations in Earth's shape. Table 1–1 presents these comparative values and shows the similarity of ground distance for a degree of latitude and one of longitude at the equator.

We have noted that latitude is easily determined by sighting the North Star, but a method of accurately determining longitude at sea remained a major difficulty in navigation until the mid-1700s. In addition, a specific determination of longitude was needed before world standard time could be established. The interesting geographic quest for longitude is the topic of Focus Study 1–2.

Prime Meridian and Standard Time

Today we take for granted the international standard time zones and an agreed-upon prime meridian. If you live in Oklahoma City and it is 3:00 P.M., you know it is 4:00 P.M. in Baltimore, 2:00 P.M. in Salt Lake City, and 1:00 P.M. in Seattle and Los Angeles. You also probably realize that it is 9:00 P.M. in London and midnight in Riyadh, Saudi Arabia. Coordination of

international trade, airline schedules, business activities, and daily living depend on a common prime meridian. But this orderly management of time zones is only a century old. Until 1884, the world lacked an international agreement on a standard prime meridian.

Because Earth revolves 360° every 24 hours, or 15° per hour (360° ÷ 24 = 15°), a time zone of one hour can be established for each 15° of longitude. Most nations in the early 1800s used their own national prime meridians for land maps, creating confusion in global mapping and clock time, as exemplified in Table 1–2. As a result, no standard time existed between or even within countries. The importance of Great Britain as a world power at the time is evident in that Greenwich, an observatory near London, was used on many nations' maritime charts as the prime meridian.

Setting time was not so great a problem in small European countries. But in North America, which covers more than 90° of longitude (the equivalent of six time zones), the problem was serious. In 1870, travelers going from Maine to San Francisco made 22

TABLE 1–2

Sample of prime meridians used in the 1800s.

Country	Prime meridians on:	
	Land Maps	Marine Charts
Austria	Ferro*	Greenwich
Belgium	Brussels	Greenwich
Brazil	Rio de Janeiro	Rio de Janeiro
France	Paris	Paris
Italy	Rome	Greenwich
Portugal	Lisbon	Lisbon
Spain	Madrid	Cádiz
United States	Washington	Greenwich

*Ferro (Spanish: Hierro) Island is the westernmost of the Canary Islands in the eastern Atlantic Ocean. The most westerly place known to early geographers, Ptolemy chose it as an initial prime meridian, ca. A.D. 150.

FIGURE 1–14

Modern international standard time zones. Numbers indicate how many hours each zone is earlier (minus sign) or later (plus sign) than Coordinated Universal Time (Greenwich Mean Time). Adapted by permission of Merrill, an imprint of Macmillan Publishing Company, from *The Cultural Landscape,* 2d ed., by James M. Rubenstein, p. 13. Copyright © 1989 by Merrill Publishing.

adjustments to their watches to stay consistent with local railroad time. Sir Sanford Fleming led the fight in Canada for standard time and an international agreement upon a prime meridian. His struggle led the United States and Canada to adopt a standard time in 1883.

In 1884 the International Meridian Conference was held in Washington, DC. After lengthy debate, most participating nations chose the Royal Observatory at Greenwich as the place for the prime meridian of 0° longitude (the Royal Observatory is pictured in the photo that opens this chapter). The Observatory was highly respected, and more than

70% of the world's merchant ships already used the London prime meridian. Thus, a world standard was established, and international time was set (Figure 1–14). Each time zone theoretically covers 7.5° on either side of a controlling meridian and represents one hour. From the map of global time zones, can you determine the present time in Moscow? London? Atlanta? Ottawa? Chicago? Winnipeg? Denver? San Francisco? Anchorage? Honolulu? Tokyo?

Today, **Greenwich Mean Time (GMT)** is called **Coordinated Universal Time (UTC)**; and although the prime meridian is still through Greenwich, UTC is based on average time calculations kept in Paris

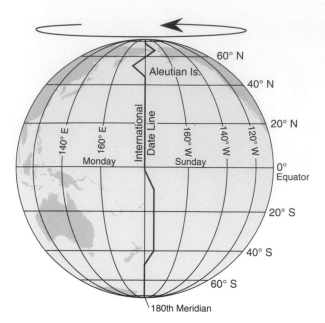

FIGURE 1–15
International Date Line location, the 180th meridian (see also Figure 1–14).

sweeps westward across Earth. This westward movement of time is created by the planet's turning eastward on its axis. At the International Date Line, the west side of the line is always one day ahead of the east side. No matter what time of day it is when the line is crossed, the calendar changes a day (Figure 1–15).

Locating the date line in the sparsely populated Pacific Ocean minimizes most local confusion. However, the consternation of early explorers before the date-line concept was understood is interesting. For example, Magellan's crew returned from the first circumnavigation of Earth, confident from their ship's log that the day of their arrival was a Wednesday. They were shocked when informed by insistent local residents that it was actually a Thursday!

and broadcast worldwide. UTC is measured today by the very regular vibrations of cesium atoms in 80 atomic clocks provided by 24 nations.

International Date Line. An important corollary of the prime meridian is the 180° meridian on the opposite side of the planet. This meridian is called the **International Date Line** and marks the place where each day officially begins (12:01 A.M.) and

Great Circles and Small Circles

Great circles and small circles are important concepts that help summarize latitude and longitude (Figure 1–16). A **great circle** is any circle of Earth's circumference whose center coincides with the center of Earth. An infinite number of great circles can be drawn on Earth. Every meridian is one-half of a great circle that passes through the poles. On flat maps, airline and shipping routes appear to arch their way across oceans and landmasses. These are great circle routes, the shortest distance between two points on Earth. Only one parallel is a great circle—the equatorial parallel. All other parallels diminish in length toward the poles and, along with any other non-great circle that one might draw,

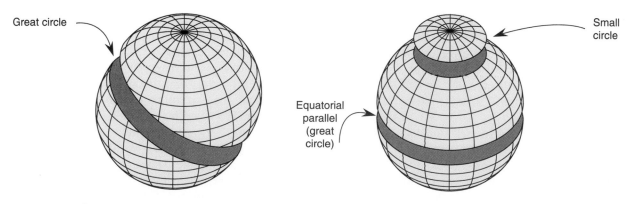

FIGURE 1–16
Great circles and small circles on Earth.

FIGURE 1–17
Three common expressions of map scale.

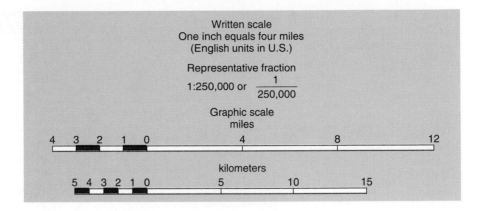

Written scale
One inch equals four miles
(English units in U.S.)

Representative fraction

$1:250,000$ or $\dfrac{1}{250,000}$

Graphic scale
miles

4 3 2 1 0   4   8   12

kilometers

5 4 3 2 1 0   5   10   15

---

constitute **small circles**—circles whose centers do not coincide with Earth's center.

MAPS AND MAP PROJECTIONS

The earliest known graphic map presentations date to 2300 B.C. when the Babylonians used clay tablets to record information about the region of the Tigris and Euphrates rivers (the area of modern-day Iraq). Much later, when sailing vessels dominated the seas, a pilot's *rutter*—a descriptive diary of locations, places, coastlines, and collected maps—became a critical reference. *Portolan* charts, describing harbor locations and coastal features and showing compass directions, also were prepared from these many navigational experiences. Today, the making of maps and charts is a specialized science as well as an art, blending aspects of geography, engineering, mathematics, graphics, computer science, and artistic specialties. It is similar in ways to architecture, in which aesthetics and utility are combined to produce an end product.

A map is a generalized view of an area, usually some portion of Earth's surface, as seen from above and greatly reduced in size. The part of geography that embodies mapmaking is called **cartography**. Maps are critical tools with which geographers depict spatial information and analyze spatial relationships. We all use maps at some time to visualize our location and our relationship to other places, or maybe to plan a trip, or to coordinate commercial and economic activities. Learning a few basics about maps is essential to our study of physical geography.

The Scale of Maps

Architects and mapmakers have something in common: they both create scale models. They both reduce real things and places to the more convenient scale of a model, diagram, or map. An architect renders a blueprint of a building to guide the contractors, selecting a scale so that one centimeter (or inch) on the drawing represents so many meters (or feet) on the proposed building. Often, the drawing is 1/50 to 1/100 of real size. The cartographer does the same thing in preparing a map. The ratio of the image on a map to the real world is called **scale**; it relates a unit on the map to a similar unit on the ground. A 1:1 scale would mean that a centimeter on the map represents a centimeter on the ground (although this certainly would be an impractical map scale, for the map would be as large as the area being mapped!). A more appropriate scale for a local map is 1:24,000, in which 1 unit on the map represents 24,000 identical units on the ground.

Map scales may be presented in several ways: a written scale, a representative fraction, or a graphic scale (Figure 1–17). A *written scale* simply states the ratio—for example, "one centimeter to one kilometer" or "one inch to one mile." A *representative fraction* (RF, or fractional scale) can be expressed with either a : or a /, as in 1:125,000 or 1/125,000. No actual units of measurement are mentioned because any unit is applicable as long as both parts of the fraction are in the same unit: 1 cm to 125,000 cm, 1 in. to 125,000 in., or even 1 armlength to 125,000 armlengths, and so on.

A *graphic scale,* or bar scale, is expressed as a bar graph with units noted to allow measurement of

TABLE 1-3

Sample representative fractions and written scales.

Units	Scale Size	Representative Fraction	Written Scale
English	Small	1:3,168,000	1 in. = 50 mi
		1:2,500,000	1 in. = 40 mi
		1:1,000,000	1 in. = 16 mi
		1:500,000	1 in. = 8 mi
		1:250,000	1 in. = 4 mi
	Medium	1:125,000	1 in. = 2 mi
		1:63,360 (or 1:62,500)	1 in. = 1 mi
		1:31,680	1 in. = 0.5 mi
		1:30,000	1 in. = 2500 ft
	Large	1:24,000	1 in. = 2000 ft

Units	Representative Fraction	Written Scale
Metric	1:1,100,000	1 cm = 10.0 km
	1:50,000	1 cm = 0.50 km
	1:25,000	1 cm = 0.25 km
	1:20,000	1 cm = 0.20 km

distances on the map. An important advantage of a graphic scale is that it changes to match the map during enlargement or reduction. If a map is "blown up" to twice its original size, the graphic scale is blown up along with it, and remains accurate. In contrast, written and fractional scales become incorrect with enlargement or reduction: you can shrink a map from 1:24,000 to 1:63,360, but the scale will still say "1 in. = 2000 ft," instead of the new correct scale of 1 in. = 1 mi.

Scales are called "small," "medium," and "large," depending on the ratio described. Thus, a scale of 1:24,000 is a large scale, whereas a scale of 1:50,000,000 is a small scale. The greater the size of the denominator in a fractional scale (or number on the right in a ratio scale), the smaller the scale and the more abstract the map must be in relation to what is being mapped. Examples of selected representative fractions and written scales are listed in Table 1-3.

If a world globe is 61 cm (24 in.) in diameter and we know Earth has an equatorial diameter of 12,756 km (7926 mi), then the scale of the globe is the ratio of 61 cm to 12,756 km. So, we divide Earth's actual diameter by the globe's diameter and determine that 1 cm of the globe's diameter equals about 20,900,000 cm of Earth's diameter. Thus, the representative fraction in centimeters is about 1:20,900,000. This representative fraction can now be expressed in *any* unit of measure, metric or English, as long as both numbers are in the same units. Thus, this 61 cm globe is a small-scale representation of Earth with a high degree of abstraction. If there is a globe available in your library or classroom, check to see the scale at which it was drawn.

Map Projections

A globe is not always a helpful representation of Earth. Travelers, for example, need more detailed information than a globe can provide, and large globes don't fit well in cars and aircraft. Consequently, to provide localized detail, cartographers prepare large-scale flat maps, which are two-dimensional representations (scale models) of our three-dimensional Earth. This reduction of the spherical Earth to a flat surface in some orderly and systematic realignment of the latitude and longitude grid is called a **map projection**.

Because a globe is the only true representation (model) of distance, direction, area, shape, and proximity, the preparation of a flat version means that decisions must be made as to the type and amount of distortion that users will accept. To understand this problem, consider these important properties of a globe: On a globe, parallels always are parallel to each other, always are evenly spaced along meridians, and always decrease in length toward the poles. On a globe, meridians converge at both poles and are evenly spaced along any individual parallel. On a globe, the distance between meridians decreases toward poles, with the spacing

FIGURE 1–18
Conversion of the globe to a flat
map. Photo courtesy of National
Aeronautics and Space
Administration.

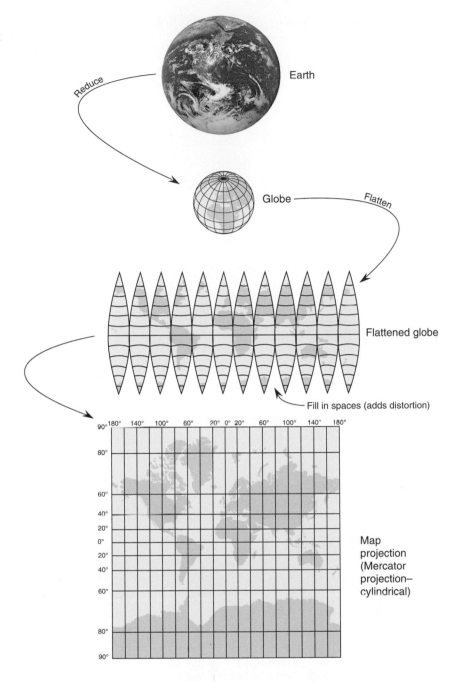

Earth

Globe — Flatten

Flattened globe

Fill in spaces (adds distortion)

90° 180° 140° 100° 60° 20° 0° 20° 60° 100° 140° 180°

80°

60°

40°

20°

0°

20°

40°

60°

80°

90°

Map
projection
(Mercator
projection–
cylindrical)

between meridians at the 60th parallel equal to one half the equatorial spacing. And on a globe, parallels and meridians always cross each other at right angles.

The problem is that all these qualities cannot be reproduced on a flat surface. Simply taking a globe apart and laying it flat on a table illustrates the problem faced by cartographers in constructing a flat map (Figure 1–18). You can see the empty spaces that open up between the sections, or gores, of the globe. Thus, no flat map projection of Earth can ever have all the features of a globe. Flat maps always possess some degree of distortion—much less for large-scale maps representing a few kilometers;

ESSENTIALS OF GEOGRAPHY ▪ **23**

much more for small-scale maps covering individual countries, continents, or the entire world.

Properties of Projections. The major decisions in selecting a map projection involve the properties of equal area (equivalence) and **true shape** (conformality). If a cartographer selects equal area as the desired trait, as for a map showing the distribution of world climates, then shape must be sacrificed by stretching and shearing. On an equal-area map, a

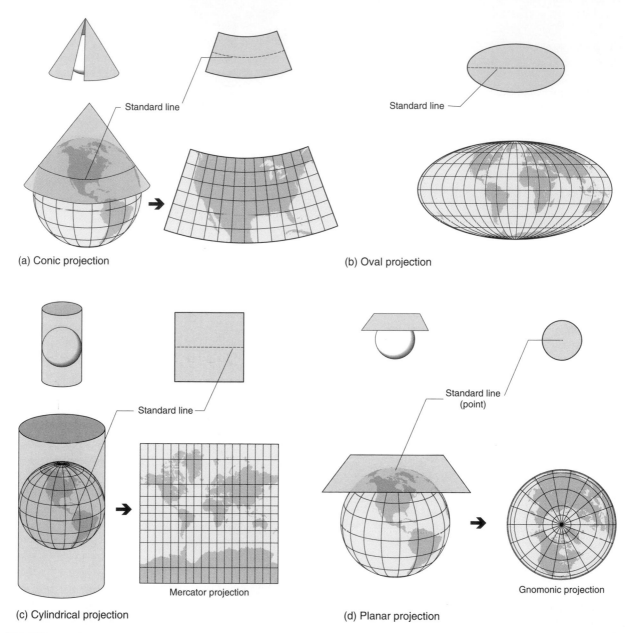

(a) Conic projection

(b) Oval projection

Mercator projection

(c) Cylindrical projection

Gnomonic projection

(d) Planar projection

FIGURE 1–19
General classes and perspectives of map projections. After American Congress on Surveying and Mapping, 1988, *Choosing a World Map,* Falls Church, VA, p. 8. Adapted by permission.

coin covers the same amount of surface area, no matter where you place it on the map. If, on the other hand, a cartographer selects the property of true shape, as for a map used for navigational purposes, then equal area must be sacrificed, and the scale will actually change from one region of the map to another. Two additional properties related to map projections are true direction (azimuth) and true distance (equidistance). *The best projection is always determined by its intended use*.

The Nature and Classes of Projections. Despite modern cartographic technology utilizing mathematical constructions and computer-assisted graphics, the word "projection" is still used. In times past, actual perspective methods were used to project the globe onto a geometric surface. A physical perspective approach uses a globe constructed of wire parallels and meridians. A light source strategically placed at varying distances then casts a pattern of latitude and longitude lines from the globe onto various geometric surfaces, such as a cone, cylinder, or plane.

Figure 1–19 illustrates the derivation of the general classes of map projections and the perspectives from which they are generated. The classes shown include the conic, cylindrical, and planar (or azimuthal). Another class of projections that cannot be derived from this physical-perspective approach is the nonperspective oval-shaped. Still others are derived from purely mathematical calculations. With all projections, the contact line or point between the wire globe and the projection surface—called a *standard line*—is *the only place where all globe properties are preserved;* thus, a *standard parallel* or *standard meridian* is a standard line true to scale along its entire length without any distortion. Areas away from this critical tangent line or point become increasingly distorted. Consequently, this area of optimum spatial properties should be centered on the region of immediate interest so that greatest accuracy is preserved there.

The commonly used Mercator projection (from Gerardus Mercator, A.D. 1569) is a cylindrical projection—Figure 1–19(c). For example, Figure 1–14 utilizes a Mercator projection for the presentation of standard time zones. The Mercator is a true-shape projection, with meridians appearing as equally spaced straight lines and parallels appearing as straight lines that come closer together near the equator. The poles are infinitely stretched, with the 84th north parallel and 84th south parallel fixed at the same length as that of the equator. Locally, the shape is accurate and recognizable for navigation; however, the scale varies with latitude. The advantage of the Mercator projection is that lines of constant direction, called **rhumb lines**, are straight and thus facilitate the task of plotting compass directions between two points. Thus, the Mercator projection is useful in navigation and has been the standard for nautical charts prepared by the National Ocean Service (formerly U.S. Coast and Geodetic Survey) since 1910.

Unfortunately, Mercator classroom maps present false notions of the size (area) of midlatitude and poleward landmasses. A dramatic example on the cylindrical projection in Figure 1–19(c) is Greenland, which looks bigger than all of South America. In reality, Greenland is only one-eighth the size of South America and is actually 20% smaller than Argentina alone!

The gnomonic or planar projection in Figure 1–19(d) is generated with a light source at the center of a globe projecting onto a plane tangent to the globe's surface. The resulting severe distortion prevents showing a full hemisphere on one projection. However, a valuable feature is derived: all great-circle routes, which are the shortest distances between two points on Earth's surface, are projected as straight lines. The great-circle routes plotted on a gnomonic projection then can be transferred to a true-direction projection, such as the Mercator, for determination of precise compass headings (Figure 1–20).

Maps Used in This Text. Two projections generally are used in this text. One is an interrupted world map designed in 1925 by Dr. J. Paul Goode of the University of Chicago and called Goode's homolosine equal-area projection (Figure 1–21). Goode's projection is a combination of two oval projections. To improve the rendering of landmass shapes, two equal-area projections are cut-and-pasted together. One type of projection is used between 40° N and S latitudes (a sinusoidal projection). The other type of projection is used from 40° N to the North Pole and from 40° S to the South Pole (a Mollweide, or homolographic, projection). Areal size relationships

(a)　　　　　　　Gnomonic projection

(b)　　　　　　　Mercator projection

FIGURE 1–20
Comparison of rhumb lines and great-circle routes on two different projections. After Fisher, 1989, *Geography and Development,* 3d ed., Columbus: Macmillan Publishing Co., p. 661. Adapted with permission.

are preserved, so the map is excellent for mapping spatial distributions when interruptions do not pose a problem. For example, the world climate map in Chapter 10, the soils map in Chapter 18, and the terrestrial ecosystems map in Chapter 20 use a Goode's homolosine projection.

Another projection used in this text is the Robinson projection, designed by Arthur Robinson in 1963 (Figure 1–22). Examples are the latitudinal geographic zones map (Figure 1–12), the world temperatures map in Chapter 5, and the world pressures

maps in Chapter 6. This projection is neither equal area nor true shape, but is a compromise between both considerations. The North and South Poles appear as lines slightly more than half the length of the equator, thus exaggerating areas in higher latitudes to a lesser degree than other oval and cylindrical projections. The National Geographic Society adopted this projection for their primary world map in 1988. Just as there are many map projections to choose from for different uses, so are there numerous surface maps available for many purposes.

Mapping and Topographic Maps

The Public Lands Survey System (1785) delineated locations in the United States. Although the Bureau of Land Management was responsible for subdividing public lands and keeping records of all related surveys, actual preparation and recording of this information fell to the U.S. Geological Survey (USGS), a branch of the Department of the Interior. The USGS depicts the information on quadrangle maps, which are rectangular maps bounded by parallels and meridians rather than by political boundaries. Detailed mapping of a variety of features is carried on continuously by the USGS. In Canada, national mapping is conducted by the Energy, Mines, and Resources Department (EMR). The EMR prepares base maps, thematic maps, aeronautical charts, federal topographic maps, and the *National Atlas of Canada.*

In mapping, a basic **planimetric map** is first prepared, showing the horizontal position of bound-

FIGURE 1–21
Goode's homolosine projection. Copyright by the University of Chicago. Used by permission of the University of Chicago Press.

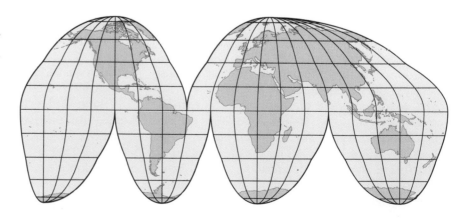

FIGURE 1–22
Robinson projection. Courtesy of American Congress on Surveying and Mapping.

FIGURE 1–23
(a) Perspective view of a hypothetical landscape; (b) depiction of that landscape on a topographic map (contours in feet). After illustrations by Dennis Tasa, Tasa Graphic Arts, Inc. Adapted by permission of Merrill, an imprint of Macmillan Publishing Company, from *The Earth,* 3d ed., by Edward J. Tarbuck and Frederick K. Lutgens, p. 620. Copyright © 1990 by Merrill Publishing.

aries, land-use aspects, and political, economic, and social features. A highway map is a common example of a planimetric map. Then, a vertical scale of physical features is added to portray the terrain. The most popular and widely used maps of this type are the detailed **topographic maps** prepared by the USGS. Topographic maps portray physical relief through the use of elevation **contour lines**. A contour line connects all points at the same elevation above or below a stated reference level, as you can see from the contour lines in the figure. This reference level—usually mean sea level—is called the *vertical datum*. The *contour interval* is the difference in elevation between two adjacent contour lines (20 ft or 6.1 m in Figure 1–23).

Slope is indicated by the pattern of these lines and the space between them. The steeper a slope or cliff, the closer together the contour lines appear—in Figure 1–23(b), note the narrowly spaced contours representing the cliffs. A more gradual slope is portrayed by a wider spacing of these contour lines, as you can see from the widely spaced lines on the level beach. Actual USGS topographic maps appear in several chapters of this text because they are so useful in depicting features of the physical landscape. The topographic map in Figure 1–23 shows a hypothetical landscape to demonstrate how contour lines and intervals indicate slope and relief.

Because a single topographic map of the United States at 1:24,000 scale would be over 200 m wide (more than 600 ft), some system had to be devised for dividing the country up into maps of manageable size. Thus, a quadrangle system based on latitude and longitude coordinates is used for map classification and layout (Figure 1–24). Because meridians converge toward the poles, the shape of quadrangles also narrows noticeably toward the poles. Each map quadrangle is referred to by its angular dimensions (Figure 1–24). Thus, a map that is one-half a degree (30') on each side is a 30-minute quadrangle, and a map one-fourth of a degree (15') on each side is a 15-minute quadrangle (the USGS standard from 1910 to 1950). A map that is one-eighth of a degree (7.5') on each side is a 7.5-minute quadrangle, the most widely produced of all USGS topographic maps, and the standard since 1950.

The USGS National Mapping Program recently completed coverage of the entire country (except Alaska) on 7.5-minute maps. The standard 7.5-minute

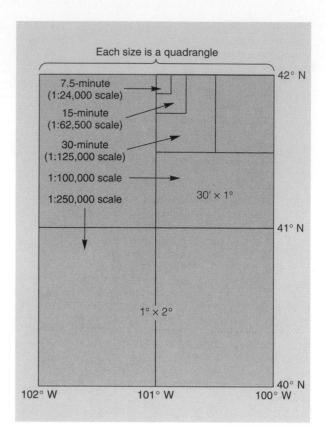

FIGURE 1–24
Quadrangle system of maps used by the USGS.

quadrangle is scaled at 1:24,000 (1 in. = 2000 ft, a large scale). It takes 53,838 separate 7.5-minute quadrangles to cover the lower 48 states, Hawaii, and U.S. territories. Alaska is covered with a series of more general 15-minute topographic maps. In Canada, the entire country is mapped at a scale of 1:250,000 (1.0 cm to 2.5 km), and about half the country also is mapped at 1:50,000, with major metropolitan areas mapped at 1:25,000. In the United States, the eventual changeover to the metric system someday will necessitate revision of the units used on all maps, with the 7.5-minute series eventually changing to a scale of 1:25,000.

The principal symbols used on USGS topographic maps are presented in Figure 1–25. Colors are standard on all USGS topographic maps: black for human constructions, blue for water features, brown for relief features and contours, pink for urbanized areas, and green for woodlands, orchards, brush, and

FIGURE 1–25

Standardized topographic map symbols used on USGS maps. English units still prevail, although some newer USGS maps are metric. From USGS, *Topographic Maps,* 1969.

Primary highway, hard surface . . . .
Secondary highway, hard surface . . . .
Light-duty road, hard or improved surface . . . .
Unimproved road . . . .
Road under construction, alinement known . . . .
Proposed road . . . .
Dual highway, dividing strip 25 feet or less . . . .
Dual highway, dividing strip exceeding 25 feet . . . .
Trail . . . .

Railroad: single track and multiple track . . . .
Railroads in juxtaposition . . . .
Narrow gage: single track and multiple track . . . .
Railroad in street and carline . . . .
Bridge: road and railroad . . . .
Drawbridge: road and railroad . . . .
Footbridge . . . .
Tunnel: road and railroad . . . .
Overpass and underpass . . . .
Small masonry or concrete dam . . . .
Dam with lock . . . .
Dam with road . . . .
Canal with lock . . . .

Buildings (dwelling, place of employment, etc.) . . . .
School, church, and cemetery . . . . Cem
Buildings (barn, warehouse, etc.) . . . .
Power transmission line with located metal tower . . . .
Telephone line, pipeline, etc. (labeled as to type) . . . .
Wells other than water (labeled as to type) . . . . oOil . . . . . . oGas
Tanks: oil, water, etc. (labeled only if water) . . . . Water
Located or landmark object; windmill . . . .
Open pit, mine, or quarry; prospect . . . .
Shaft and tunnel entrance . . . .

Horizontal and vertical control station:
Tablet, spirit level elevation . . . . BM △ 5653
Other recoverable mark, spirit level elevation . . . . △ 5455
Horizontal control station: tablet, vertical angle elevation VARM △ 95⁄9
Any recoverable mark, vertical angle or checked elevation △3775
Vertical control station: tablet, spirit level elevation . . . . . BM × 957
Other recoverable mark, spirit level elevation . . . . × 954
Spot elevation . . . . × 7369   × 7369
Water elevation 670   670

Boundaries: National . . . .
State . . . .
County, parish, municipio . . . .
Civil township, precinct, town, barrio . . . .
Incorporated city, village, town, hamlet . . . .
Reservation, National or State . . . .
Small park, cemetery, airport, etc. . . . .
Land grant . . . .
Township or range line, United States land survey . . . .
Township or range line, approximate location . . . .
Section line, United States land survey . . . .
Section line, approximate location . . . .
Township line, not United States land survey . . . .
Section line, not United States land survey . . . .
Found corner: section and closing . . . .
Boundary monument: land grant and other . . . .
Fence or field line . . . .

Index contour . . . .   Intermediate contour . .
Supplementary contour   Depression contours . .
Fill . . . .   Cut . . . .
Levee . . . .   Levee with road . . . .
Mine dump . . . .   Wash . . . .
Tailings . . . .   Tailings pond . . . .
Shifting sand or dunes   Intricate surface . . . . .
Sand area . . . .   Gravel beach . . . .

Perennial streams . . . . .   Intermittent streams . . .
Elevated aqueduct . . . .   Aqueduct tunnel . . . . . .
Water well and spring . o . . . . . . o   Glacier . . . .
Small rapids . . . .   Small falls . . . .
Large rapids . . . .   Large falls . . . .
Intermittent lake . . . . .   Dry lake bed . . . .
Foreshore flat . . . .   Rock or coral reef . . . .
Sounding, depth curve .   Piling or dolphin . . . . .
Exposed wreck . . . .   Sunken wreck . . . .
Rock, bare or awash; dangerous to navigation . . . .

Marsh (swamp) . . . .   Submerged marsh . . . . .
Wooded marsh . . . .   Mangrove . . . .
Woods or brushwood . .   Orchard . . . .
Vineyard . . . .   Scrub . . . .
Land subject to controlled inundation   Urban area . . . .

the like. The margins of a topographic map contain a wealth of information about the concept and content of the map: quadrangle name, names of adjoining quads, quad series and type, relative positioning on the latitude-longitude and other grid coordinate systems, title, legend, magnetic declination and compass information, datum plane, symbols used for roads and trails, the dates and history of the survey of that particular quad, and more. Many outdoor-product stores and state geological surveys sell topographic maps to assist people in planning their activities. These maps also may be purchased directly from the USGS or EMR.

SUMMARY

Geography is a scientific discipline that uses spatial analysis as a principal methodology. Physical geography is the spatial analysis of all the physical elements and processes that comprise the environment: energy, air, water, weather, climate, landforms, soils, animals, plants, and Earth itself. Earth's physical systems and human society are interrelated, and there is increasing concern over the state of the environment and the sustainability of certain patterns of human behavior.

A system is any ordered, interrelated set of objects, things, components or parts, and their attributes, as distinct from their surrounding environment. Earth is an open system in terms of energy and a closed system in terms of matter and physical resources. Earth is misshapen, bulging slightly through the equator, and is described as a geoid. Absolute location on Earth is described with a specific reference grid of parallels of latitude and meridians of longitude. A standardized pattern of time zones and Coordinated Universal Time (UTC) now is used internationally.

Geographers utilize maps for the spatial portrayal of Earth's physical systems. Maps are created by cartographers in a variety of projections and scales for many applications. All of the United States and Canada are mapped in a systematized format at medium and large scales.

KEY TERMS

Abiotic	Map projection
Antarctic geographic zone	Meridian
Arctic geographic zone	Midlatitude geographic zone
Atmosphere	Model
Biosphere	Movement
Biotic	Negative feedback
Cartography	Open system
Closed system	Parallel
Contour lines	Place
Coordinated Universal Time (UTC)	Planimetric map
Ecosphere	Positive feedback
Equal area	Prime meridian
Equatorial geographic zone	Process
Equilibrium	Region
Feedback loops	Rhumb lines
Geodesy	Scale
Geoid	Small circles
Great circle	Spatial analysis
Greenwich Mean Time (GMT)	Steady-state equilibrium
Human-Earth relationships	Subantarctic geographic zone
Hydrosphere	Subarctic geographic zone
International Date Line	Subtropical geographic zone
Latitude	System
Lithosphere	Topographic maps
Location	Tropical geographic zone
Longitude	True shape

REVIEW QUESTIONS

1. What is geography? Based on information in this chapter, define physical geography and give a specific instance of the geographic approach.
2. Make a list of decisions you made today or items you've read recently that involve geographic concepts.

3. Assess your geographic literacy by examining available atlases and maps. What types of maps have you used—political? physical? topographic? Do you know what projections they employed? Do you know the names and locations of the four oceans, seven continents, and individual countries?
4. Suggest a representative example for each of the five geographic themes and use that theme in a sentence.
5. Define systems theory as an organizational strategy. What are open systems and closed systems, negative feedback, and a system at a steady-state equilibrium condition? What type of system is a human body? A lake? A wheat plant?
6. Describe Earth as a system in terms of energy and of matter.
7. What are the three abiotic spheres that comprise Earth's environment? Relate these to the biosphere.
8. Describe Earth's shape and size with a diagram.
9. What are the latitude and longitude coordinates (in degrees, minutes, and seconds) of your present location? Where can you find this information?
10. Define latitude and parallel. Define longitude and meridian.
11. Draw a diagram showing Eratosthenes's method of determining Earth's circumference. Use your diagram to explain his calculations.
12. Identify the various latitudinal geographic zones. In which zone do you live?
13. What do clocks have to do with longitude? Explain this relationship. How is standard time determined on Earth?
14. What and where is the prime meridian? How was it selected? Describe the meridian opposite the prime meridian.
15. Define a great circle and a small circle. In terms of these concepts, describe the equator, other parallels, and meridians.
16. In what way is cartography an integrative discipline?
17. What is map scale? How is it expressed on a map?
18. Identify each of the following ratios as large scale, medium scale, or small scale: 1:3,168,000, 1:24,000, 1:250,000.
19. Describe the differences between the characteristics of a globe and a flat map.
20. What type of map projection is used in Figure 1–12? Figure 1–14?

SUGGESTED READINGS

Boorstin, Daniel J. *The Discoverers*. New York: Random House, 1983.

Dunn, Margery G., and Donald J. Crump. *Exploring Your World—The Adventure of Geography*. Washington: National Geographic Society, 1989.

Gaile, Gary L., and Cort J. Willmott, eds. *Geography in America*. Columbus: Macmillan Publishing Co., 1989.

Greenwood, David. *Mapping*. Chicago: The University of Chicago Press, 1964.

Howse, Derek. *Greenwich Time and the Discovery of the Longitude*. London: Oxford University Press, 1980.

Making, Joel, and Laura Bergen, eds. *The Map Catalog—Every Kind of Map and Chart on Earth*. New York: Vintage Books, 1986.

Miller, Victor C., and Mary E. Westerback. *Interpretation of Topographic Maps*. Columbus: Macmillan Publishing Co., 1989.

Odum, Howard T. *Systems Ecology—An Introduction*. New York: John Wiley & Sons, 1983.

Robinson, Arthur H., J. Morrison, and P. Muehrcke. *Elements of Cartography*. 5th ed. New York: John Wiley & Sons, 1985.

Snyder, John P. *Map Projections—A Working Manual*. U.S. Geological Survey Professional Paper 1395. Washington: U.S. Government Printing Office, 1987.

Thompson, Morris M. *Maps for America*. 3d ed. U.S. Geological Survey. Washington: U.S. Government Printing Office, 1987.

Our planet and our lives are powered by radiant energy from the star that is closest to Earth—the Sun. For more than 4.6 billion years, solar energy has traveled across interplanetary space to Earth, where a small portion of the solar output is intercepted. Because of Earth's curvature, the energy at the top of the atmosphere is unevenly distributed, creating imbalances from the equator to each pole. This unevenness of energy receipt empowers circulations in the atmosphere and on the surface below.

Earth's atmosphere acts as an efficient filter, absorbing most harmful radiation, charged particles, and space debris so that they do not reach Earth's surface. Surface energy balances are established and pulse in seasonal shifts during the year, giving rise to global patterns of temperature, winds, and ocean currents. Each of us depends on many systems that are set into motion by energy from the Sun. These systems are the subjects of Part One.

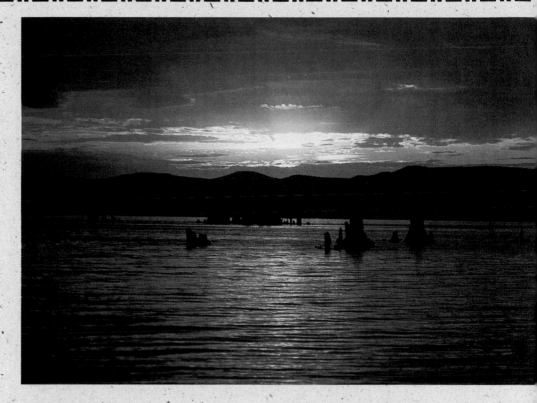

PART ONE

The Energy-Atmosphere System

Sunrise at Mono Lake, California. Photo by author.

Aurora borealis in Alaska. Photo by Lee Snyder,
Geophysical Institute, University of Alaska,
Fairbanks.

2
Solar Energy to Earth

O ur immediate home is North America, a major continent on planet Earth, third planet from a typical yellow star in a solar system. That star, our Sun, is only one of billions in the Milky Way Galaxy, which is one of millions of galaxies in the universe. This chapter examines the flow of energy from the Sun to the outer reaches of Earth's atmosphere and includes a brief section on remote sensing.

SOLAR SYSTEM: STRUCTURE AND FORMATION

Our Sun is unique to us, and yet commonplace in our galaxy. It is only average in temperature, size, and color when compared to other stars, yet it is the ultimate energy source for almost all life processes in our biosphere.

Solar System Location

Our Sun is located on a remote, trailing edge of the **Milky Way Galaxy**, a flattened, disk-shaped mass estimated to contain up to 400 billion stars. Its actual shape is thicker toward the center and thinner toward the outer edges (Figure 2–1). Our Solar System is embedded more than halfway out from the center, in one of the Milky Way's spiral gravitational arms called the Orion Arm. From our Earth-bound perspective in the Milky Way, the galaxy appears to stretch across the night sky like a narrow band of hazy light. On a clear night the unaided eye can see only a few thousand of these billions of stars.

Dimensions. The **speed of light** is 299,792 km per second (186,282 mi/sec), or more than 9 trillion km a year (9,454,000,000,000 km or 5,875,000,000,000 mi). This tremendous distance that light travels in a

FIGURE 2–1
Milky Way Galaxy, (a) viewed from above and (b) cross-section side view. Our Solar System is some 30,000 light-years from the center of the galaxy.

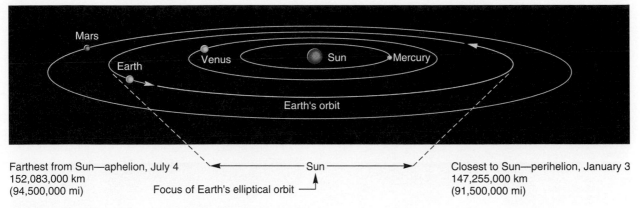

Farthest from Sun—aphelion, July 4
152,083,000 km
(94,500,000 mi)

Sun

Focus of Earth's elliptical orbit ⎤

Closest to Sun—perihelion, January 3
147,255,000 km
(91,500,000 mi)

FIGURE 2–2
Structure of Earth's elliptical orbit.

year is known as a *light-year* and is used as a unit of measurement for the vast universe. For spatial comparison, our Moon is an average distance of 384,400 km (238,866 mi) from Earth, or about 1.28 seconds in terms of light speed. Our entire Solar System is approximately 11 hours in diameter, measured by light speed. In contrast, the Milky Way is about 100,000 light-years from side to side, and the known universe stretches approximately 12 billion light-years in all directions from Earth.

Earth's Orbit. Earth's orbit around the Sun is presently elliptical—a closed, oval-shaped path (Figure 2–2). Earth's average distance from the Sun is approximately 150 million km (93 million mi), which means that light reaches Earth from the Sun in an average of 8 minutes and 20 seconds. A plane intersecting all points of Earth's orbit is termed the **plane of the ecliptic**, and Earth remains aligned with this plane as it revolves around the Sun.

Earth is at **perihelion** (its closest position to the Sun) on January 3 at 147,255,000 km (91,500,000 mi) and at **aphelion** (its farthest distance from the Sun) on July 4 at 152,083,000 km (94,500,000 mi). This seasonal difference in distance from the Sun causes a variation of 3.4% in the solar output intercepted by Earth.

The structure of Earth's orbit is not a constant but instead exhibits changes over long periods. As discussed in Chapter 17, Earth's distance from the Sun varies more than 17.7 million km (11 million mi) during a 100,000-year cycle, placing it closer or further at different periods in the cycle. This variation is thought to be one of several factors that create

Earth's cyclical pattern of glaciations and interglacial periods, which are colder and warmer times, respectively. Figure 2–3 compares Earth's orbital distance and other information with that for the other planets in the Solar System.

Solar System Formation

According to prevailing theory, our Solar System condensed from a large, slowly rotating, collapsing cloud of dust and gas called a **nebula**. As the nebular cloud organized and flattened into a disk shape, the early proto-Sun grew in mass at the center, drawing more matter to it. Small accretion (accumulation) eddies swirled at varying distances from the center of the solar nebula; these were the protoplanets. **Gravity**, the natural force exerted by the mass of an object upon all other objects, was the key organizing force in this condensing solar nebula.

The early protoplanets, called *planetesimals,* were orbiting at approximately the same distances from the Sun that the planets are today. The beginnings of the Sun and its Solar System are estimated to have occurred more than 4.6 billion years ago, with the Sun finally achieving its present brightness only after a long period of development. This same process—suns condensing from nebular clouds (the nebular hypothesis), with planetesimals forming in orbits around their central masses (the **planetesimal hypothesis**)—is actually being observed in other parts of the galaxy. The planets appear to accrete from dust, gases, and icy comets that are drawn into collision and coalescence. The development of such hypotheses and theories is an exercise of the

scientific method, a methodology important to physical geography research. Focus Study 2–1 explains this essential process of science.

Earth's Development

The development of Earth is closely related to the growth of the Sun, and the long-term future of Earth likewise is tied closely to the Sun's life cycle. The evolution of Earth's atmosphere and surface, the formation of free oxygen gas, and the development of the biosphere all represent complex interactions that were in operation from the beginning of Earth's environment. Processes set into motion as the solar nebula condensed and the planetesimals congealed proceeded at rates imperceptible on a scale of human events.

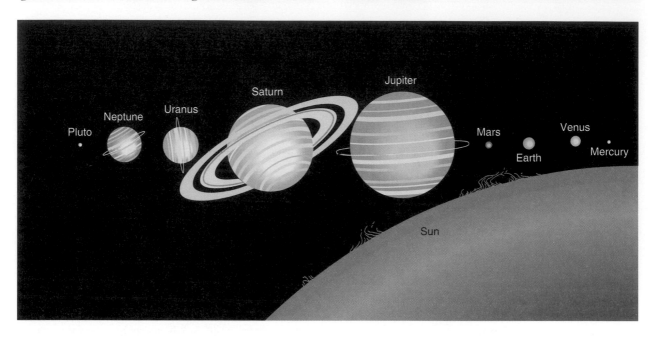

	Average Distance from Sun, millions of km(mi)		Diameter, km(mi)		Tilt of Axis	Revolution, in Earth years	Rotation, in hours, mins., or Earth days
Mercury	58	(36)	4878	(3030)	0°(?)	0.24	59d
Venus	108	(67)	12,100	(7520)	3°	0.62	243d
Earth	150	(93)	12,756	(7926)	23.5°	1.00	1d
Mars	228	(142)	6796	(4225)	25°	1.88	1d
Jupiter	778	(483)	142,800	(88,730)	3.1°	11.86	9h50m
Saturn	1427	(887)	120,660	(74,975)	26.7°	29.46	10h39m
Uranus	2870	(1783)	51,400	(31,940)	82°	84.01	17h18m
Neptune	4486	(2787)	50,950	(31,660)	28.8°	164.97	≅18h
Pluto	5900	(3666)	3500	(2170)	unknown	248.40	6.4d

FIGURE 2–3

A comparison of planets in our Solar System. All of the planets except Pluto have orbits closely aligned to the plane of the ecliptic. The size of each planet appears approximately to scale. Planetary images adapted by permission of Macmillan Publishing Company from *Earth Science,* 6th ed., by Edward J. Tarbuck and Frederick K. Lutgens. Copyright © 1991 by Macmillan Publishing Company.

FOCUS STUDY 2–1

The Scientific Method

Invoking the term "scientific method" may raise an aura of complexity that it should not. The scientific method is simply the application of common sense in an organized and objective manner. A scientist conducts observations, makes generalizations, formulates hypotheses, develops theories, and performs the tests and experiments that such formulations demand. Sir Isaac Newton (1642–1727) developed this method of discovering the patterns of nature, although the term scientific method was applied later.

The scientific method begins with a specific question to be answered by scientific inquiry. Scientists who study the physical environment turn to nature for clues that they can actually observe and measure. Then, observations and data are analyzed to identify coherent patterns that may be present. If patterns are discovered, the researcher may formulate a hypothesis—a formal generalization—deduced from a specific group of observations (e.g., planetesimal hypothesis, nuclear-winter hypothesis, moisture-benefits-from-hurricanes hypothesis). Further observations are related back to the general principles established by the hypothesis. Finally, if data are produced that fit the hypothesis, and if predictions made according to it prove accurate, the hypothesis may be elevated to the status of a theory.

A theory is a hypothesis that has been extensively and successfully tested and represents a truly broad general principle (e.g., theory of relativity, theory of evolution, atomic theory, Big Bang theory). A theory is a powerful device with which to understand both the orderliness and chaos in nature. Predictions based on a theory can be made about things not yet known, the effects of which can be tested and verified or disproved through tangible evidence. The ultimate value of a theory is the continued observation, testing, understanding, and pursuit of knowledge that the theory stimulates.

Important to consider is that pure science does not make value judgments. Instead, pure science provides people and their institutions with information on which to base their own value judgments. These social and political judgments about the applications of science are increasingly critical as Earth's natural systems respond to the impact of modern civilization.

TABLE 2–1

Summary of Earth's past atmospheres.

Name	Approximate Duration (billions of years ago)	Composition	Dominant Features
Primordial atmosphere	4.6–4.0	Water (H_2O), hydrogen cyanide (HCN), ammonia (NH_3), methane (CH_4), sulfur, iodine, bromine, chlorine	Character derived from the nebula Lighter gases of hydrogen and helium escaping to space. An unstable, hot surface with no liquid water collecting
Evolutionary atmosphere	4.0–3.3	At 4.0 billion years ago: 68% H_2O, 13% CO_2, 8% N_2, sulfurous fumes, hydrocarbons, little or no free O_2	Terrigenic origins (outgassing from Earth) Earth thought to have been shrouded in clouds Anaerobic (nonoxygen) environment Surface water accumulation Chemosynthetic bacteria (at 3.6 billion years ago)
Living atmosphere	3.3–0.6	At 3.0 billion years ago: 70% CO_2, 20% H_2O vapor, 10% N_2, <1% O_2	Continued outgassing First photosynthesis in cyanobacteria (at 3.3 billion years ago) Heavy global rains, ocean basins filling Slow evolution of gaseous constituents toward the modern atmosphere
Modern atmosphere	0.6–present	Today: 78% N_2, 21% O_2, 0.9% Argon, 0.036% CO_2, trace gases	Gradual development to the modern atmosphere Abundance of life Moderately fluctuating climate Beginning of anthropogenic atmosphere (human impact)

Earth's Past Atmospheres. A principal component of Earth's history is the evolution of its modern atmosphere. This evolution occurred in four stages. The constituents of Earth's **primordial atmosphere** were derived from the original solar nebula. This atmosphere and the second stage—**evolutionary atmosphere**—are thought to have persisted for relatively short periods. The third and fourth stages—the **living atmosphere** and the **modern atmosphere**—have existed over much greater time spans. Table 2–1 summarizes present scientific views regarding the time, duration, composition, and dominant features of all four atmospheres. The times and durations listed in the table are estimates that represent long, gradual periods of transition.

A long process of chemical evolution was necessary as a forerunner to biological evolution, beneath the shield of newly collected surface waters. In still pools at least 10 m (33 ft) deep, water provided protection from the high radiation levels that existed in the environment at that time. What a fascinating ecology was involved in the development of the atmosphere and hydrosphere, with organisms modifying their environment and in turn being modified by it! Evidence suggests that primitive cells called purple-sulfur bacteria were functioning 3.6 billion years ago, producing organic materials from inorganic elements in the nonoxygen environment of the time. That process is known as **chemosynthesis**, in which an organism uses chemical energy to synthesize organic compounds.

Approximately 3.3 billion years ago, the first organism to perform photosynthesis evolved in Earth's shallow waters. This marked the beginning of the third distinct atmosphere, the living atmosphere. Cyanobacteria (blue-green algae) released oxygen through the process of photosynthesis. However, oxygen did not reach levels comparable to today until about 500 million years ago. The modern atmosphere defined here is the setting for Chapters 3 through 10 of this text.

SOLAR ENERGY: FROM SUN TO EARTH

The Sun captured about 99.9% of the matter (dust and gas) from the original nebula. The remaining 0.1% gave rise to all the other planets, their satellites, asteroids, comets, and debris. Clearly, the dominant object in our region of space is the Sun. In the entire Solar System, it is the only object having the enormous mass needed to create the internal conditions required to produce significant energy.

Solar Operation and Fusion

The solar mass produces tremendous pressure and high temperatures deep in its dense interior. Under these conditions, pairs of hydrogen nuclei, the lightest of all the natural elements, are forced to fuse together. This process of forcibly joining positively charged hydrogen nuclei is called **fusion**. In the fusion reaction, hydrogen nuclei form helium, the second-lightest element in nature, and liberate enormous quantities of energy. During each second of operation the Sun consumes 657 million tons of hydrogen, converting it into 652.5 million tons of helium. The difference of 4.5 million tons is the quantity that is converted directly to energy—literally disappearing solar mass.

A sunny day can seem so peaceful, certainly belying the violence proceeding on the Sun. Before the lunar voyages of the late 1960s, there was serious scientific concern that humans traveling above the protective layers of the atmosphere would be killed by solar radiation. Of course, this proved not to be the case. The Sun's principal outputs consist of the solar wind and portions of the electromagnetic spectrum of radiant energy.

Solar Wind

The Sun constantly emits clouds of ionized (electrically charged) gases. Streams of these charged particles (principally electrons and protons) travel in all directions from the Sun's surface at much less than the speed of light—at about 50 million km (31 million mi) a day—taking approximately 3 days to reach Earth. The term **solar wind** was first applied to this phenomenon in 1958. Solar wind may extend from the Sun as much as 50 times the distance from the Sun to Earth, a distance beyond the planets of our Solar System.

Sunspots and Solar Activity. The most conspicuous features on the Sun are large magnetic disturbances called **sunspots**, which are caused by magnetic storms that indicate unusual solar activity. Individual sunspots may range in diameter from 10,000 to 50,000 km (6200 to 31,100 mi), with some growing as large as 160,000 km (100,000 mi), which is more than 12 times Earth's diameter (Figure 2–4). Flares, prominences, and outbreaks of charged material are associated with these surface disturbances; this material surges into space. Although sunspots were described and recorded for almost 400 years in some detail, a full explanation of their occurrence is still evolving.

Sunspot Cycles. The solar wind grows stronger during periods of increased sunspot activity and weaker during times of lowered activity. A regular cycle exists for sunspot occurrences, averaging 11 years from maximum to maximum; however, the cycle may vary from 7 to 17 years. In recent cycles, a solar minimum occurred in 1976 and a solar maximum took place in 1979, with over 100 sunspots visible. Another minimum was reached in 1986, and an extremely active solar maximum followed in 1990 with over 200 sunspots. This happened 11 years after the previous maximum, maintaining the average.

Earth

FIGURE 2–4
Sunspots appear as dark patches across the Sun. Earth is shown for scale. Photo
courtesy of The Observatories of the Carnegie Institution of Washington.

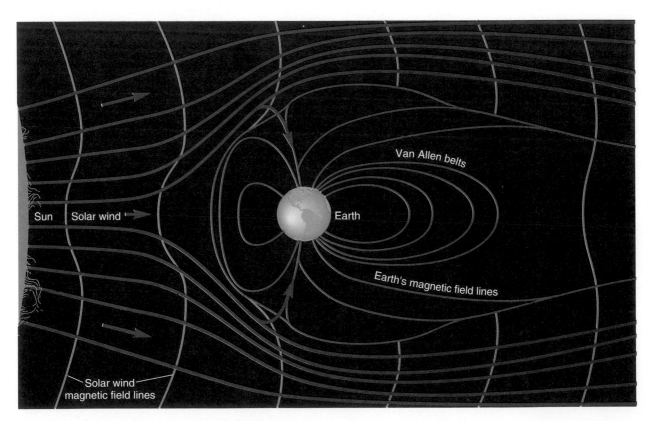

FIGURE 2–5
Earth's magnetosphere and solar wind interaction.

Earth's Magnetosphere

Earth's outer defense against the solar wind is the **magnetosphere**, which is a magnetic field surrounding Earth, generated by dynamolike motions within our planet. The magnetosphere affects the solar wind, deflecting its path (Figure 2–5). When it was first discovered and named in 1958, scientists believed the magnetosphere to be doughnut-shaped and symmetrical. Instead, sensitive instruments aboard space probes found it to be teardrop-shaped.

As the solar wind approaches Earth, the streams of charged particles are deflected by the magnetosphere and course along the magnetic-field lines. As shown in Figure 2–5, the extreme northern and southern polar regions of the upper atmosphere are the points of entry for the solar wind stream.

However, the solar wind does not completely penetrate the atmosphere and therefore does not reach Earth's protected surface. Thus, research on this phenomenon must be conducted in space. The *Apollo XI* astronauts deployed a solar wind experiment on the lunar surface on 20 July 1969 (Figure 2–6). A piece of foil was exposed to the solar wind while the astronauts worked on the Moon. When returned to Earth, the exposed foil was examined for particle impacts to confirm the presence and character of the solar wind.

Solar Wind Effects

The interaction of the solar wind and the upper layers of Earth's atmosphere produces some remarkable phenomena relevant to physical geography: the

FIGURE 2–6
Without a protective atmosphere, the lunar surface allows direct solar wind and electromagnetic radiation to reach the surface. Here a solar wind experiment is being deployed by an *Apollo XI* astronaut. Photo courtesy of National Aeronautics and Space Administration.

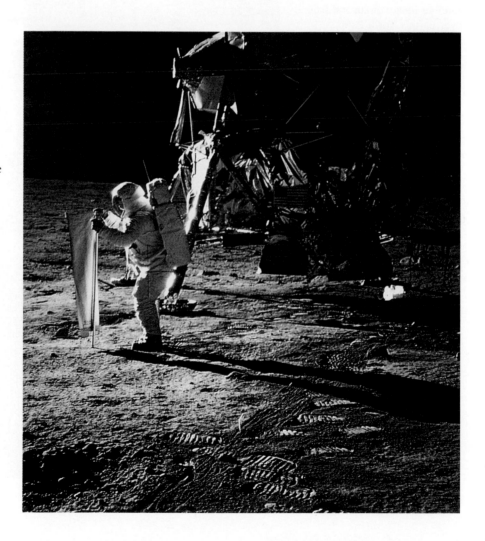

auroras that occur toward both poles, disruption of certain radio broadcasts and even some satellite transmissions, and possible effects on weather patterns.

Auroras. The solar wind creates the auroral phenomenon in the upper atmosphere, 80 to 480 km (50 to 300 mi) above Earth's surface. In this portion of the atmosphere, certain atoms and molecules are ionized, or electrically charged, by interactions with the solar wind. This absorbed energy then is radiated as light of varying colors, depending on which atoms and molecules are stimulated. These lighting effects are the **auroras**: aurora borealis (northern lights) and aurora australis (southern lights).

The auroras generally are visible poleward of 65° latitude when the solar wind is active, as shown by the dramatic polar-orbiting satellite image in Figure 2–7—the auroral halo crowning Earth. (Also see the photograph that begins this chapter). However, at times of intense solar activity, these glowing colored lights in the night sky can be seen far into the middle latitudes. In March 1989, following an enormous solar outbreak, the aurora borealis was visible as far south as Key West, Florida, and Kingston, Jamaica, below 20° N latitude.

FIGURE 2–7
Satellite image of auroral halo over Earth's North Pole. *Dynamics Explorer 1* satellite image courtesy of L. A. Frank and J. D. Craven, The University of Iowa.

Weather Effects. Another effect of the solar wind in the atmosphere involves possible influence on weather. Much scientific inquiry is in progress regarding the relationship between the Sun and Earth's weather and climate cycles. Why do wetter periods in some areas of the midlatitudes tend to coincide with every other solar maximum, and why do droughts often occur near the time of every other solar minimum? For example, sunspot cycles during the 250 years from 1740 to 1990 coincide with periods of wetness and drought, as estimated by an analysis of tree growth rings for that period throughout the western United States. These variations in weather tend to occur within two or three years after the solar maximum or minimum.

Although the correlation is interesting—and controversial, and still undergoing research—the specific causative link in the atmosphere remains to be discovered. Speculation as to the weather connection is centered on heating and wind changes in the upper atmosphere that may be attributable to a conversion of energy from the solar wind. Other explanations also have been proposed.

A remarkable failure in current planning worldwide is the nonattention given these evident cyclical patterns of drought and wetness, regardless of their cause. If such patterns were prepared for, it would be possible to accommodate their consequences. For instance, cyclical drought could be offset through implementation of widespread water conservation and efficiency practices. Wet spells might require levee strengthening and more careful reservoir management. As knowledge of the solar wind/weather phenomenon increases, the ability to forecast the consequences of solar magnetic storms may become feasible.

Electromagnetic Spectrum of Radiant Energy

The key solar input to life is electromagnetic energy. All the radiant energy produced by the Sun is a form of electromagnetic energy. This radiant energy travels at the speed of light, transmitting energy to Earth. The total spectrum of this energy is divided into wavelengths, a **wavelength** being the distance between corresponding points on any two successive waves (Figure 2–8). The number of waves passing a fixed point in one second is the frequency.

When solar radiation is examined numerically, it forms part of the **electromagnetic spectrum**. The Sun emits radiant energy composed of 8% ultraviolet, X-ray, and gamma ray wavelengths; 47% visible light wavelengths; and 45% infrared wavelengths. A portion of the electromagnetic spectrum is illustrated in Figure 2–9. Examine the spectrum and note the wavelengths at which various phenomena occur.

FIGURE 2–8

Wavelength and frequency are used to describe the electromagnetic spectrum. In electromagnetic wave motion, shorter wavelengths are higher in frequency, whereas longer wavelengths are lower in frequency.

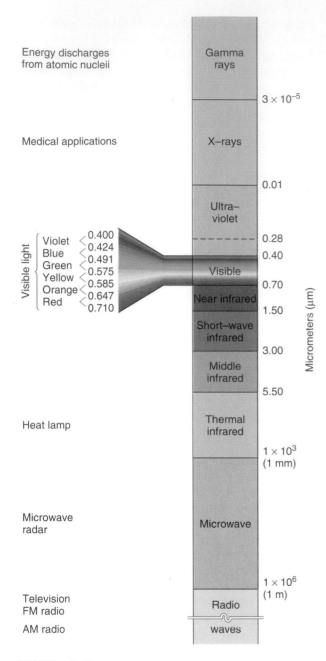

FIGURE 2-9

A portion of the electromagnetic spectrum of radiant energy.

An important physical law applicable to radiating bodies is that they radiate energy in wavelengths related to their individual surface temperatures: the hotter the body, the shorter the wavelengths emitted.

Figure 2-10 shows that the hot Sun radiates energy primarily in shorter wavelengths, concentrated around 0.4 to 0.5 micrometers (μm or microns). The Sun's surface temperature is about 6000°C (11,000°F), and its emission curve in the figure is similar to that predicted for an idealized 6000°C surface.

Earth, on the other hand, demonstrates that the cooler the radiating body, the longer the wavelengths emitted. Figure 2-10 shows that the radiation emitted by Earth occurs in longer wavelengths, centered around 10 micrometers and entirely within the infrared portion of the spectrum. In Chapter 4, we will see that Earth, clouds, sky, ground, and all that is terrestrial are cool-body radiators.

ENERGY AT THE TOP OF THE ATMOSPHERE

The region at the top of the atmosphere, approximately 480 km (300 mi) above Earth's surface, is termed the **thermopause**. It is the outer boundary of Earth's energy system and provides a useful point at which to assess the arriving solar radiation before it is diminished by passage through the atmosphere.

Intercepted Energy

Earth's distance from the Sun results in Earth's interception of only one two-billionth of the Sun's total energy output. This tiny fraction of the Sun's overall output represents an enormous amount of energy input into Earth's systems. Intercepted solar radiation is called **insolation**, a term specifically applied to radiation arriving at Earth's surface and atmosphere. Insolation at the top of the atmosphere is called the solar constant.

Solar Constant. Knowing the amount of insolation intercepted by Earth is important to climatologists and other scientists. The **solar constant** is the average value of insolation received at the thermopause (on a plane surface perpendicular to the Sun's rays) when Earth is at its average distance from the Sun. That value of the solar constant is 1370 watts per m². In more familiar calorie heat units, the solar constant

FIGURE 2–10

Solar and terrestrial energy distribution by wavelength.

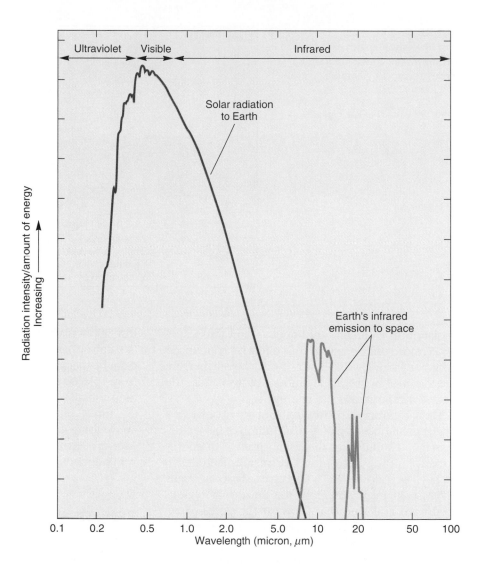

is expressed as approximately 2 calories per cm² per minute. (A **calorie** is the amount of energy required to raise the temperature of one gram of water one degree Celsius.) Another way of expressing the solar constant in terms of energy over area is to use a **langley**, which is 1 calorie per cm². Thus, the solar constant may be expressed as approximately 2 langleys per minute. (See the inside front cover of this text for more information on measurement conversions.)

The constancy of this value over time is important, for small variations of even 0.5 or 1.0% could prove dramatic for Earth's energy system. Paleontologists, who deal with the life of past geologic periods, estimate that solar energy levels have varied slightly, perhaps less than 10 to 15%, over the past several billion years. The Solar Maximum Mission (1980–1989), dubbed Solar Max, was a satellite launched to measure total solar output. Solar Max found average variations of ±0.04% in the solar constant, with the largest changes ranging up to 0.3%, essentially in correlation with sunspot activity.

Uneven Distribution of Insolation. Earth's curved surface presents a continually varied angle to the incoming parallel rays of insolation (Figure 2–11). The only point receiving insolation perpendicular to the surface (from directly overhead) is the **subsolar point**, which is intensely illuminated and occurs at lower latitudes. All other places receive

FIGURE 2–11
Solar insolation angles
determine the concentration of
energy receipts by latitude.

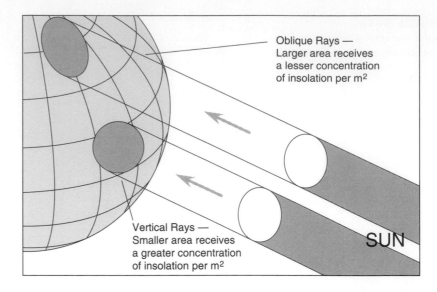

insolation at an angle—obliquely—and thus experience more diffuse energy; this effect is pronounced at higher latitudes. In Figure 2–11, the smaller the area covered by the same amount of insolation, the more concentrated is the energy.

The latitudinal variation in the angle of solar rays results in an uneven global distribution of insolation. Figure 2–12 illustrates the daily variation in energy at the top of the atmosphere for various latitudes in langleys or calories/cm². The chart shows a steady decrease in insolation from the equatorial regions northward and southward toward the poles. However, in June, the North Pole receives more than 1000 langleys per day, which is more than is received at 40° N latitude or at the equator. Such high values result from the 24-hour duration of exposure, as compared with only 15 hours of daylight at 40° N latitude and 12 hours at the equator. Even with the summer noon Sun low in the sky at the poles, a day twice as long as at the equator yields only about 200 langleys difference.

In December, the pattern reverses. Note that the top of the atmosphere at the South Pole receives even more insolation than the North Pole does in June (over 1100 langleys). This is a function of Earth's closer proximity to the Sun at perihelion.

Table 2–2 shows annual energy receipts in langleys at the thermopause and at the surface for various latitudes. This uneven global distribution of

insolation is the basis for major circulations within the atmosphere and oceans. The atmosphere is like a giant heat engine, driven by differences in insolation from place to place. During a year's time, the thermopause above the equatorial region receives 2.5 times more insolation than the thermopause above the poles. As you go about your daily activities, use the dynamic natural systems about you as reminders of this constant flow of solar energy.

REMOTE SENSING

In this era of observations from orbit outside the atmosphere and from aircraft within it, scientists are obtaining a wide array of remotely sensed data. This is an important tool for the spatial analyses of Earth's environment. When we observe the environment with our eyes, we are sensing the shape, size, and color of objects from a distance, utilizing the visible-wavelength portion of the electromagnetic spectrum. Similarly, the film in a camera senses the wavelengths for which it was designed (visible light or infrared) and is exposed by the energy that is reflected and emitted from a scene.

Our eyes and cameras both represent familiar means of **remote sensing** of information about a distant subject, without physical contact. Aerial photographs have been used for years to improve the

FIGURE 2–12
Total daily insolation received at the top of the atmosphere, charted in langleys per day by latitude and month. Adapted by permission of the Smithsonian Institution Press from *Smithsonian Miscellaneous Collections: Smithsonian Meteorological Tables,* vol. 114, 6th Edition. Robert List, ed. Smithsonian Institution, Washington, DC, 1984, p. 419, Table 134.

TABLE 2–2
Annual insolation receipts.

Latitude	Annual Insolation (calories per cm^2)	
	Thermopause	Earth's Surface
0°(equator)	320,000	186,000
40° North	250,000	152,000
90°(North Pole)	132,000	68,000

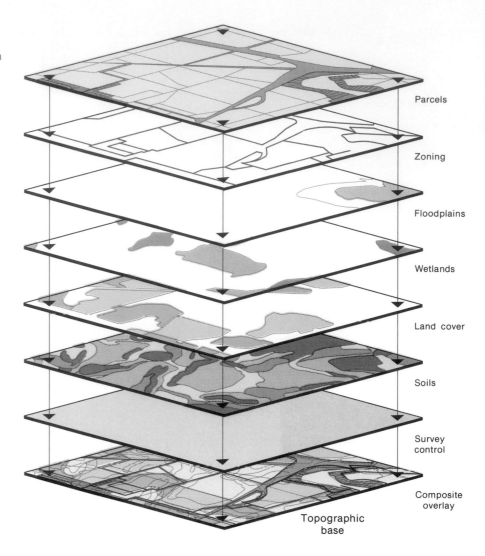

FIGURE 2–13
Layered spatial data in a geographic information system (GIS) format. After USGS.

Parcels

Zoning

Floodplains

Wetlands

Land cover

Soils

Survey control

Composite overlay

Topographic base

accuracy of surface maps at less expense and with greater ease than is permitted by actual on-site surveys. Such accurate measurements derived from photographs is the realm of **photogrammetry**, an important application of remote sensing.

Various surface materials absorb and reflect insolation in different and characteristic ways. These differences can be remotely sensed by their reflected wavelengths: ultraviolet (0.3–0.4 μm), visible light (0.4–0.7 μm), reflected infrared (0.7–3.0 μm), thermal infrared (3.0–5.0 μm and 8.0–14.0 μm), and microwave radars (0.3–300 cm) (see Figures 2–9 and 2–10).

Passive remote sensing systems record radiant energy transmitted from a surface, whereas *active*

systems direct a beam of energy at a surface and analyze the returning reflection. Side-Looking Airborne Radar (SLAR) is such an active system; its radar energy penetrates clouds and produces high-resolution images of surfaces it scans. Chapter 8 presents an analysis of Hurricane Gilbert using a SLAR image, a new application of remote-sensing techniques.

Many remote sensing-images are used throughout this text. The Japan Weather Association operates the *GMS* weather satellite, which produced the image of three typhoons shown in Chapter 8. The United States *Landsat* series of four satellites, launched between 1972 and 1984, provided a variety of visible and infrared data, as shown in images of the

Appalachians in Chapter 12, river deltas in Chapter 14, and glaciers in Chapter 17. The Geostationary Operational Environmental Satellite, known as *GOES-East,* provides the daily infrared and visible images of weather in the Western Hemisphere that you probably have seen on television (a three-day example appears in Figure 8–15). The data acquired from these systems have proven invaluable to a better understanding of Earth's air, water, land, and living systems.

Geographic Information System (GIS)

Remote sensing is an important acquisition tool that generates large volumes of spatial data. Computers have allowed an integration of geographic information from direct surveys and remote sensing in complex ways never before possible. A **geographic information system (GIS)** is a computer-based data-processing tool or methodology for gathering, manipulating, and analyzing geographic information. An example is shown in Figure 2–13. The approach involves applications of almost unlimited variety. A common terrestrial reference system, such as the latitude and longitude coordinates provided by a map, is an essential beginning component for any GIS. Such maps can be converted into digital data of areas, points, and lines, depending on the task. Remotely sensed imagery and data are then transferred to the coordinate system on the reference map.

One of the most extensive and longest-operating systems is the Canada Geographic Information System (CGIS). Environmental data about natural features, resources, and land use were taken from maps, aerial photographs, and orbital sources and reduced to map segments. These were divided into polygons and entered into the CGIS. The development of this system has progressed hand-in-hand with the ongoing Canada land-inventory project.

Prior to the advent of computers, an environmental-impact analysis required someone to gather various data and painstakingly hand-produce overlays of information to determine positive and negative aspects of a project or event. Today, this layered information can be handled by a computer-driven GIS, which assesses the complex interconnections among different components. In this way, subtle changes in one element of a landscape may be identified as having a powerful impact elsewhere. GIS applications are specifically important for the analysis of environmental events, both natural and human-caused. For example, the USGS completed a GIS to help analyze the spatial impact of the 1989 *Exxon Valdez* oil spill in Alaska.

Simulations and construction of hypothetical models are other important capabilities of a GIS. This is evident from the general circulation models used to analyze the atmosphere-ocean interaction, which is so important to forecasting future weather and climate patterns. An important planning and analytical element will be an expanded interface between GIS and remote-sensing capabilities. The entire GIS methodology represents expanding career opportunities in many fields and an important step in better understanding Earth's systems.

SUMMARY

The Solar System, planets, and Earth condensed from clouds of dust, gas, debris, and icy comets. Earth's atmosphere evolved in 4 stages—primordial, evolutionary, living, and modern. Solar energy is produced in the Sun's interior and flows to its surface, where charged particles of solar wind and the radiant energy of the electromagnetic spectrum travel outward in all directions through interplanetary space.

Solar wind is deflected by Earth's magnetosphere, producing various effects in the upper atmosphere. This interaction appears to influence Earth's weather and produces auroras in regular cycles. Electromagnetic energy passes through Earth's magnetosphere to the atmosphere. At the thermopause, the average amount of insolation received—the solar constant—is expressed as 1370 watts/m^2, or 2.0 langleys per minute (2.0 calories/cm^2/min).

Because Earth is a sphere, energy is unevenly distributed at the top of the atmosphere. The angle of the Sun's rays is more direct and thus more concentrated at lower latitudes and is more oblique and diffuse at higher latitudes. This uneven pattern of energy input, when transmitted through the atmosphere, sets in motion the atmosphere and oceans, and powers the biosphere. Much is being learned about the operation of Earth's systems through orbital and aerial remote sensing, enhanced through the application of geographic information systems.

KEY TERMS

Aphelion	Nebula
Auroras	Perihelion
Calorie	Photogrammetry
Chemosynthesis	Plane of the ecliptic
Electromagnetic spectrum	Planetesimal hypothesis
Evolutionary atmosphere	Primordial atmosphere
Fusion	Remote sensing
Geographic information system (GIS)	Scientific method
Gravity	Solar constant
Insolation	Solar wind
Langley	Speed of light
Living atmosphere	Subsolar point
Magnetosphere	Sunspots
Milky Way Galaxy	Thermopause
Modern atmosphere	Wavelength

REVIEW QUESTIONS

1. Describe the Sun's status among stars in the Milky Way Galaxy. Describe the Sun's location, size, and relationship to its planets.
2. If you have seen the Milky Way, briefly describe it. Use specifics from the text in your description.
3. Briefly describe Earth's origin as part of the Solar System.
4. How far is Earth from the Sun in terms of light speed? In terms of kilometers and miles? Relate this distance to the shape of Earth's orbit during the year.
5. How many distinct atmospheres have there been? How does Earth's present atmosphere compare to past atmospheres?
6. Within which of Earth's atmospheres did photosynthesis begin?
7. Define the scientific method and give an example of its application.
8. How does the Sun produce such tremendous quantities of energy? Write out a simple fusion reaction, using the quantities (in tons) of hydrogen involved.
9. What is the sunspot cycle? At what stage in the cycle were we in 1990?
10. Describe Earth's magnetosphere and its effects on the solar wind and the electromagnetic spectrum.
11. Summarize the presently known and hypothesized effects of the solar wind relative to Earth's environment.
12. Describe the segments of the electromagnetic spectrum, from shortest to longest wavelength.
13. Define a calorie and a langley. What is the annual total of langleys received at the top of the atmosphere above the equator? Above 40° N latitude? Above the North Pole?
14. What is the solar constant? Why is it important to know?
15. Select 40° or 50° north latitude on Figure 2–12, and describe the amount of energy in langleys per day characteristic of each month throughout the year.
16. If Earth were flat and oriented at right angles to the incoming solar radiation, what would be the latitudinal distribution of solar energy at the top of the atmosphere?

SUGGESTED READINGS

Briggs, Geoffrey, and Frederic Taylor. *The Cambridge Photographic Atlas of the Planets*. Cambridge: Cambridge University Press, 1982.

Calder, Nigel. *Time Scale—An Atlas of the Fourth Dimension*. New York: Viking Press, 1983.

Colwell, Robert N., ed. *Manual of Remote Sensing*. 2d ed. Vol. 1, *Theory, Instruments, and Techniques;* Vol. 2, *Interpretation and Applications*. Falls Church, VA: American Society of Photogrammetry, 1983.

Eddy, John A. *A New Sun—The Results from Skylab*. U.S. National Aeronautics and Space Administration. Washington: U.S. Government Printing Office, 1979.

Editors of Time-Life Books. *Stars*. Voyage Through the Universe Series. Alexandria, VA: Time-Life Books, 1988.

Frazier, Kendrick, and the editors of Time-Life Books. *Solar System*. Time-Life Planet Earth Series. Alexandria, VA: Time-Life Books, 1985.

Gallant, Roy A. *National Geographic Picture Atlas of Our Universe*. Washington: National Geographic Society, 1986.

Herman, John R., and Richard Goldberg. *Sun, Weather, and Climate*. U.S. National Aeronautics and Space Administration. Washington: U.S. Government Printing Office, 1978.

Rabenhorst, Thomas D., and Paul D. McDermott. *Applied Cartography—Introduction to Remote Sensing*. Columbus: Macmillan Publishing Co., 1989.

"The Solar System." *Scientific American* 233, no. 3 (September 1975): entire issue.

Sullivan, Walter. "Curtains of Light, Horsemen of Night." *Audubon* 89, no. 1 (January 1987): 40–51.

Photo by author.

3

Earth's Modern Atmosphere

Earth's atmosphere is a unique reservoir of gases, the product of billions of years of development. This chapter examines its structure, function, and composition, starting from outer space and descending through the various layers and regions of the atmosphere to Earth's surface. Your consideration of our modern atmosphere also must include the spatial aspects of human-induced problems that affect it, such as air pollution, the stratospheric ozone predicament, and the blight of acid deposition. We all must consider these critical topics carefully, for we are participating in producing the atmosphere of the future.

FIGURE 3–1

Earthrise over the lunar surface. Photo courtesy of National Aeronautics and Space Administration. Quotation from "The World's Biggest Membrane" from *The Lives of a Cell* by Lewis Thomas. Copyright © 1973 by the Massachusetts Medical Society. Originally published in the *New England Journal of Medicine*. Reprinted by permission of Viking Penguin, a division of Penguin Books USA, Inc.

Viewed from the distance of the Moon, the astonishing thing about Earth, catching the breath, is that it is alive. The photographs show the dry, pounded surface of the Moon in the foreground, dead as an old bone. Aloft, floating free beneath the moist, gleaming membrane of a bright blue sky, is the rising Earth, the only exuberant thing in this part of the cosmos.... It has the organized, self-contained look of a live creature, full of information, marvelously skilled in handling the Sun.... When the Earth came alive it began constructing its own membrane, for the general purpose of editing the Sun.... Taken all in all, the sky is a miraculous achievement. It works, and for what it is designed to accomplish it is as infallible as anything in nature.

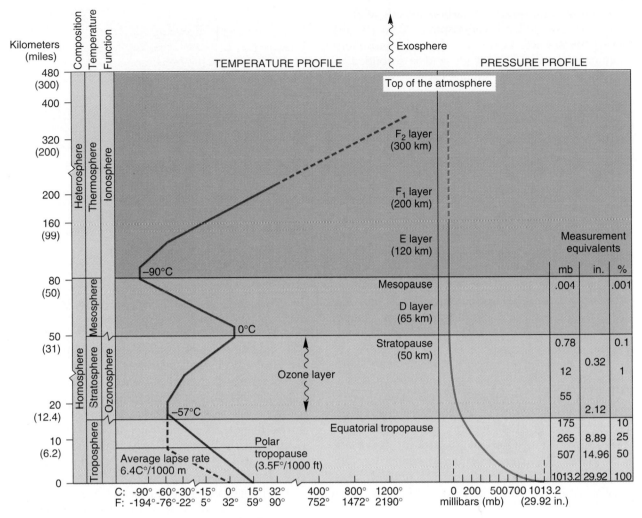

FIGURE 3–2
Modern atmosphere composite chart of composition, temperature, function, and pressure.

ATMOSPHERIC STRUCTURE AND FUNCTION

The modern atmosphere probably is the fourth general atmosphere in Earth's history. (The modern atmosphere and the other three—the primordial, evolutionary, and living atmospheres—were discussed in Chapter 2.) This modern atmosphere is in reality a gaseous mixture of ancient origins, the sum of all the exhalations and inhalations of life on Earth throughout time. The principal substance of this atmosphere is air, the medium of life as well as a major industrial and chemical raw material. Air is a simple additive mixture of gases that is naturally

odorless, colorless, tasteless, and formless, blended so thoroughly that it behaves as if it were a single gas.

In his book *The Lives of a Cell,* physician and self-styled "biology-watcher" Lewis Thomas compared the atmosphere of Earth to an enormous cell membrane. The membrane around a cell regulates the interactions between the cell's delicate inner workings and the potentially disruptive outer environment. Each cell membrane is very selective as to what it will and will not allow to pass through. The modern atmosphere acts as Earth's protective membrane, as Thomas describes so vividly in Figure 3–1.

The atmospheres of the other three inner planets bear no resemblance to Earth's atmosphere. Mars has a cold, thin atmosphere of carbon dioxide, equivalent in pressure to Earth's atmosphere 32 km (20 mi) above the surface. Venus also has an atmosphere dominated by carbon dioxide and is shrouded in clouds of sulfuric acid. Surface pressures on Venus are about 90 times those on Earth, and surface temperatures average 500°C (932°F). Mercury has no appreciable atmosphere and, in fact, appears similar to the Moon, pockmarked by innumerable craters.

Earth's modern atmosphere is arranged in a series of imperfectly shaped concentric spheres, all bound to the planet by gravity. As critical as the atmosphere is to us, it represents only a thin-skinned envelope amounting to less than one-millionth of Earth's total mass. Important to the following discussion is Figure 3–2, which illustrates the entire atmosphere and charts its essential aspects: pressure and temperature gradients, composition, and functional regions. The three criteria for classification of the atmosphere are noted on the left side of the illustration. You can use this integrative chart as a reference and guide as we explore these concepts.

Air Pressure

The gases and atoms that comprise air create pressure through their motion, size, and number. This pressure is exerted on all surfaces in contact with the air. The weight of the atmosphere, or **air pressure**, crushes in on all of us; fortunately, that same pressure also exists inside us, pushing outward. The atmosphere exerts an average force of approximately 1 kg/cm^2 (14.7 lb/in^2) at sea level. Under the acceleration of gravity, air is compressed and therefore denser near Earth's surface (Figure 3–3). It rapidly thins with increased altitude, a decrease that is measurable because air exerts its weight as pressure. Consequently, over half the total mass of the atmosphere is compressed below 5500 m (18,000 ft), 75% is compressed below 10,700 m (35,105 ft), and 90% is compressed below 16,000 m (52,500 ft). All but 0.1% is accounted for at an altitude of 50 km (31 mi), as shown in the pressure profile in Figure 3–2.

In A.D. 1643, a pupil of Galileo's named Evangelista Torricelli was working on a mine drainage problem. This led him to discover a method for measuring air pressure—Figure 3–4(a). Pumps in the mine were able to draw water upward only about 10 m (33 ft).

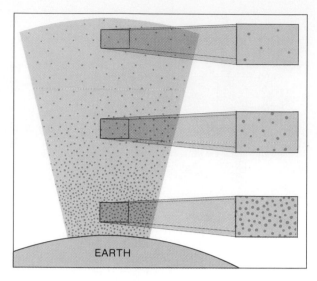

FIGURE 3–3
The atmosphere, denser near Earth's surface, rapidly decreases in density with altitude.

This limitation was not caused by the pumps but by the atmosphere itself. A pump of the type shown works by creating a vacuum above the water, which allows the pressure of the atmosphere that is pushing down on the mine water to force water up the pipe. Torricelli noted that the water level in the vertical pipe fluctuated from day to day. He surmised correctly that air pressure varied over time and with changing weather conditions.

To simulate the problem at the mine, Torricelli devised an instrument at the suggestion of Galileo. Instead of water, he used a much denser fluid, mercury (Hg), so that he could use a glass laboratory tube only 1 m high. He sealed the glass tube at one end, filled it with mercury, and then inverted it into a dish of mercury—Figure 3–4(b). Torricelli established that the average height of the column of mercury in the tube was 760 mm (29.92 in.). The column of mercury was counterbalanced by the mass of surrounding air exerting an equivalent pressure on the mercury in the dish. Using similar instruments today, normal sea level pressure is expressed as 1013.2 millibars (a way of expressing force per square meter of surface area), or 29.92 in. of mercury.

Any instrument that measures air pressure is called a barometer. Torricelli developed a **mercury barometer**. A more compact design that works without a meter-long tube of mercury is called an **aneroid barometer**—Figure 3–4(c). Inside this type of barometer, a small cell or chamber is partially emptied of air, sealed, and connected to a mechanism that is sensitive to changes in the cell. As air pressure varies, the mechanism detects the difference and relays the change to the needle on the dial. Altimeters used in aircraft are a type of aneroid barometer that is able to accurately measure altitude because of the reduced air pressure above sea level.

Figure 3–5 illustrates comparative scales in millibars and inches used to express air pressure and its relative force. Strong high pressure to deep low pressure can range from 1050 mb to 980 mb. The figure also indicates the extreme highest and lowest pressures ever recorded in the United States and on

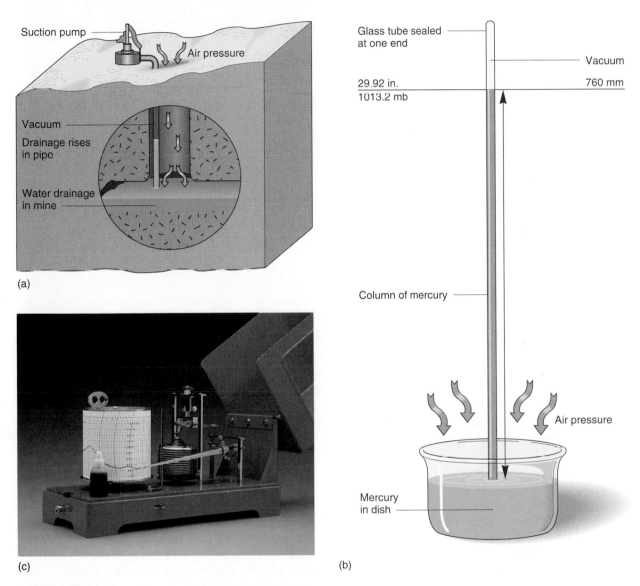

(a)

(c)

(b)

FIGURE 3–4
Evangelista Torricelli developed the barometer to measure air pressure as a by-product of trying to solve a mine drainage problem (a). Instruments used to measure atmospheric pressure are (b) a mercury barometer and (c) an aneroid barometer. (c) Courtesy of Qualimetrics, Inc., Sacramento, CA.

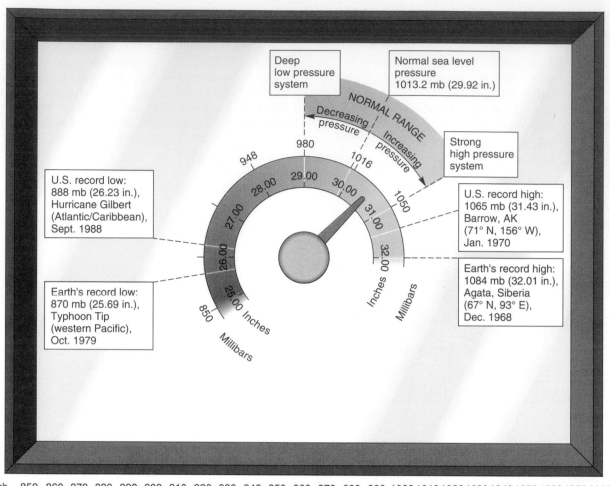

mb 850 860 870 880 890 900 910 920 930 940 950 960 970 980 990 1000 1010 1020 1030 1040 1050 1060 1070 1080

in. 25   25.5   26   26.5   27   27.5   28   28.5   29   29.5   30   30.5   31   31.5   32

FIGURE 3–5
Scales for expressing barometric air pressure in millibars and inches, with average
air pressure values and recorded pressure extremes.

Earth. The concept of air pressure, its measurement, and its application to atmospheric circulation and weather are further explained in Chapters 6 and 8. Let's now discuss the atmosphere, using several different classification criteria.

Atmospheric Composition, Temperature, and Function Classifications

This discussion of the atmosphere from the outer edges of space to Earth's surface is conveniently classified using three criteria: composition, temper-

ature, and function. Based on chemical *composition,* the atmosphere is divided into two broad regions, the heterosphere and the homosphere. Based on *temperature,* the atmosphere is divided into four distinct zones, the thermosphere, mesosphere, stratosphere, and troposphere. Finally, two specific zones are identified on the basis of *function* for their role in removing most of the harmful wavelengths of solar radiation; these zones are the ionosphere and the ozonosphere (ozone layer).

In the following sections we consider first the heterosphere, which ranges from 10,000 km (6000 mi) altitude down to 80 km (50 mi). As the prefix *hetero-* implies, this region is not uniform (not evenly mixed). Instead, distinct layers of molecules and atoms (mainly nitrogen and oxygen) are distributed according to their atomic weight. The heterosphere has a distinctive composition, thermal properties (thermosphere), and an important function (ionosphere).

Below the heterosphere we view the homosphere in similar fashion. The homosphere envelopes Earth from 80 km (50 mi) down to Earth's surface. As the prefix *homo-* implies, this region is composed of an even mixture of gases, behaving as if it were a single gas. We explore its composition, temperature regions (mesosphere, stratosphere, and troposphere) and its critical function (ozonosphere). These classifications are all shown on the composite chart in Figure 3–2.

Heterosphere

One of the criteria for classifying the atmosphere is composition. If we start at the top of the atmosphere, the first region is the **heterosphere**, which ranges from over 10,000 km altitude down to 80 km (6000 mi down to 50 mi), as shown in Figure 3–2. Although individual hydrogen and helium atoms are weakly bound by gravity as far as 32,000 km (19,900 mi) from Earth, for all practical purposes the top of our atmosphere is at roughly 480 km (300 mi)—the same as the boundary used in Chapter 2 for determining the solar constant. What lies beyond that altitude is the vacuum of space, commonly called the **exosphere**.

Thermosphere. The atmosphere also can be divided into four temperature regions. The first, dominating the heterosphere from 480 km (300 mi) down to 80 km (50 mi), is the **thermosphere** (from *thermo-,* meaning "heat"). The upper part of the thermosphere is a transition zone into space, termed the thermopause (the suffix *-pause* means "to cause to change"). High temperatures are generated in the thermosphere by the absorption of shortwave solar radiation.

The temperature profile in Figure 3–2 shows that temperatures rise sharply in the thermosphere, up to 1200°C (2200°F) and higher. However, we must use different concepts of "temperature" and "heat" to understand this effect. The intense radiation in this portion of the atmosphere excites individual molecules (nitrogen and oxygen) and atoms (oxygen) to high levels of vibration. This **kinetic energy**, the energy of motion, is the vibrational energy stated as "temperature," but the density of the molecules is so low that little actual heat is produced. Heating closer to Earth in the lower atmosphere differs because the active molecules in the denser atmosphere transmit their kinetic energy as **sensible heat** (meaning that we can sense it).

Ionosphere. A third approach to atmospheric classification identifies two functional layers, so-called because both function to filter harmful wavelengths of solar radiation, protecting Earth's surface from bombardment in any significant quantity. The upper functional layer, the **ionosphere**, extends throughout the thermosphere and beyond. It is composed of charged particles produced by the absorption of cosmic rays, gamma rays, X-rays, and shorter wavelengths of ultraviolet radiation, all of which change the atoms to positively charged ions, giving the ionosphere its name. Radiation bombards the ionosphere constantly, producing a constant flux (flow) of electrons.

Figure 3–2 shows the average daytime altitudes of four regions within the ionosphere, known as the D, E, F_1, and F_2 layers, which are important to broadcast communications. At night the D layer virtually disappears, and the F_1 and F_2 layers usually merge and reflect certain wavelengths of AM radio, returning them at predictable angles to receivers hundreds of kilometers from the transmitters. However, during the day, the D and E regions in the ionosphere actively absorb arriving radio signals, preventing such distant reception. Unaffected are FM or TV broadcast wavelengths, which pass through to space, thus necessitating use of communication satellites for retransmission of these signals to the surface. During times of intense solar wind and auroral activity, the ionosphere becomes so active that radio communications can be disrupted. The glowing auroral lights discussed in Chapter 2 occur principally within the ionosphere.

Homosphere

Below the heterosphere is the second compositional region of the atmosphere, the **homosphere**. It extends from the bottom of the heterosphere at 80 km (50 mi) to Earth's surface. Even though the atmosphere rapidly increases in density with decreasing altitude, the blend (relative proportion) of gases is nearly uniform throughout the homosphere. The only exceptions are the concentration of ozone (O_3) from 50 km down to 19 km (31–12 mi), and the variations in water vapor and pollutants in the lowest portion of the atmosphere.

The stable mixture of gases throughout the homosphere has evolved slowly. The present proportion, which includes oxygen, was attained perhaps as early as 1 billion years ago, but no more recently than 500 million years ago. Table 3–1 lists by volume the stable ingredients that comprise the homosphere.

The atmosphere is a vast reservoir of relatively inert nitrogen, principally originating from volcanic sources. Nitrogen is a key element of life, yet we exhale all the nitrogen that we inhale. This apparent contradiction is explained because the input of nitrogen to our body systems comes instead through compounds in food. In the soil, nitrogen is bound up by nitrogen-fixing bacteria and is returned to the atmosphere by denitrifying bacteria that remove nitrogen from organic materials.

Oxygen, a by-product of photosynthesis, also is essential for life processes. Along with other elements, oxygen forms compounds that compose about half of Earth's crust. Slight spatial variations occur in the atmospheric oxygen percentage because of variations in photosynthetic rates with latitude and the lag time as atmospheric circulation slowly mixes the air. Both nitrogen and oxygen reserves in the atmosphere are so extensive that, at present, they far exceed human capabilities to disrupt or deplete them.

Argon is a residue from the radioactive decay of an isotope (form) of potassium called potassium-40 (^{40}K). Slow accumulation over millions of years accounts for all the argon present in the modern atmosphere. This gas is completely inert (and thus is called a "noble" gas) and therefore is unusable in life processes. Because industry has found uses for inert argon (in light bulbs, welding, and some lasers), it is extracted from the atmosphere, along with nitrogen and oxygen, for commercial and industrial uses (Figure 3–6).

Although its percentage in the atmosphere has gradually increased over the past 200 years, **carbon dioxide** is basically a stable atmospheric component and thus is included in Table 3–1. It is a natural by-product of life processes, and the implications of its current increase are critical to society and the future. The role of carbon dioxide in the gradual warming of Earth is discussed in Chapters 5 and 10.

Mesosphere. The homosphere encompasses three regions defined by temperature criteria: the mesosphere, stratosphere, and troposphere. The **mesosphere** is the area from 80 km altitude down to 50 km (50–30 mi). The mesopause is the coldest portion of the atmosphere, averaging −90°C (−130°F), although that temperature may vary by ±25–30C° (±45–54F°). Very low pressures ranging from 0.1 mb to 0.001 mb also are found in the mesosphere. The lower limit for the auroral glow is the upper mesosphere. In addition, the mesosphere sometimes receives cosmic or meteoric dust, which acts as

TABLE 3–1
Stable components of the modern homosphere.

Gas	Symbol	% by Volume	Parts per Million (ppm)
Nitrogen	N_2	78.084	780,840
Oxygen	O_2	20.946	209,460
Argon	Ar	0.934	9,340
Carbon dioxide	CO_2	0.036	360
Neon	Ne	0.001818	18
Helium	He	0.000525	5
Methane	CH_4	0.00014	1.4
Krypton	Kr	0.00010	1.0
Ozone	O_3	variable	
Nitrous oxide	N_2O	trace	
Hydrogen	H	trace	
Xenon	Xe	trace	

FIGURE 3–6
Air is a major industrial and chemical raw material that is extracted from the atmosphere by air mining companies. Photo by author.

nuclei for forming fine ice crystals. At high latitudes an observer may see these bands of crystals glow in rare and unusual night clouds called **noctilucent clouds.**

Stratosphere and Ozonosphere. The **stratosphere** extends from 50 km down to 20 km (31–12.4 mi) above Earth's surface. Temperatures decrease throughout the stratosphere, from 0°C (32°F) at 50 km (stratopause) to −57°C (−70°F) at 18–20 km (tropopause). Meanwhile, pressures increase from only 0.78 mb at the stratopause to 175 mb at the tropopause (20 km). The heat source throughout the stratosphere is the other functional layer, called the **ozonosphere,** or **ozone layer.** The ozone layer is composed of ozone, highly reactive oxygen molecules made up of three oxygen atoms (O_3) instead of the usual two (O_2) of oxygen gas. Ozone absorbs wavelengths of ultraviolet light (0.1–0.3 micrometer) and subsequently reradiates that energy at longer wavelengths, as infrared energy. Through this process, most harmful ultraviolet radiation is effec-

tively "filtered" from the incoming solar radiation, safeguarding life at Earth's surface.

The ozone layer, presumed to be relatively stable over the past several hundred million years (allowing, of course, for daily and seasonal fluctuations), is now in a state of continuous change. Focus Study 3–1 presents a background analysis of the developing crisis in this critical section of our atmosphere. The spatial implications for humanity of even small losses in stratospheric ozone makes this an important topic for applied studies in physical geography and atmospheric sciences.

Troposphere. The home of the biosphere is the final temperature region of the atmosphere nearest Earth's surface, the **troposphere.** Approximately 90% of the total mass of the atmosphere and the bulk of all water vapor, clouds, weather, and air pollution are contained within the troposphere. The *tropopause* is defined by an average temperature of −57°C (−70°F), but its exact location varies with the seasons, latitude, and sea level temperatures and

The Stratospheric Ozone Predicament

W hy is the ozone layer so important? As noted in the text, it reduces ultraviolet light radiation levels from the Sun to relatively safe levels at Earth's surface. Even a 1% decrease in stratospheric ozone increases surface exposures to ultraviolet light from 1 to 3%, leading to a 2 to 5% increase in skin cancers. Most affected are light-skinned persons who live at higher elevations and those who work principally outdoors.

Skin cancers and related precursory ailments are steadily increasing in North America, now totaling more than 600,000 cases a year, including over 8000 malignant melanoma deaths annually. Skin cancer occurrences are increasing at an alarming 10% per year. Damage to eye tissues and even reduction in the effectiveness of certain body immunity systems are considered other possibilities. Also, potential damage to crop yields and certain aquatic life forms are probable consequences of ozone loss. An oceanographer at Texas A&M, in addition to scientists at Scripps Institution of Oceanography, completed studies in 1989 that proved the vulnerability of phytoplankton to such severe increases in ultraviolet light in the oceans and seas of polar regions. This does not bode well for essential food chains in affected oceans.

A Fragile Veil

The stratospheric ozone layer is rarefied. A sample from the densest part of the ozone layer—at an altitude of 29 km (18 mi)—contains only 1 part ozone per 4 million parts of air. Yet this layer has maintained a steady-state equilibrium for at least 500 million years. Since World War II, human activities have begun to alter this long-standing system into a dynamic condition that exhibits an overall declining ozone concentration. The possible depletion of the ozone layer by human activity was first suggested during the summer of 1974 by professors F. Sherwood Rowland and Mario J. Molina of the University of California–Irvine. Their article appeared in the British journal *Nature* and triggered the great ozone debate.*

Rowland and Molina proposed that chlorine atoms in manufactured chemicals were depleting Earth's ozone layer. Because measurement and confirmation are difficult on a scale as vast as the atmosphere, the industries producing these products successfully delayed action for 15 years, claiming that hard evidence was lacking to substantiate the ozone-depletion theory. At present, after much scientific evidence and verification of actual depletion of stratospheric ozone by chlorine atoms, even industry has had to admit the problem is real.

*Rowland, F. Sherwood, and Mario J. Molina, *Nature* 249 (1974): 810.

Chlorofluorocarbon compounds, or CFCs, are large manufactured molecules containing chlorine, fluorine, and carbon; they are marketed under many trade names. CFCs are inert, meaning that they do not readily react with anything, and they possess remarkable heat properties useful for heat-transfer systems. These two traits make them valuable as aerosol propellants and refrigerants. By 1983 some 9.1 million metric tons (10 million tons) of CFCs had been sold worldwide and subsequently released into the atmosphere. Production now averages about 910,000 metric tons (1 million tons) per year.

Rowland and Molina hypothesized that the inert CFC molecules remained intact in the atmosphere, eventually migrating upward, working their way into the stratosphere. Being stable, CFCs do not dissolve in water and do not break down biologically. However, in the stratosphere, the increased ultraviolet light dissociates (splits) the CFC molecules, releasing chlorine (Cl) atoms and forming chlorine oxide (ClO) molecules. This chlorine then acts as a catalyst, stimulating a complex reaction that results in a reduction of ozone. A single chlorine atom destroys tens of thousands of ozone molecules. Because each chlorine atom remains at ozone-layer altitudes for 40 to 100 years, the possible long-term consequences of these reactions pose an increasing threat through the next century. Since the scientists' initial proposal, hundreds of different reactions—both depleting and rehabilitating ozone—have been discovered.

How Much Damage?

Rowland and Molina initially postulated losses of ozone that ranged from 7% to 13% over the next 100 years. Four different reports from the National Academy of Sciences (NAS) postulated varying estimates of depletion, ranging from 16.5% to only 2–4% (1979 and 1984 reports, respectively). In addition, two satellites (*Nimbus 4* and *Nimbus 7*) operated by the National Oceanic and Atmospheric Administration (NOAA) recorded global decreases in stratospheric ozone of more than 3% from 1970 to 1979. By 1991 ozone losses above the midlatitudes approached 6%, much faster than scientists previously realized. With little action taking place in response to these measurements, Dr. Rowland stated his frustration, still applicable today:

> What's the use of having developed a science well enough to make predictions, if in the end all we are willing to do is stand around and wait for them to come true. . . . Unfortunately, this means that if there is a disaster in the making in the stratosphere, we are probably not going to avoid it.*

An International Response

Canada, Sweden, Norway, and the state of Oregon instituted bans on CFC propellants in aerosols between 1976 and 1979. The U.S. federal ban

*Roger B. Barry, "The Annals of Chemistry," *The New Yorker* (9 June 1986): 83.

commenced in the fall of 1978. However, even with the U.S. ban, more than half of the country's production was not outlawed, including CFCs used as blowing agents in making polyurethane foam and as a refrigerant in commercial, residential, and automotive air conditioning systems. Sales of CFCs initially dropped after the aerosol ban, but started upward again by 1981 and have continued to rise to this day (Figure 1). Foreign markets for aerosol and nonaerosol uses of CFCs have contributed to increased production, especially with the 1981 cancellation of export prohibitions for banned products.

In 1985, an initial meeting of 31 nations produced agreement on a short-term freeze of CFC production and a general consensus for a long-term phaseout. This led to the Montreal Protocol, signed in 1987 and revised in 1990. Under the auspices of the United Nations Environment Program, the nations agreed to freeze production at 1986 levels beginning in 1990, to cut production 20% by mid-1993, and to cut production another 50% by mid-1998. The 12 nations of the European Community have committed themselves to total elimination of CFCs, a commitment likely to be followed by the rest of the world. To date, 64 nations, 28 of which are developing countries, have ratified the Montreal Protocol. With new support from the United States, a fund is being set up to assist developing countries with the phaseout.

Ozone Losses Over the Poles

British scientists at their Halley Base in Antarctica have recorded steady losses of stratospheric ozone from 1970 to the present, with huge temporary decreases occurring every October. In each succeeding year of measurement, the "hole" in the ozone has widened and deepened. NOAA satellites also recorded this alarming situation (Figure 2). Chlorine freed in

FIGURE 1
CFC production 1960 to 1990. Note recovering sales in the 1980s. Adapted with permission from "Response to Ozone Depletion," *Chemical & Engineering News* (24 November 1986): 53. Copyright 1986, American Chemical Society.

FIGURE 2

Southern Hemisphere ozone observations in 1987 and 1990. The Antarctic ozone hole appears in dark red near the center. 1987, 1989, and 1990 represent the worst ozone depletions since the first images were taken in 1979. Images courtesy of National Oceanic and Atmospheric Administration and National Aeronautics and Space Administration.

the Northern Hemisphere midlatitudes evidently is redistributed by atmospheric winds, concentrating over Antarctica. The chemical model of chlorine disruption of ozone now is generally accepted by researchers.

Meanwhile, in Antarctica during October 1989 and again in 1990, stratospheric ozone concentrations approached total depletion between altitudes of 12 and 24 km (7.5–14.9 mi) over a surface area about twice the size of the Antarctic continent. The 1989 and 1990 holes exceeded the record depletion that occurred in 1987. In the Arctic, a similar ozone depletion exceeded 35% in 1989 and again in 1990. These holes in the ozone and general depletion overall provide an early warning to civilization about the spatial implications of the ozone predicament. Antarctic explorers and scientists in 1990 experienced what was described as standing under an ultraviolet lamp 24 hours a day. Any exposed skin quickly burned and swelled, and protective clothing that was not carefully selected deteriorated rapidly. Hopefully, by taking action now, this will not be as severe a hazard for life in the next century.

FIGURE 3–7
Temperature profile of the
troposphere.

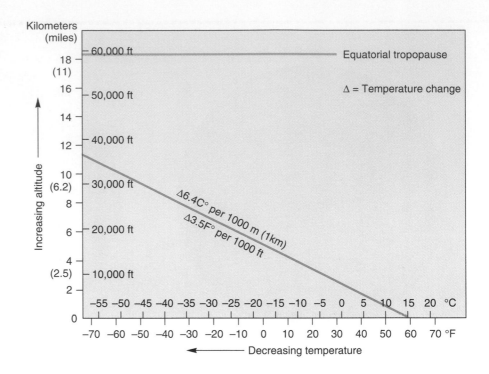

pressures. Near the equator, because of intense heating from below, the tropopause occurs at 18 km (11 mi); in the middle latitudes, it occurs at 13 km (8 mi); and at the North and South Poles it is only 8.0 km (5.0 mi) or less above Earth's surface.

The marked warming in the stratosphere causes the tropopause to act like a lid, essentially preventing whatever is in the cooler air below from mixing into the stratosphere. However, the tropopause is disrupted above the midlatitudes wherever jet streams occur (Chapter 6), and the resulting vertical turbulence does allow some interchange between the troposphere and the stratosphere. Also, hurricanes occasionally inject moisture above this inversion layer, and powerful volcanic eruptions may push ash and acid mists into the stratosphere.

Figure 3–7 illustrates the temperature profile within the troposphere during the daytime. As the graph shows, temperatures decrease with increasing altitude at an average of 6.4C° per km (3.5F° per 1000 ft), a rate known as the **normal lapse rate**. The *actual* lapse rate at any particular time and place under local weather conditions is called the **environmental lapse rate**, and may vary greatly from the normal lapse rate. This variance in temperature in the lower tropo-

sphere is important to our discussion of weather processes (Chapter 8).

Within the troposphere, both natural and human-caused variable gases, particles, and other chemicals are part of the atmosphere. The spatial aspects of these variable pollution components are a significant applied topic in physical geography.

VARIABLE ATMOSPHERIC COMPONENTS

Air pollution is not a new problem. Romans complained over 2000 years ago about the foul air of their cities. The stench of open sewers, fires, kilns, and smelters filled the air. In human experience, cities are always the place where the environment's natural ability to process and recycle waste is most taxed. English diarist John Evelyn recorded in 1684 that the air in London was so filled with smoke that ". . . one could hardly see across the streets, and this filling the lungs with its gross particles exceedingly obstructed the breast so as one could hardly breathe."* Air pollution is closely related to our production and consumption of energy, and as

---

*W. Wise, *Killer Smog: The World's Worst Air Pollution Disaster,* Chicago: Rand McNally, 1968.

TABLE 3–2
Sources of natural variable gases and materials.

Sources	Contribution
Volcanoes	Sulfur oxides, particulates
Forest fires	Carbon monoxide and dioxide, nitrogen oxides, particulates
Plants	Hydrocarbons, pollens
Decaying plants	Methane, hydrogen sulfides
Soil	Dust and viruses
Ocean	Salt spray and particulates

human occupation of a region increases in duration and complexity, the environmental impact on the troposphere also increases.

Natural sources produce a greater quantity of nitrogen oxides, carbon monoxide, and carbon dioxide than do certain human-made sources. Table 3–2 lists some of these natural sources and the substances they contribute to the air (Figure 3–8). However, any attempt to diminish the impact of human-made air pollution through a comparison with natural sources is irrelevant, for we have coevolved with and adapted to the natural ingredients in the air, and we have *not* evolved in relation to the comparatively recent concentrations of anthropogenic (human-caused) contaminants in our metropolitan areas.

Augmenting both natural and human-made atmospheric contaminants are several important natural factors. Among these are wind, local and regional landscape characteristics, and temperature inversions in the troposphere.

Winds gather and move pollutants from one area to another, sometimes reducing the concentration of pollution in one location while increasing it in

FIGURE 3–8
Natural variable pollution in the form of a dust storm. Photo by author.

another. Such air movements make the atmosphere's condition an international issue. Indeed, the movement of air pollution from the United States to Canada is the subject of much complaint and negotiation between these two governments. In Europe, the cross-boundary drift of pollution is a major issue due to the close proximity of nations, and has led in part to Europe's unification.

Local and regional landscapes are another important factor in air pollution. Surrounding mountains and hills can form barriers to air drainage or can direct pollutants from one area to another. Some of the worst incidents have resulted when local landscapes have trapped and caused a concentration of air pollution. The 1948 Donora, Pennsylvania episode is a key example and is discussed later in this chapter.

Vertical temperature distribution in the troposphere also can contribute to conditions of pollution. A **temperature inversion** occurs when the normal decrease of temperature with increasing altitude reverses at any point from ground level up to several thousand feet. Such an inversion most often results from certain weather conditions, such as when the air near the ground is radiatively cooled on clear nights, or from topographic situations, such as when cold air drains into valleys. In addition, the air above snow-covered surfaces or beneath subsiding air in a high-pressure system may cause a temperature inversion.

Figure 3–9 compares a normal temperature profile with that of a temperature inversion. The normal profile illustrated in part (a) permits lifting air to ventilate surface pollution, but the warm air inversion shown in part (b) prevents the vertical mixing of pollutants with other atmospheric gases. Thus, instead of being carried away, the pollutants are trapped beneath the inversion layer. During the winter months in eastern and midwestern regions of the United States, high-pressure areas created by subsiding cold air masses produce inversion conditions that trap air pollution. In the West, summer subtropical high-pressure systems also cause inversions which produce air stagnation.

Anthropogenic Pollution

Anthropogenic, or human-caused, air pollution remains most prevalent in urbanized regions. Table 3–3 lists the names, symbols, and principal sources of variable anthropogenic components in the air. Seven of the components in Table 3–3 (carbon monoxide, nitrogen oxides, hydrocarbons, ozone, peroxyacetyl nitrates, sulfur oxides, and particulates) result from combustion of fossil fuels in transportation (specifically automobiles) and at stationary sources such as power plants and factories. Overall, automobiles contribute about 60% of American and 40% of Canadian human-caused air pollution. Environment Canada, an agency similar to the U.S. Environmental Protection Agency, reports that automobile emissions contribute 45% of the carbon monoxide, 20% of the nitrogen oxides, and 24% of the hydrocarbons in Canada. Figure 3–10 identifies major human-caused pollutants and their proportional sources in the United States. These pollutants are described in the following sections.

The last three gases shown in Table 3–3 are discussed elsewhere in this text: water vapor with weather and the hydrologic cycle (Chapters 7 and 8), carbon dioxide with temperature and climate (Chapters 5 and 10), and methane with environmental cycles (Chapter 5).

Carbon Monoxide. Carbon monoxide (CO) is an odorless, colorless, and tasteless combination of one atom of carbon and one of oxygen. CO is produced by incomplete combustion of fuels or other carbon-containing substances. A log decaying in the woods produces CO, as does a forest fire or other organic decomposition. Natural sources produce up to 90% of existing carbon monoxide, whereas anthropogenic sources, principally transportation, produce the other 10%. However, the carbon monoxide from human sources is concentrated in urban areas, where it directly impacts human health.

The toxicity of CO is well known; one often hears of a death from carbon monoxide poisoning due to being in an enclosed space with a running automobile engine or in a tent with an unvented heater. The toxicity of CO is due to its affinity for oxygen in blood hemoglobin, which is the oxygen-carrying pigment in red blood cells. In the presence of CO, the oxygen is displaced and the blood becomes deoxygenated.

CO pollution did not become a problem until many fossil fuel–burning cars, trucks, and buses became concentrated in urban areas. Normal CO

(a)

Warmer air

Inversion layer

(b)

FIGURE 3–9

A comparison of (a) a normal temperature profile with (b) a temperature inversion in the lower atmosphere.

TABLE 3–3

Anthropogenic gases in the lower atmosphere.

Name	Symbol	Source
Carbon monoxide	CO	Incomplete combustion of fuels
Nitrogen oxides	$NO_x(NO, NO_2)$	High temperature/pressure combustion
Hydrocarbons	HC	Incomplete combustion of fuels
Ozone	O_3	Photochemical reactions
Peroxyacetyl nitrates	PAN	Photochemical reactions
Sulfur oxides	$SO_x(SO_2, SO_3)$	Combustion of sulfur-containing fuels
Particulates	—	Dust, dirt, soot, salt, metals, organics
Exotics	—	Fission products, other toxics
Carbon dioxide	CO_2	Complete combustion
Water vapor	H_2O vapor	Combustion processes, steam
Methane	CH_4	Organic processes

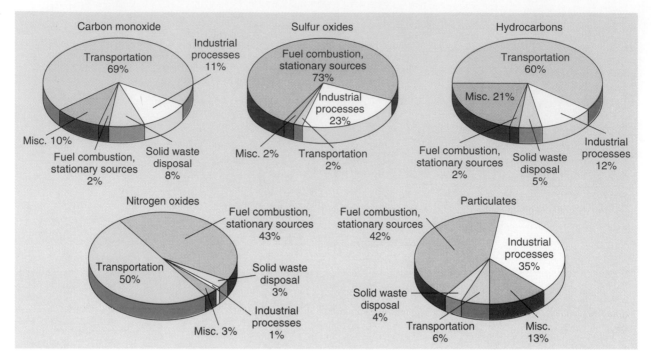

FIGURE 3–10
Major types of human-caused air pollution and their sources in the United States.

levels range from 10 to 30 parts per million (ppm) in metropolitan areas, but freeways and parking garages can reach 50 ppm CO, with an additional 30 ppm possible under a temperature inversion. These common levels in cities are dangerous, as you can see in Table 3–4, which lists various CO concentrations, the resultant reduction of oxygen in the blood, and the health symptoms experienced.

Photochemical Smog Reactions. Photochemical smog is another type of pollution that was not generally experienced in the past but developed with the advent of the automobile. Today, it is the major component of anthropogenic air pollution. **Photochemical smog** results from the interaction of sunlight and the products of automobile exhaust. Although the term "smog"—a combination of the words "smoke" and "fog"— is a misnomer, it is generally used to describe this phenomenon. Smog is responsible for the hazy sky and reduced sunlight in many of our cities (Figure 3–11).

The connection between automobile exhaust and smog was not determined until 1953, well after

TABLE 3–4
Health effects of carbon monoxide.

Concentration of CO, parts per million (ppm)	Deoxygenation of Hemoglobin	Symptoms
50	2 to 8%	Some vision and/or judgment loss
100	9 to 20%	Headache at 10%; impaired performance on simple psychological tests and in arithmetic
300–400	30 to 40%	Severe headache, nausea, collapse

FIGURE 3–11
Photochemical smog produces a yellow-white haze in the skies over Los Angeles. Photo by author.

society had established its dependence upon individualized transportation. Despite this discovery, widespread mass transit has declined, the railroads have dwindled, and the polluting individual automobile remains America's preferred transportation mode.

The higher temperatures in modern automobile engines react with fuels to produce **nitrogen dioxide** (NO_2). This nitrogen dioxide derived from automobiles, and to a lesser extent from power plants, is highly reactive with ultraviolet light, which liberates atomic oxygen (O) and a nitric oxide (NO) molecule from nitrogen dioxide (NO_2) (Figure 3–12). The free oxygen atom combines with an oxygen molecule (O_2) to form the oxidant ozone (O_3); this same gas that is so beneficial in the stratosphere is an air-pollution hazard at Earth's surface. In addition, the nitric oxide (NO) molecule reacts with hydrocarbons (HC) to produce a whole family of chemicals generally called **peroxyacetyl nitrates (PAN)**. PAN produces no known health effect in humans but is particularly damaging to plants, which provided the clue for discovery of these photochemical reactions.

Nitrogen dioxide (NO_2) is a reddish-brown gas that damages and inflames human respiratory systems, destroys lung tissue, and damages plants. Concentrations of 3 ppm are dangerous enough to require alerting parents to keep children indoors; a level of 5 ppm is extremely serious. Nitrogen dioxide interacts with water vapor to form nitric acid (HNO_3), a contributor to acid deposition by precipitation, the subject of Focus Study 3–2. Worldwide, the problem with nitrogen dioxide production is its concentration in metropolitan regions. North American urban areas may have from 10 to 100 times higher nitrogen dioxide concentrations than nonurban areas. Nitrogen dioxide is included in the nitrogen oxide emission levels shown in Figure 3–13 as characteristic of the early 1980s.

Ozone Pollution. Ozone is highly reactive and results in various negative consequences. Paint oxidizes, painted surfaces peel, elastic and rubber dry and crack, paper dries and yellows, and plants are damaged at levels of only 0.01 to 0.09 ppm. Extensive studies have found that exposure to 0.1 ppm for a few hours a day for several days to several weeks causes significant crop damage and reduces yields as much as 50% for a wide variety of agricultural crops. At 0.3 ppm eye, nose, and throat irritation is noticeable. As concentration levels rise to 1.0 to 3.0 ppm, extreme fatigue and poor coordination can affect humans.

The economics of ozone damage to plants was studied by the U.S. National Crop Loss Assessment Network (NCLAN) and reported through the U.S. Environmental Protection Agency (EPA). Similar studies have examined the damage done by photochemical smog to human health and property. The benefits of implementing air pollution abatement strategies far outweigh the implementation costs of

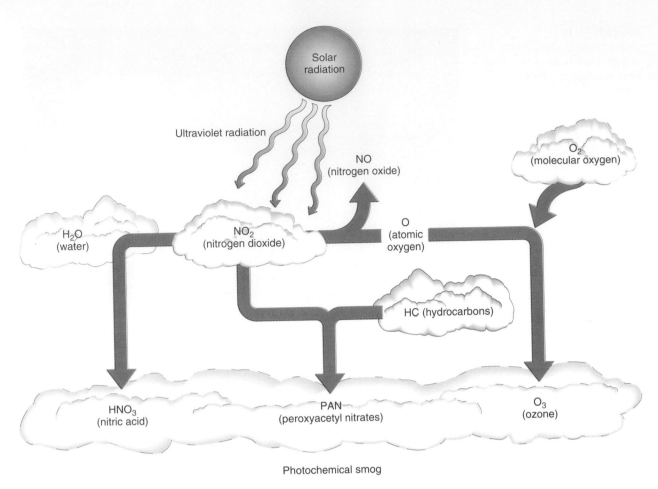

FIGURE 3–12
Photochemical reactions. After G. Tyler Miller, Jr., 1988, *Living in the Environment,*
Belmont, CA: Wadsworth, Inc., p. 428. Adapted by permission.

known solutions. Thus, taking action actually is a cheaper strategy than allowing the situation to further degrade.

Industrial Smog and Sulfur Oxides. Over the past 300 years, coal has slowly replaced wood as the basic fuel used by society. The industrial revolution required high-grade energy to run machines and to facilitate the conversion from animate energy (energy from living sources, such as animals powering farm equipment) to inanimate energy (energy from nonliving sources, such as coal, steam, and water). The air pollution associated with coal-burning industries is known as **industrial smog** (Figure 3–14). The term smog was coined by a London physician at the

turn of this century to describe the combination of fog and smoke containing sulfur, an impurity in fossil fuels. The combination of sulfur and moisture droplets forms a sulfuric acid mist that is extremely dangerous in high concentrations.

Numerous smog-related events have been reported during this century in a variety of industrial areas. Deaths occurred in five major episodes in London between 1880 and 1962, and in the Meuse Valley of Belgium during December 1930. Sickness and death also occurred in 1948 in the steelmaking town of Donora, near Pittsburgh, Pennsylvania, and in New York City in 1953, 1963, and 1966. The Donora episode resulted from a dreadful combination of factors: sulfur dioxide emissions from steel

Acid Deposition: A Blight on the Landscape

Acid deposition is a major environmental problem in some areas of the United States, Canada, Europe, and Asia. Such deposition occurs in a dry form as dust or aerosols, or wet in the form of what we call acid rain or snow. In addition, winds can carry the acid-producing chemicals many kilometers from their sources before they settle on the landscape, entering streams and lakes as runoff and groundwater flows.

Acid deposition is causally linked to declining fish populations and fish kills in the northeastern United States, southeastern Canada, Sweden, and Norway; to widespread forest damage in these same places and Germany; and to damage to buildings, sculptures, and historical artifacts. Corrective action has been delayed because of the complexity of the problem, which the U.S. General Accounting Office calls a "combination of meteorological, chemical, and biological phenomena."

The acidity of precipitation is measured on the pH scale, which expresses the relative abundance of free hydrogen ions (H^+) in a solution. (Free hydrogen ions in a solution are what make an acid corrosive, for they easily combine with other ions.) The pH scale is logarithmic: each whole number represents a 10-fold change. A pH of 7.0 is neutral (neither acidic or basic), below 7.0 is increasingly acidic, and above 7.0 is increasingly basic or alkaline. (A pH scale for soil acidity and alkalinity is graphically portrayed in Chapter 18.) Natural precipitation dissolves carbon dioxide from the atmosphere, a process that releases hydrogen ions and produces an average pH reading of 5.65, within a normal range for precipitation of 5.3 to 6.0. Thus, normal precipitation is always slightly acidic.

Certain anthropogenic gases are converted in the atmosphere into acids that are removed by wet and dry deposition processes. Nitrogen and sulfur oxides (NO_x and SO_x) released in the combustion of fossil fuels can produce nitric acid (HNO_3) and sulfuric acid (H_2SO_4) in the atmosphere. Figures 3–13(a) and 3–15(a) depict the patterns of NO_x and SO_x emissions in the United States and Canada. Precipitation as acidic as pH 2.0 has fallen in the eastern United States, Scandinavia, and Europe. By comparison, vinegar and lemon juice register slightly less than 3.0. Aquatic plant and animal life perishes when lakes drop below pH 4.8.

More than 50,000 lakes and some 96,500 km (60,000 mi) of streams in the United States and Canada are at a pH level below normal (i.e., below 5.3), with several hundred lakes incapable of supporting any aquatic life. Furthermore, acid deposition causes the release of aluminum and magnesium from clay minerals in the soil, and both of these elements negatively impact fish and plant communities. Also, relatively harmless mercury deposits in lake-bottom sediments are converted by acidified lake waters into highly toxic methyl mercury, which also is deadly to aquatic life.

Damage to forests results from the rearrangement of soil nutrients, the death of soil microorganisms, and an aluminum-induced calcium deficiency that is currently under investigation. Europe is experiencing a more advanced impact on their forests than is evidenced elsewhere, principally due to the burning of coal and density of industrial activity. In Germany, up to 50% of the forests are dead or dying; in Switzerland, 30% are afflicted. In the United States, trees at higher elevations in the Appalachians are being injured by acid-laden cloud cover. In New England, some stands of spruce are as much as 75% affected, as evidenced through analysis of tree growth-rings, which become narrower in adverse growing years. Another possible indicator of forest damage is the reduction by almost half of the annual production of U.S. and Canadian maple sugar.

The National Academy of Sciences (NAS) and the National Research Council have identified causal relationships between fossil fuel combustion and acidic deposition, citing evidence that is "overwhelming." Despite this consensus, politics and special interests have halted preventative measures. The NAS stated that a reduction of 50% in combustion emissions would reduce acid deposition by 50%. Government estimates place damage in the United States and Canada at tens of billions of dollars and in western Europe at over $50 billion annually. Because wind and weather patterns are international, efforts at reducing acidic deposition also must be international in scope.

At best, acid deposition is an unfolding international political issue of global spatial significance for which science is providing strong incentives for action. Not until March 1991 did the United States finally agree to a landmark agreement with Canada to lower sulfur dioxide and nitrogen oxide emissions—a start.

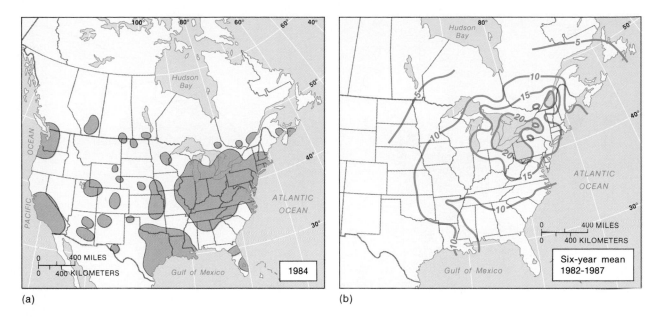

(a)   (b)

FIGURE 3–13
Spatial portrayal of (a) distribution of nitrogen oxide emissions and (b) mean annual deposition of nitrogen oxides (principally NO_3) on the landscape (in kilograms per hectare per year). Data courtesy of Robert J. Vet, Air Quality and Inter-environment Research Branch, Atmospheric Environment Service, Environment Canada.

FIGURE 3–14
Industrial smog. Pollution generated by industry differs from that produced by transportation. Industrial pollution has high concentrations of sulfur oxides, particulates, and carbon dioxide. Photo by author.

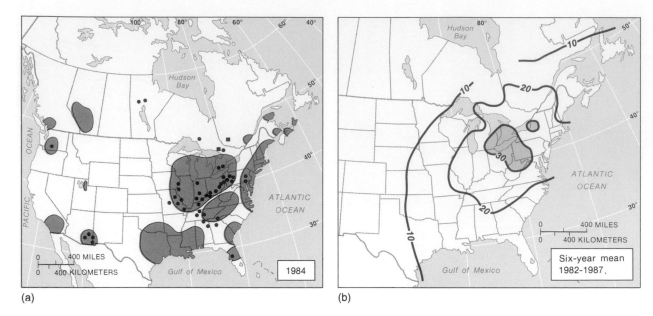

(a)   (b)

FIGURE 3–15
Spatial portrayal of sulfur dioxide (a) emissions and (b) deposition. In (a), the dots locate sources producing 100–500 metric kilotons (110,000 and 550,000 tons) per year; the squares locate sources producing 500 metric kilotons (550,000 tons) or more per year. (b) shows mean annual deposition of sulfur oxides (principally SO_4) on the landscape (in kilograms per hectare per year). Data courtesy of Robert J. Vet, Air Quality and Inter-environment Research Branch, Atmospheric Environment Service, Environment Canada.

mills and power plants, a persistent fog, a strong high-pressure weather system which caused a temperature inversion layer, the location of this town in a deep river valley, and the orientation of nearby mountains. All these factors interacted to produce a three-day siege in which 40% of the population became sick and 20 died.

Sulfur, an impurity that occurs to some degree in all fossil fuels, forms sulfur dioxide (SO_2) during combustion. Thus, sulfur dioxide is a problem wherever coal-fired power plants and heavy industry are common. **Sulfur dioxide** is colorless but can be detected by its pungent odor. At low concentrations of from 0.1 to 1 ppm, the air is dangerous to society; at 0.3 ppm the air takes on a definite taste. In addition, human respiratory systems become irritated, especially in individuals with asthma, chronic bronchitis, and emphysema.

Once in the atmosphere, SO_2 reacts with oxygen to form sulfur trioxide (SO_3), which is highly reactive and, in the presence of water or water vapor, forms

sulfuric acid (H_2SO_4). Sulfuric acid can occur in even moderately polluted air, and its synthesis does not require high temperatures. In this manner, a sulfur dioxide-laden atmosphere corrodes metals and deteriorates stone building materials at accelerated rates.

In the United States, electric utilities and steel manufacturing are the main sources of sulfur dioxide. And due to the prevailing movement of air masses, U.S. electric utilities and steel manufacturing are the main sources of sulfur dioxide in Canada, too. This conclusion comes from detailed studies that show as much as 70% of the Canadian sulfur dioxide problem is initiated within the United States. Figure 3–15 portrays representative production and deposition levels of sulfur dioxide in the early 1980s.

The oxides of sulfur and nitrogen lead to the production of airborne sulfuric acid and nitric acid, which are deposited in both wet and dry forms on the landscape and in bodies of water. This acid deposition has increased in severity since it was analyzed and described in the 1970s and is a growing blight in the biosphere. Focus Study 3–2 discusses this vital atmospheric issue.

Solutions

The distribution of human-produced carbon monoxide, nitrogen dioxide, ozone, PAN, and sulfur dioxide over North America, Europe, and Asia is definitely related to transportation and energy production. Specific studies have shown many times over that certain ailments and diseases can be generated by air pollution. Implementation of solutions is important to our individual health and agricultural productivity. Further, solutions to our growing air-pollution problem pose no mystery: technologies to eliminate or abate emissions are known and in some cases are in operation at this time.

Because energy production is at the heart of pollution production, it stands to reason that conservation (using less energy) and efficiency (using energy with less waste) are key strategies, especially considering the level of energy waste and inefficiency in the United States. However, even without such fundamental changes in our behavior, stack emissions can be cleaned with precipitators, scrubbers, filters, and separators; car exhaust can be cleaned with more efficient carburetion and emission controls; and we can reduce automobile usage with alternative transportation systems.

Action must be taken. In 1988 alone, 20,000 industrial facilities poured 1.1 billion kilograms (2.4 billion pounds) of chemicals into the atmosphere in the United States. Certainly, society cannot simply halt two centuries of industrialization; the resulting economic chaos would be devastating. However, neither can it permit pollution production to continue unabated, for catastrophic environmental feedback inevitably will result.

People have been on Earth for such a brief fragment of Earth's history, and yet our activities can be a major influence on the future of the atmosphere. We are now contributing significantly to the mix of the **anthropogenic atmosphere,** a tentative label for Earth's next (fifth) atmosphere. The urban air we breathe today may be just a preview.

SUMMARY

The modern atmosphere probably is the fourth general atmosphere in Earth's history. Our atmosphere is unique among the planets in composition, temperature, and density, and it is arranged in a series of imperfectly shaped concentric spheres. These can be defined by various criteria. By composition, we can divide the atmosphere into the heterosphere and the homosphere. By temperature, we can divide it into the thermosphere, mesosphere, stratosphere, and troposphere. (In the troposphere reside the biosphere, almost all weather and water vapor, and the bulk of pollution.) Using function as a criterion, we can distinguish two regions: the ionosphere and the ozonosphere, both of which absorb life-threatening radiation and convert it into heat energy.

The weight of the atmosphere, or air pressure, crushes in on all of us and averages around 1 kg/cm^2; we measure it with a barometer.

We coevolved with natural "pollution" and thus are adapted to it. But we are not adapted to cope with our own anthropogenic pollution, which constitutes a major health threat, particularly where people are concentrated in cities. Photochemical and industrial smog do not respect political boundaries and are carried by winds from their nations of origin to other nations. The distribution of human-produced carbon monoxide, nitrogen dioxide, ozone, PAN, and sulfur dioxide over North America, Europe, and Asia is definitely related to transportation and energy production. Energy production is at the heart of pollution production, so conservation, efficiency, and cleaning of emissions are essential strategies for abating air pollution. Earth's next atmosphere most likely will be anthropogenic in character.

Air pressure

Aneroid barometer

Anthropogenic atmosphere

Carbon dioxide

Carbon monoxide

Environmental lapse rate

Exosphere

Heterosphere

Homosphere

Industrial smog

Ionosphere

Kinetic energy

Mercury barometer

Mesosphere

Nitrogen dioxide

Noctilucent clouds

Normal lapse rate

Ozone layer

Ozonosphere

Peroxyacetyl nitrates (PAN)

Photochemical smog

Sensible heat

Stratosphere

Sulfur dioxide

Temperature inversion

Thermosphere

Troposphere

REVIEW QUESTIONS

1. What is air? Where did the components in Earth's present atmosphere originate?
2. In view of the analogy by Lewis Thomas, characterize the various functions the atmosphere performs.
3. How does air exert pressure? Describe the basic instrument used to measure air pressure. Compare the two different types of instruments discussed.
4. What is normal sea level pressure? In mm? mb? in.?
5. What three distinct criteria are employed in dividing the atmosphere for study?
6. Describe the overall temperature profile of the atmosphere.
7. Name the four most prevalent stable gases in the homosphere. Where did each originate? Is the prevalence of any of these changing at this time?
8. Why is stratospheric ozone so important? Summarize the ozone predicament and the present findings and treaties.
9. Evaluate Rowland and Molina's use of the scientific method in investigating stratospheric ozone depletion.
10. What is the difference between industrial smog and photochemical smog?
11. Describe the relationship between automobiles and the production of ozone and PAN in city air. What are the principal negative impacts of these gases?
12. How are sulfur impurities in fossil fuels related to the formation of acid in the atmosphere and acid deposition on the land?
13. Why are anthropogenic gases more significant to human health than those produced from natural sources?
14. In what ways does a temperature inversion worsen an air pollution episode?
15. In your opinion, what are the solutions to the growing problems in the anthropogenic atmosphere?

SUGGESTED READINGS

Barry, Roger B., and Richard J. Chorley. *Atmosphere, Weather, and Climate*. 5th ed. London: Methuen, 1987.

Brodeur, Paul. "Annals of Chemistry—In the Face of Doubt." *The New Yorker* (9 June 1986): 70–87.

The Conservation Foundation. *State of the Environment—A View Toward the Nineties*. Washington: The Conservation Foundation and the Charles Stewart Mott Foundation, 1987.

Ember, Lois R., Patricia L. Layman, Will Lepkowski, and Pamela S. Zurer. "The Changing Atmosphere—Implications for Mankind." *Chemical and Engineering News* (24 November 1986): entire issue.

Federal/Provincial Research and Monitoring Coordinating Committee. *Assessment of the State of Knowledge on the Long-Range Transport of Air Pollutants and Acid Deposition*. 6 parts. Ottawa: Environment Canada, 1986.

Miller, G. Tyler, Jr. *Living in the Environment*. 5th ed. Belmont, CA: Wadsworth, 1988.

Miller, John M. "Acid Rain." *Weatherwise* 37, no. 5 (October 1984): 232–49.

Office of Technology Assessment. *Acid Rain and Transported Pollutants—Implications for Public Policy*. Washington: U.S. Government Printing Office, 1984.

Stole, Richard S. "The Antarctic Ozone Hole." *Scientific American* 258, no. 1 (January 1988): 30–36.

Wayne, Richard P. *Chemistry of Atmospheres*. London: Oxford University Press, 1985.

Year-long path of the Sun inscribes an analemma
in the sky. Used by permission of Dennis diCicco.

4

The Seasons and Energy Balances

arth's biosphere pulses daily, weekly, and yearly with flows of energy. Think for a moment of the annual pace of your own life, your wardrobe, gardens, and lifestyle activities—all reflect shifting seasonal energy patterns. This chapter examines the phenomenon of Earth's seasons, looking carefully at the seasonal distribution of solar energy along various pathways as it is absorbed and redirected in the troposphere. Regions near the equator experience little seasonal variation of temperature and daylength; seasonality generally increases with latitude.

This chapter examines surface energy budgets, analyzing how net radiation is expended at different locales. We also explore the energy characteristics of urban areas, for the climates of our cities differ measurably from those of surrounding rural areas.

THE SEASONS

The periodic rhythm of warmth and cold, rain and drought, dawn and daylight, twilight and night have fascinated humans for centuries. In fact, many an-cient societies demonstrated a greater awareness of seasonal change than modern peoples and formally commemorated natural energy rhythms with festivals and monuments. At Stonehenge, rocks weighing 25 metric tons (28 tons) were hauled 480 km (300 mi) and placed in patterns that evidently mark seasonal changes, specifically the sunrise on or about June 21 at the summer solstice. Other seasonal events and eclipses of the Sun and the Moon apparently are predicted by this 3500-year-old calendar monument. Such ancient seasonal monuments and markings occur worldwide, including thousands of sites in North America (Figure 4–1).

Seasonality

Seasonality refers to both the seasonal variation of the Sun's rays above the horizon and changing daylengths during the year. Seasonal variations are a response to the change in the Sun's **altitude**, or the angular difference between the horizon and the Sun. The Sun is directly overhead (at zenith) only at the subsolar point, with all other surface points receiving

FIGURE 4–1
Stonehenge, Salisbury Plain in Wiltshire, England. Photo by Robert Llewellyn.

TABLE 4-1
Five reasons for seasons.

Factor	Description
Revolution	Orbit around the Sun; requires 365.24 days to complete at 107,280 kmph (66,660 mph)
Rotation	Earth turning on its axis; takes approximately 24 hours to complete at 1675 kmph at equator (1041 mph)
Tilt	Axis is aligned at a 23.5° angle from a perpendicular to the plane of the ecliptic (the plane of Earth's orbit)
Axis	Remains in a fixed alignment of axial parallelism, with Polaris directly overhead at the North Pole throughout the year
Sphericity	Appears as an oblate spheroid to the Sun's parallel rays; the geoid

a lower angle of more diffuse insolation. The nearer the Sun's altitude is to 90°, the more direct the Sun's rays are as they shine through the atmosphere, resulting in more concentrated solar radiation at Earth's surface. The Sun's **declination** is the angular distance from the equator to where direct overhead insolation is received. Declination annually migrates through 47° of latitude between the Tropic of Cancer at 23.5° N and the Tropic of Capricorn at 23.5° S latitude.

Seasonality also means a changing duration of exposure, or **daylength**, which varies during the year depending on latitude. The equator always receives equal hours of day and night, whereas people living along 40° N or S latitude experience about six hours' difference in daylight hours between winter and summer. Those at 50° N or S latitude experience almost eight hours of annual daylength variation. At the polar extremes, the range extends from a six-month period of no insolation to a six-month period of continuous 24-hour days.

Contributing Physical Factors

The variations in Sun angle and daylength are created by several contributing physical factors that operate together: Earth's revolution and rotation, its tilt and fixed-axis orientation, and its sphericity (Table 4–1). Of course, the essential ingredient is having a single source of illumination—the Sun.

Revolution. The structure of Earth's orbit and **revolution** about the Sun is shown in Figure 2–2. At an average distance from the Sun of 150 million km (93 million mi), Earth completes its annual orbit in 365.2422 days at speeds averaging 107,280 kmph (66,660 mph) in a counterclockwise direction when viewed from above the North Pole (Figure 4–2). This calculation is based on a tropical year, measured from March to March (or the elapsed time between two crossings of the equator by the Sun). Earth's revolution determines the length of the year and the seasons. The Earth-to-Sun distance is not a contributing seasonal factor even though it varies some 4.8 million km (3 million mi) during the year.

Rotation. Earth's **rotation**, or turning, is a complex motion that averages 24 hours in duration. Rotation

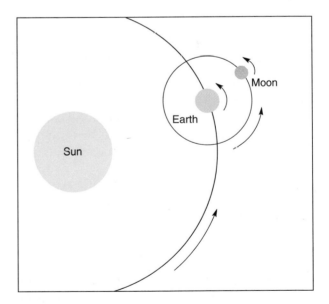

FIGURE 4-2
Earth's revolution about the Sun and rotation on its axis, viewed from above the North Pole.

TABLE 4–2
Rotational velocities at selected latitudes.

Latitude	kmph	mph	Cities at Approximate Latitudes
0°	1675	1041	Quito, Ecuador; Pontianak, Indonesia
30°	1452	902	Porto Alegre, Brazil; New Orleans
40°	1284	798	Valdivia, Chile; Columbus, Ohio; Beijing
50°	1078	670	Chibougamau, Quebec; Kiev, USSR
60°	838	521	Seward, Alaska; Oslo, Norway; Leningrad, USSR
90°	0	0	(North Pole)

determines daylength, produces the apparent deflection of winds and ocean currents, and produces the twice-daily action of the tides in relation to the gravitational pull of the Sun and the Moon. Earth's **axis** is an imaginary line extending through the planet from the geographic North Pole to the South Pole. When viewed from above the North Pole, Earth rotates counterclockwise; viewed from above the equator it moves west to east, or eastward. This west-to-east rotation creates the Sun's apparent daily journey from east to west, even though the Sun remains in a fixed position in the center of the Solar System. The Moon revolves around Earth and rotates counterclockwise on its axis.

Although points at all latitudes take the same 24 hours to complete one rotation, the *speed* of rotation at any point on Earth's surface varies with latitude. The equator is 40,075 km (24,902 mi) in length; therefore rotational speed at the equator must be approximately 1675 kmph (1041 mph) to cover that distance in one day. At 60° latitude, a parallel is only half the length of the equator, or 20,038 km (12,451 mi) long, so the rotational speed there is 838 kmph (521 mph). This variation in rotational velocity establishes the Coriolis force, discussed in Chapter 6. Table 4–2 lists the speed of rotation for several selected latitudes.

The rotation of Earth produces the diurnal pattern of day and night. The dividing line between day and night is called the **circle of illumination**. Since this day-night dividing circle of illumination is a great circle that intersects the equator, which is another great circle, daylength at the equator is always evenly divided—12 hours of day and 12 hours of night. All other latitudes experience uneven daylength through the seasons.

In reference to the Sun's position, *mean solar time* is averaged to eliminate predictable rotational and orbital (revolution) variations that cause the solar day to change in length throughout the year. In other words, it is an agreed-upon practice that a day averages 24 hours in length, or 86,400 seconds. However, an *apparent solar day* is based on observed successive passages of the Sun over a given meridian. Any difference between this observed solar time and mean solar time is called the *equation of time*. On successive days, if the Sun arrives overhead at a meridian *after* 12:00 noon local standard time (that is, taking an interval longer than the 24 hours of a mean solar day), the equation of time is negative. The Sun is described as "slow," much like an airliner arriving later than scheduled. If, on the other hand, the Sun arrives overhead *before* 12:00 noon local time on successive days (that is, taking an interval shorter than 24 hours), the equation of time is positive. The Sun then is described as "fast," like an airliner arriving ahead of schedule.

The **analemma** is useful for demonstrating the positive and negative equations of time, the periods of fast Sun times (during November and May) and slow Sun times (during February and July). It is also a convenient device to track the passage of the Sun's path and declination throughout the year (Figure 4–3). Can you identify the days of fast and slow Sun on the analemma in the figure?

The dramatic photograph at the beginning of this chapter demonstrates the changing seasons and the equation of time. To make this image, a single piece of film was exposed on 45 different dates within a year at 8:30 A.M. EST or 9:30 A.M. EDT (13:30 UTC). The result is an analemma inscribed by the Sun in the sky through multiple exposure. The portion of the analemma nearest the horizon is around December 21, whereas the upper portion is around June 21.

Tilt of Earth's Axis. To understand Earth's tilt, imagine Earth's elliptical orbit about the Sun as a

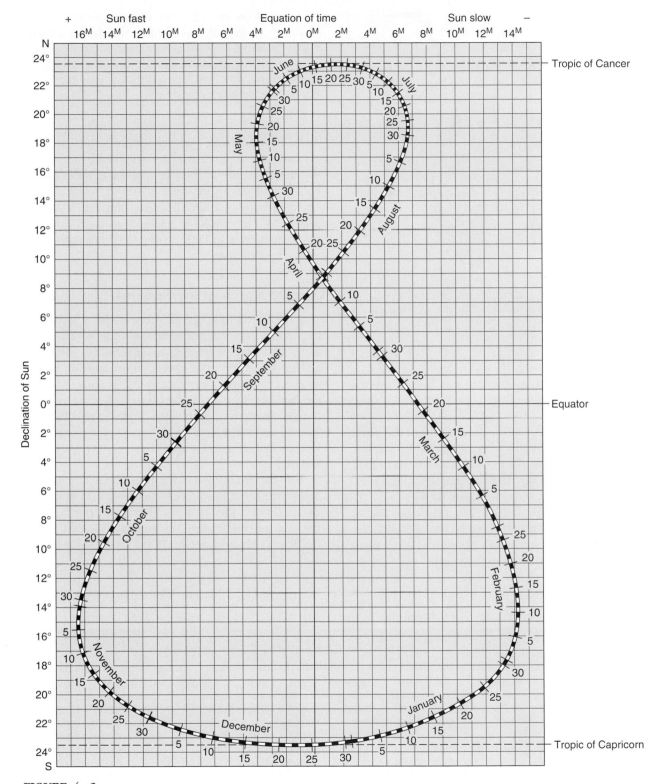

FIGURE 4–3

The analemma. Data courtesy of U.S. Coast and Geodetic Survey.

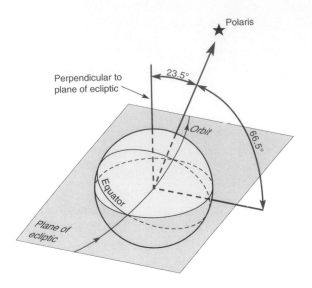

FIGURE 4–4
Earth's tilt and the plane of the ecliptic.

level plane, with half of the Sun and Earth above the plane and half below. (It may help to envision two spheres floating in water, where the water's surface forms a plane.) This level surface is termed the **plane of the ecliptic**. Now, imagine a perpendicular line passing through the plane. Earth's axis is tilted 23.5° from the perpendicular to this plane, forming a 66.5° angle from the plane itself (Figure 4–4).

The axis through Earth's two poles points just slightly off Polaris, which is appropriately called the North Star (see Figure 1–11). Hypothetically, if Earth were tilted on its side with its axis parallel to the plane of the ecliptic, we would experience a maximum variation in seasons worldwide. On the other hand, if Earth's axis were perpendicular to the plane of its orbit—that is, with no tilt—we would experience no seasonal changes, just a perpetual spring/fall season.

Axial Parallelism. Throughout our annual journey around the Sun, Earth's axis maintains the same alignment relative to the plane of the ecliptic and to Polaris and the other stars. In other words, if we compared the axis in different months, it would always appear parallel to itself, or to its orientation everywhere else in Earth's orbit, a condition known as **axial parallelism**.

Annual March of the Seasons

The combined effect of all these physical factors is the annual march of the seasons on Earth. Daylength is the most evident way of sensing changes in season at latitudes away from the equator. **Sunrise** is that moment when the disk of the Sun first appears above the horizon, whereas **sunset** is that moment when it totally disappears. Table 4–3 lists the times of sunrise and sunset and the duration of daylength for various latitudes and seasons in the Northern Hemisphere.

The extremes of daylength occur in December and June. The times around December 21 and June 21 are termed the solstices. They mark the times of the year when the Sun's declination is at one of the tropics, its farthest northerly or southerly position.

TABLE 4–3
Daylength times at selected latitudes (Northern Hemisphere).

	Winter Solstice December 21–22			Vernal Equinox March 20–21			Summer Solstice June 20–21			Autumnal Equinox September 22–23		
	AM	PM	Daylength	AM	PM	Daylength	AM	PM	Daylength	AM	PM	Daylength
0°	6:00	6:00	12:00	6:00	6:00	12:00	6:00	6:00	12:00	6:00	6:00	12:00
30°	6:58	5:02	10:04	6:00	6:00	12:00	5:02	6:58	13:56	6:00	6:00	12:00
40°	7:26	4:34	9:00	6:00	6:00	12:00	4:34	7:26	15:00	6:00	6:00	12:00
50°	8:05	3:55	8:00	6:00	6:00	12:00	3:55	8:05	16:00	6:00	6:00	12:00
60°	9:15	2:45	5:30	6:00	6:00	12:00	2:45	9:15	18:30	6:00	6:00	12:00
90°	No sunlight			Rising sun			Continuous sunlight			Setting sun		

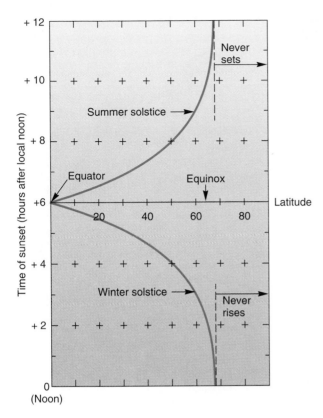

FIGURE 4–5

The time of sunset at the summer and winter solstices for different latitudes. (If you are on daylight savings time, adjust appropriately.) After Aden and Marjorie Meinel, 1983, *Sunsets, Twilights, and Evening Skies,* p. 10, Fig. 2–4. Adapted with the permission of Cambridge University Press and the authors.

Figure 4–5 illustrates how sunset times vary for these solstices at different latitudes. (For the times of sunrises, reverse the numbers on the left side of the graph.)

Table 4–4 presents the key seasonal anniversary dates, their names, and the Sun's declination. Compare the dates and terminology in this table to the analemma presented in Figure 4–3. During the year,

places on Earth go through a continuous but gradual shift in daylength, a few minutes each day, with the Sun's altitude increasing or decreasing a portion of a degree each day for any particular location. You may have noticed that these daily variations become more pronounced in spring and autumn, when the declination is changing at a faster rate (Figure 4–3).

Figure 4–6 illustrates Earth's relationship to the Sun at the four times of year described in Table 4–4. Beginning with the **winter solstice** (literally, "winter Sun stance") on December 21 or 22, the circle of illumination excludes the North Pole region from sunlight and includes the South Pole region. The subsolar point is at 23.5° S latitude at the moment of the solstice, a parallel known as the **Tropic of Capricorn**. The Northern Hemisphere is tilted away from these more direct rays of sunlight, thereby creating a lower angle for the incoming solar rays and thus a more diffuse pattern of insolation. Of lesser importance is the fact that the lower-angle solar rays must pass through a greater depth of atmosphere, resulting in greater losses of energy due to scattering, absorption, and reflection.

From 66.5° N latitude to 90° N (the North Pole), the Sun remains below the horizon the entire day. This latitude marks the Arctic Circle, the southernmost parallel (in the Northern Hemisphere) that experiences a 24-hour period of darkness. During the following three months, daylength and solar angles gradually change as Earth completes onefourth of its orbit.

The **vernal equinox** occurs on March 20–21. At that time, the circle of illumination passes through both poles so that all locations on Earth experience a 12-hour day and a 12-hour night. Those living at 40° N latitude have gained three hours of daylight since the winter solstice. At the North Pole, the Sun rises for the first time since the previous September; at the South Pole the Sun is setting. Figure 2–12 clearly illustrates these moments of beginning and ending insolation at the poles.

TABLE 4–4
Annual march of the seasons.

Approximate Date	Name	Location of the Subsolar Point
December 21	Winter solstice	23.5° S latitude (Tropic of Capricorn)
March 21	Vernal equinox	0° (Equator)
June 21	Summer solstice	23.5° N latitude (Tropic of Cancer)
September 22	Autumnal equinox	0° (Equator)

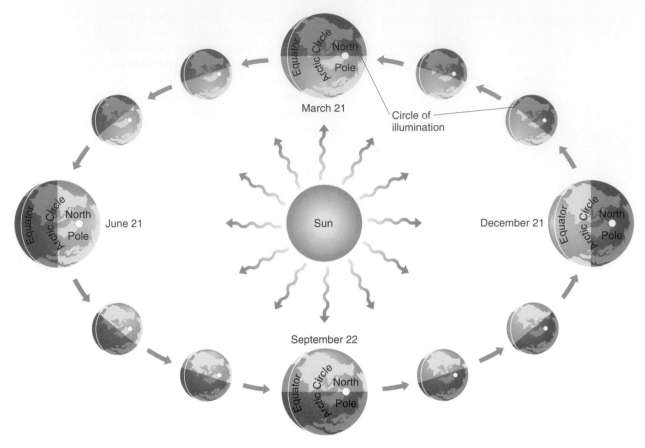

March 21

Circle of illumination

June 21

Sun

December 21

September 22

FIGURE 4–6
Annual march of the seasons as Earth revolves about the Sun. Shading indicates the changing position of the circle of illumination.

From March, the seasons move on to June 20–21 and the **summer solstice**. The subsolar point is at 23.5° N latitude, the **Tropic of Cancer**. Because the circle of illumination now includes the North Polar region, everything north of the Arctic Circle receives 24 hours of daylight. (Figure 4–7 is a multiple-image photo of the midnight Sun as seen north of the Arctic Circle.) In contrast, the region from the Antarctic Circle to the South Pole (66.5° to 90° S latitude) is in darkness the entire day.

Although the orientation of Earth's axis has remained fixed relative to the heavens and parallel to its previous position, the Northern Hemisphere in June is tilted toward the Sun. Consequently, that hemisphere receives higher Sun angles and experiences longer days—and therefore more insolation—than the Southern Hemisphere. Those living at 40° N latitude now receive more than 15 hours of Sunlight, which is 6 hours more than in December.

September 22 or 23 is the time of the **autumnal equinox**, when Earth's orientation is such that the circle of illumination again passes through both poles so that all parts of the globe receive a 12-hour day and a 12-hour night. The subsolar point has returned to the equator, with days growing shorter to the north and longer to the south. Researchers stationed at the South Pole see the disk of the Sun just rising, ending their six months of night.

Seasonal Observations. The point of sunrise migrates from the southeast in December to the northeast in June, whereas sunset moves from the southwest to the northwest during those same times. The Sun's altitude at local noon at 40° N latitude migrates from a 26° angle above the horizon at the winter solstice to a 73° angle above the horizon at the summer solstice—a range of 47°.

Dawn and Twilight Concepts. The diffused light that occurs before sunrise and after sunset represents useful work time for humans. Light is scattered by molecules and reflected by dust and moisture so that the atmosphere is illuminated. Such effects may be enhanced by the presence of pollution aerosols and other suspended particles, such as those from volcanic eruptions or forest fires. Dawn and twilight are roughly equal in extent on any given day at any particular latitude. The duration of these Sun effects is a function of latitude; that is, the angle of the Sun's path relative to the horizon determines the thickness of the atmosphere through which the Sun's rays must pass and thus the length of dawn and twilight.

At the equator, where the Sun's rays are nearly perpendicular to the horizon throughout the year, dawn and twilight are limited to 30–45 minutes each. These times increase to 1–2 hours at 40° latitude, and at 60° latitude these times range upward from 2.5 hours, with little true night in summer. The poles experience about 7 weeks of dawn and 7 weeks of twilight, leaving only 2.5 months of near darkness during the six months when the Sun is completely below the horizon.

ENERGY BALANCE OF THE TROPOSPHERE

Earth's atmosphere and surface are heated by solar energy that is unevenly distributed by latitude and that fluctuates seasonally. In terms of energy, our atmospheric budget is comprised of shortwave radiation inputs (ultraviolet light, visible light, and near-infrared wavelengths) and longwave radiation outputs (thermal infrared, or heat). **Transmission** refers to the passage of shortwave and longwave energy through the atmosphere or water. The atmosphere and surface eventually radiate heat energy back to space, and this energy, together with reflected energy, equals the initial solar input.

FIGURE 4–7
Midnight Sun north of the Arctic Circle, 9 P.M.–4 A.M., June 24. Photos © Wiancko-Braasch/Allstock.

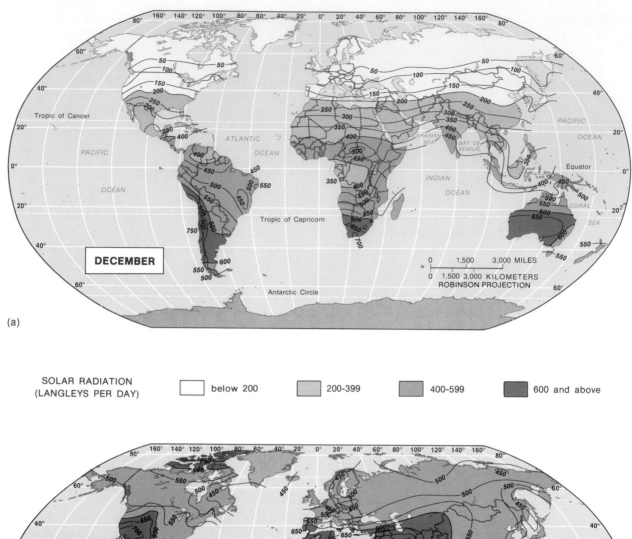

(a)

SOLAR RADIATION
(LANGLEYS PER DAY)   below 200   200-399   400-599   600 and above

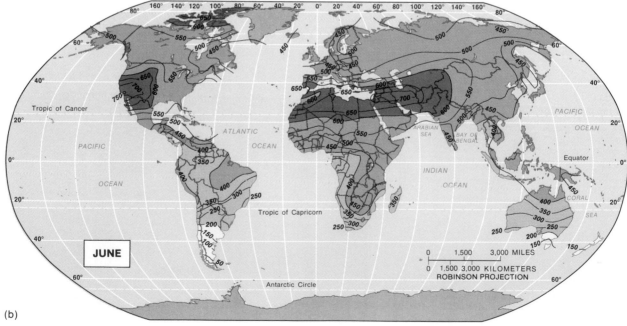

(b)

FIGURE 4–8

World solar radiation maps for (a) December and (b) June. Data used by permission
of Merrill, an imprint of Macmillan Publishing Company, from *Physical Geography* by
E. Willard Miller, p. 38. Copyright © 1985 by Merrill Publishing (after Lof, Duffie,
and Smith, 1966).

Energy in the Atmosphere: Some Basics

When viewing a photograph of Earth taken from space, the pattern of surface response to incoming insolation is clearly visible. Solar energy is intercepted by land and water surfaces, clouds, and atmospheric gases and dust. The flows of energy are evident in weather patterns, oceanic flows, and the distribution of vegetation. Specific energy patterns differ for deserts, oceans, mountain tops, rain forests, and ice-covered landscapes. In addition, clear or cloudy weather might mean a 75% difference in the amount of energy reaching the surface. (See Figure 4–10, which presents a detailed flow diagram of the energy balance between Earth and the atmosphere and summarizes the following discussion.)

Insolation Input. Insolation is the basic energy input driving the Earth-atmosphere system. The world maps in Figure 4–8 portray the distribution of average solar energy input at Earth's surface for the winter and summer months. The energy units used are calories per square centimeter (langleys) per day and represent total radiation, both direct and diffuse (or downward scattered).

Several patterns are notable on the maps. During December (winter in the Northern Hemisphere), northern latitudes experience a regular decrease of energy. Also, in general, greater insolation occurs in low-latitude deserts worldwide because of cloudless skies. The June map shows marked increases in insolation in the Northern Hemisphere; note the energy pattern in the cloudless deserts (e.g., the Sonoran, Saharan, Arabian, and Gobi Deserts). In the Southern Hemisphere, the pattern of seasonal change in insolation over Australia is especially noticeable.

Albedo and Reflection. A portion of arriving energy bounces directly back into space without being converted into heat or performing any work. This returned energy is called **reflection**, and it applies to both visible and ultraviolet light. The reflective quality of a surface is its **albedo**, or the relationship of reflected energy to insolation, expressed as a percentage. In terms of visible wavelengths, darker colors have lower albedos, and lighter colors have higher albedos. On water surfaces, the angle of the solar rays also affects albedo values; lower angles produce a greater reflection than do higher angles. In addition, smooth surfaces increase albedo, whereas rougher surfaces reduce it. Table 4–5 presents albedo values for various surfaces.

Although specific locations experience highly variable conditions during the year in response to changes in cloud and ground cover, Earth and its atmosphere reflect 31% of insolation received overall, when averaged over a year. Figure 4–10 shows that Earth's average albedo is a combination of 21% reflected by clouds, 3% reflected by the ground (combined continental and oceanic surfaces), and 7% reflected and scattered by the atmosphere. By

TABLE 4–5
Selected albedo values of various surfaces.

Type of Surface	Albedo (% Reflectivity)
Fresh snow	80–95
Snow, polluted, several days old	40–70
Sea ice	30–40
Forests	10–20
Dark soils	5–15
Dry, light, sandy soils	25–45
Crops, general	10–25
Concrete, dry	17–27
Asphalt, black top	5–10
Earth (average)	31+
Moon	6–8

SOURCE: M. I. Budyko, 1958, *The Heat Balance of the Earth's Surface,* Washington: U.S. Department of Commerce, p. 36.

comparison, a full Moon, which is bright enough to read by under clear skies, represents only a 6–8% albedo value. Thus, with earthlight being four times higher in albedo than moonlight, and with Earth four times larger than the Moon, it is no surprise that astronauts report how startling our planet looks from space.

Scattering (Diffuse Reflection). Insolation encounters an increasing density of atmospheric gases as it travels down to the surface. These molecules redirect the insolation, changing the direction of light's movement without altering its wavelengths. This phenomenon is known as **scattering** and represents 7% of Earth's albedo (Figure 4–10). Dust particles, ice, cloud droplets, and water vapor produce further scattering. A sky filled with smog and haze appears almost white because the larger particles associated with air pollution act to scatter all wavelengths of the visible light.

Why is Earth's sky blue? And why are sunsets often red? These simple questions have an interesting explanation, based upon a principle known as Rayleigh scattering (named for English physicist Lord Rayleigh, who stated the principle in 1881). This principle relates wavelength to the size of molecules or particles that cause the scattering. The general rule is: the shorter the wavelength, the greater the scattering, and the longer the wavelength, the less the scattering. Shorter wavelengths of light are scattered by small gas molecules in the air. Thus, the shorter wavelengths of visible light, the blues and violets, are scattered the most and dominate the lower atmosphere. And, because there are more blue than violet wavelengths in sunlight, a blue sky prevails.

As for red sunsets, nearer to sunrise and sunset with lower Sun altitudes, insolation must pass through an increasing mass of atmosphere. The effective scattering of shorter wavelengths leaves the residual oranges and reds to reach the observer.

Figure 4–10 shows that some incoming insolation is diffused by clouds and atmosphere and is transmitted to Earth as **diffuse radiation**, the downward component of scattered light. This light is multidirectional and thus casts shadowless light on the ground. For some outdoor sports a gray day of diffuse radiation may improve a player's ability to see the ball; however, this "flat lighting" proves poor if you are photographing the event.

Refraction. When insolation enters the atmosphere, it passes from one medium to another (from space to atmospheric gas) and is subject to a bending action called **refraction**. In the same way, a crystal or prism refracts light passing through it, bending different wavelengths to different degrees, separating the light into its component colors to display the spectrum. A rainbow is created when visible light passes through myriad raindrops and is refracted and dispersed toward the observer at a precise angle (Figure 4–9). Another example of refraction is a mirage, an image that occurs near the horizon where light waves are refracted on a hot day.

An interesting function of refraction is that it adds approximately eight minutes of daylight for us. The Sun's image is refracted in its passage from space through the atmosphere, and so, at sunrise, we see the Sun about four minutes before it actually peeks over the horizon. Similarly, at sunset, the Sun actually sets but its image is refracted from over the horizon for about four minutes afterward. To this day, modern science cannot predict the exact time of sunrise or sunset within these four minutes.

Absorption. **Absorption** is the assimilation of radiation and its conversion from one form to another. Insolation (both direct and diffuse) that is not reflected from Earth's surfaces is absorbed and converted into infrared radiation (heat) or is used by plants in photosynthesis. The temperature of the absorbing surface is raised in the process, causing that surface to radiate more total energy at shorter wavelengths.

Earth Reradiation and the Greenhouse Effect. Previously we defined Earth as a cool-body radiator emitting energy in infrared wavelengths from its surface and atmosphere back toward space. However, some of this infrared radiation is absorbed by carbon dioxide, water vapor, methane, CFCs, and other gases in the lower atmosphere and is then reradiated back toward Earth, thus delaying heat loss to space. This counterradiation process is an important factor in warming the lower atmosphere. The approximate similarity between this process and the way a greenhouse operates gives the process its name—the **greenhouse effect**.

In the greenhouse analogy, the glass is transparent to shortwave insolation, allowing light to pass through to the soil, plants, and articles inside. The

FIGURE 4–9
Double rainbow. Raindrops refract light to produce the primary rainbow, and reflection of the primary rainbow is seen as a secondary rainbow. Photo by author.

absorbed energy is then radiated as infrared energy back toward the glass, but the glass effectively traps the infrared wavelengths and warmed air inside the greenhouse. Thus, the glass acts as a one-way filter, allowing the light in but not allowing the heat out. The same process also can be observed in a dwelling that has large areas of windows or in a car parked in direct sunlight.

Opening a roof vent allows the air inside the greenhouse to mix with the outside environment, thereby removing heat by moving air physically from one place to another. Such movement is called **convection** if it involves a strong vertical motion and **advection** if a lateral (horizontal) motion is dominant. Heat in a greenhouse also can be conducted by exposed surfaces. **Conduction** is the molecule-to-molecule transfer of heat as it diffuses through a material. As molecules heat, their vibration increases, causing collisions that produce motion in neighbor-

ing molecules and thus transferring heat energy from warmer to cooler materials.

In the atmosphere, the greenhouse analogy is not fully applicable because infrared radiation is not trapped as is the case in a greenhouse. Rather, its passage to space is delayed as the heat is radiated and reradiated back and forth between Earth's surface and certain gases and particulates in the atmosphere. Today's increasing carbon dioxide concentration is causing more infrared radiation absorption in the lower atmosphere, thus forcing a warming trend and disruption of the Earth-atmosphere energy system, according to many scientists.

Earth-Atmosphere Radiation Balance

The Earth-atmosphere radiation system naturally balances in a steady-state equilibrium. But if Earth's surface and its atmosphere are considered sepa-

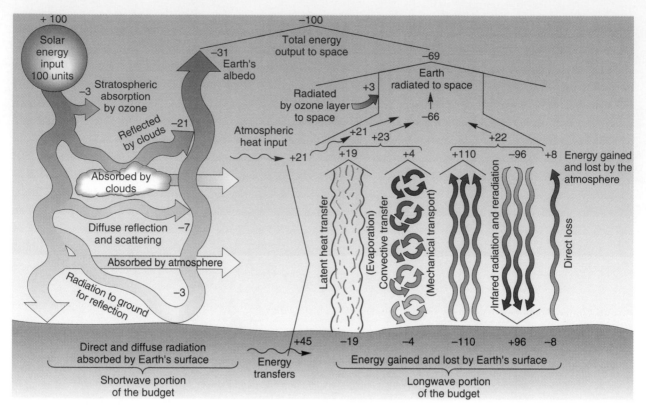

FIGURE 4–10

Earth-atmosphere energy balance. Revised and adapted by permission from Joseph E. Van Riper, *Man's Physical World,* p. 106. Copyright © 1971 by McGraw-Hill.

rately, neither exhibits a balanced radiation budget. The average annual distribution of energy demonstrates an overall positive radiation balance (energy surplus) for Earth's surface and an overall negative radiation balance (energy deficit) for the atmosphere. The natural energy balance occurs through nonradiative energy transfers in the latent heat of evaporation, convection, conduction, and through radiative transfers of infrared radiation, from the surface to the atmosphere.

Figure 4–10 presents a detailed illustration of the Earth-atmosphere radiation balance. It brings together all the elements discussed to this point in the chapter by following 100 units of energy through the system. To summarize: of 100 units of solar energy arriving, Earth's average albedo reflects 31% into space; absorption by atmospheric clouds, dust, and gases involves another 21% (which, when added to

the 3% absorbed by stratospheric ozone, totals the atmospheric heat input), and this leaves about half of the incoming insolation to actually reach Earth's surface. Earth eventually radiates 69% back into space as heat. In summary, overall Earth receives 100 units, reflects 31 units, and reradiates 69 units back to space. The 100 units of insolation received are balanced against the 100 units of reflected and radiated energy lost. However, the enhanced warming of the greenhouse effect apparently is altering this balance.

Global Net Radiation. Figure 4–11 summarizes the radiation balance for all shortwave and longwave energy by latitude. In the equatorial zone, energy surpluses dominate, for in those areas more energy is received than is lost. Sun angles are high, with consistent daylength. However, deficits exist in the polar regions, where more energy is lost than gained. At the poles the Sun is extremely low in the sky, surfaces are light and reflective, and for six months during the year no insolation is received.

This imbalance of net radiation from the equator to the poles drives vast global circulation of both energy and mass. The meridional (north-south) transfer agents are the global wind circulation, ocean currents, dynamic weather systems, and related phenomena. Tropical cyclones (hurricanes and typhoons) represent dramatic examples of such energy and mass transfers. Forming at or near the equator, they mature and migrate to higher latitudes, carrying with them surplus heat energy, water, and water vapor.

EARTH'S SURFACE ENERGY BALANCE

With only minor exceptions, solar energy is the principal heat source at Earth's surface. The direct and diffuse radiation and infrared radiation arriving at Earth's surface are shown in Figure 4–10. These daily radiation patterns at the surface demonstrate marked patterns of interest to geographers.

Daily Radiation Curves

The fluctuating daily pattern of incoming shortwave energy and outgoing longwave energy at the surface is shown in Figure 4–12. This graph represents conditions for bare soil on a cloudless summer day in the middle latitudes. Incoming energy arrives

FIGURE 4–11
Earth's energy surpluses and deficits by latitude.

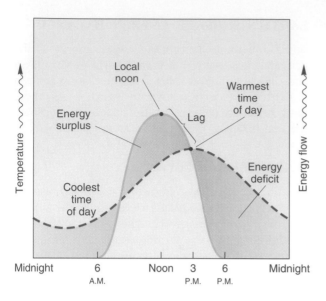

Incoming shortwave
- - - Outgoing longwave

FIGURE 4–12
Daily radiation curves demonstrate the lag between local noon and the warmest time of day.

throughout the illuminated part of the day, beginning at sunrise, peaking at local noon, and ending at sunset. The shape and height of this insolation curve varies with season and latitude. The highest trend for such a curve occurs at the time of the summer solstice (around June 21 in the Northern Hemisphere and December 21 in the Southern Hemisphere). The temperature plot (outgoing longwave energy curve) also responds to seasons and variations in input. Within a 24-hour day, temperature peaks at around 3:00–4:00 P.M. and dips to its lowest point right at or slightly after sunrise.

The relationship between the insolation curve and the temperature curve on the graph is interesting. As long as the incoming energy exceeds the outgoing energy, temperature continues to increase during the day, not peaking until the incoming energy falls below the temperature curve. The warmest time of day occurs at that moment when a maximum of insolation is absorbed. Thus, this temperature lag places the warmest time of day three to four hours after solar noon. Then, as the insolation input

decreases toward sunset, the amount of heat lost exceeds the input, and temperatures begin to drop until the surface has radiated away the maximum amount of energy, just at dawn.

The annual pattern of insolation and temperature also exhibits such a lag. For the Northern Hemisphere, January is usually the coldest month, occurring after the solstice and the shortest day in December. Similarly, the warmest months of July and August occur after the longest day of the solstice in June.

Simplified Surface Energy Balance

Earth's surface is quite active in terms of energy, and this forms the basis for varied climatic patterns. The climate at or near Earth's surface is generally termed the *boundary layer climate,* and the study of this climate is the science of **microclimatology**. The following discussion will be more meaningful if you keep in mind an actual surface—perhaps a park, a front yard, or a place on campus.

Within Figure 4–13 are the components of a surface energy balance. The column of soil diagrammed in the figure continues to a depth at which

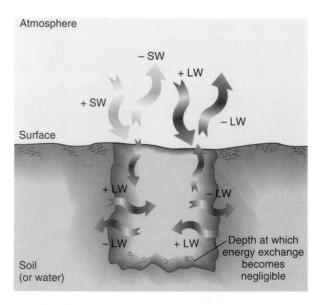

FIGURE 4–13
Input and output components of a surface energy balance for a typical soil column.

energy exchange with surrounding materials becomes negligible, usually less than a meter. The derivation of net radiation for a midlatitude location on a typical summer day is shown in Figure 4–14: direct and diffuse insolation, reflected energy (surface albedo), and infrared radiation arriving at and leaving from the surface. **Net radiation (NET R)**, or the balance of all radiation—shortwave (SW) and longwave (LW)—at Earth's surface, is derived from a consideration of these radiation inputs and outputs.

The surface receives light and heat, and it reflects light and radiates heat according to this basic scheme:

+SW ↓	−SW ↑	+LW ↓	−LW ↑	=	NET R
(Insolation, light)	(Albedo value of surface)	(Heat)	(Heat)		(Net radiation)
		(Infrared radiation)			

Energy moving toward the surface is regarded as positive (a gain), and energy moving away from the surface is considered negative (a loss). The amount of insolation to the surface (+SW ↓) varies with season, cloudiness, and latitude. The albedo dictates the amount of insolation reflected and therefore not absorbed at the surface (−SW ↑). Darker and rougher surfaces reflect less energy, therefore producing more net radiation.

Lighter and smoother surfaces reflect more shortwave energy, leading to less net radiation. Snow-covered surfaces reflect most of the light reaching them, whereas urban areas tend to be darker in color and therefore convert more insolation to infrared radiation. At night the net radiation value is negative, because the SW component ceases at sunset and the surface continues to lose radiative heat energy to the atmosphere. The surface rarely reaches a zero net radiation value—a perfect balance—at any one moment, but over time Earth's surface generally balances incoming and outgoing energies.

Net Radiation. Net radiation (NET R) is the net all-wave radiation available at Earth's surface; it is the final outcome of the entire radiation-balance process discussed in this chapter. The map in Figure 4–15 plots the distribution of mean annual net radiation. The abrupt change in radiation balance from ocean to land surfaces is quite evident. Note that all values are positive, with the only negative values probably occurring over ice-covered surfaces. The highest net

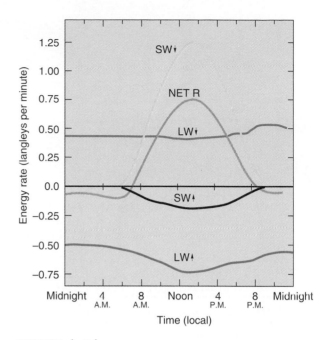

FIGURE 4–14
Radiation components and the derivation of net radiation on a typical summer day for a midlatitude location (Matador in southern Saskatchewan, about 51° N, on 30 July 1971). Adapted by permission from T. R. Oke, 1978, *Boundary Layer Climates,* New York: Methuen & Co., p. 21.

radiation occurs north of the equator in the Arabian Sea at 140 kilolangleys per year. Given the obvious interruption caused by land masses, the pattern of values appears generally zonal, or parallel to the equator.

Output paths at the surface for the principal expenditures of net radiation from a nonvegetated surface include three components:

H, or turbulent sensible heat transfer. H is the quantity of sensible heat energy transferred from air to surface and surface to air as turbulent heat. About 18% of Earth's entire NET R is reradiated as such sensible heat. Turbulent transfers can be through convection or conduction.

LE, or latent heat of evaporation. LE refers to latent heat energy that is involved in the evaporation of water. (Latent heat is absorbed or emitted when water changes state between liquid and gas.) Because the evaporation of water is the principal method of transferring and dissipating heat sur-

FIGURE 4–15
Distribution of global mean annual net radiation in kcal/cm²/year. Data from M. I.
Budyko, 1958, *The Heat Balance of the Earth's Surface,* Washington: U.S. Department
of Commerce, p. 106.

pluses vertically into the atmosphere, latent heat is
the dominant expenditure of Earth's entire NET R.
A variety of factors influence evaporation rates and
are discussed in a later chapter.

G, or ground heating and cooling. Energy that
flows into and out of the ground surface by
conduction is symbolized by G.

The highest annual values on land for LE (latent
heat of evaporation) occur between 10° N and 10° S
latitude and decrease with distance toward each of
the poles. The values for H (sensible heat) are
distributed differently, being highest in the subtropics, where nearly waterless and cloudless expanses of
subtropical deserts stretch almost vegetation-free
across vast regions. Over the oceans, the highest LE
values are at those very same subtropical latitudes
where hot, dry air comes into contact with warm
ocean water.

The overall G value for a year is zero, because the
stored energy from spring and summer is equaled by

losses in fall and winter. The input for G is derived
from absorption of insolation and infrared radiation
from above and from horizontal transfer from the
surrounding environment. Negative G values are
attributed to heat lost by radiation to the atmosphere,
sensible heat transfer when the air is cooler than the
ground, energy expended for evaporation, and horizontal transfer.

Moist surfaces covered with vegetation require
further refinement of the ground heating component. Net radiation available at the surface is decreased because incoming shortwave energy is intercepted by a forest canopy of leaves. The amount of
light reflected and absorbed depends upon tree
height and the density of the leaves. Photosynthesis
by plants absorbs approximately 8% of net radiation,
converting it into biochemical energy through the
fixation of carbon extracted from the air. The
variation in vegetation (biomass) over Earth reflects
the energy expended for photosynthesis, as discussed in Chapters 19 and 20. Those latitudes with

FOCUS STUDY 4-1

Solar Energy Collection and Concentration

Insolation not only warms Earth's surface; it provides an inexhaustible, yet essentially untapped, supply of energy for humanity. Sunlight is direct, pervasive, and renewable and has been collected for centuries through various technologies. Turn-of-the-century photographs of residences in Southern California and Florida show solar water heaters on many rooftops. Early twentieth-century ads in newspapers and merchandise catalogs featured the Climax solar water heater (1905) and the Day and Night water heater (1909). However, low-priced natural gas and oil displaced many of these early applications.

By the early 1970s, the depletion of fossil fuels, their rising prices, and our growing dependence upon foreign sources spurred a reintroduction of tried-and-true solar technologies and energy-efficiency strategies. Cities such as Davis in California passed energy-conservation and solar-utilization ordinances to guide construction and to encourage solar-energy installations. The country seemed to be rediscovering old, proven techniques. Unfortunately, during the early 1980s much of this progress was halted by political decisions and severe budget cuts made by the Reagan administration in Washington.

Any surface that receives light from the Sun is a solar collector. In fact, an average building in the United States receives 6 to 10 times more energy from the Sun than is required to heat the inside! But the diffuse nature of solar energy received at the surface requires that it be collected, transformed, and stored to be useful. Space heating is the simplest application. Windows that are carefully designed and placed allow sunlight to shine into the building, where it is absorbed and converted into sensible heat.

A *passive solar system* captures heat energy and stores it in a thermal mass, such as water-filled tanks, adobe, tile, or concrete. Three key features of passive solar design are large areas of glass facing south, a thermal storage medium, and shades or screens to prevent light entry in the warm summer months. At night, shades or other devices can cover glass areas to prevent heat loss from the structure. An *active solar system* involves the heating of water or air in a collector and then pumping it through a plumbing system to a tank where it can provide hot water for direct use or for space heating.

---

abundant insolation and water are the most photosynthetically productive.

Another item that further refines the ground heating component is energy expended at the surface to melt snow or ice. In snow- or ice-covered landscapes, almost all available energy is consumed as sensible and latent heat used in the melting and warming process.

Understanding the net radiation balance is essential to solar energy technologies that concentrate

Solar energy systems are capable of generating low-grade heat of an appropriate scale for approximately half the present domestic applications in the United States (space heating and water heating). In marginal climates, solar-assisted water and space heating is feasible as a backup; even in the New England and the Northern Plains states, solar-efficient collection systems have proven effective.

The use of reflecting surfaces, such as focusing mirrors or parabolic troughs, can attain very high temperatures to heat water or other fluids. Kramer Junction, California, about 225 km (140 mi) northeast of Los Angeles, is the location of the world's largest operating solar electric-generating facility, with 194 megawatts capacity. Long troughs of computer-guided curved mirrors concentrate sunlight to create temperatures of 390°C (735°F) in vacuum-sealed tubes filled with synthetic oil. This is then used to heat water, which produces steam that rotates turbines to generate cost-effective electricity. Thirty years of contracts for the power already are signed, and utility customers are seeking the construction of more capacity (Figure 1).

Electricity is also being produced by photovoltaic cells, a technology that has been used in spacecraft since 1958. When sunlight shines upon a semiconductor material in these cells, it stimulates a flow of electrons (an electrical current). The efficiency of these cells has gradually improved to the point where they are cost-competitive. A drawback with both solar heating and solar electric systems is that periods of cloudiness and night inhibit operations. Research is going forward to enhance energy storage techniques, such as hydrogen fuel production and improved battery technology.

Solar energy is a wise choice for the future because it is direct to the consumer, it is based upon renewable sources of an appropriate scale for end-use needs, and most solar strategies are labor intensive (rather than capital intensive). Solar seems preferable to further development of our decreasing fossil fuel reserves, oil imports, and troubled nuclear power. Large centralized power production possesses the negative traits of being indirect, nonrenewable, high-grade energy that is capital intensive. Whether

shortwave and longwave energy for use. Solar energy offers great potential worldwide and no doubt will see expanded use in the near future. Focus Study 4–1 briefly reviews solar energy utilization strategies.

Sample Stations. An energy balance equation can be structured for any time frame. This section examines daily energy balances at two locations, El Mirage in California and Pitt Meadows in British Columbia.

FIGURE 1
Luz International solar-thermal energy installation in southern California. Photo courtesy of Luz International Limited, Los Angeles.

the alternative path of solar energy is taken depends principally upon political decisions, for many of the technological innovations are presently ready for installation, requiring only leadership from the government.

El Mirage, at 35° N, is a hot desert location characterized by bare, dry soil with very little vegetation—Figure 4–16(a). The day selected for examining the energy balance is in summer, clear with a light wind in the late afternoon. The NET R value is lower than might be expected, considering the Sun's position close to zenith and no clouds. But the income of energy at this site is countered by surfaces having a higher albedo value than a forest or cropland and by extremely hot soil surfaces radiating

FIGURE 4–16
Daily net radiation expenditures for (a) El Mirage, California (near Los Angeles, about 35° N), and (b) Pitt Meadows in southern British Columbia (about 49° N). *H* = turbulent sensible heat transfer; *LE* = latent heat of evaporation; *G* = ground heating and cooling. (a) Adapted by permission from William D. Sellers, *Physical Climatology,* fig. 33, copyright © 1965 by The University of Chicago. All rights reserved. (b) Adapted by permission from T. R. Oke, 1978, *Boundary Layer Climates,* New York: Methuen & Co., p. 23.

heat back to the atmosphere throughout the afternoon.

El Mirage has little or no energy expenditure for LE. With an absence of water and plants, most of the available radiant energy is dissipated as turbulent sensible heat, warming air and soil to high temperatures. Over a 24-hour period, II consumes 90% of NET R, with the remaining 10% going to G. The G component is higher in the morning, when winds are light and turbulent transfers are reduced. In the afternoon, heated air rises off the hot ground, and convective heat expenditures are accelerated as winds increase.

The NET R at this desert location is quite similar to that of the Pitt Meadows, British Columbia site— Figure 4–16(b). However, Pitt Meadows is midlatitude (49° N), vegetated, and moist, and its energy expenditures vary greatly. The Pitt Meadows landscape is able to retain much more of its energy because of a lower albedo (less reflection), the presence of more water and plants, and lower surface temperatures. The energy balance data for Pitt Meadows are plotted for a cloudless summer day.

The higher LE values are attributed to the moist environment of rye grass and irrigated mixed-orchard ground cover for the sample area, contributing to the more moderate sensible heat levels throughout the day.

Urban-Induced Energy Balances

Most of you reading this live in or near a city. It is well established that urban microclimates generally differ from those of nearby nonurban areas. In fact, the surface energy characteristics of urban areas possess unique properties similar to energy-balance traits of desert locations. Because more than 60% of the world's population will be living in cities by the year 2000, the study of **urban heat islands** and other specific environmental alterations related to cities is an important topic for physical geographers.

At least six factors contribute to urban microclimates:

1. Urban surfaces tend to be metal, glass, asphalt, concrete, or stone. These city surfaces conduct

up to three times more heat than wet sandy soil. The heat storage capacity of these materials also exceeds that of natural surfaces. During the day, temperatures above these surfaces are higher, whereas at night such surfaces rapidly radiate this stored heat to the atmosphere. As a result, maximum and minimum temperatures are higher and lower respectively than nonurban areas, although heat-island characteristics are more profound during nighttime cooling.

2. Urban surfaces behave differently from natural surfaces in terms of energy balance. Albedo values are lower, leading to a higher net radiation value. However, urban surfaces expend more of that energy as sensible heat than nonurban areas; more than 70% of the net radiation in an urban setting is spent in this way. Figure 4–17 illustrates a generalized cross section of a typical urban heat island, showing increasing temperatures toward the center of the city.

3. Urban surfaces generally are sealed (built on and paved) so that water cannot reach the soil. Central business districts are on the average 50% sealed, whereas suburbs are about 20% sealed. Because urban precipitation cannot penetrate the soil, water runoff increases. Urban areas respond much as a desert landscape: a storm may cause a flash flood over the hard, plantless surfaces, only to be followed by a return to dry conditions a few hours later. Very little of the net radiation is expended for evaporation on such a surface.

4. The irregular geometric shapes of a modern city affect radiation patterns and winds. Incoming insolation is caught in mazelike reflection and radiation "canyons," which tend to trap energy for conduction into surface materials, thus increasing temperatures. In natural settings insolation is more readily reflected, stored in plants, or converted into latent heat through evaporation. Figure 4–18 illustrates the characteristic effects of urban structures on insolation paths and winds produced by metropolitan areas. Buildings tend to interrupt wind flows, thereby diminishing heat loss through advective movement. On the average, winds have 25% less velocity in cities than in rural areas, although the buildings do create turbulence and funneling effects. Wind flows can significantly reduce heat-island effects by increasing turbulent heat loss. Thus, maximum heat-island effects in the most built-up areas of a city occur on calm, clear days and nights.

5. Human activity alters the heat characteristics of cities. In New York City, for example, during the summer months, production of electricity and use of fossil fuels release an amount of energy equivalent to 25 to 50% of the arriving insolation. In the winter, urban-generated heat is on the average 250% greater than arriving insolation! In fact, for many northern cities in North America, winter heating requirements are actually reduced by this enhanced community heat-island effect.

6. The amount of air pollution, including gases

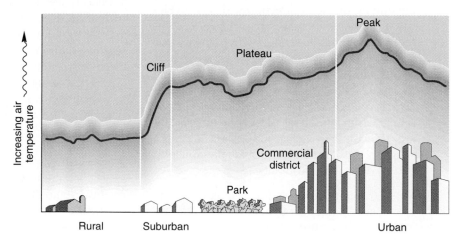

FIGURE 4–17

Generalized cross section of a typical urban heat island. The trend of the temperature gradient is described by the terms "cliff" at the edge of the city, where temperatures steeply rise, "plateau" over the suburban built-up area, and "peak" where temperature is highest at the urban core. Adapted by permission from T. R. Oke, 1978, *Boundary Layer Climates,* New York: Methuen & Co., p. 254.

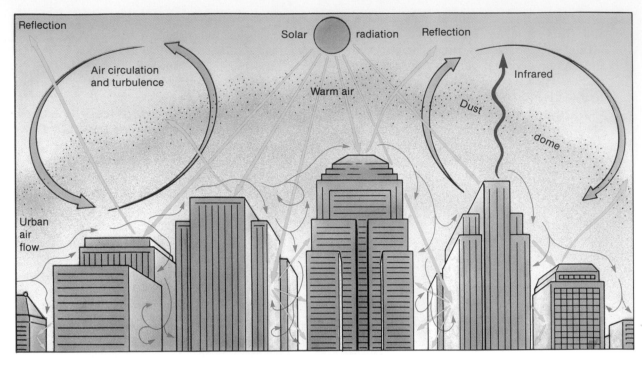

FIGURE 4–18
Insolation, wind movements, and dust dome in city environments.

and aerosols, is greater in urban areas than in comparable natural settings. Pollution actually increases reflection in the atmosphere above the city, thus reducing insolation reaching the ground. The pollution blanket also absorbs infrared radiation, radiating the heat downward. Every major city produces its own **dust dome** of airborne pollution, which can be blown from the city in elongated plumes, depending on wind direction and speed (Figure 4–18). The increased particulates present in pollution act as condensation nuclei for water vapor. Convection created by the heat island lifts the air and particulates, producing increased cloud formation and the potential for increased precipitation. Although these precipitation effects of the urban heat island are difficult to isolate and prove, research suggests that urban-stimulated precipitation occurs downwind from cities.

Table 4–6 summarizes various climatic factors and the average changes brought about by urbanization. Specific temperature differences of 6C° (11F°)

are common between urban and nonurban areas, with maximum differences of 11C° (20F°) possible on calm, clear nights.

SUMMARY

Seasonality is the change in daylength and varying altitude of the Sun above the horizon experienced over much of Earth during the year. Seasonal variations are produced by interactions of Earth's revolution, rotation, axial tilt, and alignment of the axis throughout the year. The annual march of the seasons demonstrates the varied relationship between insolation and Earth during the year.

Physical geographers utilize models that effectively budget the energy inputs, pathways, and outputs in the troposphere and at Earth's surface. Approximately 31% of insolation is reflected back to space without performing any work, as Earth's average albedo. Insolation is absorbed by atmospheric clouds, gases, and dust, with the remaining energy eventually reaching land and water surfaces.

Surface energy balances summarize the energy patterns for various locations and form an analytical

TABLE 4-6
Average climatic changes caused by urbanization.

Element	Urban Compared to Rural Environs
Contaminants:	
Condensation nuclei	10 times more
Particulates	10 times more
Gaseous admixtures	5–25 times more
Radiation:	
Total on horizontal surface	0–20% less
Ultraviolet, winter	30% less
Ultraviolet, summer	5% less
Sunshine duration	5–15% less
Cloudiness:	
Clouds	5–10% more
Fog, winter	100% more
Fog, summer	30% more
Precipitation:	
Amounts	5–15% more
Days with <5 mm (0.2 in.)	10% more
Snowfall, inner city	5–10% less
Snowfall, downwind (lee) of city	10% more
Thunderstorms	10–15% more
Temperature:	
Annual mean	0.5–3.0C° (0.3–1.7F°) more
Winter minima (average)	1.0–2C° (0.6–1.1F°) more
Summer maxima	1.0–3C° (0.6–1.7F°) more
Heating degree days	10% less
Relative humidity:	
Annual mean	6% less
Winter	2% less
Summer	8% less
Wind speed:	
Annual mean	20–30% less
Extreme gusts	10–20% less
Calm	5–20% more

SOURCE: Helmut E. Landsberg, 1981, *The Urban Climate*, Vol. 28, International Geophysics Series, p. 258. Reprinted by permission from Academic Press.

tool of microclimatology, an important aspect of physical geography. Surface energy balances produce a net radiation (NET R) value. NET R is expended through outflows of sensible heat, latent heat of evaporation, ground heating, and conversion of heat energy to biochemical energy through photosynthesis. A growing percentage of Earth's human inhabitants live in cities, which possess a unique set of altered microclimatic effects associated with urbanization: increased surface conduction, lower albedos and higher NET R values, increased runoff, complex radiation and reflection patterns, anthropogenic heating, and the gases, dusts, and aerosols that are part of urban pollution.

Absorption	Net radiation (NET R)
Advection	Plane of the ecliptic
Albedo	Reflection
Altitude	Refraction
Analemma	Revolution
Autumnal equinox	Rotation
Axial parallelism	Scattering
Axis	Summer solstice
Circle of illumination	Sunrise
Conduction	Sunset
Convection	Transmission
Daylength	Tropic of Cancer
Declination	Tropic of Capricorn
Diffuse radiation	Urban heat islands
Dust dome	Vernal equinox
Greenhouse effect	Winter solstice
Microclimatology	

REVIEW QUESTIONS

1. In what possible way does the Stonehenge monument relate to seasons?
2. The concept of seasonality refers to what specific factors? How do the variables of seasonality change with latitude? 0°? 40°? 90°?
3. Differentiate between the Sun's altitude and its declination.
4. For the latitude at which you live, how does daylength vary during the year? How does the Sun's altitude vary? Does your local newspaper publish a weather calendar containing such information?
5. List the five physical factors that operate together to produce seasons.
6. Describe revolution and rotation, and differentiate between them.
7. Why do rotational velocities vary with latitude? Describe the velocities for 0°, 40°, 60°, and 90° latitudes.
8. Define Earth's present tilt relative to our orbit about the Sun.
9. What is meant by axial parallelism? How does it influence Earth's alignment with the Sun during the year?
10. What is an analemma? How does it relate to the Sun's declination and the equation of time during the year? Relate your answer to the photograph that begins this chapter.
11. Describe seasonal conditions at each of the four key seasonal anniversary dates during the year. What are the solstices and equinoxes?
12. Do dawn and twilight vary during the year? What factors explain these seasonal changes?
13. Diagram a simple energy balance for the troposphere. Label each shortwave and longwave component and the directional aspects of related flows.
14. List several types of surfaces and their albedos. Explain the differences. What determines the reflectivity of a surface?

15. Why is the lower atmosphere blue? What would you expect the sky color to be at 50 km (30 mi) altitude? Why?
16. Define conduction, convection, absorption, transmission, and diffuse radiation.
17. What are the similarities and differences between an actual greenhouse and the gaseous atmospheric greenhouse? Why is Earth's greenhouse changing?
18. Why is there a temperature lag between the highest Sun altitude and the warmest time of day? Relate your answer to the insolation and temperature curves.
19. Generalize the pattern of global net radiation. How might this pattern drive the atmospheric weather machine?
20. In terms of surface energy balance, explain the term net radiation.
21. What are the expenditure pathways for surface net radiation? What kind of work is accomplished?
22. Compare the daily surface energy balances of El Mirage and Pitt Meadows. Explain the differences.
23. What is the basis for the urban heat-island concept? Describe the climatic effects attributable to urbanization as compared to nonurban environments.

SUGGESTED READINGS

Ackerman, B., S. A. Changnon, Jr., and others. *Summary of METROMEX*. Vol. 2, *Causes of Precipitation Anomalies*. State of Illinois, Department of Registration and Education Bulletin 63. Urbana: Illinois State Water Survey, 1978.

Budyko, M. I. *The Heat Balance of the Earth's Surface*. Translated by Nina A. Stepanova. Washington: U.S. Department of Commerce, U.S. Weather Bureau, 1958.

Budyko, M. I. *Climate and Life*. Edited by David H. Miller. International Geophysics Series, vol. 18. New York: Academic Press, 1974.

Gates, David M. "The Flow of Energy in the Biosphere." *Scientific American* 224, no. 3 (September 1971): 88–100.

Hobbs, John E. *Applied Climatology—A Study of Atmospheric Resources*. Kent, England: Dawson Westview Press, 1980.

Landsberg, Helmut E. *The Urban Climate*. International Geophysics Series, vol. 28. New York: Academic Press, 1981.

Meinel, Aden, and Marjorie Meinel. *Sunsets, Twilights, and Evening Skies*. Cambridge: Cambridge University Press, 1983.

Miller, David H. *Energy at the Surface of the Earth—An Introduction to the Energetics of Ecosystems*. International Geophysics Series, vol. 27. New York: Academic Press, 1981.

Oke, T. R. *Boundary Layer Climates*. 2d ed. London: Methuen, 1987.

Sellers, William D. *Physical Climatology*. Chicago: University of Chicago Press, 1965.

United States Naval Observatory. *The Air Almanac*. Washington: U.S. Government Printing Office, published annually.

Photo by author.

5
World
Temperatures

What temperature is it now (both indoors and outdoors) as you read these words? How is it measured and what does the value mean? How is today's air temperature controlling your plans for the day and your choice of clothing? Air temperature has a remarkable influence upon our lives, both at the microlevel just indicated and at the macrolevel. A variety of temperature regimes worldwide affect entire lifestyles, cultures, decision making, and resources spent. Global temperature patterns presently are changing in a warming trend that is affecting us all and is the subject of much scientific, geographic, and political interest.

Our bodies sense temperature and subjectively judge comfort, reacting to changing temperatures with predictable responses. Focus Study 5–1 introduces some essential temperature concepts to begin our study of world temperatures.

PRINCIPAL TEMPERATURE CONTROLS

Air temperature, a measure of sensible heat energy present in the atmosphere, indicates the average kinetic energy of individual molecules within the atmosphere. Principal controls and influences upon temperature patterns include latitude, altitude, land-water heating differences, ocean currents, cloud cover, and surface conditions. Of course, Earth's temperatures have not always been as they are today. In fact, humans are too recent inhabitants of Earth to have experienced the planet's "normal" climate—those climate patterns characteristic of Earth's first 3 billion years. Chapter 17 examines past Earth temperatures and some of the temperature controls that influence such long-term patterns. Let's first examine each of the principal temperature controls.

Latitude

Insolation received is the single most important influence on temperature variations. The distribution of intercepted solar energy at the top of the atmosphere exhibits an overall imbalance among equatorial, midlatitude, and polar locations. As discussed previously (Figures 2–11 and 2–12), insolation intensity decreases as surface distance increases from the subsolar point, a point that migrates between the Tropics of Cancer and Capricorn from 23.5° N to 23.5° S during the year. In addition, daylength and Sun angle change throughout the year, increasing in seasonal effect with increasing latitude. From equator to poles, Earth ranges from continually warm, to seasonally variable, to continually cold.

Altitude

Air temperatures decrease with increasing elevation above Earth's surface in the troposphere (recall that the normal lapse rate of temperature change with altitude is 6.4C°/1000 m or 3.5F°/1000 ft). Thus, worldwide, mountainous areas experience lower temperatures than do regions nearer sea level, even at similar latitudes. The density of the atmosphere also diminishes with increasing altitude, as discussed in Chapter 3. In fact, air pressure at an elevation of 5500 m (18,000 ft) is about half of that at sea level. As the atmosphere thins, its ability to absorb and radiate heat is reduced.

The consequences are that average air temperatures at higher elevations are lower, nighttime cooling increases, and the temperature range between day and night and between areas of sunlight and shadow also increases. Temperatures may decrease noticeably in the shadows and shortly after sunset. Surfaces both heat rapidly and lose their heat rapidly. Also, at higher elevations, the insolation received is more intense because of the reduced mass of atmospheric gases. As a result of this intensity, the ultraviolet energy component makes sunburn a distinct hazard.

The snowline seen in mountain areas indicates where winter snowfall exceeds the amount of snow lost through summer melting and evaporation. The snowline's location is a function both of altitude and latitude, which means that glaciers can exist even at equatorial latitudes if the altitude is great enough. In equatorial mountains, the snowline occurs at approximately 5000 m (16,400 ft), and so, because of this altitude, permanent ice fields and glaciers exist on equatorial mountain summits in the Andes and East Africa. With increasing latitude toward the poles, snowlines gradually descend in elevation from 2700 m (8850 ft) in the midlatitudes to lower than 900 m (2953 ft) in southern Greenland.

Temperature Concepts, Terms, and Measurements

The amount of heat energy present in any substance can be measured and expressed as its temperature. Temperature actually is a reference to the speed and movement of the atoms and molecules that make up a substance. Changes in temperature are caused by the absorption or emission (gain or loss) of energy. The temperature at which all motion in a substance stops is called 0° absolute temperature. Its equivalent in different temperature-measuring schemes is −273° Celsius (C), −459.4° Fahrenheit (F), or 0 Kelvin.

The Celsius scale (formerly called centigrade) is named after Swedish astronomer Anders Celsius, 1701–1744. It divides the difference between the freezing and boiling temperatures of water into 100 degrees on a decimal scale.

The Fahrenheit scale places the freezing point of water at 32°F (0°C, 273 K) and boiling point of water at 212°F (100°C, 373 K). The United States remains the only major country still using the Fahrenheit scale, named after the German physicist who invented the alcohol and mercury thermometers and who developed this thermometric scale in the early 1700s.

The Kelvin scale, named after British physicist Lord Kelvin (born William Thomson, 1824–1907) who proposed the scale in 1848, is used in science because temperature readings are proportional to the actual kinetic energy in a material.

Most countries use Celsius as part of the International System of Units, or SI. This text presents Celsius (with Fahrenheit equivalents in parentheses) throughout in an effort to bridge this transitional era. The continuing pressure of the scientific community and corporate interests, not to mention the rest of the world, makes adoption of the SI system inevitable for the United States. Figure 1 compares these two scales. Formulas for converting between SI and English units are on the inside front cover of this text.

Outdoor temperature usually is measured with a mercury thermometer or alcohol thermometer in a sealed glass tube. Alcohol is preferred in cold climates because it freezes at −112°C (−170°F), whereas mercury freezes at −39°C (−38.2°F). The principle of these thermometers is simple: when these fluids are heated, they expand; upon cooling, they contract. A thermometer stores fluid in a small reservoir at one end and is marked with calibrations to measure the expansion or contraction of the fluid, which reflects the temperature of the thermometer's environment.

Figure 2 shows a minimum-maximum thermometer, designed to record the day's highest and lowest temperatures on a mercury thermometer and preserve the readings until it is reset. Resetting is accomplished by moving the markers with a magnet. A recording

Degrees

Water boils (at normal air pressure at sea level) 100°C (212°F)

Highest air temperatures recorded:
World—
Al 'Azīzīyah, Libya (32° N, 13° E) 58°C (136°F) September 13, 1922

North America—
Death Valley, CA (37°N, 117°W) 57°C (134°F) July 10, 1913

Average normal body temperature 37°C(98.6°F)

Normal room temperature 20°C(68°F)

Freezing point of water 0°C(32°F)

Degrees

°F | °C
9F° | 5C°

Note the distinction between °C and C°. Actual temperature measurements give results in degrees Celsius, or °C— "it is 15 degrees Celsius (°C) outside." But when you are describing a *change* in temperature, you give the number of degrees of change in Celsius degrees, or C°— "the temperature went up by 3 Celsius degrees (C°)."

Freezing point of mercury −39°C(−38°F)

Lowest air temperatures recorded:
Northern Hemisphere—
Verkhoyansk, USSR (67°N, 133°E) −68°C (−90°F) February 7, 1892

World—
Vostok, Antarctica (78°S, 106°E) −89°C (−129°F) July 21, 1983

FIGURE 1
Scales for expressing temperature in degrees Celsius and Fahrenheit.

FIGURE 2
Minimum-maximum thermometer. Photo by author.

FIGURE 3
Standard weather instrument
shelter. Photo by author.

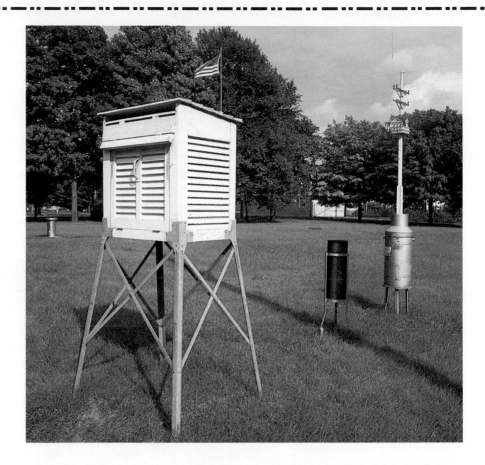

thermometer creates an inked record on a turning drum; it is usually set for one full rotation every 24 hours or every 7 days.

Thermometers for standardized official readings are deployed inside white (high albedo), louvered (ventilated) weather stations (Figure 3), designed to avoid excessive heating of the instruments inside. They are placed at least 1.2–1.8 m (4–6 ft) above a surface, usually on turf; in the United States 1.2 m (4 ft) is standard. Official temperatures are measured in the shade to prevent the effect of direct insolation. Temperature readings are taken daily, sometimes hourly, at more than 10,000 weather stations worldwide. Some stations with recording equipment also report the duration of temperatures, rates of rise or fall, and variation over time throughout the day and night.

Daily minimum-maximum readings are averaged to get the daily mean temperature. The monthly mean temperature is the total of daily mean temperatures for the month, divided by the number of days in the month. An annual temperature range expresses the difference between the lowest and the highest monthly mean temperatures for a given year. If you install a thermometer for outdoor temperature reading, be sure to avoid direct sunlight on the instrument and place it in an area of good ventilation.

Two cities in Bolivia illustrate the interaction of the two temperature controls, latitude and altitude. Figure 5–1 displays temperature data for the cities of Concepción and La Paz, which are at the same latitude (about 16° S). Note the elevation, average annual temperature, and precipitation for each location. The hot, humid climate of Concepción at its much-lower elevation stands in marked contrast to the cool, dry climate of highland La Paz. People living around La Paz actually grow wheat, barley, and potatoes—crops characteristically grown in the cooler midlatitudes—despite the fact that La Paz is at 4103 m (13,461 ft) above sea level. (For comparison, the summit of Pikes Peak in Colorado is at 4301 m or

14,111 ft.) The combination of elevation and equatorial location guarantee La Paz nearly constant daylength and moderate temperatures, averaging about 9°C (48°F) for every month. Such moderate temperature and moisture conditions lead to the formation of more fertile soils than those found in the warmer, wetter climate of Concepción.

Cloud Cover

Local temperatures may be affected by cloud cover. Orbital surveys reveal that approximately 50% of Earth is cloud-covered at any one time. Clouds are the most variable factor influencing Earth's radiation

Station: Concepción, Bolivia **Elevation:** 490 m (1607.6 ft)
Lat/long: 16°15'S, 62°03'W **Population:** 768,000
Avg. Ann. Temp.: 24°C (75.2°F) **Ann. Temp. Range:**
Total Ann. Precip.:   5C° (9F°)
 121.2 cm (47.7 in.)

Station: La Paz, Bolivia **Elevation:** 4103 m (13,461 ft)
Lat/long: 16°30'S, 68°10'W **Population:** 993,000
Avg. Ann. Temp.: 9°C (48.2°F) **Ann. Temp. Range:**
Total Ann. Precip.:   3C° (5.4F°)
 55.5 cm (21.9 in.)

FIGURE 5–1
Comparison of temperature patterns in La Paz and Concepción, Bolivia.

budget, and they are the subject of much investigation and effort to improve computer models of atmospheric behavior. The International Satellite Cloud Climatology Project, part of the World Climate Research Program, is presently in the midst of such research.

Clouds are moderating influences on temperature, producing lower daily maximums and higher nighttime minimums. Acting as insulation, clouds hold heat energy below them at night, preventing more rapid radiative losses, whereas during the day, clouds reflect insolation as a result of their high albedo values. The moisture in clouds both absorbs and liberates large amounts of heat energy, yet another factor in moderating temperatures at the surface.

Land-Water Heating Differences

The irregular arrangement of landmasses and water bodies on Earth contributes to the overall pattern of temperature. The physical nature of the substances themselves—solid rock and soil vs. water—is the reason for these **land-water heating differences**. More moderate temperature patterns are associated with water bodies compared to more extreme temperatures inland. Figure 5–2 visually summarizes the following discussion.

Evaporation. More of the energy arriving at the ocean's surface is expended for evaporation than is expended over a comparable area of land. An estimated 84% of all evaporation on Earth is from the oceans. When water evaporates and thus changes to water vapor, heat energy is absorbed in the process and is stored in the water vapor. This stored heat energy is called *latent heat* and is discussed fully in Chapter 7. You can experience this evaporative heat loss (cooling) by wetting the back of your hand and then blowing on the moist skin. Sensible heat energy is drawn from your skin to supply some of the energy for the water evaporation, and you feel the cooling. Similarly, as surface water evaporates, substantial energy is absorbed, resulting in a lowering of nearby air temperatures.

Transmissibility. The transmission of light obviously differs between soil and water; solid ground is opaque, water is transparent. Consequently, light striking a soil surface does not penetrate but is absorbed, heating the ground surface. That heat is accumulated during times of exposure and is rapidly lost at night or in shadows. Figure 5–3 shows the profile of diurnal temperatures for a column of soil and the atmosphere above it at a midlatitude location. You can see that maximum and minimum temperatures generally are experienced right at ground level. Below the surface, even at shallow depths, temperatures remain relatively static throughout the day. This same situation often exists at a beach, where surface sand may be painfully hot to your feet but the sand a few centimeters below the surface feels cooler and offers relief.

In contrast, when light reaches a body of water it penetrates the surface because of water's **transparency**, transmitting light to an average depth of 60 m (200 ft) in the ocean. This illuminated zone is known as the *photic layer* and has been recorded in some ocean waters to depths of 300 m (1000 ft). This characteristic results in the distribution of available heat energy over a much greater depth and volume, forming a larger heat reservoir than the surface layers of the land.

Specific Heat. When similar volumes of water and land are compared, water requires far more heat to raise its temperature. In other words, water can hold more heat than can soil or rock, and therefore water is said to have a higher **specific heat**. A given volume of water represents a more substantial heat reservoir than an equal volume of land, so that changing the temperature of the oceanic heat reservoir is a slower process than changing the temperature of land.

Movement. Land is a rigid, solid material, whereas water is a fluid and capable of movement. Differing temperatures and currents result in a mixing of cooler and warmer waters, and that mixing spreads the available heat over an even greater volume. Surface water and deeper waters mix, redistributing heat energy. Both ocean and land surfaces radiate heat at night, but land loses its heat energy more rapidly than does the moving mass of the oceanic heat reservoir.

Ocean Currents. Although ocean currents are discussed in greater detail in the next chapter, the influence of currents on temperature requires men-

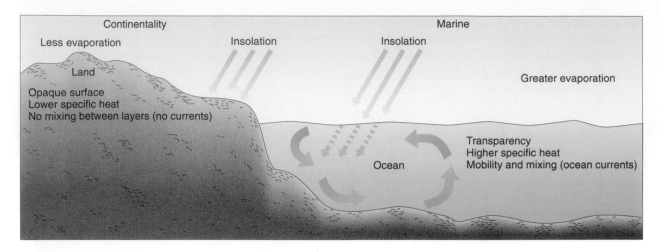

FIGURE 5-2
The differential heating of land and water.

FIGURE 5-3
Profile of air and soil temperatures in Seabrook, New Jersey. From John R. Mather, 1974, *Climatology: Fundamentals and Applications,* New York: McGraw-Hill, p. 36. Adapted by permission.

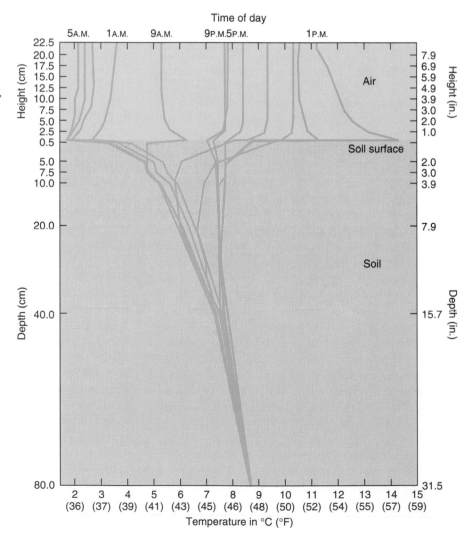

FIGURE 5–4

Satellite image of the warm Gulf Stream as it flows northward along the U.S. East Coast. Instruments sensitive to infrared wavelengths produced this remote-sensing image. Imagery courtesy of National Aeronautics and Space Administration and National Oceanic and Atmospheric Administration.

tion here. As a specific example, the **Gulf Stream** moves northward off the east coast of North America, carrying warm water far into the North Atlantic (Figure 5–4). As a result, the southern third of Iceland experiences much milder temperatures than would be expected for a latitude of 65° N, just below the Arctic Circle at 66.5°. In Reykjavik, on the southwest coast of Iceland, monthly temperatures average above freezing during all months of the year. The Gulf Stream affects Scandinavia and northwestern Europe in the same manner.

In the western Pacific Ocean, the *Kuroshio* or Japan Current, similar to the Gulf Stream, functions much the same in its warming effect on Japan, the Aleutians, and the northwestern margin of North America. In contrast, along midlatitude west coasts, cool ocean currents influence air temperatures. When conditions in these regions are warm and moist, fog frequently forms in the chilled air over the cooler currents.

Summary of Marine vs. Continental Conditions. Figure 5–2 summarizes the operation of all

these land-water temperature controls: evaporation, transmissibility, specific heat, movement, and ocean currents. The term **marine**, or maritime, is used to describe locations that exhibit the moderating influences of the ocean, usually along coastlines or on islands. **Continentality** refers to the condition of areas that are less affected by the sea and therefore have a greater range between maximum and minimum temperatures diurnally and yearly.

The Canadian cities of Vancouver, British Columbia, and Winnipeg, Manitoba, exemplify these marine and continental conditions (Figure 5–5). Both cities are at approximately 49° N latitude. Respectively, they are at sea level and 248 m (814 ft) elevation. However, Vancouver has a more moderate pattern of average maximum and minimum temperatures than Winnipeg. Vancouver's annual range of 16.0C° (28.8F°) is far below the 38.0C° (68.4F°) temperature range in Winnipeg. In fact, Winnipeg's continental temperature pattern is more extreme in every aspect than that of maritime Vancouver.

A similar comparison of San Francisco and Wichita, Kansas, is presented in Figure 5–6. Both cities are

Station: Vancouver, British Columbia
Lat/long: 49°11'N, 123°10'W
Avg. Ann. Temp.: 10°C (50°F)
Total Ann. Precip.:
    104.8 cm (41.3 in.)

Elevation: sea level
Population: 431,000
Ann. Temp. Range:
    16C° (28.8F°)

Station: Winnipeg, Manitoba
Lat/long: 49°54'N, 97°14'W
Avg. Ann. Temp.: 2°C (35.6°F)
Total Ann. Precip.:
    51.7 cm (20.3 in.)

Elevation: 248 m (813.6 ft)
Population: 561,000
Ann. Temp. Range:
    38C° (68.4F°)

FIGURE 5–5
Comparison of temperatures in coastal Vancouver, British Columbia, and continental Winnipeg, Manitoba.

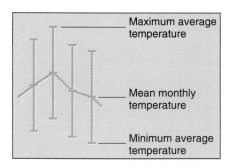

at approximately 37° 40′ N latitude and are at 5 m (16.4 ft) and 403 m (1321 ft) in elevation, respectively. Summer fog plays a role in delaying until September the warmest summer month in San Francisco. In 90 years of weather records there, summer maximum temperatures have risen above 32.2°C (90°F) an average of only once per year and have dropped below freezing an average of once per

year. In contrast, Wichita is susceptible to freezes from late October to mid-April, with diurnal variations slightly enhanced by its elevation. Wichita has temperatures of 32.2°C (90°F) or higher at least 65 days each year. West of Wichita, winters increase in severity with increasing distance from the moderating influences of invading air masses from the Gulf of Mexico.

EARTH'S TEMPERATURE PATTERNS

Both the pattern and variety of temperatures on Earth result from the global energy system. The interaction of the principal temperature control factors is portrayed in the temperature maps in this section, which show worldwide mean monthly air temperatures for January (Figure 5–7) and July (Figure 5–9). These maps, along with Figure 5–11 of January–July ranges, also are useful in identifying areas that experience the greatest annual extremes. We have used January and July maps instead of maps of the solstice months (December and June) because, as explained in Chapter 4, a lag occurs between insolation received and maximum or minimum temperatures experienced.

The lines on temperature maps are known as isotherms. An **isotherm** connects points of equal

Station: San Francisco, California
Lat/long: 37°37'N, 122°23'W
Avg. Ann. Temp.: 14°C (57.2°F)
Total Ann. Precip.:
   47.5 cm (18.7 in.)

Elevation: 5 m (16.4 ft)
Population: 750,000
Ann. Temp. Range:
   9C° (16.2F°)

Station: Wichita, Kansas
Lat/long: 37°39'N, 97°25'W
Avg. Ann. Temp.: 13.7°C (56.6°F)
Total Ann. Precip.:
   72.2 cm (28.4 in.)

Elevation: 402.6 m (1321ft)
Population: 280,000
Ann. Temp. Range:
   27C° (48.6F°)

FIGURE 5–6
Comparison of temperatures in coastal San Francisco and continental Wichita, Kansas.

FIGURE 5–7

Worldwide mean sea level temperatures for January. Temperatures are in Celsius, which can be converted to Fahrenheit by means of the scale.

temperature and portrays the temperature pattern, just as a contour line on a topographic map portrays points of equal elevation. Geographers are concerned with spatial analysis of temperatures, and isotherms facilitate this process.

With each map, begin by finding your own city or town and noting the temperatures indicated by the isotherms (the small scale of these maps will permit only a general determination). Record the information from these maps in your notebook. As you work through the different maps throughout this text, note atmospheric pressure and winds, precipitation, climate, landforms, soil orders, vegetation, and terrestrial biomes. By the end of the course you will have recorded a complete profile of your specific locale.

January Temperature Map

Figure 5–7 is a map of mean temperatures for January. In the Southern Hemisphere, the higher Sun altitude means longer days and summer weather

conditions, whereas in the Northern Hemisphere, January marks winter's shortened daylength and lower Sun angles. Isotherms generally trend east-west, parallel to the equator, marking the general decrease in insolation and net radiation with distance from the equator. The **thermal equator**, a line connecting all points of highest mean temperature (red line on the map), trends southward into the interior of South America and Africa, indicating higher temperatures over landmasses. In the Northern Hemisphere, isotherms trend equatorward as cold air chills the continental interiors. The oceans, on the other hand, appear more moderate, with warmer conditions extending further north than over land at comparable latitudes.

The coldest area on the map is in northeastern Siberia in the Soviet Union. The cold experienced there relates to consistent clear, dry, calm air, small insolation input, and an inland location far from any moderating maritime effects. Verkhoyansk and Oymyakon, USSR, each have experienced a minimum temperature of −68°C (−90°F) and a daily average of −50.5°C (−58.9°F) for January. Verkhoyansk experiences at least seven months of temperatures below freezing, including at least four months below −34°C (−30°F)! People do live and work in Verkhoyansk, which has a population of 1400; it has been continuously occupied since 1638 and is today a minor mining district. In contrast, these same towns experience maximum temperatures of +37°C (+98°F) in July.

Trondheim, Norway, is near the latitude of Verkhoyansk and at a similar elevation. However, Trondheim's coastal location moderates its annual temperature regime: January minimum and maximum temperatures range between −17° and +8°C (+1.4° and +46°F), and the minimum/maximum

Station: Trondheim, Norway
Lat/long: 63°25'N, 10°27'E
Avg. Ann. Temp.: 5°C (41°F)
Total Ann. Precip.:
  85.7 cm (33.7 in.)
Elevation: 115 m (377.3 ft)
Population: 135,000
Ann. Temp. Range:
  17C° (30.6F°)

Station: Verkhoyansk, USSR
Lat/long: 67°33'N, 135°23'E
Avg. Ann. Temp.: −15°C (5°F)
Total Ann. Precip.:
  15.5 cm (6.1 in.)
Elevation: 137 m (449.5 ft)
Population: 1400
Ann. Temp. Range:
  63C° (113.4F°)

FIGURE 5-8
Comparison of temperatures in coastal Trondheim, Norway, and continental Verkhoyansk, USSR.

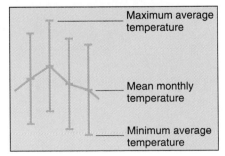

range for July is +5° to +27°C (+41° to +81°F). The most extreme minimum and maximum temperatures ever recorded in Trondheim are −30° and +35°C (−22° and +95°F)—quite a difference from the extremes at Verkhoyansk (Figure 5–8).

July Temperature Map

Average July temperatures are presented in Figure 5–9. The longer days of summer and higher Sun altitude now are in the Northern Hemisphere. Winter dominates the Southern Hemisphere, although it is milder there because large continental landmasses are absent and oceans and seas dominate. The thermal equator shifts northward with the high summer Sun and reaches the Persian Gulf–Pakistan–Iran area. The Persian Gulf is the site of the highest recorded sea-surface temperature of 36°C (96°F), difficult to imagine for a sea or ocean body.

July is a time of 24-hour-long nights in Antarctica. The lowest natural temperature reported on Earth occurred on 21 July 1983 at the Russian research base at Vostok, Antarctica (78° 27′ S, elevation 3420 m or 11,220 ft): a frigid −89.2°C (−128.56°F). For comparison, such a temperature is 11C° (19.8F°) colder than dry ice (solid carbon dioxide)! During July in the Northern Hemisphere, isotherms trend poleward over land, as higher temperatures dominate continental interiors. July temperatures in Verkhoyansk average more than 13°C (56°F), which represents a 63C° (113F°) seasonal variation between winter and summer averages. The Verkhoyansk region of Siberia is probably the greatest example of continentality on Earth. Figure 5–10 shows maximum and minimum average temperatures and mean monthly temperatures for a comparable interior station in North America—Baker Lake, Northwest Territories (about 64° N).

The hottest places on Earth occur in Northern Hemisphere deserts during July. These deserts are areas of clear and dry skies and strong surface heating, with virtually no surface water and few plants. Locations such as portions of the Sonora Desert area of North America and the Sahara of Africa

FIGURE 5–9
Worldwide mean sea level temperatures for July. Temperatures are in Celsius.

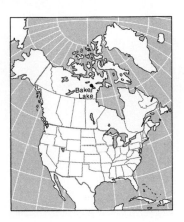

Station: Baker Lake, N.W. Territories **Elevation:** sea level
Lat/long: 64°18'N, 96°00'W **Population:** 1000
Avg. Ann. Temp.: −11°C (12.2°F) **Ann. Temp. Range:**
Total Ann. Precip.:   44C° (79.2F°)
  20.8 cm (8.2 in.)

FIGURE 5–10
Temperature graph for Baker Lake, Northwest
Territories.

are prime examples. Africa has recorded shade temperatures in excess of 58°C (136°F), such as a record set on 13 September 1922 at Al 'Azīzīyah, Libya (32° 32′ N; 112 m or 367 ft elevation). The highest maximum and annual average temperatures in North America occurred in Death Valley, California, where the Greenland Ranch Station (37° N; −54.3 m or −178 ft below sea level) reached 57°C (134°F) in 1913. Such arid and hot lands are discussed further in Chapter 15.

Annual Range of Temperatures

The pattern of marine vs. continental influence emerges dramatically when the difference, or range in averages, is mapped for a full year (Figure 5–11). As you might expect, the largest temperature ranges occur in subpolar locations in North America and Asia, where average ranges of 64C° (115F°) are recorded. The Southern Hemisphere, on the other hand, produces little seasonal variation in mean temperatures, although the smaller land masses do produce some increase in temperature ranges.

For example, in January (see Figure 5–7) Australia is dominated by isotherms of 20° to 30°C (68° to 86°F), whereas in July (see Figure 5–9) Australia experiences isotherms of 10° to 20°C (50° to 68°F). Generally speaking, Southern Hemisphere patterns are more marine, whereas Northern Hemisphere patterns are more continental. The Northern Hemisphere overall registers a slightly higher average surface temperature than does the Southern Hemisphere: 15°C (59°F) as opposed to 12.7°C (55°F).

A logical question to ask about these patterns of world temperatures is what effect they have on the human body. What would you experience if you went to some of these locations?

Air Temperature and the Human Body

We humans are capable of sensing small changes in temperature in the environment around us. Our perception of temperature is described by the terms **apparent temperature** or sensible temperature. This perception of temperature varies among individuals and cultures. Through complex mechanisms, our

D E G R E E S	F°	5	9	18	27	36	45	54	63	72	81	90	99	108	F°	D E G R E E S
	C°	3	5	10	15	20	25	30	35	40	45	50	55	60	C°	

FIGURE 5–11
Annual range of global temperatures in °C (°F).

bodies maintain an average internal temperature of 37°C (98.6°F), slightly lower in the morning or in cold weather and slightly higher at emotional times or during exercise and work.

The water vapor content of air, the wind speed, and air temperature taken together affect each individual's sense of comfort. Higher temperatures, high humidity, and low winds produce the most heat discomfort, whereas low humidity and strong winds enhance cooling rates. Although modern heating and cooling systems can reduce the impact of uncomfortable temperatures indoors, the danger to human life from excessive heat or cold persists outdoors. When changes occur in the surrounding air, the human body reacts in various ways to maintain its core temperature and to protect the brain at all cost. Table 5–1 summarizes the human body's response to stress induced by low and high temperatures.

The wind chill index is important to those who experience winters with freezing temperatures (Figure 5–12). The **wind chill factor** indicates the enhanced rate at which body heat is lost to the air. As wind speeds increase, heat loss from the skin increases. For example, if the air temperature is −1°C (30°F) and the wind is blowing at 32 kmph (20 mph), skin temperatures will be −16°C (4°F). The colder wind chill values present a serious danger to exposed flesh.

A measured index of the human body's reaction to air temperature and water vapor is called the *heat index (HI)*. (The water vapor in air is expressed as relative humidity, a concept presented in Chapter 7.) For now, we simply can assume that the amount of water vapor in the air affects the evaporation rate of perspiration on the skin, because the more water vapor in the air (the higher the humidity), the less water from perspiration the air can absorb through evaporation. The heat index indicates how the air *feels to an average person*—in other words, its apparent temperature.

Figure 5–13 is an abbreviated version of the heat index used by the National Weather Service (NWS) and now included in their daily weather summaries during appropriate months. The inset describes the

TABLE 5-1
The human body's response to temperature stress.

At Low Temperatures	At High Temperatures
Temperature Regulation Methods	
Heat-gaining mechanism	Heat-dissipation mechanism
Constriction of surface blood vessels	Dilation of surface blood vessels
Concentration of blood	Dilution of blood
Flexing to reduce surface exposure	Extending to increase exposure
Increased muscle tone	Decreased muscle tone
Decrease in sweating	Sweating
Inclination to increase activity	Inclination to decrease activity
Shivering	
Increased cell metabolism	
Consequent Disturbances	
Increased urine volume	Decreased urine volume
Danger of inadequate blood supply to exposed parts; frostbite	Reduced blood supply to brain; dizziness, nausea, fainting
Discomfort leading to neuroses	Discomfort leading to neuroses
Increased appetite	Decreased appetite
	Mobilization of tissue fluid
	Thirst and dehydration
	Reduced chloride balance; heat cramps
Failure of Regulation	
Falling body temperature	Rising body temperature
Drowsiness	Impaired heat-regulating center
Cessation of heartbeat and respiration	Failure of nervous regulation; cessation of breathing

Data from CLIMATOLOGY—Arid Zone Research X. © UNESCO 1958. Reproduced by permission of UNESCO.

effects of heat-index categories on higher-risk groups, because a combination of high temperature and humidity can severely reduce the body's natural ability to regulate internal temperature. There is a distinct possibility that future humans may experience greater temperature-related challenges due to complex changes now underway in the lower atmosphere.

FUTURE TEMPERATURE TRENDS

Significant climatic change has occurred on Earth in the past and most certainly will occur in the future. There is nothing society can do about long-term influences that cycle Earth through swings from ice ages to warmer periods. However, our global society must address possible short-term changes that are influencing global temperatures within the life span of present generations.

A cooperative global network of nations, under the auspices of the United Nations Environment Programme (UNEP), participate in the World Weather Watch System and the Global Change Program for the gathering of temperature and other weather and climatic information. Evidence of this cooperation is the recent study completed by the Intergovernmental Panel on Climate Change (IPCC). Computers make possible the processing of global data with which to assess trends in average temper-

Actual Air Temperatures in °C (°F)

Wind speed, kmph (mph)	10° (50°)	4° (40°)	−1° (30°)	−7° (20°)	−12° (10°)	−18° (0°)	−23° (−10°)	−29° (−20°)	−34° (−30°)	−40° (−40°)	−46° (−50°)	−51° (−60°)
						Temperature Effects on Exposed Flesh						
Calm	10° (50°)	4° (40°)	−1° (30°)	−7° (20°)	−12° (10°)	−18° (0°)	−23° (−10°)	−29° (−20°)	−34° (−30°)	−40° (−40°)	−46° (−50°)	−51° (−60°)
8 (5)	9° (48°)	3° (37°)	−2° (28°)	−9° (16°)	−14° (6°)	−21° (−5°)	−26° (−15°)	−32° (−26°)	−38° (−36°)	−44° (−47°)	−49° (−57°)	−56° (−68°)
16 (10)	4° (40°)	−2° (28°)	−9° (16°)	−16° (4°)	−23° (−9°)	−29° (−21°)	−36° (−33°)	−43° (−46°)	−50° (−58°)	−57° (−70°)	−64° (−83°)	−71° (−95°)
24 (15)	2° (36°)	−6° (22°)	−13° (9°)	−21° (−5°)	−28° (−18°)	−38° (−36°)	−43° (−45°)	−50° (−58°)	−58° (−72°)	−65° (−85°)	−73° (−99°)	−74° (−102°)
32 (20)	0° (32°)	−8° (18°)	−16° (4°)	−23° (−10°)	−32° (−25°)	−39° (−39°)	−47° (−53°)	−55° (−67°)	−63° (−82°)	−71° (−96°)	−79° (−110°)	−87° (−124°)
40 (25)	−1° (30°)	−9° (16°)	−18° (0°)	−26° (−15°)	−34° (−29°)	−42° (−44°)	−51° (−59°)	−59° (−74°)	−64° (−83°)	−76° (−104°)	−81° (−113°)	−92° (−133°)
48 (30)	−2° (28°)	−11° (13°)	−19° (−2°)	−28° (−18°)	−36° (−33°)	−44° (−48°)	−53° (−63°)	−62° (−79°)	−70° (−94°)	−78° (−109°)	−87° (−125°)	−96° (−140°)
56 (35)	−3° (27°)	−12° (11°)	−20° (−4°)	−29° (−20°)	−37° (−35°)	−45° (−49°)	−53° (−64°)	−63° (−82°)	−72° (−98°)	−84° (−119°)	−89° (−129°)	−98° (−145°)
64 (40)	−3° (26°)	−12° (10°)	−21° (−6°)	−29° (−21°)	−38° (−37°)	−47° (−53°)	−56° (−69°)	−65° (−85°)	−74° (−102°)	−82° (−116°)	−91° (−132°)	−100° (−148°)

Low danger with proper clothing   Exposed flesh in danger of freezing

FIGURE 5–12

Wind chill factor for various temperatures and wind velocities.

atures and sea level. What follows is an overview of current scientific efforts to determine future short-term temperature trends.

Global Warming

Human activities are enhancing the greenhouse effect described in Chapter 4. Average air temperatures worldwide are increasing. Several world conferences on global warming and climate change now are held each year. Representatives at the Villach Conference, held as an initial assessment by the United Nations Environment Programme in 1985 at Villach, Austria, offered some direction:

> The rate and degree of future warming [of Earth's climates] could be profoundly affected by governmental policies on energy conservation, use of fossil fuels, and the emission of some greenhouse gases. . . . understanding of the greenhouse

question is sufficiently developed that scientists and policy makers should begin an active collaboration to explore the effectiveness of alternative policies and adjustments.*

Five years later these findings were mirrored in a pledge agreement to reduce carbon dioxide emissions, signed in November 1990. This historic agreement was reached by 130 nations, or about two-thirds of Earth's governments. We are in a unique position to alter, perhaps reverse, Earth's probable anthropogenic warming trend by controlling the production of certain radiatively active gases. But first, we must understand the problem and its roots.

Background on Global Warming. Radiatively active gases are atmospheric gases, such as carbon

---

*United Nations Environment Programme, 1985, *An Assessment of the Role of Carbon Dioxide and of Other Greenhouse Gases in Climate Variations and Associated Impacts,* Geneva: World Meteorological Organization. Reprinted by permission.

FIGURE 5-13

Heat index graph for various temperatures and relative humidity levels. Data courtesy of National Weather Service.

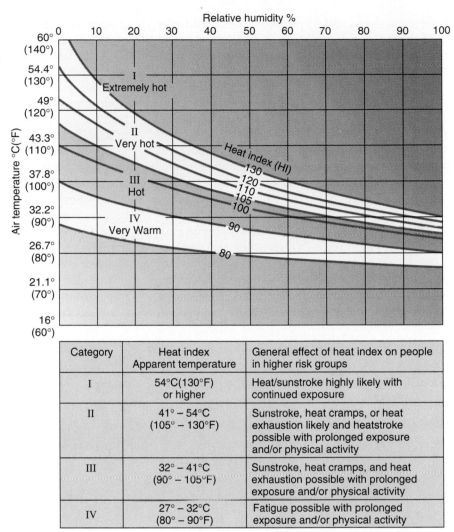

Category	Heat index Apparent temperature	General effect of heat index on people in higher risk groups
I	54°C(130°F) or higher	Heat/sunstroke highly likely with continued exposure
II	41° – 54°C (105° – 130°F)	Sunstroke, heat cramps, or heat exhaustion likely and heatstroke possible with prolonged exposure and/or physical activity
III	32° – 41°C (90° – 105°F)	Sunstroke, heat cramps, and heat exhaustion possible with prolonged exposure and/or physical activity
IV	27° – 32°C (80° – 90°F)	Fatigue possible with prolonged exposure and/or physical activity

dioxide, methane, CFCs, and water vapor, that absorb and radiate infrared wavelengths. Carbon dioxide is the principal radiatively active gas causing Earth's natural greenhouse effect. CO_2 is transparent to light but opaque to the infrared wavelengths radiated by Earth, and thus its presence delays heat loss to space. While detained, that heat energy is absorbed and reradiated over and over in the lower atmosphere, causing average temperatures to increase. The Industrial Revolution, beginning in the mid-1700s, has prompted a tremendous surge in the burning of fossil fuels. This, coupled with the destruction and inadequate replacement rate of harvested forests, is causing atmospheric CO_2 levels to increase 7.3–9.1 billion metric tons (8–10 billion tons) per year

(Figure 5–14). CO_2 alone is thought to be responsible for almost 60% of the global warming trend.

Table 5–2 shows the increasing percentage of CO_2 in the lower atmosphere from past to present, and gives estimates for the future. CO_2 is presently increasing in concentration at the rate of 0.4% per year. This increase appears sufficient to override any natural climatic tendencies toward a cooling trend, as well as to produce possible unwanted global warming in the next century.

Figure 5–15 shows past and projected sources of excessive (non-natural) carbon dioxide by country or region, clearly identifying developing nations as the sector with the greatest probable growth in fossil-fuel consumption and new CO_2 production.

However, national and corporate policies could alter this forecast by actively steering developing countries toward more appropriately scaled alternative energy sources.

Especially important in the United States are policies to eliminate energy waste and inefficiency, both in terms of reducing carbon dioxide production and in setting a positive example. Yet relatively simple solutions are not being promoted, and throughout the 1980s the United States was uncooperative in worldwide efforts to halt this human-induced increase in CO_2. The United States was not a signatory to the November 1990 agreement mentioned, despite the fact that our nation alone produces 27% of the excess CO_2 annually! And again, at an April 1991 conference, the United States failed to pledge emission cutbacks with other nations of the world. U.S. emissions instead are predicted to rise 15% over 1990 levels by the year 2000.

Another radiatively active gas contributing to the overall greenhouse effect is **methane** (CH_4), which also is increasing in concentration at more than 1% per year. Air bubbles in ice show that past concentrations of methane—500 to 27,000 years ago—were approximately 0.7 ppm, whereas today that figure is 1.6 ppm. Methane is generated by organic processes, such as burning, digestion, and rotting in the presence of oxygen. About 50% of the excess methane being produced comes from bacterial action in the intestinal tracts of livestock and from underwater bacteria in rice paddies. Burning of vegetation causes another 20% of the excess, and bacterial action inside the digestion systems of increasing termite populations also is a significant source. Methane is now believed responsible for at least 12% of the total atmospheric warming, complementing the warming caused by the buildup of CO_2 and equaling about one-half the contribution of CFCs.

FIGURE 5–14
Industrial landscapes exemplify the increasing output of CO_2 into the atmosphere.
Photo by author.

TABLE 5–2
Lower atmosphere concentration of CO_2.

Date	Percentage (%) of CO_2 Among Atmospheric Gases	Parts per Million
1825	0.021	210
1888	0.028	280
1970	0.032	320
1985	0.035	350
1992 (estimate)	0.036	360
2020 (estimate)	0.055	550
2050 (estimate)	0.060	600

Chlorofluorocarbons (CFCs) are thought to contribute about 25% of the global warming. CFCs absorb infrared in wavelengths missed by carbon dioxide and water vapor in the lower troposphere. As radiatively active gases, CFCs enhance the greenhouse effect, in addition to their other negative role in stratospheric ozone depletion.

Present Indications and the Future. The year 1987 was the warmest year in recorded history—until it was exceeded by the year 1988, which then was exceeded by 1990—the single warmest year in the history of instrumented weather data, at a global average temperature of 15.4°C (59.8°F). As an exam-

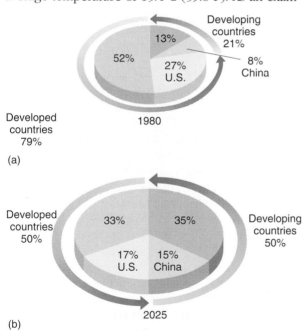

(a)

(b)

FIGURE 5–15
Countries and regions of origin (a) for excessive CO_2 in 1980 and (b) forecast for 2025.

ple, the winter of 1987–1988 was one of the mildest ever observed on the Great Lakes. The other warmest years were, in order of severity, 1983, 1981, 1980, 1989, and 1986.

Comparing annual temperatures and 5-year mean temperatures gives a sense of overall trends. Figure 5–16 shows observed temperatures from 1880 through 1990. The overall pattern is clear—we are entering a period of record warmth. The 1990 Scientific Assessment of Climate Change, an international working group of 170 scientists from 30 nations, formed a consensus that the warming will be large and socially significant, and that warming rates will accelerate in the near future.

The consequences of this potential warming are complex. During the decade of the 1980s, average ocean surface temperatures rose at an annual rate of 0.11C° (0.2F°). Of course, any change in ocean temperature has a profound effect on weather and, indirectly, on agriculture and soil moisture. The rate of increase in ocean temperatures as measured by satellites is twice the rate measured by previous methods, which were far less accurate and reliable. Specific biological effects presently thought to be a result of these warmer waters are discussed in Chapter 16.

Perhaps the most pervasive effect to monitor is high-latitude and polar warming, which could lead to a rapid escalation of ice melting. In the past century alone, sea-level rise is estimated at 10–20 cm (4–8 in.). Additional meltwater, especially from grounded ice masses (those that rest entirely on land) in Antarctica and Greenland, could raise global sea level in the future. Even the warming of the ocean itself will contribute about 25% of the sea-level rise due to thermal expansion of the water. Further discussion of the overall climatic effects of higher temperatures is presented in Chapter 10.

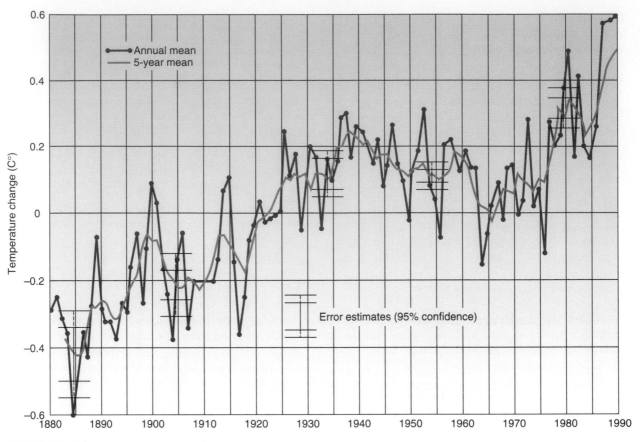

FIGURE 5–16

Global temperature trends from 1880 to 1990. The 0 baseline represents the 1950–1980 global average temperature. Data courtesy of Dr. James E. Hansen, Goddard Institute, National Aeronautics and Space Administration.

Global Cooling

In the 1980s several volcanic eruptions also affected global temperatures. The 1982 El Chichón eruption in south-central Mexico sent a cloud of sulfuric acid and ash 25 km (15.5 mi) into the atmosphere. Within 21 days the cloud had circled the globe, increasing the atmospheric albedo and lowering temperatures by 0.5C° (0.9F°). Some of the finer particles remained in the stratosphere through 1985. Such volcanic effects are well documented in history. In addition, worldwide air pollution has increased the turbidity, or murkiness, of the atmosphere (Chapter 3). Perhaps a combination of increasing air pollution and several volcanic eruptions explains why the warming trend during this century was interrupted slightly between 1940 and 1970, only to be overwhelmed after 1970 by warming factors (see Figure 5–16).

Another potential cooling effect was postulated by Paul Crutzen (a Dutch atmospheric scientist with the Max Planck Institute of West Germany) and John Birks (a chemist at the University of Colorado) in "The Atmosphere After a Nuclear War: Twilight at Noon." Their study, published in the Royal Swedish Academy of Sciences publication *Ambio* in June 1982, launched an avalanche of scientific analyses, assessment, and general confirmations (see Suggested Readings).

The **nuclear winter hypothesis**, as it has come to be known, now encompasses a whole range of ecological, biological, and climatic impacts associated with the detonation of a relatively small number of nuclear warheads within the biosphere. Resulting urban fires and fire storms would produce such a pall of insolation-obscuring smoke that surface heating would be drastically reduced, with temperatures dropping at least 25C° (45F°) and perhaps more in the midlatitudes. Scientists are anxiously studying the localized temperature effects caused by the oil-well fires in Kuwait and Iraq which were started in early 1991. This potential global impact of modern warfare affects literally all aspects of Earth systems and therefore is appropriate to our study of physical geography. Further discussion of this issue is in the concluding chapter.

SUMMARY

Worldwide, a variety of temperature regimes affect individuals, lifestyles, and cultures. The principal factors that control the pattern of world temperatures include latitude, altitude, cloud cover, and the numerous physical properties that create differential heating and cooling rates for land and water. The pattern of present Earth temperatures for January and July is best portrayed on isothermal maps, which permit geographic analysis. The annual range of mean temperatures between January and July demonstrates the interaction of the principal temperature controls and isolates regions of greatest continentality.

As humans, we need to be aware of our sensitivity to temperature differences and the physiological responses produced in our bodies by exposure to various temperature extremes. The heat index and wind chill charts are two indicators of apparent temperature we can use. Human activities apparently are affecting short-term temperature trends by producing increases in certain radiatively active gases. An increase in carbon dioxide concentration over the past two centuries is probably enhancing the natural atmospheric greenhouse effect. Further international cooperation seems mandatory, based on existing evidence of this escalating condition. The pattern of world temperatures produced by specific natural and unnatural factors knows no boundaries and is truly an international issue.

Apparent temperature

Continentality

Gulf Stream

Isotherm

Land-water heating differences

Marine

Methane

Nuclear winter hypothesis

Radiatively active gases

Specific heat

Thermal equator

Transparency

Wind chill factor

REVIEW QUESTIONS

1. Distinguish between sensible heat and sensible temperature.
2. What does air temperature indicate about energy in the atmosphere?
3. Explain the effect of altitude upon air temperature.
4. What noticeable effect does air density have on the absorption and radiation of heat? What role does altitude play in that process?
5. How is it possible to grow moderate-climate crops such as wheat, barley, and potatoes at an elevation of 4103 m (13,461 ft) near La Paz, Bolivia, so near the equator?
6. Describe the effect of cloud cover with regard to Earth's temperature patterns.
7. List the physical aspects of land and water that produce their different responses to heating. What is the specific effect of transmissibility?
8. Differentiate between marine and continental temperatures. Give geographical examples of each from the text.
9. What is the thermal equator? Describe the location of the thermal equator in January and in July.
10. Explain the extreme temperature range experienced in north-central Siberia between January and July.
11. Where are the hottest places on Earth? Are they near the equator or elsewhere? Explain. Where is the coldest place on Earth?
12. From the maps in Figures 5–7, 5–9, and 5–11, determine the average temperature values and annual range of temperatures for your location.
13. Identify the different responses of the human body to low-temperature and high-temperature stress.
14. Describe the interaction between air temperature and wind speed and their affect on skin chilling. Select several temperatures and wind speeds from Figure 5–12 and determine the wind chill index.
15. Define the heat index. What characteristic patterns are experienced throughout the year where you live relative to the graph in Figure 5–13?
16. Explain the factors that are creating the global warming trend. What are the indications that such a pattern is occurring?
17. Briefly, what strategies might be implemented to reduce the buildup of radiatively active gases?
18. What could cause a global cooling trend? What are some real and potential cooling factors that might affect global temperatures?

SUGGESTED READINGS

Begley, Sharon, with Mark Miller and Mary Hager. "The Greenhouse Effect." *Newsweek* (11 July 1988): 16–24.

Budyko, M. I. *The Earth's Climate: Past and Future*. International Geophysics Series, vol. 29. New York: Academic Press, 1982.

Graedel, Thomas E., and Paul J. Crutzen. "The Changing Atmosphere." *Scientific American* 261, no. 3 (September 1989): 58–68.

Hansen, James E., and others. "Prediction of Near-Term Climate Evolution: What Can We Tell the Decision-Makers Now?" *Proceedings of the First North American Conference on Preparing for Climate Changes*. Washington: Government Institutes, 1987.

Hansen, James E., and Sergej Lebedeff. "Global Trends of Measured Surface Air Temperature." *Journal of Geophysical Research* 92, no. D11 (20 November 1987): 13,345–72.

Houghton, Richard A., and George M. Woodwell. "Global Climate Change." *Scientific American* 260, no. 4 (April 1989): 36–44.

Kerr, Richard A. "New Greenhouse Report Puts Down Dissenters." *Science* 249 (3 August 1990): 481.

Lemonick, Michael D. "The Heat Is On." *Time* 130, no. 16 (19 October 1987): 58–67.

Mather, John R. *Climatology—Fundamentals and Applications*. New York: McGraw-Hill, 1974.

Riordan, Pauline and Paul G. Bourget. *World Weather Extremes*. Fort Belvoir, VA: U.S. Army Corp of Engineers, 1985.

Rudloff, Willy. *World Climates*. Stuttgart, Germany: Wissenschaftliche Verlagsgesellschaft, 1981.

Schneider, Stephen H. "The Changing Climate." *Scientific American* 261, no. 3 (September 1989): 70–79.

Turco, R. P., O. B. Toon, T. P. Ackerman, J. B. Pollack, and C. Sagan. "Climate and Smoke: An Appraisal of Nuclear Winter." *Science* 247, no. 4939 (12 January 1990): 166–76.

———. "Nuclear Winter: Global Consequences of Multiple Nuclear Explosions." *Science* 222, no. 4630 (23 December 1983): 1283–92.

———. "The Climatic Effects of Nuclear War." *Scientific American* 251, no. 2 (August 1984): 33–43.

Wernstedt, Frederick L. *World Climatic Data*. Lemont, PA: Climatic Data Press, 1972.

Willmott, Cort J., John R. Mather, and Clinton M. Rowe. *Average Monthly and Annual Surface Air Temperature and Precipitation Data for the World*. Part 1, "The Eastern Hemisphere," and Part 2, "The Western Hemisphere." Elmer, NJ: C. W. Thornthwaite Associates and the University of Delaware, 1981.

Photo by author.

6

Atmospheric and Oceanic Circulations

Early in April 1815, on an island named Sumbawa in present-day Indonesia, the volcano Tambora erupted violently, releasing an estimated 150 km³ (36 mi³) of material, an 80-times greater volume than produced by the 1980 Mount Saint Helens eruption in Washington State. Some material from Tambora—the aerosols and acid mists—were carried worldwide by global atmospheric circulation, creating a stratospheric dust veil. The result was both a higher atmospheric albedo and absorption of energy by the particulate materials injected into the stratosphere. Remarkable optical and meteorological phenomena resulting from the spreading dust were noted for months and years after the eruption. Beautifully colored sunsets and periods of twilight were enjoyed in London, Paris, New York, and elsewhere.

In the summer of 1816, one year later, farmers in New England and Europe were shocked to experience frosts in every month, and in some places, every week. Mean temperatures in the Northern Hemisphere apparently had dropped 0.4–0.7C° (0.72–1.26F°). Fascinatingly, scientists can still study the eruption debris nearly two centuries later, for it was carried by the winds and embedded in the layers of the ice sheets in Greenland and Antarctica. Although Tambora's eruption has not been conclusively tied to the cold summer in 1816, the impact of the eruption appears so large that, despite measurement uncertainties, it remains as powerful evidence of global circulation.

More recently, during April 25–26, 1986, a nuclear power reactor in the town of Pripyat, 16 km (10 mi) northwest of Chernobyl in the Soviet Ukraine, experienced one of the worst meltdown accidents in the history of nuclear reactors. Tons of radioactive inventory explosively spewed from the reactor during the accident. As winds passed over the accident site, they spread radioactive debris across the countryside. By the 28th, only two days later, atmospheric circulation had carried the hazard over Sweden and the rest of Scandinavia, and in later days it spread to Italy, France, and the rest of Europe. Trace amounts drifted around the planet and were detected in North America. Permanent evacuation of over 125,000 people eventually occurred in the region surrounding the plant site.

In this chapter we examine the dynamic circulation of Earth's atmosphere that carried Tambora's debris and Chernobyl's fallout worldwide, and which carries the common ingredients oxygen, carbon dioxide, and water vapor around the globe. We also consider Earth's wind-driven oceanic currents.

WIND ESSENTIALS

Earth's atmospheric circulation is an important transfer mechanism for both energy and mass. In the process, the energy imbalance between equatorial surpluses and polar deficits is partly resolved, Earth's weather patterns are generated, and ocean currents are produced. Human-caused pollution also is spread worldwide by this circulation, far from its points of origin—certainly an important reason why

FIGURE 6–1
Instruments used to measure wind direction (wind vane) and wind speed (anemometer). Photo by author.

the United States, the Soviet Union, and Great Britain signed the 1963 Limited Test Ban Treaty. That treaty banned above-ground testing of nuclear weapons because atmospheric circulation was spreading radioactive contamination worldwide. This is an illustration of how the atmosphere socializes humanity more than any other natural or cultural factor. Our atmosphere makes all the world a spatially linked society—one person's or nation's exhalation is another's breath.

Atmospheric circulation is generally categorized at three levels: primary (general) circulation, secondary circulation of migratory high-pressure and low-pressure systems, and tertiary circulation that includes local winds and temporal weather patterns. Winds that move principally north or south along meridians are known as *meridional flows* or meridional circulations. Winds moving east or west along parallels of latitude are called *zonal flows*, or zonal circulations.

Wind: Description and Measurement

Simply stated, **wind** is the horizontal motion of air relative to Earth's surface. It is produced by differences in air pressure from one location to another and is influenced by several variables. The two principal variables are speed and direction, and instruments have been developed to measure each. Wind speed is measured with an **anemometer**, and may be expressed in kilometers per hour (kmph), miles per hour (mph), meters per second (m/sec), or knots. (A *knot* is a nautical mile per hour, covering 1 minute of Earth's arc in an hour, equivalent to 1.85 kmph or 1.15 mph.) Wind direction is determined with a **wind vane**; the standard measurement is taken 10 m (33 ft) above the ground to avoid, as much as possible, local effects of topography upon wind direction (Figure 6–1).

Winds are named for the direction *from which they originate*. For example, a wind from the west is a westerly wind (it blows eastward); a wind out of the south is a southerly wind (it blows northward). Figure 6–2 illustrates a simple wind compass, naming the 16 principal wind directions.

A descriptive scale useful in visually estimating winds is the traditional **Beaufort wind scale**. Originally established in 1806 by Admiral Beaufort of the British Navy for use at sea, the scale was expanded to

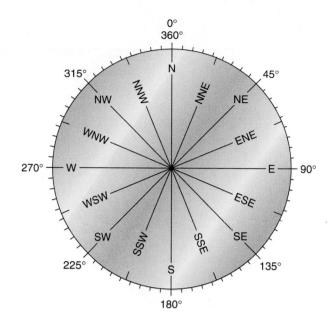

FIGURE 6–2
A wind compass.

include wind speeds by G. C. Simpson in 1926 and standardized by the National Weather Service (Weather Bureau) in 1955. The scale presented in Table 6–1 is modernized and adapted from these earlier versions. It still is referenced on ocean charts and is presented here with descriptions of visual wind effects on land and sea. This observational scale makes possible the estimation of wind speed without instruments, although most modern ships use sophisticated equipment to perform such measurements.

Global Winds

The primary circulation of winds across Earth has fascinated travelers, sailors, and scientists for centuries, although only in the modern era is a true picture emerging of the pattern and cause of global winds. With the assistance of scientific breakthroughs in space-based observations and Earthbound computer technology, models of total atmospheric circulation are becoming more refined. A NASA *SEASAT* satellite image, derived from 350,000 wind measurements over the world's oceans during September 1978, is presented in Figure 6–3. The patterns you see are the response to specific forces at work in the atmosphere, and these forces are our next topic.

TABLE 6–1

Beaufort wind scale.

Wind Speed			Beaufort Number	Wind Description	Observed Effects at Sea	Observed Effects on Land
kmph	mph	knots				
<1	<1	<1	0	Calm	Glassy calm, like a mirror	Calm, no movement of leaves
1–5	1–3	1–3	1	Light air	Small ripples, wavelet scales, no foam on crests	Slight leaf movement, smoke drifts, wind vanes still
6–11	4–7	4–6	2	Light breeze	Small wavelets; glassy look to crests, which do not break	Leaves rustling; wind felt; wind vanes moving
12–19	8–12	7–10	3	Gentle breeze	Large wavelets, dispersed whitecaps as crests break	Leaves and twigs in motion, small flags and banners extended
20–29	13–18	11–16	4	Moderate breeze	Small, longer waves; numerous whitecaps	Small branches moving; raising dust, paper litter, and dry leaves
30–38	19–24	17–21	5	Fresh breeze	Moderate, pronounced waves; many whitecaps; some spray	Small trees and branches swaying, wavelets forming on inland waterways
39–49	25–31	22–27	6	Strong breeze	Large waves, white foam crests everywhere; some spray	Large branches swaying, overhead wires whistling, difficult to control an umbrella
50–61	32–38	28–33	7	Moderate (near) gale	Sea mounding up, foam and sea spray blown in streaks in the direction of the wind	Entire trees moving, difficult to walk into wind
62–74	39–46	34–40	8	Fresh gale (or gale)	Moderately high waves of greater length, breaking crests forming seaspray, well-marked foam streaks	Small branches breaking, difficult to walk, moving automobiles drifting and veering
75–87	47–54	41–47	9	Strong gale	High waves, wave crests tumbling and the sea beginning to roll, visibility reduced by blowing spray	Roof shingles blown away, slight damage to structures, broken branches littering the ground
88–101	55–63	48–55	10	Whole gale (or storm)	Very high waves and heavy, rolling seas; white appearance to foam-covered sea; overhanging waves; visibility reduced	Uprooted and broken trees, structural damage, considerable destruction, seldom occurring
102–116	64–73	56–63	11	Storm (or violent storm)	White foam covering a breaking sea of exceptionally high waves, small and medium-sized ships lost from view in wave troughs, wave crests frothy	Widespread damage to structures and trees, a rare occurrence
>117	>74	>64	12–17	Hurricane	Driving foam and spray filling the air, white sea, visibility poor to nonexistent	Severe to catastrophic damage, devastation to affected society

FIGURE 6–3
SEASAT imagery of global winds. Courtesy of National Aeronautics and Space Administration.

DRIVING FORCES WITHIN THE ATMOSPHERE

Several forces act on the wind in terms of both speed and direction:

- Earth's *gravitational force* is virtually uniform, equally compressing the atmosphere near the ground worldwide, with decreasing density occurring with increasing altitude.
- The **pressure gradient force** drives air from areas of higher barometric pressure to areas of lower barometric pressure.
- The **Coriolis force**, a deflective force, apparently deflects wind from a straight path in relation to Earth's rotating surface—to the right in the Northern Hemisphere and to the left in the Southern Hemisphere.
- The **friction force** drags on the wind as it moves across surfaces; it decreases with height above the surface.

All four of these forces operate on moving air and water at Earth's surface and affect the circulation patterns of global winds. The following sections describe the actions of the pressure gradient, Coriolis, and friction forces.

Pressure Gradient Force

High- and low-pressure areas exist on Earth principally as a result of unequal heating at Earth's surface and as a result of certain dynamic forces in the atmosphere. A pressure gradient is the difference in atmospheric pressure over distance between areas of higher pressure and lower pressure. Air pressure is measured with a barometer, as described in Chapter 3. Lines plotted on a weather map connecting points of equal pressure are called **isobars**; the distance between isobars indicates the degree of pressure difference.

Just as closer contour lines on a topographic map mark a steeper slope, so do closer isobars denote steepness in the pressure gradient. A steep gradient causes faster air movement from a high-pressure area to a low-pressure area. Isobars spaced at greater distances from one another mark a more gradual pressure gradient, one that creates a slower air flow. Along a horizontal surface, the pressure gradient force acts at right angles to the isobars (Figure 6–4).

Figure 6–5 illustrates the combined effect of the forces that direct the wind. The pressure gradient force acting alone is shown in Figure 6–5(a) from two perspectives. As air descends in the high-

FIGURE 6–4
Pressure gradient.

FIGURE 6–5
Three physical forces that integrate to produce wind patterns at the surface and aloft: (a) the pressure gradient force, (b) the Coriolis force, and (c) the friction force. (The gravitational force is a given.)

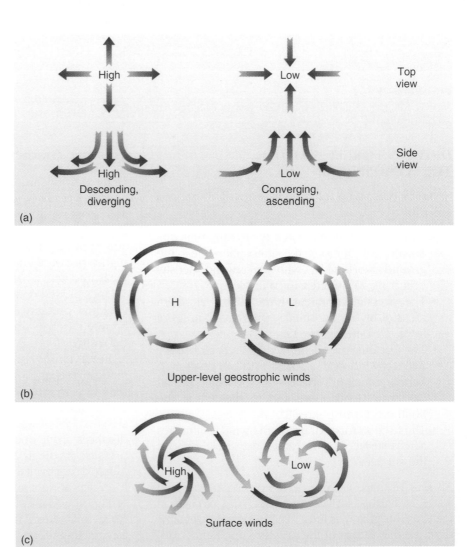

pressure area, a field of subsiding, or sinking, air develops. Air moves out of the high-pressure area in a flow described as diverging. High-pressure areas feature *descending, diverging* air flows. On the other hand, air moving into a low-pressure area does so with a converging flow. Thus, low-pressure areas feature *converging, ascending* air flows.

Coriolis Force

The Coriolis force applies to objects moving across Earth's surface that appear to deflect from a straight path. Because the physicist Sir Isaac Newton stated that, if something is accelerating over a space, a force is in operation, the label "force" is appropriate here. This apparent force is named for Gaspard Coriolis, a French mathematician and researcher of applied mechanics, who described the phenomenon in 1831.

The deflection produced by the Coriolis force is caused by the fact that Earth's rotational speed varies with latitude, decreasing from 1675 kmph (1041 mph) at the equator to 0 at the poles (see Table 4–2). The Coriolis effect increases as the speed of moving objects increases; thus, the faster the movement, the greater the apparent deflection. And, because Earth rotates eastward, objects that move in an absolute straight line over a distance for some time (such as winds and ocean currents) appear to curve to the right in the Northern Hemisphere and to the left in the Southern Hemisphere. Winds, ocean currents, and long-range missiles—all are affected by the Coriolis force.

Viewed from an object passing over Earth's surface, the surface rotates beneath the object. But viewed from the surface, the object appears to curve off-course. The object does not actually deviate from a straight path, but it *appears* to because of the rotational movement of Earth's surface beneath it. The deflection to the right or left occurs regardless of the direction in which the object is moving. The Coriolis force is zero along the equator, increasing to half the maximum deflection at 30° N and S latitudes, and to maximum deflection at the poles (Figure 6–6).

As air rises through the lowest levels of the atmosphere, leaving the drag of surface friction behind, it accelerates in speed, and thus increases the Coriolis force, spiraling the winds to the right in

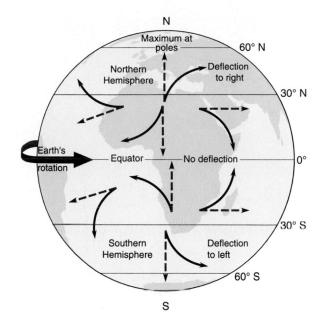

FIGURE 6–6
Distribution of the Coriolis force on Earth: apparent deflection to the right in the Northern Hemisphere; apparent deflection to the left in the Southern Hemisphere.

the Northern Hemisphere or left in the Southern Hemisphere. In the upper troposphere the Coriolis force just balances the pressure gradient force. Consequently, the winds between high-pressure and low-pressure areas aloft flow parallel to the isobars.

Figure 6–5(b) illustrates the combined effect of the pressure gradient force and the Coriolis force on the atmosphere, producing **geostrophic winds**. Geostrophic winds are characteristic of upper tropospheric circulation. The air does not flow directly from high to low, but around the pressure areas instead, remaining parallel to the isobars and producing the characteristic pattern shown on the upper-air weather chart in Figure 6–7.

Friction Force

The effect of surface friction extends to a height of about 500 m (1640 ft) and varies with surface texture, wind speed, time of day and year, and atmospheric conditions. Generally, rougher surfaces produce more friction. Figure 6–5(c) adds the effect of friction to the Coriolis and pressure gradient forces

The wind speed table:

wind speed symbol	miles (statute) per hour	knots
◎	Calm	Calm
—	1-2	1-2
⊥	3-8	3-7
⌐	9-14	8-12
⌐	15-20	13-17
⌐	21-25	18-22
⌐	26-31	23-27
⌐	32-37	28-32
⌐	38-43	33-37
⌐	44-49	38-42
⌐	50-54	43-47
◣	55-60	48-52
◣	61-66	53-57
◣	67-71	58-62
◣	72-77	63-67
◣	78-83	68-72
◣	84-89	73-77
◣	119-123	103-107

FIGURE 6–7

Isobaric chart for 25 April 1977, 7:00 A.M. EST, showing the 500 mb contours (in feet) and geostrophic wind patterns in the troposphere at approximately 5500 m (18,000 ft) altitude. Data courtesy of National Weather Service.

on wind movements. Because surface friction decreases wind speed, it reduces the effect of the Coriolis force and causes winds to move across isobars at an angle, spiraling out from a high-pressure area clockwise to form an **anticyclone** and spiraling into a low-pressure area counterclockwise to form a **cyclone**. In the Southern Hemisphere these circulation patterns are reversed, flowing out of high-pressure cells counterclockwise and into low-pressure cells clockwise.

ATMOSPHERIC PATTERNS OF MOTION

With these forces and motions in mind, we are ready to understand the *SEASAT* image of winds (see Figure

6–3) and build a general model of total atmospheric circulation. If Earth did not rotate, then the warmer, less-dense air of the equatorial regions would rise, creating low pressure at the surface, and the colder and denser air of the poles would sink, creating high pressure at the surface. The net effect would be a meridional flow of winds established by this simple pressure gradient. But in reality, because Earth does rotate, a much more complex system exists for the transfer of energy and mass from equatorial surpluses to polar deficits—one with waves, streams, and eddies on a planetary scale. And instead of this hypothetical meridional flow, the flow is predominantly zonal (latitudinal) at the surface and aloft: westerly (eastward) in the middle and high latitudes and easterly (westward) in the low latitudes.

TABLE 6–2
Four hemispheric pressure areas.

Name	Cause	Location	Air temperature/ moisture
Polar high-pressure cells	Thermal	90° N, 90° S	Cold/dry
Subpolar low-pressure cells	Dynamic	60° N, 60° S	Cool/wet
Subtropical high-pressure cells	Dynamic	20° to 35° N and S	Hot/dry
Equatorial low-pressure trough	Thermal	10° N to 10° S	Warm/wet

Primary High-Pressure and Low-Pressure Areas

Four identifiable pressure areas cover both the Northern and Southern Hemispheres. Two are specifically stimulated by thermal (temperature) factors: the **equatorial low-pressure trough** and the weak **polar high-pressure cells**, both north and south. The other two areas are formed by dynamic factors: the **subtropical high-pressure cells** and **subpolar low-pressure cells**. Table 6–2 summarizes the characteristics of these pressure areas.

The following discussion of Earth's pressure and wind patterns refers often to Figure 6–8, which shows January and July maps of average surface barometric pressure. These maps indicate prevailing surface winds, suggested by the isobars. The primary high- and low-pressure areas on Earth appear on these isobaric maps as interrupted cells or uneven belts of similar pressure that stretch across the face of the planet. Between these areas are the primary winds, which have been noted in adventure and myth throughout human experience. Secondary highs and lows, from a few hundred to a few thousand kilometers in diameter and hundreds to thousands of meters high, are formed within the primary pressure areas. These pressure systems seasonally migrate to produce changing weather and climate patterns in the regions over which they pass.

Equatorial Low-Pressure Trough. The equatorial low-pressure trough is an elongated, narrow band of low pressure that nearly girdles Earth, following an undulating linear axis. Constant high Sun altitude and consistent daylength make large amounts of energy available at this region of Earth's surface throughout the year. The warming creates lighter, less-dense, ascending air, with winds converging all along the extent of the trough. This converging air is extremely moist and full of latent heat energy. Vertical cloud columns frequently reach the tropopause, and precipitation is heavy throughout this zone. The combination of heating and convergence forces air aloft and forms the **intertropical convergence zone (ITCZ)**. Figure 6–9 is a satellite image of the equatorial low-pressure trough, identified by bands of clouds associated with the successive lows of the ITCZ along the equator.

During summer, a marked wet season accompanies the shifting ITCZ over various regions. The maps in Figure 6–8 show the location of the ITCZ in January and July. In January, the zone touches northern Australia and dips southward in eastern Africa and South America, whereas in July the zone shifts northward in the Americas and as far north as Pakistan and southern Asia. The zone is fairly constant in the eastern Pacific and Atlantic oceans, varying more in the western Pacific and Atlantic and the Indian oceans.

Figure 6–10 shows two views of the **Hadley cell**, named for the eighteenth-century English scientist who described the tradewinds. Such a cell is generated by this equatorial low-pressure system of converging, ascending air. These winds converging on the equatorial low-pressure trough are known as the northeast tradewinds (in the Northern Hemisphere) and southeast tradewinds (in the Southern Hemisphere), or generally as **tradewinds**. The tradewinds pick up large quantities of moisture as they return through the Hadley circulation cell for

(a)

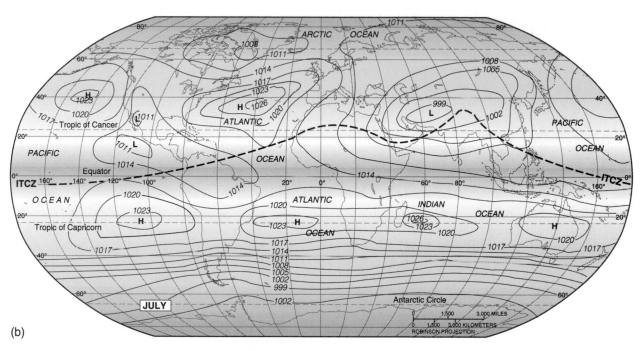

(b)

FIGURE 6-8
Average surface barometric pressure (millibars) for (a) January and (b) July.

60° N
50° N
40° N
30° N
20° N
10° N
0°
10° S
20° S

FIGURE 6–9
Satellite image of Earth, showing the ITCZ (the band of clouds) surrounded on either side by the cloudless subtropical high-pressure systems. Courtesy of National Oceanic and Atmospheric Administration/NESDIS/NCDC/SDSD.

another cycle of uplift and condensation. Within the ITCZ, winds are calm or mildly variable because of the even pressure gradient and the strong vertical ascent of air. These equatorial calms are the **doldrums**, a name formed from an older English word meaning foolish, because of the difficulty sailing ships encountered when attempting to move through this zone.

The rising air from the equatorial low-pressure area spirals upward into a geostrophic flow to the north and south. These upper-air winds turn eastward, flowing from west to east, beginning at about 20° N and S, forming descending anticyclonic flows above the subtropics. In each hemisphere, this circulation pattern appears most vertically symmetrical near the equinoxes.

Subtropical High-Pressure Cells. Between 20° and 35° latitude in both hemispheres, a broad high-pressure zone of hot, dry air is evident across the globe. It is represented by the clear, cloudless skies of the satellite imagery in Figure 6–9 (particularly western North America and the eastern Atlantic). The dynamic cause of these subtropical anticyclones is too complex to detail here, but generally they form as air in the Hadley cell descends in these latitudes, constantly supported by a dynamic mechanism in the upper-air circulation. Air above the subtropics is mechanically pushed downward and is heated as it is compressed on its descent to the surface, as illustrated in Figure 6–10. The air is dry because heavy precipitation along the equatorial circulation removed much of its moisture.

FIGURE 6–10
Two views of the equatorial
circulation pattern and the
Hadley cell in each hemisphere.

(a)

(b)

Surface air diverging from the subtropical high-pressure cells generates Earth's principal surface winds: the westerlies and the tradewinds. The **westerlies** are the dominant surface winds from the subtropics to high latitudes. They diminish somewhat in summer and are stronger in winter.

If you examine Figure 6–8, you will find several high-pressure areas. In the Northern Hemisphere, the Atlantic subtropical high-pressure cell is called the **Bermuda high** (in the western Atlantic) or the **Azores high** (when it migrates to the eastern Atlan-

tic). The area under this subtropical high features clear, warm waters and large quantities of *Sargassum* (a seaweed), which gives the area its name—the Sargasso Sea. The **Pacific high**, or Hawaiian high, dominates the Pacific in July, retreating southward in January. In the Southern Hemisphere, three large high-pressure centers dominate the Pacific, Atlantic, and Indian oceans and tend to move along parallels in shifting zonal positions.

The equatorward boundary of the subtropical highs is generally stable and cloud-free, but the poleward boundaries are quite variable. The entire high-pressure system migrates with the summer high Sun, fluctuating about 5° to 10° in latitude. The eastern side of these anticyclonic systems are drier, more stable, and have cooler ocean currents beneath. Thus, subtropical and midlatitude west coasts are influenced and experience dry summer conditions. In fact, Earth's major deserts generally occur within the subtropical belt and extend to the west coast of each continent—for example, the Sahara, Sonora, Great Australian, Kalahari, Namib, and Atacama deserts (see Figure 15–13).

The western sides of subtropical high-pressure cells tend to be moist and unstable, as characterized in Figure 6–11. This causes warm, moist conditions in Hawaii, Japan, southeastern China, and the southeastern United States.

Because the core of the subtropical belts is near 25° N and S latitudes, these areas sometimes are known as the calms of Cancer and the calms of Capricorn. The name **horse latitudes** also is used for these zones of windless, hot, dry air, so deadly in the era of sailing ships. This label is popularly attributed to becalmed sailing crews of past centuries who destroyed the horses on board, not wanting to share food or water with the livestock. The term's true origin may never be known; the *Oxford English Dictionary* calls it "uncertain."

Subpolar Low-Pressure Cells. In Figure 6–8(a), the January map shows two low-pressure cyclonic cells over the oceans around 60° N latitude. The North Pacific **Aleutian low** is near the Aleutian Islands; the North Atlantic **Icelandic low** is near Iceland. Both cells are dominant in winter and weaken significantly or disappear altogether in the summer—see Figure 6–8(b). The area of contrast between cold and warm air masses forms a contact zone known as the **polar front**, which encircles Earth at this latitude and is focused in these low-pressure areas.

Figure 6–12 illustrates this confrontation between warmer, moist air from the westerlies and colder, drier air from the polar and Arctic regions. The warmer air is displaced upward by the heavier cold air, forcing cooling and condensation in the lifted air. Low-pressure cyclonic storms migrate out of the Aleutian and Icelandic frontal areas and may produce precipitation in North America and Europe, respectively. Northwestern sections of North America and Europe generally are cool and moist as a result of the passage of these cyclonic systems onshore—consider the weather in Washington State and Ireland.

In the Southern Hemisphere, a noncontinuous belt of subpolar cyclonic pressure systems surrounds Antarctica and is especially present during the Antarctic winter (July). The spiraling cloud patterns

FIGURE 6–11

Characteristic circulation and climate conditions related to Atlantic subtropical high-pressure anticyclone in the Northern Hemisphere.

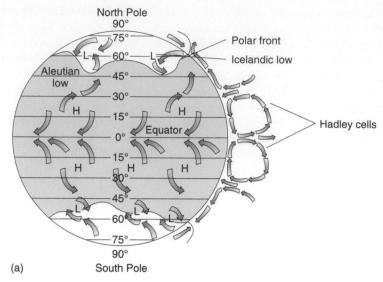

FIGURE 6–12
Polar front and the subpolar low-pressure cells, as related to the general atmospheric circulation. (a) equatorial view; (b) cross-section. Cross-section adapted from Joseph E. Van Riper, *Man's Physical World,* p. 211. Copyright © 1971 by McGraw-Hill. Adapted by permission.

produced by these cyclonic systems are clearly visible in Figure 6–13. Cyclonic storms can even cross Antarctica in great severity, producing strong winds and new snowfall.

Polar High-Pressure Cells. The weakness of the polar high-pressure cells seems to argue against their inclusion among pressure areas on Earth. The polar atmospheric mass is small, receiving little energy to put it into motion. The variable cold, dry winds moving away from the polar region are anticyclonic,

descending and diverging clockwise in the Northern Hemisphere (counterclockwise in the Southern Hemisphere) and forming weak, variable winds called **polar easterlies**. Of the two polar regions, the **Antarctic high** predominates in terms of both strength and persistence. In the Arctic, a polar high-pressure cell is less pronounced, and when it does form, it tends to locate over the colder northern continental areas in winter (Canadian and Siberian highs) rather than directly over the relatively warmer Arctic Ocean.

FIGURE 6–13
Satellite imagery showing
swirling subpolar lows in the
Southern Hemisphere. Courtesy
of National Oceanic and
Atmospheric Administration.

Upper Atmospheric Circulation

Middle and upper tropospheric circulation is an important component of the atmosphere's general circulation. As described previously, these upper-atmosphere winds tend to blow west to east from the subtropics to the poles. The isobars that mark the paths of upper-air winds are based upon elevation above sea level; these form an **isobaric surface** along which all points have the same pressure (Figure 6–14). Just as sea level is a reference datum for evaluating air pressure at the surface, this isobaric surface is a reference datum used to evaluate air-pressure characteristics at altitude.

Closer spacing of the isobars indicates greater winds; wider spacing indicates lesser winds. In this isobaric pressure surface, altitude variations from the reference datum are called *ridges* for high pressure

(with isobars on the map bending poleward) and *troughs* for low pressure (with isobars on the map bending equatorward). Looking at Figure 6–7, can you identify such ridges and troughs on the isobaric chart of geostrophic winds?

Within the westerly flow of geostrophic winds are great waving undulations called **Rossby waves**, named for meteorologist C. G. Rossby, who first described them mathematically. The polar front is the contact between colder air to the north and warmer air to the south (Figure 6–15). The Rossby waves bring tongues of cold air southward, with warmer tropical air moving northward. The development of Rossby waves in the upper-air circulation is shown in the figure. As these disturbances mature, distinct cyclonic circulations form, with warmer air and colder air mixing along distinct fronts. The development of cyclonic storm systems at the surface

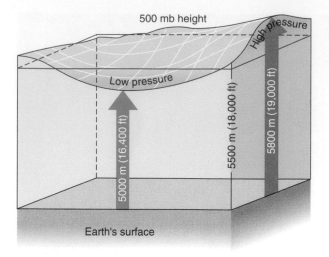

FIGURE 6–14

Representation of the 500 mb isobaric surface is used for analyzing pressure patterns at this height above the surface. After *Meteorology: The Atmosphere in Action* by Joe R. Eagleman, © 1980 by Litton Educational Publishing, Inc. Adapted by permission of Wadsworth, Inc.

is supported by these wave-and-eddy formations and other upper-air flows. These Rossby waves develop along the flow axis of a jet stream.

Jet Streams. The most prominent movement in these upper-level westerly wind flows is the **jet stream**, an irregular, concentrated band of wind occurring at several different locations. Rather flattened in vertical cross section, the jet streams normally are 160–480 km wide by 900–2150 m thick (100–300 mi wide by 3000–7000 ft thick), with core speeds which can exceed 300 kmph (190 mph).

Airline schedules reflect the presence of these upper-level winds, for they allot shorter flight times from west to east and longer flight times from east to west. Also important to both military and civilian aircraft is the effect the jet streams have on fuel consumption and the existence of air turbulence. Jet streams in both hemispheres tend to weaken during each hemisphere's summer and strengthen during winter as the streams shift closer to the equator. A stylized view of a polar jet stream flow is shown in Figure 6–16.

The **polar jet stream** is located at the tropopause along the polar front, at altitudes between 7600 m and 10,700 m (24,900 to 35,100 ft), meandering between 30° and 70° N latitude. The polar jet stream can migrate as far south as Texas, steering colder air masses into North America, influencing surface storm paths traveling eastward. In the summer, the guiding influence of the polar jet stream stays far to the north.

In subtropical latitudes, near the boundary between tropical and midlatitude air, another jet stream flows near the tropopause. This **subtropical jet stream** ranges from 9100 to 13,700 m (29,850 to 45,000 ft) in altitude, and although it is generally weaker, it can reach greater speeds than those of the polar jet stream. The subtropical jet stream meanders from 20° to 50° latitude and may occur over North America simultaneously with the polar jet stream. The cross section of the troposphere in Figure 6–12(b) depicts the general location of these two jet streams.

Global Circulation Model

The general circulation of the atmosphere is schematically presented in Figure 6–17, which summarizes the chapter to this point. You can use this idealized model to locate and identify each of the four principal pressure and three wind systems we have discussed. In this ideal model, the differing effects of land and water and seasonal variations are generalized. Atmospheric circulation is drawn in dimensional perspective, with the vertical dimension exaggerated to better exhibit the anatomy of the circulation. Surface wind patterns are seen through the cutaway view.

Local Winds

Several winds that form in response to local terrain deserve mention in our discussion of atmospheric circulation. These local effects can, of course, be overwhelmed by weather systems passing through an area.

Land-sea breezes occur on most coastlines (Figure 6–18). The differential heating characteristics of land and water surfaces create these winds. During the day, land heats faster and becomes warmer than the water offshore. Because warm air is less dense, it rises and triggers an onshore flow of cooler marine air, usually stronger in the afternoon. At night, inland areas cool radiatively faster than offshore waters. As a result, the cooler air over the land subsides and flows offshore over the warmer water, where the air is lifted. This night pattern reverses the process that developed during the day.

Well inland from the Pacific Ocean—160 km or 100 mi—Sacramento, California, demonstrates the sea-breeze effect. The city is at 39° N and 5 m (17 ft) elevation. The average July maximum-minimum temperatures are 34°C and 14°C (93°F and 57°F). This evening cooling by the natural flow of marine air establishes a monthly mean of only 24°C (75°F), despite high daytime temperatures. In some locations (for example, Hyderabad, Pakistan), large de-

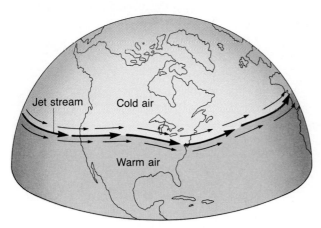

(a) Gently undulating upper airflow

(b) Meanders form in jet stream

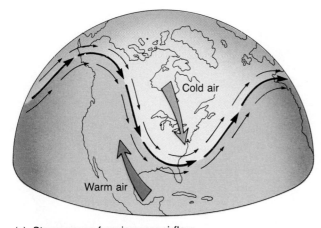

(c) Strong waves form in upper airflow

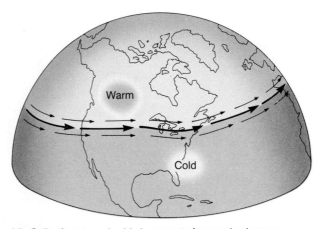

(d) Cells of warm and cold air separate from main air mass

FIGURE 6–15
Development of Rossby waves in the upper-air circulation. Reprinted by permission of Merrill, an imprint of Macmillan Publishing Company, from *Earth Science,* 5th ed., by Edward J. Tarbuck and Frederick K. Lutgens, p. 395. Copyright © 1988 by Merrill Publishing. Adapted from J. Namias, National Oceanic and Atmospheric Administration.

FIGURE 6–16 (above) Stylized portrait of a polar jet stream.

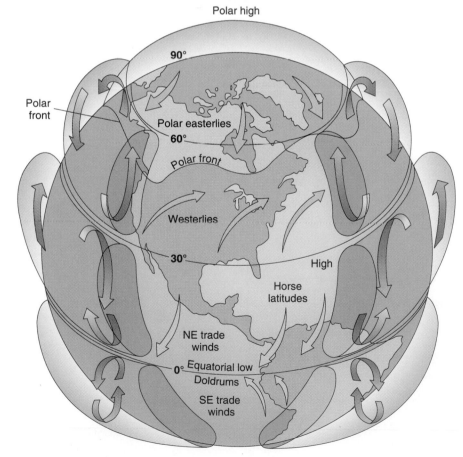

FIGURE 6–17 (right) Idealized global circulation model. Adapted by permission of Merrill, an imprint of Macmillan Publishing Company, from *Earth Science,* 5th ed., by Edward J. Tarbuck and Frederick K. Lutgens, p. 392. Copyright © 1988 by Merrill Publishing.

FIGURE 6-18

Land-sea breezes characteristic of day and night. Reprinted by permission of Merrill, an imprint of Macmillan Publishing Company, from *Earth Science,* 5th ed., by Edward J. Tarbuck and Frederick K. Lutgens, p. 396. Copyright © 1988 by Merrill Publishing

flectors are placed over roof openings to direct onshore sea-to-land breezes into dwellings to take advantage of the natural cooling.

Mountain-valley breezes are created in a somewhat similar exchange. Mountain air cools rapidly at night, and valley air heats rapidly during the day (Figure 6-19). Thus, warm air rises upslope during the day, particularly in the afternoon; at night, cooler air subsides downslope into the valleys. Other downslope winds are discussed in Chapter 8.

Katabatic winds, or gravity drainage winds, are significant on a larger regional scale, under certain conditions. These winds usually are stronger than mountain-valley winds. An elevated plateau or highland is essential, where layers of air at the surface cool, become denser, and flow downslope. The ferocious winds that can blow off the ice sheets of Antarctica and Greenland are katabatic in nature. Worldwide, a variety of terrains produce such winds and bear many local names. The *mistral* of the Rhône Valley in southern France can cause frost damage to vineyards as the cold north winds move over the region to the Gulf of Lion and the Mediterranean Sea. The frequently stronger *bora,* driven by the cold air

FIGURE 6–19
Pattern of mountain-valley
breezes during day and night.

of winter high-pressure systems inland, flows across the Adriatic Coast to the west and south. In Alaska such winds are called the *Taku*.

Santa Ana winds are another local wind type, generated by high pressure over the Great Basin of the western United States. A strong, dry wind is produced that flows out across the desert to southern California coastal areas. The air is heated by compression as it flows from higher to lower elevations and with increasing speed it moves through constricting valleys to the southwest. These winds irritate the population with their dust, dryness, and heat.

Locally, wind represents a potential source of renewable energy. Focus Study 6–1 briefly explores the potential for development of wind resources.

Monsoonal Winds. Regional wind systems that seasonally change direction are important in some regions. Examples of intense, seasonally shifting wind systems occur in the tropics over Southeast Asia, Indonesia, India, northern Australia, and equatorial Africa. These winds involve an annual cycle of returning precipitation with the summer Sun and are

Wind Power: An Energy Resource

The principles may be old, but the technology is modern and the prospects favorable. More than half of the world's population relies on renewable energy, derived from resources that are not depleted in the span of a human lifetime. That energy in developing countries—for cooking, heating, and pumping—comes principally from small hydroelectric plants, wind-power systems, and wood resources. In developed nations, energy sources are focused more upon nonrenewable fuels, although some electricity is generated at wind farms, where groups of wind machines are massed, and at solar installations.

More than 50,000 wind turbines have been installed worldwide since 1974, with 13,000 of them in California alone (Figure 1). Denmark and California lead the world in this technology, with Denmark moving forward under its own official National Wind Strategy. Sweden is implementing plans to dismantle nuclear power plants and no doubt will increase development of its appreciable wind resources.

Power generation from wind is site-specific, meaning that the localized conditions that produce adequate winds are confined to certain areas. In places where wind is reliable less than 25–30% of the time, only small-scale uses are economically feasible. Wind resources are greatest along coastlines influenced by tradewinds and westerly winds; where mountain passes constrict the flow of air and interior valleys develop thermal low-pressure areas, thus drawing air across the landscape; or where localized winds such as katabatic and monsoonal flows occur. In addition, the mountains of North America, northwestern Europe, southwestern Australia, and the Arctic coastlands of the Soviet Union all offer favorable sites.

Problems with further deployment of wind farms in the United States arose because of political decisions in the 1980s. All alternative energy programs, including wind, were cut or canceled, and tax incentives eliminated, even though tax incentives and research for nonrenewable fossil fuels and nuclear power were increased.

One factor that may accelerate wind applications to replace oil-fired electrical generation at appropriate sites is an estimate of remaining domestic oil reserves made by the U.S. Geological Survey in 1989. With 51 billion barrels of recoverable oil left, including expected undiscovered reserves, and with the rate of domestic consumption at over 5.4 billion barrels per year, the United States has only a 16-year supply left. Imports at the 50% level will stretch this reserve only to 32 years, with unknown economic consequences. And the domestic natural gas supply estimate is not much better, with a 35-year supply left until depletion, at present consumption rates.

These economic realities no doubt will override any further delaying actions, especially in regions where peak winds are in concert with peak electrical demand for air cooling, space heating, or agricultural water pumping. Long-term and marginal costs also favor wind-energy deployment,

FIGURE 1
Wind farm at Gorgonio Pass in southern California. Photo by California Energy Commission.

for it does not produce carbon dioxide, radioactivity, sulfur dioxide, toxic wastes, or ash, nor does it require mining or the importation of foreign fuels—all of which represent significant cost with nonrenewable fuels. The utility of wind power will be enhanced by improvements in energy-storage capability. In addition, wind farm sites can have multiple uses, such as pasturelands and farming.

History is rich with the accomplishments of wind-powered sails. Experiments are going forward today to see whether the wind can still drive ships across the sea as in days past. In an experiment more than 10 years ago, the Japanese oil tanker *Shin Aitoku Maru* installed two computer-guided sails of 300 m^2 (3228 ft^2) each. Fuel consumption was reduced by 50% in early tests. If we consider the world's cargo fleet of more than 25,000 vessels, consuming 4 to 5 million barrels of oil a day, a savings of just 10 to 15% would be significant.

Cash-poor, energy-poor developing countries already have a decentralized demand for energy, so for them large, centralized, capital-intensive electrical generation seems inappropriate. The key issue appears to be whether societies will develop nonrenewable-centralized or renewable-decentralized energy resources. And, as for sails on ships: wind is a renewable resource, waiting to be realized as in the sailing days of old.

FIGURE 6–20
Asian monsoon patterns: (a) winter pressure and wind patterns over Asia; (b) summer pressure and wind patterns over Asia and the location of the ITCZ. From Joseph E. Van Riper, *Man's Physical World,* p. 215. Copyright © 1971 by McGraw-Hill. Adapted by permission.

named after the Arabic word for season, *mausim,* or **monsoon**. (Specific monsoonal weather, associated climate types, and vegetation regions are discussed in Chapters 8, 10, and 20.)

The monsoons of southern and eastern Asia are driven by the location and size of the Asian landmass and its proximity to the Indian Ocean. Also important to the generation of monsoonal flows are wind and pressure patterns in the upper-air circulation. The extreme temperature range from summer to winter over the Asian landmass is due to its continentality, reflecting its isolation from the modifying effects of the ocean. This continental landmass is dominated by an intense high-pressure anticyclone in winter, whereas the central area of the Indian Ocean is dominated by the equatorial low-pressure trough.

Resultant cold, dry winds blow from the Asian interior over the Himalayas, downslope, and across India, producing average temperatures between 15°

and 20°C (60° and 68°F) at lower elevations. These dry winds desiccate (dehydrate) the landscape and then give way to hot weather from March through May. When the monsoonal rains arrive from June to September, they are welcome relief from the dust, heat, and parched land of Asia's springtime (Figure 6–20).

During the June–September wet period, the Sun shifts northward to the Tropic of Cancer, near the mouths of the Indus and Ganges Rivers. The Asian continental interior develops a thermal low pressure, associated with higher average temperatures, and the intertropical convergence zone shifts northward over southern Asia. Meanwhile, the Indian Ocean, with surface temperatures of 30°C (86°F), is under the influence of subtropical high pressure. As a result, hot, dry subtropical air sweeps over the warm ocean, producing extremely high evaporation rates.

FIGURE 6–21
Major ocean currents.

By the time this mass of air reaches India and the convergence zone, it is laden with moisture in thunderous, dark clouds. The warmth of the land lends additional lifting to the incoming air, as do the Himalayas, forcing the air mass to higher altitudes. These conditions produce the wet monsoon of India, where world-record rainfalls have occurred: both the second-highest average rainfall (1143 cm or 450 in.) and the highest single-year rainfall total (2647 cm or 1042 in.) were measured at Cherrapunji, India. This city's record rainfall is discussed further in Chapter 8.

The annual monsoon is an integral part of Indian culture and life, as writer Khushwant Singh describes:

> To know India and her peoples, one has to know the monsoon. . . . It has to be a personal experience because nothing short of living through it can fully convey all it means . . . . What the four seasons of the year mean to the European, the one season of the monsoon means to the Indian. It is preceded by

> desolation; it brings with it the hopes of spring; it has the fullness of summer and the fulfillment of autumn all in one . . . . much of India's art, music, and literature is concerned with the monsoon . . . . First it falls in fat drops; the earth rises to meet them. She laps them up thirstily and is filled with fragrance. It brings the odor of the earth and of green vegetation to the nostrils.*

OCEANIC CURRENTS

Earth's atmospheric and oceanic circulations are interrelated. The driving force for ocean currents is the frictional drag of the winds. Also important in shaping these currents is the interplay of the Coriolis force, density differences associated with temperature and salinity, the configuration of the continents and ocean floor, and astronomical forces (the tides).

---

*Khushwant Singh, 1959, *I Hear the Nightingale,* New York: Grove Press, 101–2. Reprinted by permission.

You may be surprised to learn that the ocean surface is not level, for it features small height differences in response to these currents, atmospheric pressure differences, slight variations in gravity, and the presence of waves and swells.

The general patterns of major ocean currents are shown in Figure 6–21. Because ocean currents flow over distance and through time, they come under the influence of the deflective Coriolis force. However, their pattern of deflection is not as tightly circular as that of the atmosphere. If we compare this map with that of Earth's pressure and wind systems (Figure 6–8), we find the ocean currents driven by the circulation around subtropical high-pressure cells in both hemispheres. These circulation systems are known as **gyres** and generally appear offset toward the western side of each ocean basin. To understand these gyres and their western margins, which feature slightly stronger currents, we must understand a process known as **western intensification**.

Along the full extent of areas adjoining the equator, the tradewinds drive the oceans westward in a concentrated channel. These currents are kept near the equator by the weaker Coriolis force influence. As the surface current approaches the western margins of the oceans, the water actually piles up an average of 15 cm (6 in.). From this western edge, ocean water then spills northward and southward in strong currents, flowing in tight channels along the western edges of the ocean basins (eastern shorelines of continents). Additionally, a strong countercurrent also is generated that flows through the full extent of the Pacific, Atlantic, and Indian oceans. This occasionally strong, somewhat sporadic eastward **equatorial countercurrent** may be alongside or just beneath the surface current, at depths of 100 m (300 ft).

In the Northern Hemisphere, the Gulf Stream and the Kuroshio move forcefully northward as a result of western intensification, with their speed and depth increased by the constriction of the area they occupy. The warm, deep-blue water of the ribbonlike Gulf Stream (Figure 5–4) usually is 50–80 km wide and 1.5–2.0 km deep (30–50 mi wide and 0.9–1.2 mi deep), moving at 3–10 kmph (1.8–6.2 mph). In 24 hours, ocean water can move 70–240 km (40–150 mi) in the Gulf Stream, although a complete circuit around an entire gyre may take a year.

Where surface water is swept away from a coast, either by surface divergence (induced by the Coriolis force) or by offshore winds, an **upwelling current** occurs. This cool water generally is nutrient-rich and rises from great depth to replace the vacating water. Such cold upwelling currents occur off the Pacific coasts of North and South America and the subtropical and midlatitude west coast of Africa. In other portions of the sea where there is an accumulation of water—as at the western end of an equatorial current or along the margins of Antarctica—excess water is thrust downward in a **downwelling current**. Important mixing currents along the ocean floor are generated from such downwelling zones and travel the full extent of the ocean basins, carrying heat energy and salinity. A complete circuit may require 1000 years from downwelling in the Labrador Sea off Greenland to its reemergence in the southern Indian or North Pacific Oceans.

SUMMARY

Atmospheric circulation facilitates important transfers of energy and mass on Earth. More than any other spatial factor, natural or cultural, the atmosphere socializes humanity—one person's or nation's exhalation is another's breath. Wind is the horizontal movement of air across Earth's surface, directed and driven by pressure differences, the Coriolis force, and frictional drag. Earth's atmospheric and oceanic circulations represent a vast heat engine powered by the Sun.

The patterns of high and low pressures on Earth in four generalized belts provide the basis for the distribution of specific wind systems. A generalized model of the atmospheric circulation system is helpful in visualizing both the surface and upper tropospheric wind circulation. Constructing an accurate multidimensional computer model of the atmospheric and oceanic circulation remains an elusive and actively pursued goal for climatologists.

Ocean currents are primarily caused by the frictional drag of wind and occur worldwide at varying intensity, temperature, and speed. Wind itself is an important energy source and represents a renewable resource with potential for increasing development in the future.

KEY TERMS

Aleutian low	Isobars
Anemometer	Jet stream
Antarctic high	Katabatic winds
Anticyclone	Monsoon
Azores high	Pacific high
Beaufort wind scale	Polar easterlies
Bermuda high	Polar front
Coriolis force	Polar high-pressure cells
Cyclone	Polar jet stream
Doldrums	Pressure gradient force
Downwelling current	Rossby waves
Equatorial countercurrent	Subpolar low-pressure cells
Equatorial low-pressure trough	Subtropical high-pressure cells
Friction force	Subtropical jet stream
Geostrophic winds	Tradewinds
Gyres	Upwelling current
Hadley cell	Westerlies
Horse latitudes	Western intensification
Icelandic low	Wind
Intertropical convergence zone (ITCZ)	Wind vane
Isobaric surface	

REVIEW QUESTIONS

1. What is a possible explanation for the cold summer of 1816 experienced by New England farmers and others? Relate your answer to global circulation.
2. Explain the statement, ". . . the atmosphere socializes humanity, making all the world a spatially linked society." Illustrate your answer with some examples.
3. Define wind. How is it measured? How is its direction determined?
4. Distinguish among primary, secondary, and tertiary classifications of global atmospheric circulation.
5. What is the purpose of the Beaufort wind scale? Characterize winds given Beaufort numbers of 4, 8, and 12, relative both to water and land effects.
6. Describe the horizontal and vertical air motions in a high-pressure cell and in a low-pressure cell.
7. Describe the effect of the Coriolis force. How does it apparently deflect atmospheric and oceanic circulations? Explain.
8. What are geostrophic winds, and where are they encountered in the atmosphere?
9. Construct a simple diagram of Earth's general circulation, including the four principal pressure belts or zones and the three principal wind systems that are produced.
10. How does the intertropical convergence zone (ITCZ) relate to the equatorial low-pressure trough? How might it appear on a satellite image?

11. What is one source of migratory low-pressure cyclonic storms in North America? In Europe?

12. Relate the jet-stream phenomenon to general upper-air circulation. How does the presence of this circulation relate to airline schedules from New York to San Francisco and the return trip to New York?

13. People living along coastlines generally experience variations in day-night winds. Explain the factors that produce these changing wind patterns.

14. Describe the seasonal pressure patterns that produce the Asian monsoonal wind and precipitation patterns.

15. What is the relationship between global atmospheric circulation and ocean currents? Specifically relate oceanic gyres to patterns of subtropical high pressure.

16. Define the western intensification. How is it related to the Gulf Stream and Kuroshio currents?

17. Where on Earth are upwelling currents experienced? Why? Characterize such currents.

18. What is the present status of wind-energy technologies? Describe deployment delays. How do these delays relate to the nature of the technology?

SUGGESTED READINGS

Eagleman, Joe R. *Meteorology—The Atmosphere in Action*. New York: D. Van Nostrand, 1980.

Marr, John C., ed. *The Kuroshio—A Symposium on the Japan Current*. Honolulu: East-West Center Press, 1970.

Martz, Larry, with Mark Miller, Nikki Finke Greenberg, and Karen Springen. "Chernobyl Meltdown." *Newsweek* (12 May 1986): 20–49.

McBoyle, Geoffrey, ed. *Climate in Review*. Boston: Houghton Mifflin, 1973.

McDonald, James E. "The Coriolis Effect." *Scientific American* 186 (May 1952): 72–76.

Neiburger, Morris, James G. Edinger, and William D. Bonner. *Understanding Our Atmospheric Environment*. San Francisco: W. H. Freeman, 1973.

Starr, Victor P. "The General Circulation of the Atmosphere." *Scientific American* 195 (December 1956): 40–45.

Thurman, Harold V. *Introductory Oceanography*. 5th ed. Columbus: Macmillan Publishing Co., 1988.

Vesilind, Priit J. "Monsoons—Life Breath of Half the World." *National Geographic* 166, no. 6 (December 1984): 712–47.

Whipple, A. B. C., and the editors of Time-Life Books. *Restless Oceans*. Alexandria, VA: Time-Life Books, 1983.

Wiebe, Peter H. "Rings of the Gulf Stream." *Scientific American* 246 (March 1982): 60–70.

Earth is the water planet. Its surface waters are unique in the solar system. Answers to why water is here in such quantity, what qualities and properties it possesses, and how it circulates over Earth are the topics that begin this section. The dynamics of daily weather phenomena follow: the effects of moisture and energy in the atmosphere, the interpretation of cloud forms, conditions of stability or instability, the interaction of air masses, and the occurrence of violent weather. The specifics of the hydrologic cycle are explained through the water-balance concept, which is useful in understanding water-resource relationships, whether global, regional, or local. Important water resources include rivers, lakes, groundwater, and oceans. The spatial implications over time of this water-weather system leads to the final topic in Part Two—world climate patterns.

PART TWO

The Water, Weather, and Climate System

Photo by author.

Burney Falls, McArthur–Burney Falls State Park,
California. Photo by author.

7

Water and Atmospheric Moisture

Walden is blue at one time and green at another, even from the same point of view. Lying between the earth and the heavens, it partakes of the color of both. . . . A lake is the landscape's most beautiful and expressive feature. It is earth's eye; looking into which the beholder measures the depth of his own nature. . . . Sky water. It needs no fence. Nations come and go without defiling it. It is a mirror which no stone can crack . . . Nature continually repairs . . . a mirror in which all impurity presented to it sinks, swept and dusted by the sun's hazy brush. A field of water . . . is continually receiving new life and motion from above. It is intermediate in its nature between land and sky.*

Thus did Thoreau speak of the water so dear to him—Walden Pond in Massachusetts, along whose shore he lived. Water is the essential medium of our daily lives and a principal compound in nature. It covers 71% of Earth (by area) and in the solar system occurs in such significant quantities only on our planet. It constitutes nearly 70% of our bodies by weight and is the major ingredient in plants, animals, and our food. A human being can survive 50 to 60 days without food, but only 2 or 3 days without water. The water we use must be adequate in quantity as well as quality for its many tasks—everything from personal hygiene to vast national water projects. Water indeed occupies that place between land and sky, mediating energy and shaping both the lithosphere and atmosphere, which Thoreau revealed in his thoughtful observations.

This chapter examines the dynamics of water and atmospheric moisture in particular. The clouds that form as air becomes water-saturated are more than whimsical, beautiful configurations; they are important indicators of overall atmospheric conditions.

BEGINNINGS

Earth's hydrosphere contains about 1,359,208,000 km^3 (326,074,000 mi^3) of water. All water on Earth was formed within the planet, reaching Earth's surface in an ongoing process called **outgassing**, by which water and water vapor emerge from layers

*Reprinted by permission of Merrill, an imprint of Macmillan Publishing Company, from *Walden* by Henry David Thoreau, p. 192, 202, 204. Copyright © 1969 by Merrill Publishing. Originally published 1854.

deep within and below the crust—25 km (15.5 mi) or more below Earth's surface (Figure 7–1). Various geophysical factors explain the timing of this outgassing over the past 4 billion years. In the early atmosphere, massive quantities of outgassed water vapor condensed and fell in torrents, only to vaporize again because of high temperature at Earth's surface. For water to remain on Earth's surface, land temperatures had to drop below the boiling point of 100°C (212°F), something that occurred about 3.8 billion years ago.

The lowest places across the face of Earth then began to fill with water: first ponds, then lakes and seas, and eventually ocean-sized bodies of water. (A sea is generally smaller than an ocean and is near a landmass, or sometimes refers to a large, inland, salty body of water.) Massive flows of water washed over the landscape, carrying both dissolved and undissolved elements to these early seas.

Quantity Equilibrium

Today, water is the most common compound on Earth, having achieved the present volume of 1.36 billion km^3 (326 million mi^3) approximately 2 billion years ago. This quantity has remained relatively constant, even though water is continuously being lost from the system, escaping to space or breaking down and forming new compounds with other elements. Lost water is replaced by pristine water not previously at the surface, water that emerges from within Earth. The net result of these inputs and outputs to water quantity is a steady-state equilibrium in Earth's hydrosphere.

Despite this overall net balance in quantity, worldwide changes in sea level do occur, called **eustasy**—specifically related to changes in volume of water and not movement of land. These changes are explained by **glacio-eustatic** factors (see Chapter 17). Glacio-eustatic factors are based on the amount of water stored on Earth as ice. As more water is bound up in glaciers (in mountains worldwide) and in ice sheets (Greenland and Antarctica), sea level lowers, whereas a warmer era reduces the quantity of water stored as ice and thus raises sea level. Over the past 100 years, mean sea level has steadily risen and is still rising worldwide at this time. Apparent changes in sea level also are related to actual physical changes in landmasses called isostatic change, such as continen-

FIGURE 7–1
Outgassing of water from Earth's crust in Chilean Andes. Photo by Stephen Cunha.

tal uplift or subsidence (*isostasy* is discussed in Chapter 11).

Distribution of Earth's Water Today. Most of Earth's continental land is in the Northern Hemisphere, whereas the Southern Hemisphere is dominated by water. The present location of all of Earth's water, in terms of fresh vs. saline, surface vs. underground, and frozen vs. liquid, is illustrated in Figure 7–2. Data are first divided between ocean water and freshwater, with oceans containing 97.22% of all water, or about 1.321 billion km^3 (317 million mi^3). The remaining 2.78% is classified as fresh, or nonoceanic, water.

Table 7–1 divides this freshwater into two categories—surface and subsurface water. The greatest single repository of surface freshwater is ice. Ice sheets and glaciers account for 77.14% of all fresh-

water on Earth. Add to this the subsurface groundwater, and 99.36% of all freshwater is accounted for. The remaining freshwater, although very familiar to us, and present in seemingly huge amounts in lakes, rivers, and streams, actually represents but a small quantity, less than 1%. All the freshwater lakes in the world total only 125,000 km^3 (30,000 mi^3), with 80% of this in just 40 of the largest lakes, and about 50% contained in just 7 lakes (Table 7–2).

Saline lakes and inland seas, which are neither fresh nor connected to the ocean, contain 104,000 km^3 (25,000 mi^3) of water. Think of the weather occurring in the atmosphere worldwide at this moment, and the many flowing rivers and streams, which combined amount to only 14,250 km^3 (3400 mi^3), or only 0.033% of freshwater, or 0.0011% of all water! Yet, this small amount is very dynamic. A water molecule traveling along atmospheric and surface-

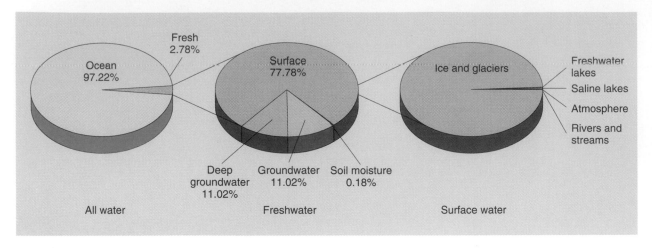

FIGURE 7–2
Ocean and freshwater distribution on Earth.

water paths moves through the entire hydrologic system in less than two weeks, whereas a water molecule located in deep ocean circulation, groundwater, or a glacier moves slowly in transit, taking thousands of years.

Quality (Composition) Equilibrium

Water acts as a solvent, dissolving at least 57 of the elements found in nature. In fact, most natural elements and the compounds they form are found in the seas as dissolved solids, or solutes. Thus, seawater is a solution, and the concentration of dissolved solids is called **salinity**. Long ago, during the evolutionary and living atmosphere periods, heavy precipitation delivered enormous quantities of dissolved solids to the ocean. Later, about a billion years ago and after much accumulation, the salinity of the oceans achieved a steady-state equilibrium. Any **influx** of material from physical, chemical, and

TABLE 7–1
Distribution of freshwater.

Location	Amount km³	(mi³)	Percent of Freshwater	Percent of Total Water
Surface Water—				
Ice sheets and glaciers	29,180,000	(7,000,000)	77.14	2.146
Freshwater lakes	125,000	(30,000)	0.33	0.009
Saline lakes and inland seas	104,000	(25,000)	0.28	0.008
Atmosphere	13,000	(3100)	0.03	0.001
Rivers and streams	1,250	(300)	0.003	0.0001
Total surface water	29,423,250	(7,058,400)	77.78	2.164
Subsurface Water—				
Groundwater—surface to 762 m (2500 ft) depth	4,170,000	(1,000,000)	11.02	0.306
Groundwater—762 m (2500 ft) to 3962 m (13,000 ft) depth	4,170,000	(1,000,000)	11.02	0.306
Soil moisture storage	67,000	(16,000)	0.18	0.005
Total subsurface water	8,407,000	(2,016,000)	22.22	0.617
TOTAL (rounded)	37,800,000	(9,070,000)	100.00%	2.78%

TABLE 7–2
Major freshwater lakes.

Lake	Volume in km³ (mi³)	Surface Area in km² (mi²)	Depth in m (ft)
Baikal (USSR)	22,000 (5280)	31,500 (12,160)	1620 (5315)
Tanganyika (Africa)	18,750 (4500)	39,900 (15,405)	1470 (4823)
Superior (U.S./Canada)	12,500 (3000)	83,290 (32,150)	397 (1301)
Michigan (U.S.)	4920 (1180)	58,030 (22,400)	281 (922)
Huron (U.S./Canada)	3545 (850)	60,620 (23,400)	229 (751)
Ontario (U.S./Canada)	1640 (395)	19,570 (7550)	237 (777)
Erie (U.S./Canada)	485 (115)	25,670 (9910)	64 (210)

biological processes was countered by an **efflux**, or removal, in the form of sediments, sulfides, silicas, clays, and animal shells (calcium carbonate).

Salinity. The oceans remain a remarkably homogeneous mixture, and the ratio of individual salts does not change, despite minor fluctuations in salinity level. In 1874 the British ship *HMS Challenger* sailed around the world, taking surface and depth measurements and collecting samples of seawater. Careful analyses of those samples demonstrated its uniform composition.

Seven elements comprise more than 99% of the dissolved solids in seawater: chlorine (Cl), sodium (Na), magnesium (Mg), sulfur (S), calcium (Ca), potassium (K), and bromine (Br). Seawater also contains dissolved gases (such as carbon dioxide, nitrogen, and oxygen), solid and dissolved organic matter, and a multitude of trace elements. Only

sodium chloride (common table salt), magnesium, and a little bromine are commercially extracted in any significant amount. Future mining of minerals from the seafloor is technically feasible, although it remains uneconomical.

There are several ways to express salinity in seawater. Here are examples, using the worldwide average value:

3.5% (% = parts per hundred)

35,000 ppm (parts per million)

35 g/kg

35‰ (‰ = parts per thousand), the most common notation

Salinity worldwide normally varies between 34‰ and 37‰; variations are attributable to atmospheric conditions above the water and to the quantity of freshwater inflows. In equatorial water, precipitation

FIGURE 7–3
The variation in salinity by latitude. Salinity (green line) is principally a function of climatic conditions, specifically the difference between evaporation and precipitation (E−P), shown by the purple line. From G. Wüst, 1936, p. 347–59.

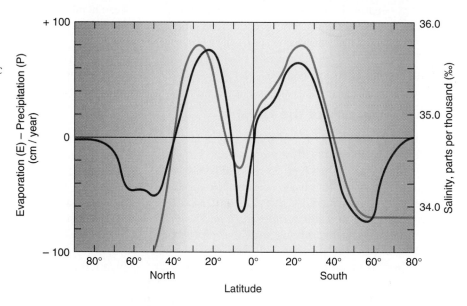

is high throughout the year, diluting salinity values to slightly lower than average (34.5‰). In subtropical oceans—where evaporation rates are high due to the influence of hot, dry subtropical high-pressure cells—salinity is more concentrated, increasing to 36.5‰. Figure 7–3 plots evaporation/precipitation and salinity by latitude to illustrate this slight spatial variability.

The term **brine** is applied to water that exceeds the average of 35‰ salinity, whereas **brackish** applies to water that is less than 35‰. Generally, oceans are lower in salinity near landmasses, because of river discharges and runoff. Extreme examples include the Baltic Sea (north of Poland and Germany) and the Gulf of Bothnia (between Sweden and Finland), which average 10‰ or less salinity because of heavy freshwater runoff and low evaporation rates. On the other hand, the Sargasso Sea, within the North Atlantic subtropical gyre, averages 38‰, and the Persian Gulf is at 40‰ as a result of high evaporation rates in an almost-enclosed basin. Deep pockets near the floor of the Red Sea register a very salty 225‰.

GLOBAL OCEANS AND SEAS

The ocean is one of Earth's last great scientific frontiers and is of great interest to oceanographers and geographers. From a geographical point of view, ocean and land surfaces are not evenly distributed. In fact, from certain perspectives, Earth appears to have a distinct oceanic hemisphere and a distinct land hemisphere—Figure 7–4. The present distribution of the world's oceans and major seas is illustrated in Figure 7–5. (Names and locations of major ocean currents were shown in Figure 6–21.)

Physical Structure of the Oceans

The basic physical structure of the ocean is layered (Figure 7–6). The graph illustrates average temperature, salinity, and dissolved carbon dioxide and oxygen levels with increasing depth. The ocean's surface layer is warmed by the Sun and is wind-driven. Variations in water temperature and solutes are blended rapidly in a **mixing zone** that represents only 2% of the oceanic mass. Below this is the **thermocline transition zone**, a region of strong temperature gradient that lacks the motion of the

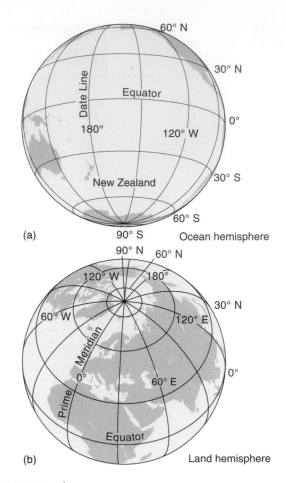

FIGURE 7–4
Two perspectives that divide Earth's surface into (a) an ocean hemisphere and (b) a land hemisphere.

surface. Friction dampens the effect of surface currents, with colder water temperatures at the lower margin tending to inhibit any convective movements.

Starting at a depth of 1–1.5 km (0.62–0.93 mi) and going down to the bottom, temperature and salinity values are quite uniform. Temperatures in this **deep cold zone** are near 0°C (32°F); but, due to its salinity, seawater freezes at about −2°C (28.4°F). The coldest water is at the bottom, except near the poles where cold water may be near or at the surface.

This physical structure exposes the inaccuracy of the supposition that pollution pumped into the sea is mixed throughout its mass (the solution-to-pollution-is-dilution rationale). In reality, only the surface

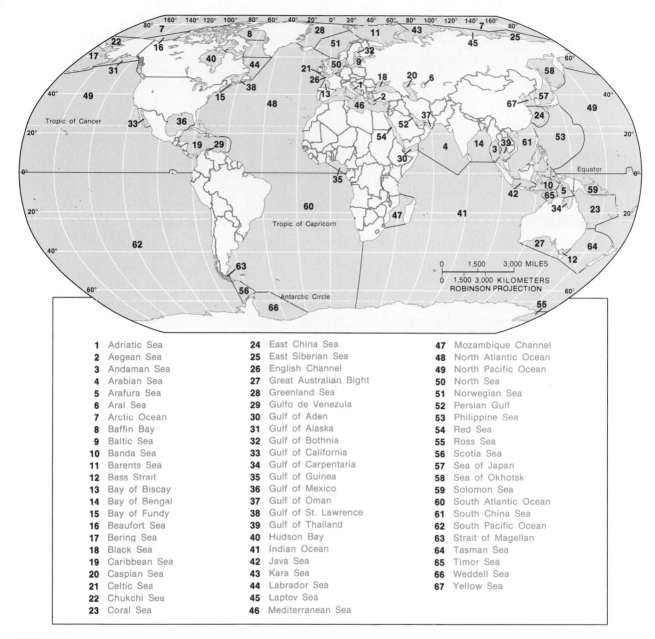

FIGURE 7–5
The names of principal oceans and seas of the world.

1	Adriatic Sea	24	East China Sea
2	Aegean Sea	25	East Siberian Sea
3	Andaman Sea	26	English Channel
4	Arabian Sea	27	Great Australian Bight
5	Arafura Sea	28	Greenland Sea
6	Aral Sea	29	Gulfo de Venezuela
7	Arctic Ocean	30	Gulf of Aden
8	Baffin Bay	31	Gulf of Alaska
9	Baltic Sea	32	Gulf of Bothnia
10	Banda Sea	33	Gulf of California
11	Barents Sea	34	Gulf of Carpentaria
12	Bass Strait	35	Gulf of Guinea
13	Bay of Biscay	36	Gulf of Mexico
14	Bay of Bengal	37	Gulf of Oman
15	Bay of Fundy	38	Gulf of St. Lawrence
16	Beaufort Sea	39	Gulf of Thailand
17	Bering Sea	40	Hudson Bay
18	Black Sea	41	Indian Ocean
19	Caribbean Sea	42	Java Sea
20	Caspian Sea	43	Kara Sea
21	Celtic Sea	44	Labrador Sea
22	Chukchi Sea	45	Laptev Sea
23	Coral Sea	46	Mediterranean Sea

47	Mozambique Channel
48	North Atlantic Ocean
49	North Pacific Ocean
50	North Sea
51	Norwegian Sea
52	Persian Gulf
53	Philippine Sea
54	Red Sea
55	Ross Sea
56	Scotia Sea
57	Sea of Japan
58	Sea of Okhotsk
59	Solomon Sea
60	South Atlantic Ocean
61	South China Sea
62	South Pacific Ocean
63	Strait of Magellan
64	Tasman Sea
65	Timor Sea
66	Weddell Sea
67	Yellow Sea

layers actively mix. Deep currents are slow-moving masses. It takes thousands of years for deep currents to flow among the North Atlantic, Indian, and North Pacific Oceans, or for the colder, saltier flows that sink near Antarctica to flow northward along the ocean floor past the equator.

UNIQUE PROPERTIES OF WATER

Earth's distance from the Sun places it within a most remarkable temperate zone, compared to the other planets. This temperate location allows all states of water—ice, liquid, and vapor—to occur naturally

FIGURE 7–6

Schematic of average physical characteristics observed throughout the ocean's vertical structure.

on Earth. Water is composed of two atoms of hydrogen and one of oxygen, which readily bond. The resulting water molecule exhibits a unique stability, is a versatile solvent, and possesses extraordinary heat characteristics. As the most common compound on Earth, water exhibits the most uncommon of properties.

Once hydrogen and oxygen atoms combine, they are quite difficult to separate, thereby producing a water molecule that remains stable in Earth's environment. The nature of the hydrogen-oxygen bond results in the hydrogen side having a positive charge and the oxygen side a negative charge (see inset in Figure 7–7). This polarity of the water molecule explains why water "acts wet" and dissolves so many other molecules and elements. Because of this solvency ability, pure water is rare in nature.

Water molecules are attracted to each other because of their polarity; the positive (hydrogen) side of one water molecule is attracted to the negative (oxygen) side of another. This bonding between water molecules is called *hydrogen bonding*. The effects of hydrogen bonding in water are observable in everyday life. Hydrogen bonding creates the *surface tension* that allows you to float a steel needle on the surface of water, even though

steel is denser than water. This surface tension also allows you to slightly overfill a glass with water, so that the water surface actually is above the rim of the glass, retained by millions of hydrogen bonds. In chemistry laboratory classes, students learn to read the convex meniscus that hydrogen bonding forms in a cylinder or a test tube.

Hydrogen bonding also is the cause of capillarity, which you observe when you use a paper towel. The towel draws water upward through its fibers, because each molecule is pulling on its neighbor. Capillary action is an important component of soil moisture processes, discussed in Chapters 9 and 18. It is important to note that, without hydrogen bonding, water would be a gas at ambient surface temperatures.

Heat Properties

For water to change from one state to another, heat energy must be added to it or released from it. The amount of heat energy must be sufficient to affect the hydrogen bonds between the molecules. Figure 7–7 presents the three states of water and the terms used to describe each **phase change**. The term **sublimation** refers to the direct change of water vapor to ice or ice to water vapor, whereas *melting* and *freezing* describe the phase change between solid and liquid. The terms *condensation* and *evaporation,* or *vaporization* at boiling temperature, apply to the change between liquid and vapor.

Ice, the Solid Phase. As water cools, it behaves like most compounds and contracts in volume, reaching its greatest density at 4°C (39°F). But below that temperature, water behaves very differently from most compounds, and begins to expand as more hydrogen bonds form among the slower-moving molecules, creating the hexagonal structures shown in Figure 7–7. This expansion continues to a temperature of −29°C (−20°F), with up to a 9% increase in volume possible. As shown in Figure

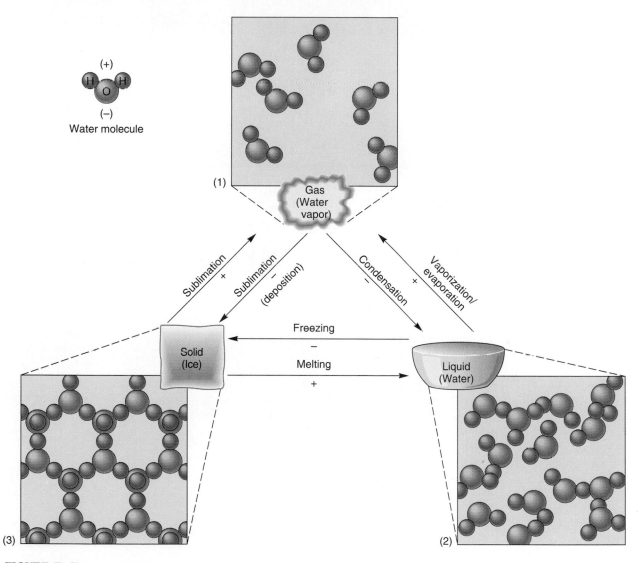

FIGURE 7–7
The three physical states of water: (1) water vapor, (2) liquid water, and (3) ice. Note the molecular arrangement in each state and the terms that describe the changes from one phase to another. Also note how the polarity of water molecules bonds them to one another, loosely in the liquid state and firmly in the solid state.

FIGURE 7–8
(a) Ice-crystal patterns are dictated by the internal structure between water molecules. This structure also explains the lower density of ice and why ice floats, as illustrated by the icebergs (b). (a) Photo by American Museum of Natural History; (b) photo courtesy of Jim Ferry.

(a)

(b)

7–8(a), the rigid internal structure of ice dictates the six-sided appearance of all ice crystals, which can loosely combine to form snowflakes.

The expansion in volume that accompanies the freezing process results in a decrease in density. Specifically, ice has 0.91 times the density of water, and so it floats. Without this change in density, much of Earth's freshwater would be bound in masses of ice on the ocean floor. Instead, we have floating icebergs, with approximately 1/11 (9%) of their mass exposed and 10/11 (91%) hidden beneath the ocean's surface—Figure 7–8(b).

Winter damage to highways is often the result of precisely this phase change from water to ice. Rainwater seeps into roadway cracks and then expands as it freezes, thus breaking up the pavement. Historically, this physical property of water was put to useful work in quarrying rock for building materials. Holes were drilled and filled with water before winter, so that when cold weather arrived the water would freeze and expand, cracking the rock into manageable shapes. The freezing action of ice as an important physical weathering process is discussed in Chapter 13.

Water, the Liquid Phase. As a liquid, water assumes the shape of its container and is a noncompressible fluid, quite different from solid, rigid ice. For ice to melt, heat energy must increase the motion of the water molecules to break some of the hydrogen bonds. The heat characteristics for all three states are illustrated in Figure 7–9. Despite the fact

FIGURE 7–9
Significant latent heat energy is involved in the phase changes of water.

that there is no change in sensible temperature between ice at 0°C and water at 0°C, 80 calories are required for the phase change of 1 gram of ice to 1 gram of water. This heat, called **latent heat**, is hidden within the water and is liberated whenever a gram of water freezes. These 80 calories are the *latent heat of fusion,* or the *latent heat of melting* and of *freezing.* To raise the temperature of 1 gram of water from freezing at 0°C (32°F) to boiling at 100°C (212°F), we must add 100 additional calories, gaining an increase of 1C° (1.8F°) for each calorie added.

Water Vapor, the Gas Phase. Water vapor is an invisible and compressible gas in which each molecule moves independently (Figure 7–7). To accomplish the phase change of liquid to vapor at boiling, under normal sea-level pressure, 540 calories must be added to 1 gram of boiling water to achieve a phase change to water vapor. Those calories are the **latent heat of vaporization**. When water vapor condenses to a boiling liquid, each gram gives up its hidden 540 calories as the **latent heat of condensation**.

To summarize, taking 1 gram of ice at 0°C and raising it to water vapor at 100°C—changing it from a solid, to a liquid, to a gas—*absorbs* 720 calories (80 cal + 100 cal + 540 cal). Or, reversing the process, changing from 1 gram of water vapor at 100°C to ice at 0°C *liberates* 720 calories.

Heat Properties of Water in Nature. In a lake, stream, or in soil water, at 20°C (68°F) every gram of water that breaks away from the surface through evaporation must absorb from the environment approximately 585 calories as the *latent heat of evaporation* (Figure 7–9). This is slightly more energy than would be required if the water were boiling (540 cal). You can feel this absorption of latent heat as evaporative cooling on your skin when it is wet. As air that contains water vapor is cooled, the vapor eventually condenses back into the liquid state, forming moisture droplets and thus liberating 585 calories as the *latent heat of condensation* for every gram of water.

When you realize that a small, puffy, fair-weather cumulus cloud holds 500–1000 metric tons (550–1100 tons) of moisture droplets, you can appreciate the tremendous energy available in the atmosphere! Weather events such as Hurricane Camille (1969), Hurricane Gilbert (1988), and Hurricane Hugo

(1989) involve a staggering amount of energy. Government meteorologists estimate that the moisture in Camille weighed 27 trillion metric tons (30 trillion tons) at its maximum power and mass. The latent heat liberated each day from those hurricane clouds approximately equalled the total energy consumption in the United States for six months! These unique heat properties and latent heat exchanges in water, empowered by solar energy, drive weather systems worldwide.

THE HYDROLOGIC CYCLE

Vast currents of water, water vapor, ice, and energy are flowing about us continuously in an elaborate open global plumbing system. Together they form the **hydrologic cycle**, or hydrologic system, which has operated for billions of years, involving the lower atmosphere down to several kilometers beneath Earth's surface, circulating and transforming water throughout Earth's various spheres.

A Hydrologic Cycle Model

A simplified model of this complex system is useful to our study of the hydrologic cycle. In the following discussion, refer to Figure 7–10 to follow each flow path in the cycle. The ocean provides a starting point, where more than 97% of all water is located and most evaporation and precipitation occur. If we assume that the mean annual global evaporation equals 100 units, we can trace 86 of them to the ocean. The other 14 units come from the land, including water moving from the soil into plant roots and passing through their leaves (a process called transpiration, described further in Chapter 9). Of the ocean's evaporated 86 units, 66 combine with 12 advected from the land to produce the 78 units of precipitation that fall back into the ocean. The remaining 20 units of moisture evaporated from the ocean, plus 2 units of land-derived moisture, produce the 22 units of precipitation that fall over land. Clearly, the bulk of continental precipitation derives from the oceanic portion of the cycle.

Precipitation that reaches Earth's surface follows a variety of pathways. The process of precipitation striking vegetation or other groundcover is called **interception**. Precipitation that falls directly to the ground, coupled with that which drips onto the

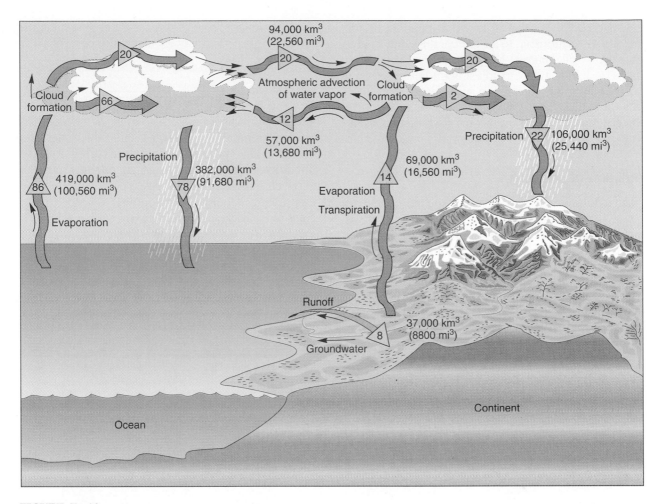

FIGURE 7–10
The hydrologic cycle model with percentages and volumes and directional arrows denoting flow paths.

ground from vegetation, constitutes **throughfall**. Intercepted water that drains across plant leaves and down plant stems is termed **stem flow** and can represent an important moisture route to the surface. Water reaches the subsurface through **infiltration**, or penetration of the soil surface. It then permeates soil or rock through vertical movement called **percolation**.

The atmospheric advection of water vapor from sea to land and land to sea in Figure 7–10 appears to be unbalanced: 20 units moving inland, with only 12 units moving out to sea. However, this exchange is balanced by 8 units of runoff that also flow from land to sea. Most of this runoff—about 95%—comes from surface waters washing across land as overland flow and streamflow, leaving only 5% of the runoff

attributable to slow-moving subsurface groundwater. From Table 7–1 you can see that the small amount of water in rivers and streams is very dynamic, whereas the large quantity of subsurface water is sluggish in comparison and represents only a small portion of total runoff. Now that you are acquainted with the hydrologic cycle and surface water, let's examine atmospheric moisture.

HUMIDITY

The water vapor content of air is termed **humidity**. The capacity of air to hold water vapor is primarily a function of temperature: warmer air has a greater capacity for water vapor, whereas cooler air has a lesser capacity. We are all aware of humidity in the air, for its relationship to air temperature determines our sense of comfort. North Americans spend billions of dollars a year to adjust humidity, either with air conditioning (extracting water vapor and cooling) or with air humidifying (adding water vapor). To determine the energy available for weather activities, it is essential to know the water vapor content of air.

Relative Humidity

Next to temperature and pressure, the most common piece of information in local weather broadcasts is **relative humidity**. Relative humidity is not a direct measurement of water vapor, rather it is a ratio expressed as a percentage of the amount of water vapor that is actually in the air (content), compared to the maximum water vapor the air could hold at a given temperature (capacity). If the air is relatively

dry in comparison to its capacity, the percentage is lower; if the air is relatively moist, the percentage is higher.

$$\text{relative humidity} = \frac{\text{actual water vapor content of the air}}{\text{maximum water vapor capacity of the air}} \times 100$$

Relative humidity varies due to evaporation, condensation, or temperature changes, all of which affect the overall ability of the air to hold water vapor. Relative humidity is an expression of an ongoing process between air and moist surfaces, for condensation and evaporation operate continuously—water molecules move back and forth between air and water. Air temperature is the principal regulator of those processes and thus of relative humidity.

Relative humidity indicates the nearness of air to a saturated condition and indicates when active condensation begins. Air is said to be **saturated**, or full, if it is holding all the water vapor that it can hold at

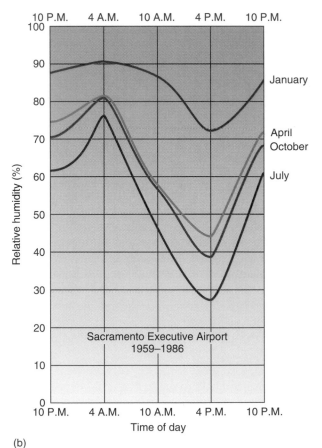

(a)

FIGURE 7–11
(a) Typical daily variations in relative humidity and temperature and (b) seasonal daily variations in relative humidity for Sacramento, California.

a given temperature; under such conditions the net transfer of water molecules between surface and air achieves a saturation equilibrium. Saturation indicates that any further addition of water vapor (change in content) or any decrease in temperature (change in capacity) will result in active condensation.

The temperature at which a given mass of air becomes saturated is termed the **dew-point temperature**. In other words, air is saturated when the dew-point temperature and the air temperature are the same. A cold drink in a glass provides a common example of these conditions: the water droplets that form on the outside of the glass condense from the air because the air layer next to the glass is chilled to its dew-point temperature and thus is saturated.

During a typical day, air temperature and relative humidity relate inversely—Figure 7–11(a). Relative humidity is highest at dawn, when air temperature is lower and the capacity of the air to hold water vapor is less. Relative humidity is lowest in the late afternoon, when higher air temperatures increase the capacity of the air to hold water vapor. The actual water vapor content in the air may have remained the same throughout the day, but because the temperature varied, relative humidity percentages changed from morning until afternoon.

Think of the capacity of air to hold water vapor as being like sponges of varying sizes: warmer air has in it a larger sponge; cooler air a smaller sponge. A cooler body of air can be saturated—filled to capacity—by an amount of water vapor that will only partially fill the capacity of the warmer air.

Weather records for Sacramento, California, show the seasonal variation in relative humidity by time of day, confirming the relationship of temperature and relative humidity—Figure 7–11(b). Beyond the expected similarity in overall daily pattern, January readings are higher than July readings because air temperatures are lower overall in winter. Similar relative humidity records at most weather stations demonstrate the same relationship among season, temperature, and relative humidity.

Expressions of Relative Humidity

Three ways are used to express humidity and relative humidity, each with its own utility and application: vapor pressure, specific humidity, and absolute humidity.

Vapor Pressure. One way of describing humidity is related to air pressure. As free water molecules evaporate from a surface into the atmosphere, they become water vapor, one of the gases in air. That portion of total air pressure that is made up of water vapor molecules is termed **vapor pressure** and is expressed in millibars (mb). Water vapor molecules continue to evaporate from a moist surface, slowly diffusing into the air, until the increasing vapor pressure in the air causes some molecules to return to the surface. Saturation is reached when the movement of water molecules between surface and air is in equilibrium. The maximum capacity of the air at a given temperature is termed the *saturation vapor pressure* and indicates the maximum pressure that water vapor molecules can exert. Any increase or decrease in temperature causes the saturation vapor pressure to change. Figure 7–12 graphs the saturation vapor pressure at varying air temperatures.

As the graph shows, air at 20°C (68°F) has a saturation vapor pressure of 24 mb; that is, the air is

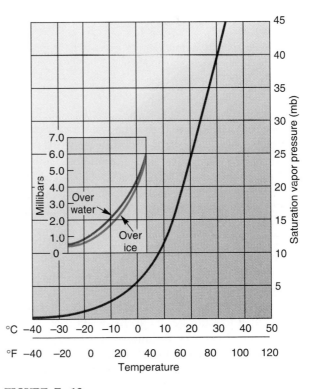

FIGURE 7–12
Saturation vapor pressure of air at various temperatures. Inset compares saturation vapor pressure over water and over ice at subfreezing temperatures.

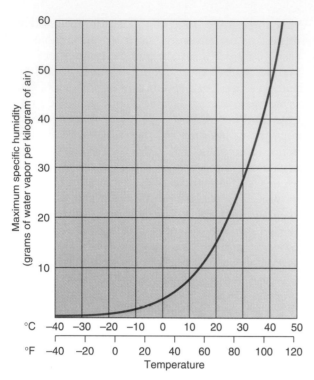

FIGURE 7–13
Maximum specific humidity for a mass of air at various temperatures.

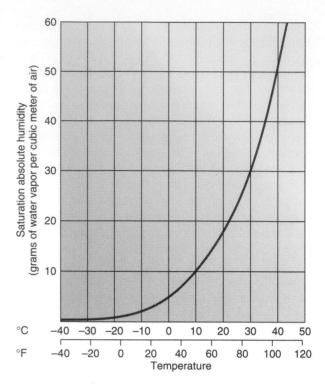

FIGURE 7–14
Saturation absolute humidity for a volume of air at various temperatures.

saturated if the water vapor portion of the air pressure is at 24 mb. Thus, if the water vapor content actually present is exerting a vapor pressure of only 12 mb in 20°C air, the relative humidity is only 50% (12 mb ÷ 24 mb = 0.50 × 100 = 50%). The graph illustrates that, for every temperature increase of 10C° (18F°), the vapor pressure capacity of air nearly doubles.

The inset in Figure 7–12 compares saturation vapor pressure over water and over ice at subfreezing temperatures. You can see that saturation vapor pressure is greater above a water surface than over an ice surface—that is, it takes more water vapor molecules to saturate air above water than it does above ice. This fact is important to condensation processes and rain droplet formation, both of which are discussed with clouds later in this chapter.

Specific Humidity. A humidity measure that remains constant as temperature and pressure change is useful. **Specific humidity** refers to the mass of water vapor (in grams) per mass of air (in kilograms)

at any specified temperature. Because it is measured in mass, specific humidity is not affected by changes in temperature or pressure such as occur when an air parcel rises to higher elevations, and is therefore a more valuable term in weather considerations. Specific humidity stays constant despite volume changes.

The maximum mass of water vapor that a kilogram of air can hold at any specified temperature is termed the *maximum specific humidity* and is plotted in Figure 7–13. The graph shows that a kilogram of air could hold a maximum specific humidity of 47 g of water vapor at 40°C (104°F), 15 g at 20°C (68°F), and 4 g at 0°C (32°F). Therefore, if a kilogram of air at 40°C has a specific humidity of 12 g, its relative humidity is 25.5% (12 g ÷ 47 g = 0.255 × 100 = 25.5%). Specific humidity is useful in describing the moisture content of large air masses that are interacting in a weather system.

Absolute Humidity. Absolute humidity is the actual amount of water vapor in the air, expressed as the weight of water vapor (grams) per unit volume (cubic meter) of air—that is, an expression of weight per volume. Figure 7–14 plots the *saturation absolute humidity* at various temperatures. Air at 30°C (86°F) has the capacity to hold 30 g of water vapor per cubic meter; at 20°C (68°F), 17 g; and at 10°C (50°F), 10 g.

A distinct limitation of using absolute humidity is that volume changes with pressure. Absolute humidity declines as air rises (expands), even though no water vapor has been removed; it increases as air falls (compresses), even though no new vapor has been added. Absolute humidity, therefore, has limited utility in weather analysis, but it is used for engineering and air-conditioning calculations.

Instruments for Measurement. Relative humidity is measured with various instruments. The **hair hygrometer** uses the principle that human hair changes as much as 4% in length between 0 and 100% relative humidity. The instrument connects a standardized bundle of human hair through a mechanism to a gauge and a graph to indicate relative humidity.

Another instrument used to measure relative humidity is a **sling psychrometer**. Figure 7–15 shows this device, which has two thermometers mounted side-by-side on a metal holder. One is called the dry-bulb thermometer; it simply records the ambient (surrounding) air temperature. The other thermometer is called the wet-bulb thermometer; it is set lower in the holder and has a cloth wick over the bulb, which is moistened with distilled water. The psychrometer is then spun by its handle. (Other versions mechanically spin the thermometers or place them where a fan forces air over them.)

The rate at which water evaporates from the wick depends on the relative saturation of the surrounding air. If the air is dry, water evaporates quickly, absorbing the latent heat of evaporation from the wet-bulb thermometer, causing the temperature to drop. In an area of high humidity, much less water evaporates from the wick. After spinning or ventilating the psychrometer for a minute or two, the temperature on each bulb is noted and compared on a relative humidity (psychrometric) chart, from which relative humidity can be determined.

Global Distribution of Relative Humidity

Relative humidity varies across Earth's surface according to temperature and moisture variability. Figure 7–16(a) plots the approximate mean distribution of relative humidity and Figure 7–16(b) plots the mean distribution of vapor pressure by latitude for January and July. The graphs illustrate the relative dryness of the subtropical regions and the relative wetness of equatorial areas and higher latitudes.

In terms of vapor pressure, the water vapor actually in the air above subtropical deserts at 30° latitude is about double the amount at higher latitudes such as 50°, yet the relative humidity in the subtropical regions is lower because the warmer air has a greater capacity to hold water vapor. Thus, the relatively dry air over the cloudless Sahara (warmer air) actually contains more water vapor than the relatively moist air in the midlatitudes (cooler air).

FIGURE 7–15
Sling psychrometer used for measurement of relative humidity. Photo by author.

(a)

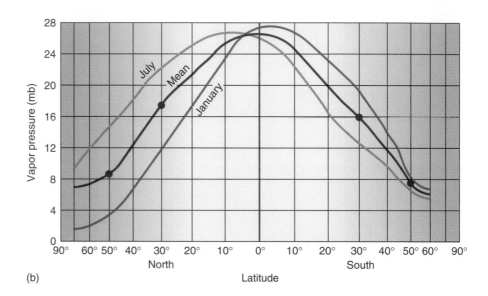

(b)

CLOUDS AND FOG

Clouds are beautiful indicators of atmospheric moisture and weather conditions. Although cloud types are too numerous to describe all of them here, the classification schemes for cloud types are quite simple.

Cloud Formation Processes

Clouds are not initially composed of raindrops. Instead they are made up of a multitude of **moisture droplets,** each individually invisible to the human eye without magnification. An average raindrop, at 2000 μm diameter (0.2 cm or 0.078 in.), is made up of a million or more moisture droplets, each approximately 20 μm diameter (0.002 cm or 0.0008 in.). After more than a century of debate, science has determined how these droplets form during condensation and what processes cause them to coalesce into raindrops.

Under unstable conditions (discussed at the beginning of the next chapter), a parcel of air may rise to where it becomes saturated—i.e., the air cools to the dew-point temperature and 100% relative humidity (although under certain conditions condensation may occur at slightly less or more than 100% relative humidity). Further cooling (lifting) of the air parcel produces active condensation of water vapor to water. This condensation requires microscopic particles called **condensation nuclei,** which always are present in the atmosphere.

In continental air masses, which average 10 billion nuclei per cubic meter, condensation nuclei are derived from ordinary dust, volcanic and forest-fire soot and ash, and particles from fuel combustion.

Given the air composition over cities, much higher concentrations of such nuclei are available. In maritime air masses, which average 1 billion nuclei per cubic meter, the nuclei are formed principally from sea salts derived from ocean sprays. Salts are particularly attracted to moisture and thus are called **hygroscopic nuclei**. The atmosphere never lacks such nuclei.

Raindrop Formation. Given the preconditions of saturated air, availability of condensation nuclei, and the presence of cooling mechanisms in the atmosphere, active condensation occurs. Two principal processes account for the majority of the world's raindrops. The **collision-coalescence process** predominates in clouds that form at above-freezing temperatures, principally in the warm clouds of the tropics. Initially, simple condensation takes place on small nuclei, some of which are larger and produce larger water droplets. As those larger droplets fall through a cloud, they combine with smaller droplets, gradually coalescing in size until their weight is beyond the ability of the cloud circulation to hold them aloft. Towering tropical clouds keep a growing raindrop aloft longer, thus producing more collisions and more drops of greater diameter—Figure 7–17(a).

The existence of ice crystals and supercooled water droplets also is a significant mechanism for raindrop production. Water in bulk freezes at 0°C (32°F), but fine, pure (distilled) water droplets don't freeze until the temperature drops to −40°C (−40°F). Consequently, **supercooled droplets** may be found in clouds at temperatures from −15° to −35°C (+5° to −31°F).

Figure 7–12 showed that the saturation vapor pressure near an ice surface is lower than that near a water surface; therefore supercooled water droplets will evaporate more rapidly near ice crystals, which then absorb the vapor. In this **ice-crystal process**, first proposed in 1928 by Swedish scientist Tor Bergeron, the ice crystals feed on the supercooled cloud droplets, grow in size, and eventually fall as snow or rain—Figure 7–17(b). Precipitation in middle and high latitudes begins as ice and snow high in the clouds, then melts and gathers moisture as it falls through the warmer portions of the cloud.

FIGURE 7–17
Principal processes for raindrop formation: (a) the collision-coalescence process; (b) the ice-crystal process. After Frederick K. Lutgens and Edward J. Tarbuck, *The Atmosphere: An Introduction to Meteorology*, 3d. ed., copyright © 1986, p. 127. Reprinted by permission of Prentice Hall, Inc., Englewood Cliffs, NJ.

TABLE 7–3
Cloud classes and types.

Class	Altitude/Composition at midlatitudes	Type	Symbol	Description
Low clouds (C_L)	Up to 2000 m (6500 ft). Water	Stratus (St)		Uniform, featureless, gray, like high fog
		Stratocumulus (Sc)		Soft gray globular masses in lines, groups, or waves, heavy rolls, irregular overcast patterns
		Nimbostratus (Ns)		Gray, dark, low, with drizzling rain
Middle clouds (C_M)	2000–6000 m (6500–20,000 ft). Ice and water	Altostratus (As)		Thin to thick, no halos, Sun's outline just visible, gray day
		Altocumulus (Ac)		Patches of cotton balls, dappled, arranged in lines or groups, rippling waves, the lenticular clouds associated with mountains
High Clouds (C_H)	6000–13,000 m (20,000–43,000 ft). Ice	Cirrus (Ci)		Mare's-tails, wispy, feathery, hairlike, delicate fibers, streaks, or plumes
		Cirrostratus (Cs)		Veil of fused sheets of ice crystals, milky, with Sun and Moon halos
		Cirrocumulus (Cc)		Dappled, "mackerel sky," small white flakes, tufts, in lines or groups, sometimes in ripples
Vertically developed clouds	Near surface to 13,000 m (43,000 ft). Water below, ice above	Cumulus (Cu)		Sharply outlined, puffy, billowy, flat-based, swelling tops, fair weather
		Cumulonimbus (Cb)		Dense, heavy, massive, dark thunderstorms, hard showers, explosive top, great vertical development, towering, cirrus-topped plume blown into anvil-shaped head

FIGURE 7–18 (right)
Principal cloud types, classified by form and altitude. Reprinted by permission of Merrill, an imprint of Macmillan Publishing Company, from *Earth Science,* 5th ed., by Edward J. Tarbuck and Frederick K. Lutgens, p. 370–71. Copyright © 1988 by Merrill Publishing, after Ward's Natural Science Establishment, Inc., Rochester, NY.

Cloud Types and Identification

A cloud is simply an aggregation, or grouping, of moisture droplets and ice crystals that are suspended in air and are great enough in volume and density to be visible to the human eye. In 1803 an English biologist named Luke Howard established a classification system for clouds and coined Latin names for them that we still use today.

Clouds usually are classified by altitude and by shape. They come in three basic forms—flat, puffy, and wispy—which are found in four primary classes and ten basic types. Clouds that are developed horizontally—flat and layered—are called **stratiform** clouds. Those that are developed vertically—puffy and globular—are termed **cumuliform**. Wispy clouds usually are quite high, composed of ice crystals, and are labeled **cirroform**. These three basic forms occur in four altitudinal classes: low, middle, high, and those that are vertically developed across all altitudes. Table 7–3 presents the basic cloud classes and types. Figure 7–18 illustrates the general appearance of each type, and Figure 7–19 includes representative photographs.

Cloud Classes. Low clouds, ranging from the surface up to 2000 m (6500 ft) in the middle latitudes, are simply called **stratus** or **cumulus**. Stratus clouds appear dull, gray, and featureless. When they yield precipitation, they are called **nimbostratus** (*nimbo*- denotes precipitation), and their showers fall as drizzling rain. Cumulus clouds appear bright and puffy, like cotton balls. When they do not cover the sky, they float by in infinitely varied shapes.

Sometimes near the end of the day, lumpy, grayish, low-level clouds called **stratocumulus** may fill the sky in patches. Near sunset, these spreading puffy stratiform remnants may catch and filter the Sun's rays, sometimes indicating clearing weather.

Stratus and cumulus middle-level clouds are denoted by the prefix *alto*-. They are made of water droplets and, when cold enough, can be mixed with ice crystals. Altocumulus clouds, in particular, represent a broad category that occurs in many different styles.

Clouds occurring above 6000 m (20,000 ft) are composed principally of ice crystals in thin concentrations. These wispy filaments, usually white except when colored by sunrise or sunset, are termed **cirrus**

FIGURE 7–19
Principal cloud types: (a) stratus, (b) nimbostratus, (c) altostratus, (d) altocumulus, (e) cirrus, (f) cirrostratus, (g) cumulus, (h) cumulonimbus. Photos by author.

FIGURE 7–20
Structure and form of a cumulonimbus cloud.

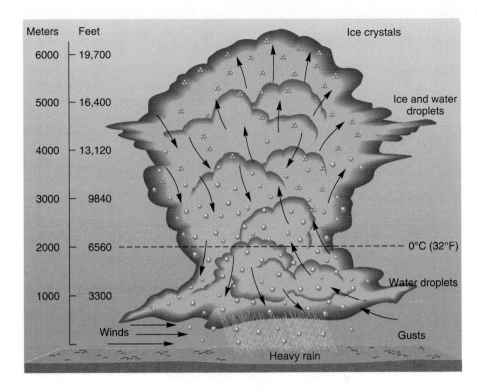

clouds, sometimes dubbed marc's-tails. Cirrus clouds look as if an artist has taken a brush and placed delicate feathery strokes high in the sky. Often, cirrus clouds are associated with an oncoming storm, especially if they thicken and lower in elevation. The prefix *cirro-* (cirrostratus, cirrocumulus) is used for other high clouds.

A cumulus cloud can develop into a towering giant called **cumulonimbus** (again, *-nimbus* denotes precipitation). Such clouds generally are referred to as thunderheads because of their shape and their associated lightning and thunder (Figure 7–20). When the rising cloud column enters a region of temperatures well below freezing, ice crystals form and consume available moisture droplets. It is possible to witness this transformation from a water-droplet cloud to an ice-crystal cloud, a process called **glaciation**, because the sharply delineated form of the cloud top softens in appearance. High-altitude winds may then shear the top of the cloud into the characteristic anvil shape of the mature thunderhead.

Fog

A cloud in contact with the ground is commonly referred to as **fog**. The presence of fog tells us that the air temperature and the dew-point temperature at ground level are nearly identical, producing saturated conditions. Generally, fog is capped by an inversion layer, with as much as 30C° (50F°) difference in air temperature between the ground under the fog and the clear, sunny skies above.

By international agreement, fog is officially described as a cloud layer on the ground, with visibility restricted to less than 1 km (3300 ft). Almost all fog is warm—that is, its moisture droplets are above freezing. Supercooled fog, which occurs when the moisture droplets are below freezing, is special because it can be dispersed through artificial seeding with ice crystals or other crystals that mimic ice, following the principles of the ice-crystal formation process described earlier.

Two principal forms of fog related to cooling are advection fog and radiation fog. As the name implies, **advection fog** forms when air in one place migrates to another place where saturated conditions exist. When warm, moist air overlays cooler ocean currents, lake surfaces, or snow masses, the layer of air directly above the surface is chilled to the dew point. Off all subtropical west coasts in the world, summer fog forms in this manner (Figure 7–21). Organisms have adapted to this special condition, as they adapt

FIGURE 7–21
San Francisco's Golden Gate Bridge shrouded by an advection fog. Photo by author.

to all Earthly phenomena. For example, sand beetles that have adapted to summer advection fog live in the Namib Desert in extreme southwestern Africa. In an otherwise dry environment, they harvest water from the fog by holding up their wings so condensation runs down to their mouths. In the Atacama Desert of Chile and Peru, residents stretch large nets on frames in the air to intercept the fog; moisture condenses on the netting and drips into barrels.

Another type of fog forms when cold air flows over the warm water of a lake, ocean surface, or even a swimming pool. An **evaporation fog**, or steam fog, may form as the water molecules evaporate from the water surface into the cold overlying air. When visible at sea, the term *sea smoke* is applied to this shipping hazard.

A type of advection fog involving the movement of air forms when moist air is forced to higher elevations along a hill or mountain. This upslope lifting leads to cooling by expansion as the air rises. The resulting **upslope fog** forms a stratus cloud at the level of saturation. Along the Appalachians and the eastern slopes of the Rockies, such fog is common in winter and spring. Another fog associated with topography is called **valley fog**. Because cool air is denser, it settles in low-lying areas, producing a fog in the chilled, saturated layer near the ground.

Radiation fog forms when radiative cooling of a surface chills the air layer directly above that surface to the dew-point temperature, creating saturated conditions and fog. This fog occurs especially on clear nights over moist ground; it does not occur over water, because water does not cool appreciably overnight. Slight movements of air deliver even more moisture to the cooled area for more fog formation of greater depth.

Fog poses a hazard to various human activities, including driving automobiles, flying aircraft, navigating ships, and for pedestrians or cyclists negotiating traffic. In addition, fog droplets can combine with atmospheric pollutants to form airborne acid pollution and even a "killer fog" in extreme cases, with stagnant air leading to unhealthy concentrations and possible fatalities. The prevalence of fog throughout the United States is shown in Figure 7–22. The spatial implications of fog occurrence on a local scale should represent a key component in any location analysis for a proposed airport or harbor facility.

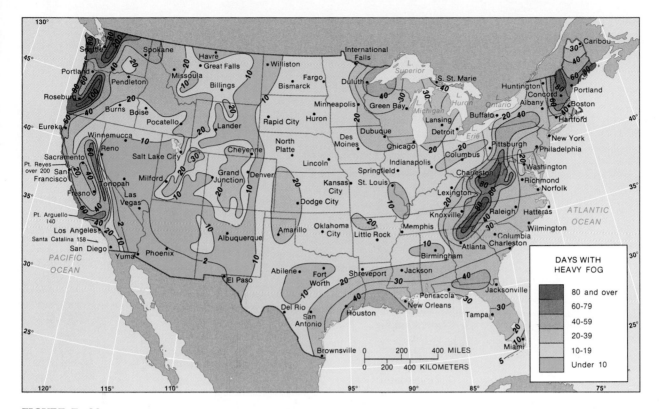

FIGURE 7–22
Mean annual number of days with heavy fog in the United States. The foggiest spot in the United States is the mouth of the Columbia River where it enters the Pacific Ocean at Cape Disappointment, Washington. Data courtesy of National Weather Service.

SUMMARY

The next time it rains where you live, pause and reflect on the journey each of those water molecules has made. Water molecules came from within Earth over billions of years, in a process called outgassing, and began endless cycling through the hydrologic system. Water is the most common compound on Earth and, in addition to its solvency, possesses great stability and unusual heat characteristics. The present salinity of the sea is a result of water's great dissolving ability. Water naturally exists in all three states—solid, liquid, and gas. The measurement of water vapor in the air is expressed as humidity and relative humidity. Saturated air and the presence of condensation nuclei create conditions for condensation and cloud formation, which occur in a variety of classes and types. Fog is a cloud that occurs at the surface and can prove hazardous to human activities.

Absolute humidity	Latent heat
Advection fog	Latent heat of condensation
Brackish	Latent heat of vaporization
Brine	Mixing zone
Cirroform	Moisture droplets
Cirrus	Nimbostratus
Collision-coalescence process	Outgassing
Condensation nuclei	Percolation
Cumuliform	Phase change
Cumulonimbus	Radiation fog
Cumulus	Relative humidity
Deep cold zone	Salinity
Dew-point temperature	Saturated
Efflux	Sling psychrometer
Eustasy	Specific humidity
Evaporation fog	Stem flow
Fog	Stratiform
Glaciation	Stratocumulus
Glacio-eustatic	Stratus
Hair hygrometer	Sublimation
Humidity	Supercooled droplets
Hydrologic cycle	Thermocline transition zone
Hygroscopic nuclei	Throughfall
Ice-crystal process	Upslope fog
Infiltration	Valley fog
Influx	Vapor pressure
Interception	

REVIEW QUESTIONS

1. Approximately where and when did Earth's waters originate?
2. If the quantity of water on Earth has been quite constant in volume for at least 2 billion years, how can sea level have fluctuated? Explain.
3. Describe the location of Earth's water, both oceanic and fresh. What is the largest repository of freshwater at this time? In what ways is this significant to modern society?
4. Evaluate the salinity of seawater: its composition, amount, and distribution.
5. Analyze the latitudinal distribution of salinity shown in Figure 7–3. Why is salinity lower along the equator and higher in the subtropics?
6. What are the three general zones relative to physical structure within the ocean? Characterize each in terms of temperature, salinity, oxygen, and carbon dioxide.
7. Describe the three states of matter as they apply to ice, water, and water vapor.
8. What happens to the physical structure of water as it cools below 4°C (39°F)? What are some of the visible indications of these physical changes?

9. What is latent heat? How is it involved in the phase changes of water? What amounts of energy are involved?
10. Explain the simplified model of the complex flows of water on Earth—the hydrologic cycle.
11. What are the possible routes that a raindrop may take on its way to and into the soil surface?
12. Where does the bulk of evaporation and precipitation on Earth take place?
13. Define relative humidity. What does the concept represent? What is meant by saturation and dew-point temperature?
14. Using several different measures of humidity in the air, derive relative humidity values (vapor pressure/saturation vapor pressure; specific humidity/maximum specific humidity).
15. How do the two instruments described in this chapter measure relative humidity?
16. Evaluate the latitudinal distribution of relative humidity as shown in Figure 7–16.
17. How does this compare with the actual water vapor content of air?
18. Specifically, what is a cloud? Describe the droplets that form a cloud.
19. Explain the condensation process: what are the necessary requirements? What two principal processes are discussed in this chapter?
20. What are the basic forms of clouds? Using Table 7–3, describe how the basic cloud forms vary with altitude.
21. What type of cloud is fog? List and define the principal types of fog.
22. Why might you describe Earth as the water planet? Explain.

SUGGESTED READINGS

Allen, Oliver E., and the editors of Time-Life Books. *Atmosphere*. Planet Earth Series. Alexandria, VA: Time-Life Books, 1983.

Blair, Thomas A., and Robert C. Fite. *Weather Elements—A Text in Elementary Meteorology*. 5th ed. Englewood Cliffs, NJ: Prentice Hall, 1965.

Buswell, Arthur M., and Worth H. Rodebush. "Water." *Scientific American* 194 (April 1956): 76–80.

Knight, Charles, and Nancy Knight. "Snow Crystals." *Scientific American* 228 (January 1973): 100–107.

Leopold, Luna B. *Water—A Primer*. San Francisco: W. H. Freeman, 1974.

Leopold, Luna B., Kenneth S. Davis, and the editors of Time-Life Books. *Water*. Life Science Library. New York: Time-Life Books, 1966.

Miller, Albert, Jack C. Thompson, Richard E. Peterson, and Donald R. Haragan. *Elements of Meteorology*. Columbus: Macmillan Publishing Co., 1983.

Neiburger, Morris, James G. Edinger, and William D. Bonner. *Understanding Our Atmospheric Environment*. San Francisco: W. H. Freeman, 1973.

Pringle, Laurence, and the editors of Time-Life Books. *Rivers and Lakes*. Planet Earth Series. Alexandria, VA: Time-Life Books, 1985.

Runnels, L. K. "Ice." *Scientific American* 215 (December 1966): 118–24.

Scorer, Richard and Arjen Verkaik. *Spacious Skies*. London: David & Charles, 1989.

Scorer, Richard and Harry Wexler. *Cloud Studies in Colour*. London: Pergamon Press, 1967.

Photo by author.

8
Weather

Every day a vast drama plays out on Earth's stage, involving the stability or instability of interacting air mass actors, and featuring powerful dialogue and beautiful special effects in the lower atmosphere. Large air masses come into conflict, moving and shifting to dominate different regions in response to the controls of insolation, latitude, patterns of atmospheric temperature and pressure, the arrangement of land, water, and mountains, and migrating cyclonic systems that cross the middle latitudes. Tropical cyclones travel northward and southward from equatorial latitudes, hitting susceptible coastal lowlands as hurricanes and typhoons.

Temperature, air pressure, relative humidity, wind speed and direction, daylength, and Sun angle are important measurable elements that contribute to the weather. We tune to a local station for the day's weather report from the National Weather Service (in the United States) or Atmospheric Environment Service (in Canada) to see the current satellite images and to hear tomorrow's forecast.

We begin our study of weather with a discussion of atmospheric stability and instability. We will follow huge air masses across North America, observe powerful lifting mechanisms in the atmosphere, examine cyclonic systems, and conclude with a portrait of the violent and dramatic weather that occurs in the atmosphere.

ATMOSPHERIC STABILITY

Stability refers to the tendency of a parcel of air to either remain as it is or change its initial position by

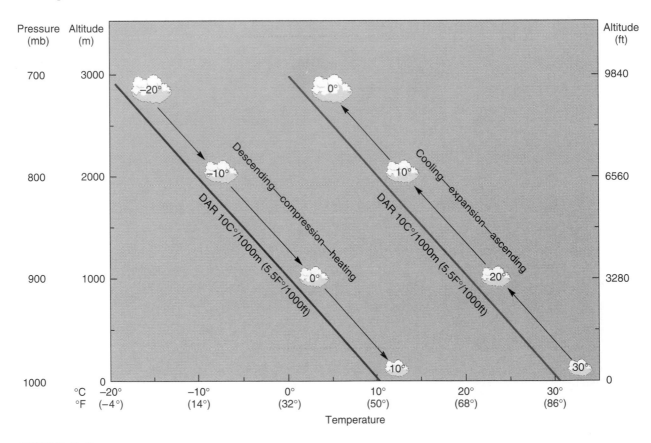

FIGURE 8–1
Vertically moving air parcels that are not saturated expand when they rise and are compressed when they descend. An ascending parcel is cooled by expansion; a descending parcel is heated by compression. Temperature change in both expansion and compression occur at the dry adiabatic rate (DAR).

lifting or falling. An air parcel is termed *stable* when it resists displacement upward or, if disturbed, it tends to return to its starting place. On the other hand, an air parcel is considered *unstable* when it continues to rise until it reaches an altitude where the surrounding air has a density similar to its own. Determining the degree of stability or instability involves measuring simple temperature relationships between the air parcel and the surrounding air. Such temperature measurements are made daily with balloon soundings (instrument packages called radiosondes) at thousands of weather stations.

Adiabatic Processes

The normal lapse rate, as introduced in Chapter 3, is the average decrease in temperature with increasing altitude, a value of 6.4C° per 1000 m (3.5F° per 1000 ft). This rate of temperature change is for still, calm air, but it can differ greatly under varying weather conditions, and so the actual lapse rate at a particular place and time is labeled the *environmental lapse rate*. Within this environment, an ascending (lifting) parcel of air tends to cool by expansion, responding to the reduced pressure at higher altitudes. Descending (subsiding) air tends to heat by compression.

Both ascending and descending temperature changes are assumed to occur without any heat exchange between the surrounding environment and the moving parcel of air. The warming and cooling rates for a parcel of expanding or compressing air are termed **adiabatic** and are measured with one of two dynamic rates. (*Diabatic* means to occur *with* an exchange of heat; *adiabatic* means to occur *without* a loss or gain of heat.)

Dry Adiabatic Rate (DAR). The dry adiabatic rate is the rate at which "dry" air cools by expansion (if ascending) or heats by compression (if descending). "Dry" air is less than saturated, with a relative humidity less than 100%. The DAR is 10C° per 1000 m (5.5F° per 1000 ft), as illustrated in Figure 8–1. For example, consider an unsaturated parcel of air at the surface, whose temperature measures 27°C (80°F). It is lifted, expands, and cools at the DAR as it rises from the ground to 2500 m (approximately 8000 ft). What happens to the temperature of the parcel? Its temperature at that altitude is 2°C (36°F), cooling by expansion as it is lifted.

$$10C° \text{ per } 1000 \text{ m} \times 2500 \text{ m} = 25C° \text{ of total cooling}$$
$$(5.5F° \text{ per } 1000 \text{ ft} \times 8000 \text{ ft} = 44F°)$$

Subtracting the 25C° of cooling from the starting temperature of 27°C gives the temperature at 2500 m of 2°C.

Moist Adiabatic Rate (MAR). The moist adiabatic rate is the average rate at which air that is moist (saturated) cools by expansion (if ascending) or heats by compression (if descending). The *average* MAR is 6C° per 1000 m (3.3F° per 1000 ft), or roughly 4C° (2F°) less than the DAR. However, the MAR varies with moisture content and temperature, and can range from 4C° to 10C° per 1000 m (2F° to 6F° per 1000 ft). The reason is that, in a saturated air parcel, latent heat of condensation is liberated as sensible heat, which reduces the adiabatic rate of cooling. The release of latent heat may vary, which affects the MAR. The MAR is much lower than the DAR in warm air, whereas the two rates are more similar in cold air.

Stable and Unstable Atmospheric Conditions

The relationship among the dry adiabatic rate (DAR), moist adiabatic rate (MAR), and the environmental (actual) lapse rate is complex and determines the stability of the atmosphere over an area. You can see the range of possible relationships in Figure 8–2.

Let's apply this concept to a specific situation. Figure 8–3 illustrates three temperature relationships in the atmosphere that lead to three different conditions: unstable, conditionally unstable, and stable. For the sake of illustration, all three examples begin with a parcel of air at the surface at 25°C (77°F). In each of the three examples, examine the temperature relationships between the parcel of air and the surrounding environment, assuming that a lifting mechanism is present.

Under the unstable conditions in Figure 8–3(a), the air parcel continues to rise through the atmosphere because it is warmer than the surrounding environment, which is cooler by an environmental lapse rate of 12C° per 1000 m (6.6F° per 1000 ft). By 1000 m (3280 ft), the parcel has cooled by expansion to 15°C (59°F), but the surrounding air is at 13°C (55.4°F). So, the temperature in the parcel is 2C° (3.6F°) warmer.

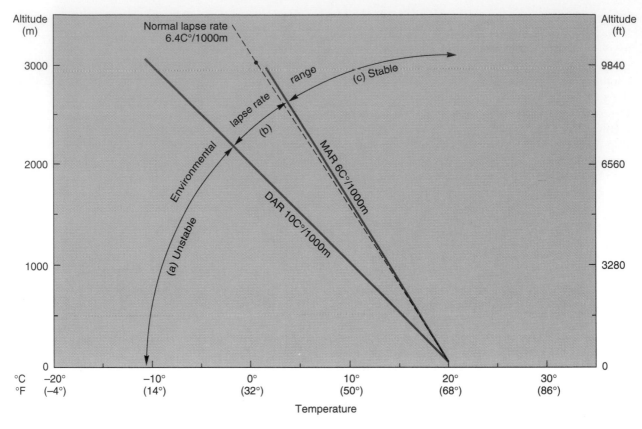

FIGURE 8–2

The relationship between dry and moist adiabatic rates and environmental lapse rates produces three conditions: (a) unstable atmosphere—environmental lapse rate exceeds the DAR; (b) conditionally unstable atmosphere—environmental lapse rate is between the DAR and MAR; (c) stable atmosphere—environmental lapse rate is less than the DAR and MAR.

On the other hand, the stable conditions in Figure 8–3(c) are caused by an environmental lapse rate of 5C° per 1000 m (3F° per 1000 ft), which is less than both the DAR and the MAR, thus forcing the parcel of air to settle back to its original position because it is cooler (denser) than the surrounding environment.

In example 8–3(b), the environmental lapse rate is between the DAR and the MAR at 7C° per 1000 m. Under these conditions, a parcel resists upward movement if it is less than saturated. However, if it is saturated and cools at the MAR, it eventually becomes unstable and continues to rise.

With these stability relationships in mind, let's examine the specific actors in the weather drama—

the air masses—each with their own specific humidity, temperature, and stability characteristics.

AIR MASSES

Earth's surface imparts its varying characteristics to the air it touches. Such a distinctive body of air is called an **air mass**, and it initially reflects the characteristics of its source region. The longer an air mass remains stationary over a region, the more definite its physical attributes become. Within each air mass there is a homogeneity of temperature and humidity that sometimes extends through the lower half of the troposphere.

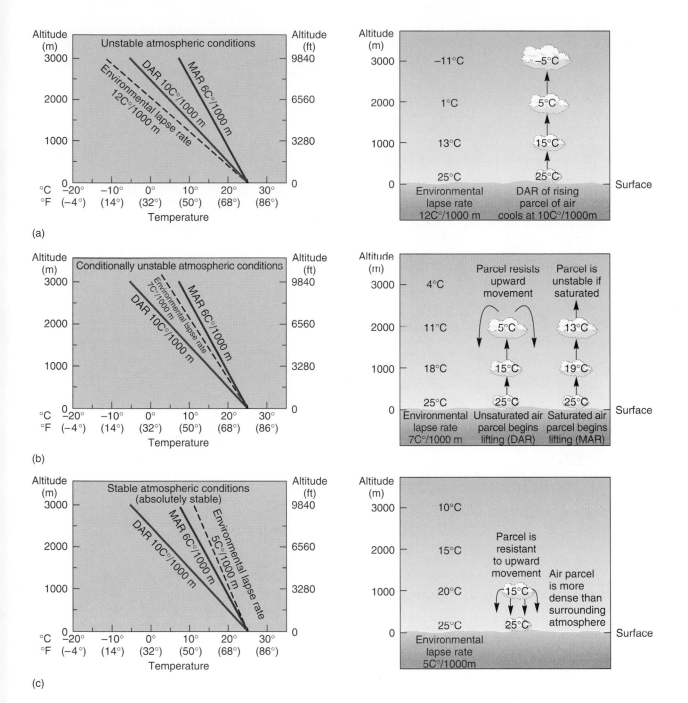

FIGURE 8–3

Specific examples of (a) unstable, (b) conditionally unstable, and (c) stable conditions in the lower atmosphere.

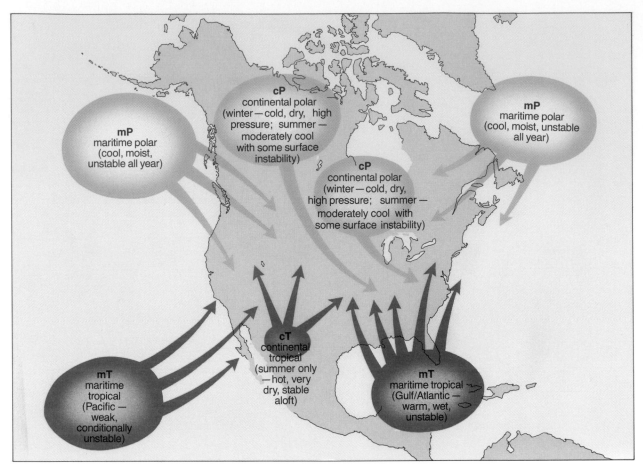

FIGURE 8–4

Air masses that influence North America and their characteristics. Adapted by permission of Merrill, an imprint of Macmillan Publishing Company, from *Earth Science,* 5th ed., by Edward J. Tarbuck and Frederick K. Lutgens, p. 403. Copyright © 1988 by Merrill Publishing.

Air Masses Affecting North America

Air masses generally are classified according to the moisture and temperature characteristics of their source regions:

1. Moisture—designated by the letters **m** for maritime and **c** for continental.
2. Temperature (latitude)—designated by the letters **A** (arctic), **P** (polar), **T** (tropical), **E** (equatorial), and **AA** (antarctic).

The principal air masses that affect North America are mapped in Figure 8–4 and detailed in Table 8–1.

Continental polar (**cP**) air masses form only in the Northern Hemisphere and are most developed in winter. The Southern Hemisphere lacks the necessary continental masses (continentality) at high latitudes to produce the required source-region characteristics.

Marine polar (**mP**) air masses in the Northern Hemisphere are northwest and northeast of the North American continent, and within them cool, moist, unstable conditions prevail throughout the year. The Aleutian and Icelandic subpolar low-pressure cells reside within these mP air masses, especially in their well-developed winter state.

North America also is influenced by two maritime tropical (**mT**) air masses—the **mT Gulf/Atlantic** and the **mT Pacific**. The humidity and precipitation experienced in the East and Midwest is created by

TABLE 8–1
Principal air masses.

Latitudinal Source Region	Moisture Source Region	
	c: Continental	m: Marine
A: Arctic, 60°–90° N	cA: Continental Arctic— *Winter:* very cold, very dry, stable (avg. SH = 0.1 g/kg) *Summer:* cold, dry	(not applicable)
P: Polar, 40°–60° N or S	cP: Continental polar— *Winter:* (N.H. only) cold, dry, stable, with high pressure (avg. SH = 1.4 g/kg) *Summer:* warm to cool, dry, moderately stable	mP: Maritime polar— Cool, moist, unstable all year (avg. SH = 4.4 g/kg)
T: Tropical, 15°–35° N or S	cT: Continental tropical— hot, very dry, stable aloft, unstable at surface, turbulent in summer (avg. SH = 10 g/kg	mT: Maritime tropical— mT Gulf *Winter:* warm, wet, unstable *Summer:* warm, wet, very unstable (avg. SH = 17 g/kg) mT Pacific Weak, conditionally unstable
E: Equatorial, 15° N–15° S	(not applicable)	mE: Maritime equatorial— unstable, very warm, very wet, winterless (avg. SH = 19 g/kg)
AA: Antarctic, 60°–90° S	cAA: Continental Antarctic— very cold, very dry, very stable	(not applicable)

SH = specific humidity, g/kg

the mT Gulf/Atlantic air mass, which is particularly unstable and active from late spring to early fall. In contrast, the mT Pacific is stable-to-conditionally unstable and generally much weaker. As a result, the western United States, influenced by this weaker Pacific air mass, receives lower average precipitation than the rest of the country. (Please refer to Figure 6–11 and the discussion of subtropical high-pressure cells off the coast of North America—the moist, unstable conditions on the western edge (east coast), and the drier, stable conditions on the eastern edge (west coast). These conditions, coupled respectively with warmer and cooler ocean currents, produce the characteristics of each source region.)

Air Mass Modification

As air masses migrate from their source regions, their temperature and moisture characteristics are modified. For example, an mT Gulf/Atlantic air mass may carry humidity to Chicago and on to Winnipeg, but gradually will become more stable with each day's passage northward. Similarly, temperatures below freezing occasionally reach into southern Texas and Florida, influenced by an invading winter **cP** air mass from the north. However, that air mass will have warmed from the −50°C (−58°F) of its source region in central Canada. In winter, as a **cP** air mass moves southward over warmer land, it is modified, especially warming after it crosses the snowline. If it passes over the warmer Great Lakes, it will absorb heat and moisture and produce heavy lake-effect snowfall downwind as far east as New York, Pennsylvania, New England, and the Maritime Provinces of Canada.

On the opposite side of the globe, winter monsoon winds bring cold, dry air from the interior of Siberia over the surrounding lands and oceans. The air mass is modified as it passes southward and eastward. Figure 8–5 shows a satellite image of such an air mass as it moves out over the Sea of Japan and the China Sea. Over warmer water the moisture content of the mass is increased, and its clouds form elongated bands. The result is a cool, moist **cP** air

FIGURE 8–5
Air mass modification over the Sea of Japan and the North Pacific, as evidenced by cloud patterns. GMS satellite image for 24 January 1988, courtesy of the Japan Weather Association.

mass that is humidified, warmed, and destabilized. It has thus become a **secondary air mass** to the initial **cA** or **cP** mass that left Siberia.

ATMOSPHERIC LIFTING MECHANISMS

If air masses are to cool adiabatically (by expansion), and if they are to reach the dew-point temperature and saturate, condense, form clouds, and perhaps precipitate, then they must be lifted. Three principal lifting mechanisms operate in the atmosphere: convectional lifting (stimulated by local surface heating), orographic lifting (when air is forced over a barrier such as a mountain range), and frontal lifting (along the leading edges of contrasting air masses). Descriptions of all three mechanisms follow.

Convectional Lifting

When an air mass passes from a maritime source region to a warmer continental region, heating from the warmer land causes lifting in the air mass. If conditions are unstable, initial lifting is sustained and clouds develop. Figure 8–6 illustrates convectional action stimulated by local heating, with unstable conditions present in the atmosphere. The rising parcel of air continues its ascent because it is warmer than the surrounding environment.

As an example of local heating and convectional lifting, the land mass of Florida was warmer than the surrounding Gulf of Mexico and Atlantic Ocean on the day shown in Figure 8–7. Because the land gradually heats throughout the day, convectional showers tend to form in the afternoon and early evening. The land is covered by clouds formed by this process. Towering cumulonimbus clouds are an expected summertime feature in the regions of North America affected by mT Gulf/Atlantic air masses, and to a lesser extent by the weaker mT Pacific air mass. Convectional precipitation also dominates along the intertropical convergence zone (ITCZ), over tropical islands, and anywhere that moist, unstable air is heated from below, or where dynamic convergence piles up air.

Orographic Lifting

The physical presence of a mountain acts as a topographic barrier to migrating air masses. **Orographic lifting** occurs when air is lifted upslope as it is pushed against a mountain and cools adiabatically. Stable air that is forced upward in this manner may produce stratiform clouds, whereas unstable or conditionally unstable air usually forms a line of cumulus and cumulonimbus clouds. An orographic barrier enhances convectional activity and causes additional lifting during the passage of weather fronts and cyclonic systems, thereby extracting more moisture from passing air masses. Figures 8–8 and 8–9 illustrate the operation of orographic lifting.

The wetter intercepting slope is termed the **windward slope**, as opposed to the drier far-side slope, known as the **leeward slope**. Moisture is condensed from the lifting air mass on the windward side of the mountain; on the leeward side the descending air mass is heated by compression, causing evaporation of any remaining water in the air. Thus, air can begin its ascent up a mountain warm and moist but finish its descent hot and dry. In North America, **chinook winds** (called föhn or foehn winds in Europe) are the warm, downslope air flows characteristic of the leeward side of mountains. Such winds can bring a 20C° (36F°) jump in temperature and greatly reduced relative humidity.

The term **rain shadow** is applied to dry regions leeward of mountains. East of the Cascade Range, Sierra Nevada, and Rocky Mountains, such rain-shadow patterns predominate. In fact, the precipitation pattern of windward and leeward slopes persists worldwide, as confirmed by the precipitation maps for North America (Figure 9–2) and the world (Figure 10–2).

Orographic Precipitation World Records. Orographic precipitation is limited in areal extent because it can occur only where a topographic barrier exists, thus making it the least dominant

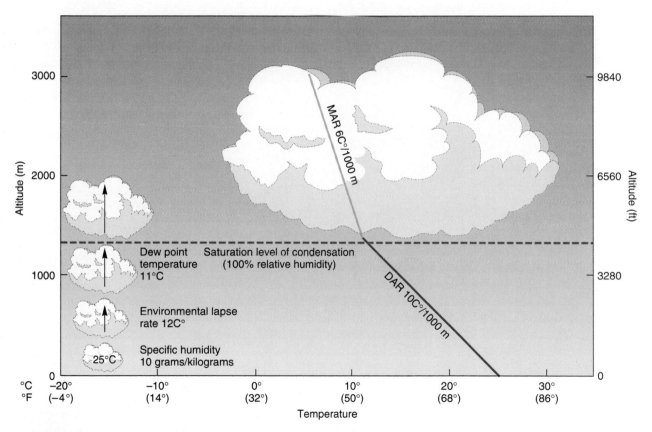

FIGURE 8–6
Local heating and convection under unstable atmospheric conditions. The air parcel has a specific humidity of 10 g/kg and thus an initial dew-point temperature of 11°C (51.8°F). The environmental lapse rate is 12C° per 1000 m (7F° per 1000 ft) so that the parcel is warmer and continues to rise. At 1400 m (4590 ft) the parcel reaches the dew point and becomes saturated.

lifting mechanism in terms of total worldwide precipitation. Nonetheless, because mountains are fixed features on the landscape, they are the most consistent precipitation-inducing mechanism. Thus, the greatest average annual precipitation and maximum annual precipitation on Earth occur on the windward slopes of mountains that intercept tropical tradewinds.

The world's greatest average annual precipitation occurs in the United States on the windward slopes of Mount Waialeale, on the island of Kauai, Hawaii, which rises 1569 m (5147 ft) above sea level. It averages 1234 cm (486 in. or 40.5 ft) a year (records for 1941–present). In contrast, the rain-shadow side of Kauai receives only 50 cm (20 in.) of rain annually. If no high islands existed at this location, this portion of the Pacific Ocean would receive only an average 63.5 cm (25 in.) of precipitation a year.

Another place receiving world-record precipitation is Cherrapunji, India, 1313 m (4309 ft) above sea level at 25° N latitude, in the Assam Hills south of the Himalayas. It is located just north of Bangladesh, where tropical cyclones, torrential rains, and flooding in 1970 and 1991 killed hundreds of thousands.

FIGURE 8–7
Convectional activity over Florida. Cumulus clouds outline the land, with several cells developing into cumulonimbus thunderheads. Project Gemini photograph courtesy of National Aeronautics and Space Administration.

FIGURE 8–8
Orographic barrier and precipitation patterns given unstable conditions. Warm moist air is forced upward by a mountain range, producing adiabatic cooling and eventual saturation. On the leeward slopes, compressional heating leads to hot, relatively dry air at the foot of the mountain.

Because of the summer monsoons that pour in from the Indian Ocean and the Bay of Bengal, Cherrapunji has received 930 cm (366 in. or 30.5 ft) of rainfall in one month and a total of 2647 cm (1042 in. or 86.8 ft) in one year. Not surprisingly, Cherrapunji is the all-time precipitation record holder for a single year and for every other time interval from 15 days to 2 years. The average annual precipitation there is 1143 cm (450 in.), placing it second only to Mount Waialeale.

Frontal Lifting

The leading edge of an advancing air mass is called its **front**. That term was applied by Vilhelm Bjerknes and a team of meteorologists working in Norway during World War I, because it seemed to him that migrating air-mass "armies" were doing battle along their fronts. A front is a place of atmospheric discontinuity, of considerable difference in temperature, pressure, humidity, wind direction and speed, and cloud development. The leading edge of a cold air mass is a **cold front**, whereas the leading edge of a warm air mass is a **warm front**.

Cold Front. On weather maps, such as those illustrated in Figure 8–15, a cold front is identified by a line marked with triangular spikes pointing in the direction of frontal movement along an advancing cP air mass in winter. The steep face of the cold air mass suggests its ground-hugging nature, caused by its density and unified physical character (Figure 8–10).

Warmer, moist air in advance of the cold front is lifted upward abruptly and is subjected to the same adiabatic rates and concepts of stability/instability that pertain to all lifting air parcels. A day or two ahead of the cold front's passage, high cirrus clouds appear, telling observers that a lifting mechanism is on the way. The cold front's passage is marked by a wind shift, temperature drop, and lowering barometric pressure. Air pressure reaches a local low-pressure reading as the line of most intense lifting passes, usually just ahead of the front itself. Clouds

FIGURE 8–9
Orographic lifting along the Pacific Coast. Photo by author.

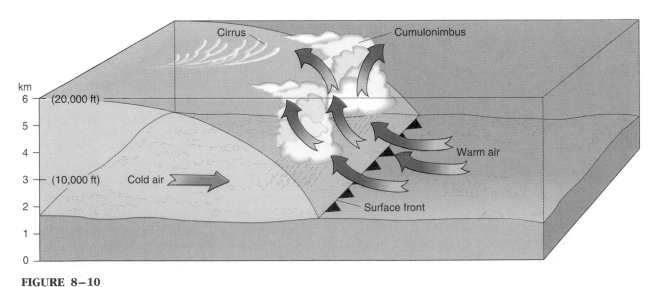

FIGURE 8–10
Cold front. Adapted by permission of Merrill, an imprint of Macmillan Publishing Company, from *Earth Science,* 5th ed., by Edward J. Tarbuck and Frederick K. Lutgens, p. 406. Copyright © 1988 by Merrill Publishing.

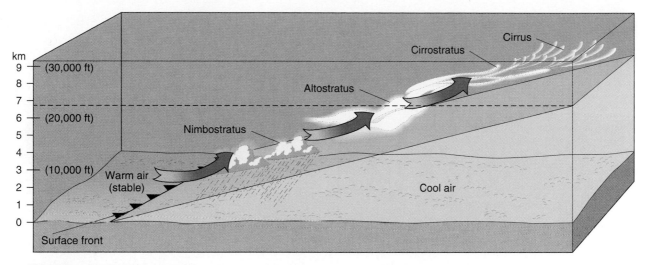

FIGURE 8–11

Warm front. Note the sequence of cloud development as the warm front approaches.
Adapted by permission of Merrill, an imprint of Macmillan Publishing Company,
from *Earth Science,* 5th ed., by Edward J. Tarbuck and Frederick K. Lutgens, p. 405.
Copyright © 1988 by Merrill Publishing.

may build up along the cold front into characteristic cumulonimbus types and may appear as an advancing wall of clouds. Precipitation usually is hard, containing large droplets, and can be accompanied by hail, lightning, and thunder. The aftermath usually brings northerly winds, lower temperatures, and broken sky cover.

The particular shape and size of the North American landmass and its latitudinal position represent conditions where cP and mT air masses are best developed and have the most direct access to each other. The resulting contrast can lead to dramatic weather, particularly in late spring, with sizable temperature differences from one side of a cold front to the other. A fast-advancing cold front can cause violent lifting and create a zone slightly ahead of the front called a **squall line**, where wind patterns are turbulent and wildly changing and precipitation is intense. Tornadoes also may develop along such a squall line.

Warm Front. A warm front is denoted on weather maps by a line marked with semicircles facing the direction of frontal movement. The leading edge of an advancing warm air mass is unable to displace cooler, passive air, which is more dense. Instead, the warm air tends to push the cooler, underlying air into a characteristic wedge shape, with the warmer air sliding up over the cooler air. Thus, in the cooler-air region, a temperature inversion is present, sometimes causing poor air drainage and stagnation.

Figure 8–11 illustrates a typical warm front in which mT air is gently lifted, leading to stratiform cloud development and characteristic nimbostratus clouds and drizzly precipitation. Ahead of the warm front are the high cirrus and cirrostratus clouds that announce the advancing frontal system. Closer to the front, the clouds lower and thicken to altostratus, and then to stratus within several hundred kilometers of the front.

MIDLATITUDE CYCLONIC SYSTEMS

The conflict between contrasting air masses along fronts leads to conditions appropriate for the development of a midlatitude **wave cyclone**, a migrating center of low pressure, with converging, ascending air, spiraling inward counterclockwise in the Northern Hemisphere (or converging clockwise in the Southern Hemisphere). This cyclonic motion is generated by the pressure gradient force and the Coriolis force, discussed in Chapter 6.

Wave cyclones form a dominant type of weather pattern in the middle and higher latitudes of both

(a) Cyclogenesis

(b) Open Stage

Open Stage Cross Section

Cold air

Warm air

Cool air

A   B

Cold front
(abrupt lifting)

Warm front
(gentle lifting)

(c) Occluded Stage

(d) Dissolving Stage

FIGURE 8–12

Life cycle of a midlatitude cyclone. Adapted by
permission of Merrill, an imprint of Macmillan
Publishing Company, from *Earth Science,* 5th ed., by
Edward J. Tarbuck and Frederick K. Lutgens, p. 406, 408.
Copyright © 1988 by Merrill Publishing.

hemispheres and act as a catalyst for air-mass conflict. A *midlatitude cyclone*, or extratropical cyclone, is born along the polar front, particularly in the region of the Icelandic and Aleutian subpolar low-pressure cells in the Northern Hemisphere. Strengthening and development of a wave cyclone is known as cyclogenesis. In addition to the polar front, certain other areas are associated with wave cyclone development and intensification: the eastern slope of the Rockies and other north-south mountain barriers, and the North American and Asian east coasts.

Figure 8–12 shows the birth, maturity, and death of a typical midlatitude cyclone. On the average, a midlatitude cyclone takes from 4 to 9 days to progress through these stages and cross the continent.

Cyclogenesis: The Birth of a Cyclone

Cyclogenesis is the atmospheric process in which these low-pressure systems originate and strengthen—Figure 8–12(a). Along the polar front, cold and warm air masses converge and conflict. The polar front forms a discontinuity of temperature, moisture, and winds that establishes potentially unstable conditions. However, for a wave cyclone to form along the polar front, a point of *convergence* at the surface must be matched by a compensating area of *divergence* aloft. Even a slight disturbance along the polar front, perhaps a small change in the path of the jet stream, can initiate the converging, ascending motion of a surface low-pressure system (wave cyclone).

The growing circulation system then is vented into the streams of upper-level winds. To the east of the developing low-pressure center, warm air begins to move along an advancing front, while cold air advances to the west of the center—Figure 8–12(b). As the midlatitude cyclone matures, the counterclockwise flow in the Northern Hemisphere draws the cold air mass from the north and west and the warm air mass from the south.

These cyclonic storms—1600 km (1000 mi) wide—and their attendant air masses move across the continent along storm tracks, which shift latitu-

FIGURE 8–13
Cyclonic storm tracks for North America vary seasonally.

FIGURE 8–14
Balloon tracks follow prevailing surface and tropospheric air patterns. Data courtesy
of Geography Education Program, National Geographic Society Alliances Program.

dinally with the Sun, crossing North America farther
to the north in summer and farther to the south in
winter (Figure 8–13). As the tracks begin to shift
northward in the spring, cP and mT air masses are
brought into their clearest conflict. This is the time of
strongest frontal activity, with associated thunderstorms and tornadoes.

During Geography Awareness Week in November
1987, thousands of geography students throughout
the United States launched helium-filled balloons.
They generally followed the same upper-air winds
that direct storm systems across the country—in
other words, the balloon tracks shown in Figure
8–14 depicted the prevailing storm tracks for those
days.

Cyclonic circulation also frequently develops on
the lee side of mountain ranges, as along the Rockies
from Alberta to Colorado. As air moves downslope,
the vertical axis of the air column extends, shrinking

the system horizontally, intensifying wind speed. As
the air travels downslope, it is deflected in a cyclonic
flow, thus developing new cyclonic systems or
intensifying existing ones. By crossing the mountains, such systems gain access to the moisture- and
energy-rich mT air mass from the Gulf of Mexico.

Occluded Fronts and a Dissolving Cyclone

Because the cP air mass is more homogeneous in
temperature and pressure than the mT air mass, the
cold front leads a denser, more unified mass, and
therefore moves faster than the warm front. Cold
fronts can average 40 kmph (25 mph), whereas warm
fronts average 16–24 kmph (10–15 mph). Thus, a
cold front may overrun the cyclonic warm front,
producing an **occluded front** within the overall
cyclonic circulation—Figure 8–12(c). Precipitation

may be moderate to heavy initially and then taper as the warmer air wedge is lifted higher by the advancing cold air mass.

The final dissolving stage of the midlatitude cyclone occurs when its lifting mechanism is completely cut off from the warm air mass, which was its source of energy and moisture—Figure 8–12(d). Remnants of the cyclonic system then dissipate in the atmosphere.

Daily Weather Map and the Midlatitude Cyclone

The four-stage sequence in Figure 8–12 illustrates the evolution in an ideal model, but the actual pattern of cyclonic passage over North America demonstrates a great variety of shape, form, and duration. However, we can apply this model along with our understanding of warm and cold fronts to the actual midlatitude cyclone shown in the weather-map sequence for April 1–3, 1988 (Figure 8–15).

Preparing a weather report and forecast requires analysis of a daily weather map and satellite images that depict atmospheric conditions. The figure presents weather maps and satellite images for a typical three-day period and explains the standard weather station symbol. Weather data from these sources include the following:

- Barometric pressure
- Pressure tendency
- Surface air temperature
- Dew-point temperature
- Wind speed and direction
- Type and movement of clouds
- Current weather
- State of the sky
- Precipitation since last observation

On the map for April 2 in Figure 8–15(b), you can identify the temperatures reported by various stations, the patterns created, and the location of warm and cold fronts. The distribution of air pressure is defined by the use of isobars, the lines connecting points of equal pressure on the map. Although the low-pressure center identified is not intense, it is well defined over eastern Nebraska, with winds following in a counterclockwise, cyclonic path

around the low pressure. The maps and satellite images also illustrate the evolving pattern of precipitation over the three-day period as the warm and cold fronts sweep eastward. The mT air mass, advancing northward from the Gulf/Atlantic source region, provides the moisture for afternoon and evening thundershowers. Can you identify and plot the storm track for the three days shown in the sequence?

VIOLENT WEATHER

Weather provides a continuous reminder of nature's power. The flow of energy across the latitudes can at times set into motion violent and potentially destructive conditions. Thunderstorms, tornadoes, and tropical cyclones are such forms of violent weather.

Thunderstorms

Tremendous energy is liberated by the condensation of large quantities of water vapor. This process is accompanied by violent updrafts and downdrafts. As a result, giant cumulonimbus clouds can create dramatic weather moments—squall lines of heavy precipitation, lightning, thunder, hail, blustery winds, and tornadoes. Thunderstorms may develop within an air mass, along a front (particularly a cold front), or where mountain slopes cause orographic lifting.

Thousands of thunderstorms occur on Earth at any given moment. Equatorial regions and the ITCZ have many of them, as exemplified by the city of Kampala in Uganda, East Africa (north of Lake Victoria), which sits virtually on the equator and averages 242 days a year of thunderstorms—a record. Figure 8–16 shows the distribution of annual days with thunderstorms across the United States. You can see that in North America most thunderstorms occur in areas dominated by mT air masses.

Lightning and Thunder. Lightning refers to flashes of light caused by enormous electrical discharges—tens to hundreds of millions of volts—which briefly ignite the air to temperatures of 15,000° to 30,000°C (27,000° to 54,000°F). The violent expansion of this abruptly heated air sends shock waves through the atmosphere—the sonic bangs known as

FIGURE 8–15

Weather maps and *GOES-East* satellite images for April 1, 2, and 3, 1988. Courtesy of National Weather Service and National Oceanic and Atmospheric Administration, National Climatic Center, Satellite Data Services Division.

(a) Weather map for April 1, 1988

WEATHER STATION SYMBOL

COLD FRONT	WARM FRONT
STATIONARY FRONT	OCCLUDED FRONT

PRECIPITATION TYPE

Symbol	Type
,	DRIZZLE
•	RAIN
✳	SNOW
▽	SHOWERS
⊤	THUNDERSTORMS
≡	FOG
∞	DRY HAZE
◠	FREEZING RAIN
△	HAIL
△	SLEET

WIND SPEED

(kph/mph)

Speed	Symbol	Speed
CALM	◎	
5–13	\	3–8
14–23	\	9–14
24–32	\\	15–20
33–40	\\\	21–25
89–97	◣	55–60
116–124	◣\\\	72–77

CLOUD COVER

Symbol	Description	Symbol	Description
○	NO CLOUDS	◑	SIX-TENTHS
①	ONE-TENTH OR LESS	◕	SEVEN-TENTHS TO EIGHT-TENTHS
◔	TWO-TENTHS TO THREE-TENTHS	◕	NINE-TENTHS OR OVERCAST WITH OPENINGS
◓	FOUR-TENTHS	●	COMPLETELY OVERCAST
◑	FIVE-TENTHS	⊗	SKY OBSCURED

Polar Stereographic projection true at 60°N

Scale at different latitudes

0   400   800 km
0   250   500 mi

SURFACE WEATHER MAP AND STATION WEATHER AT 7:00 A.M., E.S.T.

(b) Weather map for April 2, 1988

(c) Weather map for April 3, 1988

(a) Satellite image for April 1, 1988

(b) Satellite image for April 2, 1988

(c) Satellite image for April 3, 1988

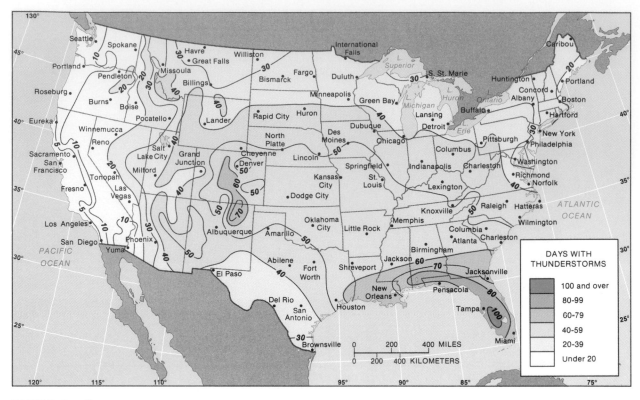

FIGURE 8–16

Mean annual number of days with thunderstorms. Data courtesy of National Weather Service.

thunder. The greater the distance a lightning stroke travels, the longer the thunder echoes. Lightning is created by a buildup of electrical energy between areas within a cumulonimbus cloud or between the cloud and the ground, with sufficient electrical potential to overcome the resistance of the atmosphere and leap from one surface to the other—a giant spark.

The beginning of a lightning discharge, called the *leader stroke*, goes from the cloud to the surface, following the route of least resistance, usually in a succession of forked steps. The path of charged ions remaining in the air behind this relatively slow-moving stroke then becomes the pathway for the main stroke, which surges in an enormous rush of energy along that path a fraction of a second later. This *return stroke,* traveling at a tenth the speed of light, produces most of the light seen as lightning. From 2 to 40 repetitions of this sequence occur in a second or two along the same path and produce *forked lightning*. A discharge of forked lightning

from one portion of a cumulonimbus cloud to another, usually masked by the cloud and appearing as a broad flash of light, is termed *sheet lightning*. Lightning poses a hazard, both to aircraft and to people and structures at the surface. Individuals should acknowledge that certain precautions are mandatory when the threat of a discharge is near. When such lightning is imminent, agencies such as the National Weather Service issue severe storm warnings and caution people to remain indoors.

Hail. Annual hail damage in the United States tops $750 million. **Hail** generally forms within a cumulonimbus cloud when a raindrop circulates repeatedly above and below the freezing level in the cloud, adding layers of ice until its weight can no longer be supported by the circulation in the cloud. Pea-sized and marble-sized hail are common, although hail the size of golf balls and baseballs also is reported at least once or twice a year. For larger stones to form, the frozen pellets must stay aloft for longer periods.

The largest authenticated hailstone in the world landed in Kansas in 1970 and measured 14 cm (5.6 in.) in diameter and weighed 758 grams (1.67 pounds)! The pattern of hail occurrences across the United States is similar to that of thunderstorms shown in Figure 8–16.

Tornadoes

The updrafts associated with a cumulonimbus cloud appear on satellite images as pulsing bubbles of clouds. Because winds in the troposphere blow stronger above Earth's surface than they do at the surface, a body of air pushes forward faster at altitude than at the surface, thus creating a rotation in the air along a horizontal axis that is parallel to the ground. When that rotating air encounters the strong updrafts associated with frontal activity, the axis of rotation is shifted to a vertical alignment, perpendicular to the ground—as described in one current model.

It is this spinning, cyclonic, rising column of mid-troposphere-level air that forms a **mesocyclone**. A mesocyclone can range up to 10 km (6 mi) in diameter and rotate over thousands of feet vertically within the parent cloud. As a mesocyclone extends vertically and contracts horizontally, wind speeds accelerate in an inward vortex (much as ice skaters accelerate while spinning by pulling their arms in closer to their bodies). A well-developed mesocyclone most certainly will produce heavy rain, large hail, blustery winds, and lightning; some mature mesocyclones will generate tornado activity.

As more moisture-laden air is drawn up into the circulation of a mesocyclone, more energy is liberated, and the rotation of air becomes more rapid. The narrower the mesocyclone, the faster the spin of converging parcels of air being sucked into the rotation. The swirl of the mesocyclone itself is visible, as are dark gray **funnel clouds** that pulse from the bottom side of the parent cloud. The terror of this stage of development is the lowering of such a funnel to Earth (Figure 8–17).

A **tornado** can range from a few meters to a few hundred meters in diameter and can last anywhere from a few moments to tens of minutes. Pressures inside a tornado funnel usually are about 10% less than those in the surrounding air. This disparity is similar to the pressure difference between sea level and an altitude of 1000 m (3280 ft). The in-rushing convergence created by such a horizontal pressure gradient causes severe wind speeds that can exceed 483 kmph (300 mph), with at least a third of all tornadoes exceeding 182 kmph (113 mph). When tornado circulation occurs over water, a **waterspout** forms, and the sea or lake surface water is drawn up into the funnel some 3–5 m (10–16 ft). The rest of the waterspout funnel is made visible by the rapid condensation of water vapor.

Of the 50 states, 49 have experienced tornadoes, as have all the Canadian provinces and territories. May is the peak month. A small number of tornadoes are reported in other countries each year, but North America receives the greatest share because its latitudinal position and shape permit conflicting and contrasting air masses access to each other. In the United States, 25,318 tornadoes were recorded in the 31 years between 1959 and 1990, with 723 causing 2577 deaths. In addition, these tornadoes resulted in almost 50,000 injuries and damages over $15 billion (Figure 8–18). Interestingly, 1990 set a new record for the most tornadoes in a single year. Only six states escaped tornado activity during the year. Of these, a record 14 had winds exceeding 333 kmph (207 mph).

It is not enough to account statistically for past tornadoes; the need is for accurate forecasting to avoid loss of life and promote safety measures in vulnerable areas. The National Severe Storms Forecast Center in Kansas City, Missouri, is the center for such activities. Increasing use of Doppler radar is enabling scientists to detect the specific direction and flow of moisture droplets in mesocyclones, allowing a warning period of 30 minutes to 1 hour in areas at risk. A severe storm watch sometimes is possible a few hours in advance. But at other times, this most intense phenomenon of the atmosphere can strike with destructive suddenness and little warning. On April 27, 1991, a four-state area was devastated when over 70 tornadoes touched down from Texas to Nebraska, killing over 30 and causing millions in property damage. Compare this to the 53 casualties that occurred during the entire 1990 record year.

Tropical Cyclones

A powerful manifestation of Earth's energy and moisture systems is the **tropical cyclone**, which originates entirely within tropical air masses. The tropics extend from the Tropic of Cancer at 23.5° N to the Tropic of Capricorn at 23.5° S, encompassing the equatorial zone between 10° N and 10° S. Approximately 80 tropical cyclones occur annually worldwide. Table 8–2 identifies the varying stages of cyclonic development designated by wind speeds and notes some characteristic effects.

The cyclonic systems forming in the tropics are quite different from midlatitude cyclones because the air of the tropics is essentially homogeneous, with no fronts or conflicting air masses. In addition, the warm air and warm seas ensure an abundant supply of water vapor and thus the necessary latent heat to fuel these storms. Tropical meteorologists now think that the cyclonic motion begins to form as disturbances associated with slow-moving easterly waves of low pressure in the tradewind belt of the tropics. It is along the eastern (leeward) side of these migrating troughs of low pressure, a place of convergence and rainfall, that tropical cyclones are formed. Surface air flow then spins in toward the low-pressure area, ascends, and flows outward aloft. This important divergence aloft acts as a chimney, pulling more moisture-laden air into the developing system. Tropical cyclones also may result from extratropical cyclonic flows extending equatorward as well as along the ITCZ.

FIGURE 8–17
A fully developed tornado locks onto the ground. Photo courtesy of National Severe Storms Laboratory, National Oceanic and Atmospheric Administration.

(a)   Average tornadoes per year (1959–1989)

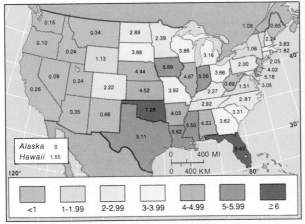

(b)   Tornado frequency per 10,000 mi² (1959–1989)

JAN FEB MAR APR MAY JUN JUL AUG SEP OCT NOV DEC
(c)

STATE	RANK BY 30-YR TOTAL	30 YEAR TOTAL	AVERAGE PER YEAR	FREQUENCY PER 10,000 MI² PER YR.	RANK BY FREQUENCY PER 10,000 MI²
TX	1	4025	134	5.11	8
FL	2	1533	51	9.42	1
OK	3	1488	50	7.28	2
KS	4	1099	37	4.52	10
NE	5	1020	34	4.44	11
IA	6	995	33	5.89	3
SD	7	839	28	3.68	18
MO	8	800	27	3.92	15
IL	9	790	26	4.67	9
MS	10	766	26	5.50	6
LA	11	757	25	5.62	4
CO	12	677	23	2.22	30
AL	13	672	22	4.33	12
GA	14	633	21	3.62	20
AR	15	627	21	4.03	13
WI	16	626	21	3.86	16
ND	17	597	20	2.89	25
IN	18	587	20	5.56	5
MN	19	573	19	2.39	27
MI	20	527	18	3.16	22
OH	21	457	15	3.66	19
NC	22	409	14	2.87	26
TN	23	346	12	2.92	24
WY	24	315	11	1.13	36
SC	25	302	10	3.31	21
PA	26	280	9	2.00	32
KY	27	274	9	2.27	28
NM	28	246	8	0.66	40
VA	29	178	6	1.51	35
NY	30	152	5	1.06	38
MT	31	144	5	0.34	43
CA	32	129	4	0.26	44
AZ	33	106	4	0.35	42
MA	34	89	3	3.83	17
MD	35	81	3	3.05	23
NJ	36	75	3	4.02	14
WV	37	66	2	0.69	39
ID	38	64	2	0.24	45
NH	39	60	2	2.24	29
ME	40	56	2	0.65	41
UT	41	50	2	0.24	46
WA	42	42	1	0.15	47
NV	43	38	1	0.09	49
CT	44	37	1	2.05	31
DE	45	30	1	5.18	7
OR	46	30	1	0.10	48
VT	47	26	1	1.08	37
HI	48	25	1	1.55	34
RI	49	6	0.2	1.82	33
AK	50	0	0	0	50

(d)   30 years of tornadoes in the United States

FIGURE 8–18
(a) Average tornadoes per year during 1959–1989; (b) tornado frequency per 10,000 km² during 1959–1989; (c) average tornadoes per month 1959–1990; (d) tornado rankings and frequency by state. During an average year, 778 tornadoes strike throughout the United States. Data prepared by Leo A. Grenier, National Severe Storm Forecast Center, National Weather Service.

TABLE 8–2
Tropical cyclone classification.

Designation	Winds	Features
Tropical disturbance	Variable, low	Definite area of surface low pressure; patches of clouds
Tropical depression	Up to 30 knots (63 kmph/39 mph)	Gale force, organizing circulation, light-to-moderate rain
Tropical storm	30 to 64 knots (63–117.5 kmph/39–73 mph)	Closed isobars, definite circular organization, heavy rain, assigned a name
Hurricane (Atlantic and E. Pacific) Typhoon (W. Pacific) Cyclone (Indian Ocean, Australia) Baguios (Philippines)	Greater than 65 knots (119 kmph/74 mph)	Circular, closed isobars; heavy rain, storm surges, tornadoes in right-front quadrant

The map in Figure 8–19 shows tropical cyclone formation areas, some of the characteristic storm tracks traveled, and indicates the range of months during which tropical cyclones are most likely to appear. As the map shows, tropical cyclones tend to occur when the equatorial low-pressure trough is farthest from the equator, that is, during the months that follow the summer solstice in each hemisphere.

H —Hurricane
T —Typhoon
C —Cyclone
B —Baguios (Philippine Is.)

FIGURE 8–19
Worldwide pattern of the most intense tropical cyclones, with principal months of occurrence and regional names.

FIGURE 8–20
Hurricane Gilbert, 13
September 1988: (a) *GOES-East*
satellite image; (b) side-view
radar image (Side-Looking
Airborne Radar—SLAR) taken
from an aircraft flying through
the central portion of the storm.
National Oceanic and
Atmospheric Administration and
the National Hurricane Center,
Miami.

(a)

(b)

About 10% of all tropical disturbances have the right ingredients to intensify into a full-fledged **hurricane** or **typhoon**.

Hurricanes and Typhoons. Tropical cyclones are potentially the most destructive storms experienced by humans, costing an average of several hundred million dollars in property damage and thousands of lives each year. The tropical cyclone that struck Bangladesh in 1970 killed an estimated 300,000 people and the one in 1991 claimed over 200,000 initially. The Galveston, Texas, hurricane of 1900 killed 6000; Hurricane Audry (1957), 400; Hurricane Gilbert (1988), 318; Hurricane Camille (1969), 256; and Hurricane Agnes (1972), 117.

Before the 1989 California earthquake, the largest dollar losses in the United States due to a natural disaster were from Hurricane Agnes in 1972 and Hurricane Camille in 1969, totaling $3.1 billion and $1.5 billion respectively. Most recently, Hurricane Hugo in 1989 caused several hundred deaths and more than $4 billion in damages. Yet, despite these losses, hurricanes can have beneficial aspects in some cases, as Focus Study 8–1 explains.

Fully organized tropical cyclones possess an intriguing physical appearance. They range in diameter from a compact 160 km (100 mi), to 960 km (600 mi), to some western Pacific typhoons that reach 1300 to 1600 km (800 to 1000 mi). Vertically, these storms dominate the full height of the troposphere. The inward-spiraling clouds form dense rain bands, with a central area designated the *eye,* around which a thunderstorm cloud called the *eyewall* swirls, producing the area of most intense precipitation. The eye remains an enigma, for in the midst of devastating winds and torrential rains the eye has quiet, warm air with even a glimpse of blue sky or stars possible.

The entire storm travels at 16–40 kmph (10–25 mph). When it moves ashore, or makes **landfall**, it pushes **storm surges** of seawater inland. The strongest winds of a tropical cyclone are usually recorded in its right-front quadrant (relative to the storm's directional path), where dozens of fully developed tornadoes may be embedded at the time of landfall. For example, Hurricane Camille in 1969 had up to 100 tornadoes embedded in that quadrant.

Hurricane Gilbert (September 1988) illustrates these general characteristics. For a Western Hemisphere hurricane, Gilbert achieved the largest-ever size (1600 km or 1000 mi in diameter) and the lowest-ever barometric pressure (888 mb or 26.22 in. of mercury). In addition, it sustained winds of 298 kmph (185 mph) with peak winds of exceeding 320 kmph (200 mph). Figure 8–20 portrays Gilbert from overhead and in cross section. From these images you can see the vast counterclockwise circulation, the central eye, eyewall, rain bands, and overall hurricane appearance.

Typhoons form in the vast region of the western Pacific, where the annual frequency of tropical cyclones is greatest. The magnitude and intensity of typhoons generally exceed those of hurricanes. In fact, the lowest sea-level pressure recorded on Earth was 870 mb (25.69 in.) in the center of Typhoon Tip, in October 1979, 837 km (520 mi) northwest of Guam (17° N 138° E). On September 9, 1987, the *Geostationary Meteorological Satellite* (GMS), operated by the Japan Weather Association, caught three well-developed typhoons in a single image, stretched across the western Pacific (Figure 8–21).

SUMMARY

The stability of the lower atmosphere determines whether active lifting of air parcels will occur. The interaction of air masses of differing temperature and moisture characteristics is essential to weather phenomena. Unstable conditions create lifting air masses that cool, saturate and condense, and form clouds. These masses can be lifted by convection heating, topographic barriers, and frontal contrasts. Midlatitude wave cyclones organize air masses in conflicting patterns as they migrate across the continents and oceans from west to east. The violent power of some weather phenomena are displayed in thunderstorms, lightning, tornadoes, hurricanes, and typhoons. Analyzing and understanding the daily weather map is essential in forecasting the weather. As forecasting and the public's perception of hazards improves, losses of life and property could stabilize rather than continue to increase.

FIGURE 8–21
Typhoons Gerald, Freda, and Holly in a remarkable image from *Geostationary Meteorological Satellite* (GMS), 9 September 1987. Courtesy of the Japan Weather Association.

Hurricane Camille: Disaster and Benefit

One of the most important hurricanes to strike the United States coastline this century was Hurricane Camille. It was significant not only for the disaster it brought (256 dead, $1.5 billion damage), but also for the moisture benefit it delivered and the drought it abated. Figure 1 presents two satellite images from August 17 and 18, 1969, as Camille made landfall on the Gulf Coast west of Biloxi, Mississippi, and then moved north. Hurricane-force winds ceased shortly after landfall, leaving an elaborate rainstorm that traveled from the Gulf Coast through Mississippi, western Tennessee, Kentucky, and into central Virginia. Severe flooding occurred along the Gulf Coast near the point of landfall and in the James River basin of Virginia, where torrential rains produced record floods. However, beyond this flooding and immediate coastal damage, Camille's effect was beneficial, ending drought conditions along major portions of its track.

(a)

(b)

FIGURE 1
Hurricane Camille (a) made landfall on the night of August 17, 1969, and (b)
continued inland on August 18. These two images were made exactly 24 hours apart.
Courtesy of National Oceanic and Atmospheric Administration.

FIGURE 2

A water resource view of Hurricane Camille's impact on local water budgets: (a) precipitation attributable to the storm (moisture supply); (b) resulting deficit abatement (avoided moisture shortages) attributable to Camille.

PRECIPITATION

cm	in.
14.0-28.8	5.50-11.33
11.5-13.9	4.50-5.49
8.9-11.4	3.50-4.49
6.4-8.8	2.50-3.49
3.8-6.3	1.50-2.49
1.3-3.7	0.50-1.49
0-1.2	0.00-0.49

(a)

DEFICIT ABATEMENT

cm	in.
5.4-8.1	2.10-3.20
2.8-5.3	1.10-2.09
0.1-2.7	0.04-1.09
Under 0.1	Under 0.04

(b)

According to the weather records of the past 40 years, about one-third of all hurricanes making landfall in the United States have created beneficial conditions inland in terms of the precipitation contributed to local water budgets. Figure 2 shows two maps of the precipitation attributable to Camille (moisture supply) and the moisture deficits abated (moisture shortages avoided) due to rains generated during Camille's passage. Dry soil conditions and low reservoirs were recharged by the rains. Camille's monetary benefits inland outweighed costs by 2 to 1, it is estimated. The tragic loss of life must not be forgotten, of course. But, in terms of overall effect, hurricanes should be viewed as normal and natural meteorological events that certainly have destructive potential but also contribute to the precipitation regimes of the southern and eastern United States.

A closing thought on our society's response to these powerful weather events: despite the increased damage from hurricanes, we see evidence of increased construction and development in vulnerable coastal lowlands. Even after Camille's devastation, a building boom hit the region. Ironically, new buildings, apartments, and governmental offices opened right next to still-visible rubble and bare foundation pads. Unfortunately, careful hazard planning to guide the settlement of these high-risk areas has never been policy. Thoughtful hazard perception is rarely practiced by public or private decision makers, whether along coastal lowlands, river floodplains, or earthquake fault zones.

The consequence is that our entire society bears the financial cost of poor perception, no matter where in the country it occurs, not to mention those victims who directly shoulder the physical hardship of the event. This recurrent, yet avoidable cycle—construction, devastation, reconstruction, devastation—was recognized two decades ago in *Fortune:*

> Before long the beachfront is expected to bristle with new motels, apartments, houses, condominiums, and office buildings. Gulf Coast businessmen, incurably optimistic, doubt there will ever be another hurricane like Camille, and even if there is, they vow, the Gulf Coast will rebuild bigger and better after that.*

Fortune, October 1969, p. 62.

Adiabatic	Occluded front
Air mass	Orographic lifting
Chinook winds	Rain shadow
Cold front	Secondary air mass
Cyclogenesis	Squall line
Dry adiabatic rate (DAR)	Stability
Front	Storm surges
Funnel clouds	Thunder
Hail	Tornado
Hurricane	Tropical cyclone
Landfall	Typhoon
Leeward slope	Warm front
Lightning	Waterspout
Mesocyclone	Wave cyclone
Moist adiabatic rate (MAR)	Windward slope

REVIEW QUESTIONS

1. Differentiate between stability and instability relative to a parcel of air rising vertically in the atmosphere.
2. How do the adiabatic rates of heating or cooling differ from the normal lapse rate and environmental lapse rate?
3. Why is there a difference between the dry adiabatic rate (DAR) and the moist adiabatic rate (MAR)?
4. How does a source region influence the type of air mass that forms over it? Give specific examples of each basic classification.
5. Of all the air masses, which are of greatest significance to the United States and Canada? What happens to them as they migrate to locations different from their source regions? Give an example of air-mass modification.
6. Explain why it is necessary for an air mass to rise if there is to be precipitation. What are the three main lifting mechanisms described in this chapter?
7. When an air mass passes across a mountain range, many things happen to it. Describe each aspect of a mountain crossing by a moist air mass. What is the pattern of precipitation that results?
8. Differentiate between a cold front and a warm front.
9. How does a wave cyclone act as a catalyst for conflict between air masses?
10. What is meant by cyclogenesis? Where does it occur and why?
11. What did the 1987 student balloon launch, during Geography Awareness Week, detect about cyclonic storm tracks?
12. Diagram a midlatitude cyclonic storm during its open stage. Label each of the components in your illustration, and add arrows to indicate wind patterns in the system.
13. What constitutes a thunderstorm? What type of cloud is involved? What type of air masses would you expect in an area of thunderstorms in North America?
14. Lightning and thunder are powerful phenomena in nature. Briefly describe how they develop.

15. Describe the formation process of a mesocyclone. How is this development associated with that of a tornado?
16. Evaluate the pattern of tornado activity in the United States from Figure 8–18. What generalizations can you make about the distribution and timing of tornadoes?
17. What are the different classifications for tropical cyclones? List the various names used worldwide for hurricanes.
18. Was Hurricane Camille detrimental, or beneficial, or both? Explain your answer and give specific support for your assessment of the storm's impact.

SUGGESTED READINGS

Department of the Air Force. *Weather for Aircrews*. Air Force Manual 51–12. Washington: U.S. Air Force, 1982.

Ferguson, Edward, and Frederick Ostby. "Tornadoes of 1990." *Weatherwise* 44, no. 2 (April 1991): 19–28.

Fincher, Jack. "Summer's Light and Sound Show: A Deadly Delight." *Smithsonian* 19, no. 4 (July 1988): 110–21.

Gedzelman, Stanley David. *The Science and Wonders of the Atmosphere*. New York: John Wiley & Sons, 1980.

Grenier, Leo A., John Halmstad, and Preston Leftwich, Jr. *Severe Local Storm Warning Verification: 1989*. National Oceanic and Atmospheric Administration Technical Memorandum NWS NSSFC–27. Kansas City: Severe Storms Forecast Center, May 1990.

Hughes, Patrick. *American Weather Stories*. Environmental Data Service and National Oceanic and Atmospheric Administration. Washington: U.S. Government Printing Office, 1976.

Lockhart, Gary. *The Weather Companion*. New York: John Wiley & Sons, 1988.

Miller, Peter. "Tracking Tornadoes!" *National Geographic* 171, no. 6 (June 1987): 690–715.

Oliver, John E., and Rhodes W. Fairbridge, eds. *The Encyclopedia of Climatology*. New York: Van Nostrand Reinhold Co., 1987.

Petterssen, Sverre. *Introduction to Meteorology*. New York: McGraw-Hill, 1969.

Roberts, Walter Orr. "We're Doing Something About the Weather." *National Geographic* 141, no. 4 (April 1972): 518–55.

Schaeffer, Vincent J., and John A. Day. *A Field Guide to the Atmosphere*. Peterson Field Guide Series. Boston: Houghton Mifflin, 1981.

Snow, John T. "The Tornado." *Scientific American* 250, no. 4 (April 1984): 86–96.

Thompson, Philip D., Robert O'Brien, and the editors of Time-Life Books. *Weather*. Life Science Library. New York: Time-Life Books, 1965.

Whipple, A. B. C., and the editors of Time-Life Books. *Storm*. Planet Earth Series. Alexandria, VA: Time-Life Books, 1982.

Williams, Earle R. "The Electrification of Thunderstorms." *Scientific American* 259, no. 5 (November 1988): 88–99.

Desert farmlands on the Hopi Reservation,
Arizona. Photo by author.

9

Water Balance and Water Resources

Water is not always naturally available when and where it is needed. From the maintenance of a houseplant to the distribution of local water supplies, from an irrigation program on a farm to the rearrangement of river flows—all involve aspects of the water balance and water-resource management. The availability of water is controlled by atmospheric processes and climate patterns.

We begin this chapter by examining the water balance, which is an accounting of the hydrologic cycle for a specific area with emphasis on plants and soil moisture. We discuss the nature of groundwater and look at several examples of this generally abused resource. Groundwater resources are closely tied to surface-water budgets. We also consider the water we withdraw and consume from available resources, in terms of both quantity and quality.

THE WATER-BALANCE CONCEPT

A water balance can be established for any area of Earth's surface by calculating the total precipitation input and the total of various outputs. The water-balance approach allows an examination of the hydrologic cycle, including estimation of streamflow, at a specific site or area for any period of time. Charles Warren Thornthwaite (1899–1963), a pioneer geographer in the field of applied water-resource analysis, worked with others to develop a water-balance methodology. He and others related these water-balance concepts to various geographical problems, especially to accurately determining irrigation quantity and timing. Thornthwaite also developed methods for estimating evaporation and transpiration from a variety of groundcovers and surfaces, and he identified the relationship between a given supply of water and the local demand as an important climatic element. In fact, his initial application of the techniques discussed here was to develop a new climatic classification system.

The Water-Balance Equation

The purpose of the water balance is to describe the various ways in which the water supply is expended. To understand Thornthwaite's water-balance methodology and "accounting" or "bookkeeping" procedures, we must first understand the terms and concepts in the simple water-balance equation:

$$\underset{\text{Supply}}{\text{PRECIP}} = \underset{\underset{\text{ACTET}}{\underbrace{\text{(Demand} - \text{Shortage)}}}}{\text{(POTET} - \text{DEFIC)}} + \underset{\text{Oversupply}}{\text{SURPL}} \pm \underset{\substack{\text{Soil moisture utilization} \\ \text{or recharge}}}{\Delta \text{ STRGE}}$$

Actual evapotranspiration

in which:

- PRECIP (precipitation) is rain, sleet, snow, and hail—*the moisture supply*.
- POTET (potential evapotranspiration) is the amount of moisture that would evaporate and transpire if the moisture were available; the amount lost under optimum moisture conditions—*the moisture demand*.
- DEFIC (deficit) is the amount of unmet, or unsatisfied, POTET; the amount of demand that is not met either by PRECIP or by soil moisture storage—*the moisture shortage*.
- SURPL (surplus) is the amount of moisture that exceeds POTET, when soil moisture storage is at field capacity (full)—*the moisture oversupply*.
- ±ΔSTRGE (soil moisture storage change) is the utilization (decrease) or recharge (increase) of soil moisture, snow pack, or lake and surface storage or detention of water—*the moisture savings*.
- ACTET (actual evapotranspiration): the actual amount of evaporation and transpiration that occurs; derived from POTET − DEFIC; thus, if all the demand is satisfied, POTET will equal ACTET—*the actual satisfied demand*.

Precipitation. The supply of moisture to Earth's surface is **precipitation** (PRECIP)—the rain, sleet, snow, and hail that fall toward the surface. The instrument used to measure precipitation is the **rain gauge**. A standard rain gauge is cylindrical, 20.3 cm (8.0 in.) in diameter and 58 cm (23 in.) deep. Modern styles include the tipping-bucket and weighing-type gauges, which receive precipitation through an open top so that precipitation can be measured in terms of depth or weight or volume (Figure 9–1).

Various conditions affect the performance of a rain gauge. For instance, winds can cause an undercatch of precipitation; a wind of 37 kmph (23 mph) produces an undercatch as high as 40%. Consequently, windshields are installed above the opening of a gauge to reduce error. In addition, when excessively heavy rains hit an area, various containers such as buckets or pans may be used to augment the rain gauge.

According to the World Meteorological Organization, more than 40,000 weather monitoring stations are operating worldwide, with over 100,000 places actually measuring precipitation. Precipitation patterns over the United States and Canada are shown in

FIGURE 9–1
Standard rain gauge installation. The water is guided by a funnel into a measuring tube. The device is designed to minimize evaporation, which could lead to inaccurate readings. Photo by author.

Figure 9–2. The highest amounts are in the East, Southeast, Northwest, and western mountain areas. Lower precipitation is found in the southwestern and western interiors and in the extreme North. A map displaying the pattern of world precipitation is included in the next chapter as Figure 10–2.

Potential Evapotranspiration. **Evaporation** is the movement of free water molecules away from a wet surface into air that is less than saturated. **Transpiration** takes place in plants. It is the outward movement of water through small openings called stomata in the underside of leaves. Both evaporation and transpiration directly respond to climatic conditions of temperature and humidity. Additionally, transpiration rates are partially controlled by the plants themselves through the functioning of small cells around the stomata. Transpired quantities can be significant: on a hot day, a single tree may transpire hundreds of liters of water. The overall coincidence of these two rates makes it convenient to combine evaporation and transpiration into one term—**evapotranspiration**.

The demand for water at Earth's surface is called **potential evapotranspiration** (POTET), which represents the amount of water that would evaporate and transpire under optimum moisture conditions (adequate precipitation and full soil moisture capacity). Filling a bowl with water and letting it evaporate might help illustrate this concept: when the bowl becomes dry, is there still an evaporation demand? We can conclude that a demand is present, even when the bowl is dry. Similarly, the amount of water that would evaporate and/or transpire if water always were available is the POTET, or that amount that would evaporate from the bowl if it were constantly supplied with water.

Determining POTET. For the empirical measurement of POTET, probably the easiest method employs an **evaporation pan**, or evaporimeter (Figure 9–3). As evaporation occurs, water in measured amounts is automatically replaced in the pan. Screens of various-sized mesh are used to protect against overmeasurements created by wind.

A **lysimeter** is a relatively elaborate device for measuring POTET, for an actual portion of a field is isolated so that the moisture moving through it can

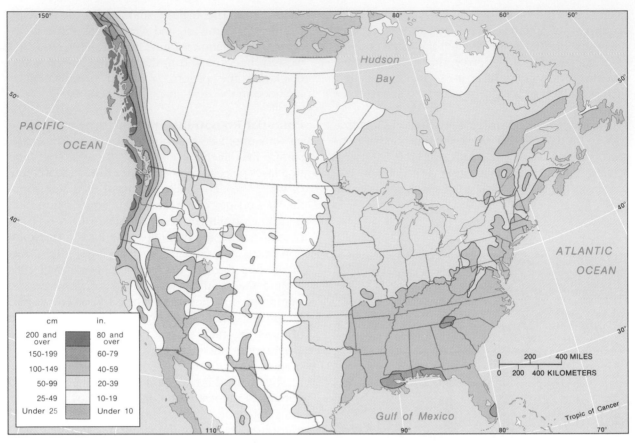

FIGURE 9–2
Annual precipitation in the United States and Canada. Adapted from U.S. National
Weather Service, U.S. Department of Agriculture, and Environment Canada.

FIGURE 9–3
Standard evaporation pan. Photo
by author.

be measured. In its simplest form, a lysimeter is a buried tank, open at the surface and approximately a cubic meter in size. Deeper-rooted crops require a larger tank so that root growth is not inhibited. Care must be taken to create similar conditions in the lysimeter and the surrounding field to maximize accuracy. A rain gauge next to the lysimeter measures the precipitation input. Some of the water remains as soil moisture or is incorporated into the vegetation cover, or is measured as gravitational water which drains out of the bottom of the lysimeter. The remainder is credited to evapotranspiration. Given natural conditions, the lysimeter measures actual evapotranspiration. If adequate water is supplied through irrigation and the soil moisture available to the plant is optimum, the rate of water loss will be equivalent to POTET.

A more complex device is the weighing lysimeter. It isolates a large portion of a field in an embedded tank, with the entire mass on a scale, sometimes constructed to allow access through an underground room where weight is recorded. The utility of lysimeters and even evaporation pans is greatly limited by their scarcity across North America. They do, however, provide a necessary database with which to develop other methods of estimating POTET.

The Thornthwaite method of estimating POTET is based on empirical measurements from watersheds and lysimeters in the central and eastern United States. Thornthwaite found that POTET is best approximated using mean monthly air temperature (measured with a thermometer) and daylength (a function of the measuring station's latitude). Both factors are easily determined, making it relatively simple to calculate POTET and the other water-balance components in the equation indirectly, with fair accuracy for most midlatitude locations.

The limitations of Thornthwaite's method, and why it is more useful in some climates than others, stem from its lack of consideration of some other factors. For example, Thornthwaite's method does not consider wind movements of warm or cool air across a surface, sublimation effects and surface-water retention when temperatures are below freezing, and problems associated with areas of frozen soil (*permafrost,* discussed in Chapter 13) or poor drainage. Nevertheless, we use the method here because of its great utility as a teaching tool and its

overall ease of application in geographic studies to analyze water budgets with acceptable results over large areas.

Figure 9–4 presents POTET values derived by the Thornthwaite method for the United States. Higher values are found in the South, with highest readings in the Southwest. Lower POTET values are found at higher latitudes and elevations. Please compare this POTET (demand) map with Figure 9–2, which is the PRECIP (supply) map for the same region. The relationship between PRECIP supplies and POTET demands determines the remaining components of the water-balance equation. Can you identify from the two maps regions where PRECIP is higher than POTET? Or where POTET is higher than PRECIP?

Deficit. POTET can be satisfied either by PRECIP or by moisture derived from the soil, or it can be artificially met through the application of irrigation water. If these sources are inadequate, the site experiences a moisture shortage. This unmet, or unsatisfied, POTET is referred to as the **deficit** (DEFIC). By subtracting DEFIC from POTET, we determine the **actual evapotranspiration**, or ACTET, that takes place. Under ideal conditions POTET and ACTET are about the same, so that plants do not experience a water shortage, for prolonged deficits would lead to drought conditions.

Surplus. If POTET is satisfied and soil moisture is full, then additional water input becomes **surplus** (SURPL), or water oversupply. This excess water may sit on the surface in puddles, a situation known as **water detention**, or flow through the soil to groundwater storage. Surplus water that flows across the surface toward stream channels is termed **overland flow**, which together with other precipitation and subsurface flows into river channels is the **total runoff** of the area. Because streamflow, or runoff, is generated mostly from surplus water, the water-balance methodology can be thought of as a useful technique for indirectly estimating streamflow.

Soil Moisture Storage. Soil moisture storage represents a savings account of water that can receive deposits and allow withdrawals as conditions change in the water balance. Soil moisture storage (ΔSTRGE) refers to the amount of water that is stored in the soil and is accessible to plant roots, or the

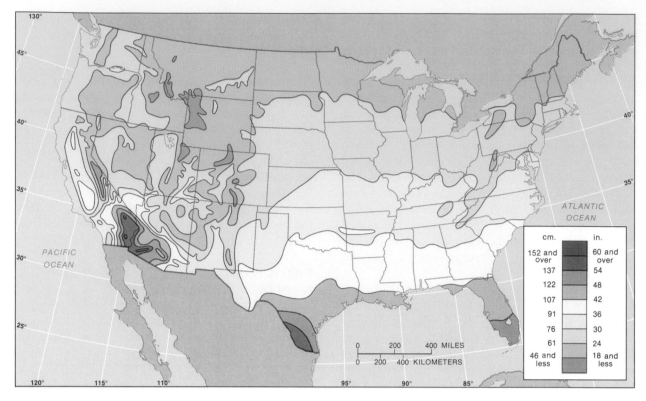

FIGURE 9–4
Potential evapotranspiration for the United States. From Charles W. Thornthwaite, 1948, "An Approach Toward a Rational Classification of Climate," *Geographical Review* 38, p. 64. Adapted by permission from the American Geographical Society.

effective rooting depth of plants in a specific soil. This water is held in the soil against the pull of gravity and is discussed in more detail in Chapter 18.

The moisture retained in the soil is comprised of two forms of water, only one of which is accessible to plants (Figure 9–5). The inaccessible form is **hygroscopic water**, the thin molecular layer of water that is bonded tightly to each soil particle by the natural hydrogen bonding of water molecules. Such water exists even in the desert, and is not available to meet POTET demands. Soil is said to be at the **wilting point** when plant roots are unable to extract water; in other words, plants will wilt and eventually die after prolonged moisture stress.

The soil moisture that is generally accessible to plant roots is **capillary water**, held in the soil by surface tension and cohesive forces between the water and the soil. Almost all capillary water is **available water** in soil moisture storage and is removable for POTET demands through the action of

plant roots and surface evaporation; some capillary water remains adhered to soil particles along with hygroscopic water. When capillary water is full in a particular soil, that soil is said to be at **field capacity**, an amount determined by actual soil surveys. Water surpluses beyond field capacity and moisture demands form **gravitational water** that percolates downward from the capillary zone to the groundwater zone. "Available water" is generally the preferred term for soil moisture available to plants.

A simplified graph of soil water capacity shows the relationship of soil texture to soil moisture content (Figure 9–6). The lower line on the graph plots the wilting point; the upper line plots field capacity. The space between the two lines represents the amount of water available to plants, given various soil textures. Various plant types growing in various types of soil send roots to different depths and therefore are exposed to varying amounts of soil moisture. For example, shallow-rooted crops such as spinach,

Soil particles with forms of soil moisture

Soil moisture availability (increasing ⟶)

← H₂O unavailable for plants → | ← H₂O available → | ← H₂O →

Hygroscopic H₂O*

Capillary H₂O

Gravitational H₂O

Wilting point   Field capacity   Gravitational water

*Note: Some capillary water is bound to hygroscopic water on soil particle and is also unavailable.

FIGURE 9–5
Forms of soil moisture: hygroscopic, capillary, available, and gravitational. After Donald Steila, *The Geography of Soils,* © 1976, p. 45. Adapted by permission of Prentice Hall, Inc., Englewood Cliffs, NJ.

beans, and carrots send roots down 65 cm (25 in.) in a silt loam, whereas deep-rooted crops such as alfalfa and shrubs exceed a depth of 125 cm (50 in.) in such a soil. A soil blend that maximizes available water is best for supplying plant water needs. Based on Figure 9–6, can you determine the soil texture with the greatest quantity of available water?

As plant roots work the soil for their water needs, they exert a tensional force on soil moisture. When soil moisture is at field capacity, plant roots are able to obtain water with less effort, and water is thus rapidly available to them. As the soil water is reduced

by *soil moisture utilization,* the plants must exert greater effort to extract the same amount of moisture. As a result, even though a small amount of water may remain in the soil, plants may be unable to exert enough pressure to utilize it. The unsatisfied demand resulting from this situation is calculated as a deficit. Avoiding a deficit and reducing plant growth inefficiencies are the goals of a proper irrigation program, for the harder plants must work to get water, the less their yield and growth will be.

Whether naturally occurring or artificially applied, water infiltrates soil and replenishes available water

FIGURE 9–6
The relationship between soil moisture availability and soil texture. From Donald Steila, *The Geography of Soils,* © 1976, p. 46. Adapted by permission of Prentice Hall, Inc., Englewood Cliffs, NJ, as adapted from U.S. Department of Agriculture, *1955 Yearbook of Agriculture— Water,* p. 120.

in./ft   cm/m

Available water per unit of rooting depth

Field capacity

Available water

Unavailable water

Wilting point

Sand   Sandy loam   Loam   Silt loam   Clay loam   Clay

TABLE 9–1

PRECIP − POTET for Kingsport, Tennessee.

	Jan	Feb	Mar	Apr	May	Jun	Jul	Aug	Sep	Oct	Nov	Dec
PRECIP	97	99	97	84	104	97	132	112	66	66	66	99
	(3.8)	(3.9)	(3.8)	(3.3)	(4.1)	(3.8)	(5.2)	(4.4)	(2.6)	(2.6)	(2.6)	(3.9)
POTET	7	8	24	57	97	132	150	133	99	55	12	7
	(0.3)	(0.3)	(0.9)	(2.2)	(3.8)	(5.2)	(5.9)	(5.2)	(3.9)	(2.2)	(0.5)	(0.3)
Net supply (+) or demand (−):	+90	+91	+73	+27	+7	−35	−18	−21	−33	+11	+54	+92
	(3.5)	(3.6)	(2.9)	(1.1)	(0.3)	(−1.4)	(−0.7)	(−0.8)	(−1.3)	(0.4)	(2.1)	(3.6)

All quantities in millimeters (inches).

content, a process known as *soil moisture recharge*. The texture and the structure of the soil dictate available pore spaces, or **porosity**; and the flow of the water through the soil, or **permeability**, determines the rate of soil moisture recharge. Permeability depends on particle sizes and the shape and packing of soil grains; it is expressed as a rate, generally lower for clays and higher for sands and gravels. (Chapter 18 presents more on soil texture, structure, and moisture characteristics.)

Thornthwaite assumed that the soil moisture recharge rate is 100% efficient as long as the soil is below field capacity and above a temperature of −1°C (30.2°F). Under real conditions we know that infiltration actually proceeds rapidly in the first minutes of the precipitation event, slowing down as

the upper layers of soil become saturated even though the soil below is still dry. Various agricultural practices can improve both soil permeability and the depth to which moisture can efficiently penetrate to recharge a deeper soil storage.

Sample Water Balances

Using all of these concepts, we can now apply the water-balance equation to a specific station. Table 9–1 presents the long-term average water supply and demand conditions for the city of Kingsport in the extreme northeast corner of Tennessee. Kingsport is at 36.6° N 82.5° W, with an elevation of 390 m (1280 ft), and has maintained weather records for more than 55 years. The monthly values in Table 9–1

TABLE 9–2

Water balance for Kingsport, Tennessee.

	Jan	Feb	Mar	Apr	May	Jun	Jul	Aug	Sep	Oct	Nov	Dec	Total
PRECIP	97	99	97	84	104	97	132	112	66	66	66	99	1119
	(3.8)	(3.9)	(3.8)	(3.3)	(4.1)	(3.8)	(5.2)	(4.4)	(2.6)	(2.6)	(2.6)	(3.9)	(44.1)
POTET	7	8	24	57	97	132	150	133	99	55	12	7	781
	(0.3)	(0.3)	(0.9)	(2.2)	(3.8)	(5.2)	(5.9)	(5.2)	(3.9)	(2.2)	(0.5)	(0.3)	(30.7)
PRECIP − POTET	+90	+91	+73	+27	+7	−35	−18	−21	−33	+11	+54	+92	—
	(+3.5)	(+3.6)	(+2.9)	(+1.1)	(+0.3)	(−1.4)	(−0.7)	(−0.8)	(−1.3)	(+0.4)	(+2.1)	(+3.6)	
STRGE	100	100	100	100	100	67	51	34	9	20	74	100	—
	(4.0)	(4.0)	(4.0)	(4.0)	(4.0)	(2.7)	(2.1)	(1.4)	(0.4)	(0.8)	(2.9)	(4.0)	
Δ STRGE	0	0	0	0	0	−33	−16	−17	−25	+11	+54	+26	±0
						(−1.3)	(−0.6)	(−0.7)	(−1.0)	(+0.4)	(+2.1)	(+1.1)	
ACTET	7	8	24	57	97	130	148	129	91	55	12	7	765
	(0.3)	(0.3)	(0.9)	(2.2)	(3.8)	(5.1)	(5.8)	(5.1)	(3.6)	(2.2)	(0.5)	(0.3)	(30.6)
DEFIC	0	0	0	0	0	2	2	4	8	0	0	0	16
						(0.1)	(0.1)	(0.2)	(0.3)				(0.6)
SURPL	90	91	73	27	7	0	0	0	0	0	0	66	354
	(3.5)	(3.6)	(2.9)	(1.1)	(0.3)							(2.6)	(14.0)

All quantities in millimeters (inches).

assume that PRECIP and POTET are evenly distributed throughout each month, smoothing out actual daily and hourly conditions.

Let's first compare PRECIP with POTET by month to determine whether there is a net supply (+) or net demand (−) for water. We can see that Kingsport experiences a net supply from January through May and from October through December. However, the warm days and months from June through September result in net demands for water. Assuming a soil moisture storage capacity of 100 mm (4.0 in.), typical of shallow-rooted plants, those months of net demand for moisture are satisfied through soil moisture utilization.

Table 9–2 presents the annual water balance for Kingsport, and Figure 9–7 graphs the water-balance components of these same monthly averages. The following discussion elaborates on both this table and graph.

Table 9–2 shows that soil moisture estimates are at field capacity through May, with the excess supply crediting surplus each month. In June the net demand is −35 mm (−1.4 in.), most of which is withdrawn from soil moisture storage, bringing soil moisture capacity down to 67 mm (2.7 in.) at the end of June. The other 2 mm (0.1 in.) of net demand appears as DEFIC; that amount remains in the soil, reflecting the increasing inefficiency as soil moisture drops below field capacity. From June through September the ACTET (actual evapotranspiration) is progressively less than the optimum POTET, and slight deficits continue to be recorded.

July, August, and September feature soil moisture utilization of −16 mm (−0.6 in.), −17 mm (−0.7 in.), and −25 mm (−1.0 in.), so that by the end of September shallow-rooted soil moisture is near the wilting point. During October, however, 11 mm of net supply is generated and is credited as soil moisture recharge. Recharge occurs again in Novem-

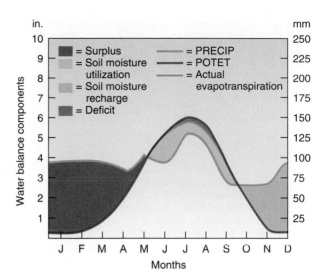

FIGURE 9–7
Annual average water balance for Kingsport, Tennessee.

ber and December, bringing the soil moisture back up to field capacity by the end of the year, with a surplus of 66 mm (2.6 in.) estimated for December. The Kingsport water-balance equations for the year and for the months of March and September are isolated in Table 9–3 to show the interaction of these various components. Check these three functions against Table 9–2 and perform the functions indicated to see whether we have accounted for all the PRECIP received.

Obviously, not all stations experience the surplus moisture patterns of this humid continental city and surrounding region; different climatic regimes experience different relationships among their water-balance components. Also, medium- and deep-rooted plants produce different results. Figure 9–8 presents water-balance graphs for several other cities in North America and the Caribbean Sea region. (Appendix A presents PRECIP, POTET, and tempera-

TABLE 9–3
Annual, March, and September water balances for Kingsport.

	Precip	=	Potet	−	Defic	+	Surpl	±	Δ Strge
Annual	1119	=	(781	−	16)	+	354	±	0
	(44.1)	=	(30.7	−	0.6)	+	(14.0)	±	0
March	97	=	(24	−	0)	+	73	±	0
	(3.8)	=	(0.9	−	0.0)	+	(2.9)	±	0
September	66	=	(99	−	8)	+	0	−	25
	(2.6)	=	(3.9	−	0.3)	+	0	−	(1.0)

All quantities in millimeters (inches)

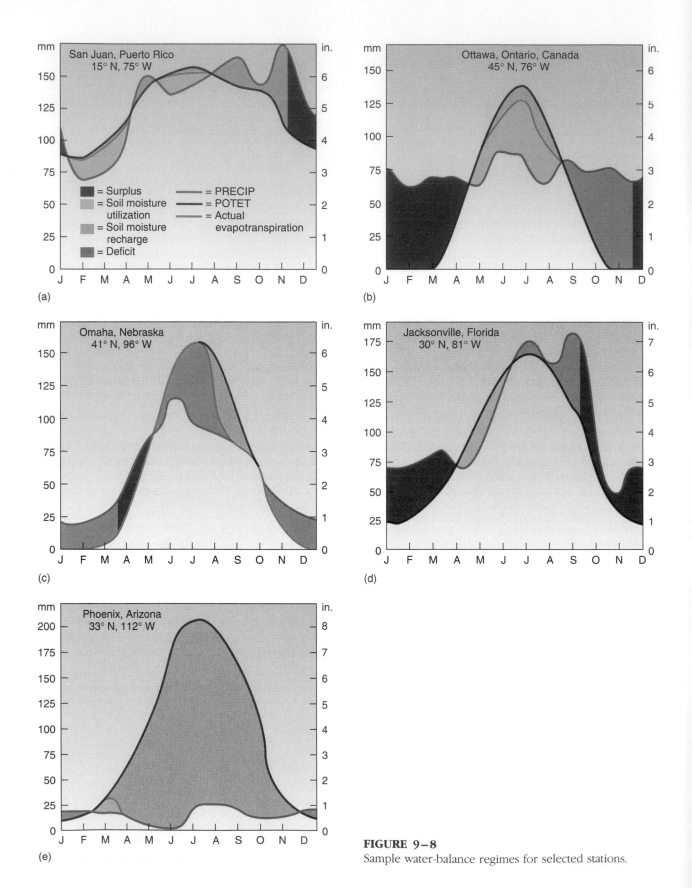

FIGURE 9–8
Sample water-balance regimes for selected stations.

ture values, along with other information, for a wide variety of stations worldwide, grouped by climatic classification.)

Water Management: An Example

Most water management projects attempt to rearrange water resources either geographically or across the calendar. Because water is distributed unevenly over space and time, substantial human effort has been expended to override natural water-balance regimes and reoccurring drought. In North America, several large water-management projects are already in operation: the Bonneville Power Authority in Washington State, the Tennessee Valley Authority in the Southeast, the California Water Project, the Central Arizona Project, and the Churchill Falls and Nelson River Projects of Manitoba, to name only a few. Controversy and battles among conflicting interests invariably follow water projects of this magnitude, especially as most ideal sites already are taken.

A particularly ambitious regional water project is the Snowy Mountains Hydro-electric Authority Scheme of Australia, where water-balance variations created the need to relocate water. The precipitation patterns in southeastern Australia are portrayed on the world precipitation map in Figure 10–2. In the Snowy Mountains, part of the Great Dividing Range in extreme southeastern Australia, precipitation ranges from 100 to 200 cm (40 to 80 in.) a year, whereas interior Australia receives under 50 cm (20 in.), and drops to less than 25 cm (10 in.) further inland. POTET values are high throughout the Australian outback and lower in the higher elevations of the Snowy Mountains.

The Snowy Mountain Scheme was begun in the 1950s and completed over 20 years later. The plan was to take surplus water that flowed down the Snowy River eastward to the Tasman Sea and reverse the flow to support newly irrigated farmland in the interior of New South Wales and Victoria. Through some of the longest tunnels ever built, vast pumping systems, numerous reservoirs, and power plants, this plan was completed and even expanded to include many tributary streams and drainage systems high in the mountains. The westward-flowing rivers—the Murray, Tumut, and Murrumbidgee—have their flows augmented, and as a result, new acreage is now

in production in what was dry outback, formerly served only by wells drawing on meager groundwater supplies (Figure 9–9).

GROUNDWATER RESOURCES

Groundwater is a part of the hydrologic cycle that lies beneath the ground and is therefore tied to surface supplies. Groundwater is the largest potential source of freshwater in the hydrologic cycle—larger than all surface reservoirs, lakes, rivers, and streams combined. Between Earth's surface and a depth of 4 km (13,100 ft) worldwide, some 8,340,000 km^3 (2,000,000 mi^3) of water resides. Despite this volume and its obvious importance, groundwater is subject to abuse. It is polluted in many areas and is consumed in quantities beyond natural recharge rates.

About 50% of the U.S. population derives a portion of its freshwater from groundwater sources. Between 1950 and 1990 the annual groundwater withdrawal increased 160%. In Canada, about 4% of all water withdrawals come from groundwater sources—4.2 million m^3/day, 1.5 billion m^3/year (148.3 million ft^3/day, 53 billion ft^3/year). Figure 9–10 shows potential groundwater resources in the United States and Canada.

Groundwater Description

Figure 9–11 provides a helpful integrative reference for this entire section and illustrates the following discussion. Groundwater is fed by surplus water, which percolates downward from the zone of capillary water as gravitational water. This excess surface water moves through the **zone of aeration** and eventually reaches an area of collected subsurface water known as the **zone of saturation**. Like a sponge made of sand, gravel, and rock, this zone stores water within its structure, filling all available pores and voids. The *porosity* of this layer is dependent on the arrangement, size, and shape of individual component particles, the cement between them or how compacted they are, and the presence of noncompressed bits of rock. Subsurface structures are referred to as permeable or impermeable, depending on whether they permit or obstruct water flows.

An **aquifer** is a rock layer that is permeable to groundwater flow in significant amounts, whereas an

FIGURE 9–9
Snowy Mountain irrigation water makes possible the opening of new lands in
Australia's eastern interior. Photo by Snowy Mountains Hydro-electric Authority.

aquiclude is a body of rock that does not conduct water in usable amounts. Technically, the zone of saturation is an **unconfined aquifer**, a water-bearing stratum that is not confined by an impermeable overburden. The upper limit of the water that collects in the zone of saturation is called the **water table**; it is the contact point between the zones of saturation and aeration. Water movement is controlled by the slope of the water table, which generally follows the contours of the land surface. To optimize their potential flow, water wells must penetrate the water table. When the water table intersects the surface, it creates springs or feeds lakes or riverbeds (Figure 9–12). Ultimately, groundwater may enter stream channels and flow as surface water. During dry periods the water table can act to sustain river flows.

A confined aquifer differs from an unconfined aquifer in several important ways. A **confined aquifer** is bounded above and below by impermeable layers of rock or sediment, while an unconfined aquifer is

not (Figure 9–11). In addition, the two aquifers differ in **aquifer recharge area**, which is the surface area where water enters an aquifer to recharge it. For an unconfined aquifer, that area generally extends above the entire aquifer, but in the case of a confined aquifer the recharge area is far more restricted, as you can see in the figure. Once these recharge areas are identified and mapped, the appropriate government body should zone them to prohibit pollution discharges, septic and sewage system installations, or hazardous-material dumping. Of course, an informed population would insist on such action.

Aquifers also differ in pressure. A well drilled into an unconfined aquifer must be pumped to make the water level rise above the water table. In contrast, the water in a confined aquifer is under the pressure of its own weight, creating a pressure level called the **piezometric surface**, which actually can be above ground level. Under this condition, **artesian water** may rise up in wells and even flow out at the surface if the head of the well is below the piezometric

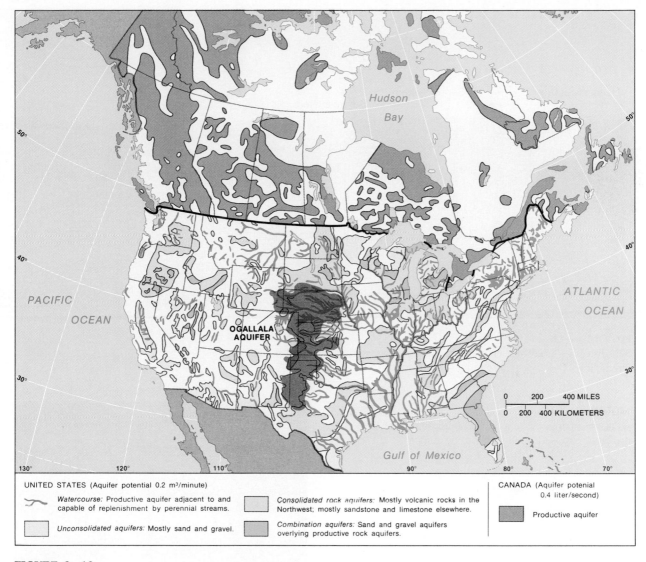

FIGURE 9–10
Groundwater resource potential for the United States and Canada. For the United States, we highlight areas underlain by productive aquifers capable of yielding freshwater to wells at 0.2 m³/minute or more (for Canada, 0.4 liter/second). Data courtesy of Water Resources Council for the United States and the Inquiry on Federal Water Policy for Canada.

surface. In other wells, however, pressure may be inadequate, and the artesian water must be pumped the remaining distance to the surface.

Groundwater Utilization

Specific yield is the groundwater available for utilization at the surface. When specific yield can be determined, it is expressed as a percent of the total volume of the saturated aquifer. As water is pumped from a well, the adjacent water table within an unconfined aquifer will **drawdown**, or become lower, if the rate of pumping exceeds the horizontal flow of water in the aquifer. The resultant shape of the water table around the well is called a **cone of depression** (Figure 9–11).

WATER BALANCE AND WATER RESOURCES • **241**

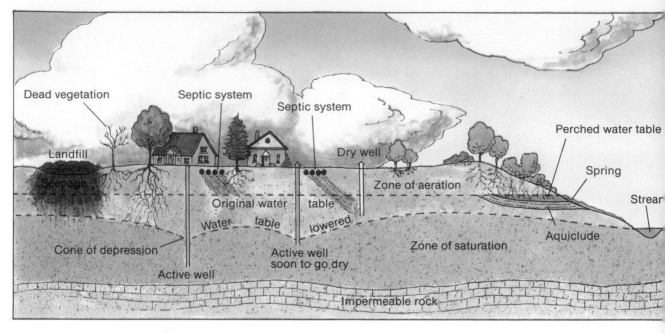

FIGURE 9–11
Subsurface groundwater characteristics, processes, water flows, and human interactions.

FIGURE 9–12
An active spring, evidence of groundwater and a probable exposed aquifer. Photo by author.

Aquifers frequently are pumped beyond their flow and recharge capacities; **groundwater mining** refers to this overutilization of groundwater resources. Large tracts of the Midwest, West, lower Mississippi Valley, and Florida experience chronic groundwater overdrafts. In many places the water table or artesian water level has declined more than 12 m (40 ft). Groundwater mining is of special concern today in the Ogallala aquifer, which is the topic of Focus Study 9–1.

Another possible effect of water removal from an aquifer is that the stratum loses its internal support—the water in the pore spaces between the rock grains—and overlying rock may crush the aquifer, resulting in land subsidence and lower surface elevations. For example, within an 80 km (50 mi) radius of Houston, Texas, land has subsided more than 3 m (10 ft). In the Fresno area of the San Joaquin Valley in central California, after years of intensive irrigation, land levels have dropped almost 10 meters (32 ft) because of water removal and soil compaction. Unfortunately, collapsed aquifers may not be rechargeable even if surplus gravitational water becomes available, because pore spaces may be permanently closed. Aquifers also are destroyed by strip mining for coal and lignite in such places as eastern Texas and northeastern Louisiana, Wyoming, and Arizona, among other locales.

An additional problem arises when aquifers are overpumped near the ocean. In such a circumstance the groundwater interface between freshwater and seawater often retreats inland. As a result, wells may become contaminated with saltwater, and the aquifer may become useless as a freshwater source (Figure 9–11 illustrates this on the far right margin). Aquifers also can be invaded by ancient seawater that was trapped in sediments at the time they were deposited (known as *connate water*). The contamination may be halted by pumping freshwater back into the aquifer, although the contaminated aquifer usually cannot be reclaimed. This problem plagues the Persian Gulf states, which soon may run out of freshwater. Their vast groundwater resource is being overpumped to such an extent that saltwater is filling aquifers tens of kilometers inland. By the year 2000, water on the Arabian Peninsula may be undrinkable. Imagine having the main issue in the Middle East become water, not oil!

Remedies for groundwater overutilization are neither easy nor cheap. In the Persian Gulf area additional freshwater is being obtained through desalinization plants along the coasts, as well as a very costly water pipeline that will carry water overland from Turkey some 1500 km (930 mi) away (the distance from New York City to St. Louis). Estimates predict the cost of the piped water to be

Ogallala Aquifer Overdraft

Probably the largest confined and unconfined aquifer in the world lies below a six-state area extending from southern South Dakota and Nebraska to Texas, a region known as the American High Plains (Figure 1). The sand and gravel of the Ogallala aquifer (High Plains aquifer) were charged with meltwaters from retreating glaciers for several hundred thousand years. However, for the past 100 years, Ogallala groundwater has been heavily mined. The mining has been especially intense since World War II, when center-pivot irrigation was introduced. These large circular devices facilitate production of wheat, sorghums, cotton, corn, and about 40% of the grain fed to cattle in the United States.

The Ogallala aquifer is used to irrigate an area representing about one-fifth of all U.S. cropland: some 5.7 million hectares (14 million acres) are provided with water from 150,000 wells. In this process, water is being removed from the aquifer at 26 billion m^3 (21 million acre-feet) a year, an increase of more than 300% since 1950. In the past four decades the water table in the aquifer has dropped more than 30 m (100 ft), and throughout the 1980s it has averaged a 2 m (6 ft) drop per year because of extensive groundwater mining. U.S. Geological Survey scientists estimate that recharging the recoverable portions of the Ogallala aquifer (those portions that have not been crushed or subsided) would take at least 1000 years if mining stopped today!

The aquifer is most impacted in Texas, where the saturated layer is shallower and some 75,000 wells are in operation. The declining water table in Floyd County, Texas, illustrates the overdraft problem (Figure 2). Between 1952 and 1982, pumping costs in Floyd County have increased 221% (adjusted for increases in the price index for crops), even though irrigated acreage dropped during that same time by almost 15%.

Obviously billions of dollars of agricultural activity cannot be abruptly halted, but neither can profligate water mining continue. This issue raises tough questions, involving the management of cropland, reliance on extensive irrigation, the demand to produce exports, and the need to assess the importance of certain crops that are in chronic oversupply, as well as various crop subsidies. Present agricultural practices, if continued, will result in the loss of about half the total Ogallala aquifer resource, and a two-thirds loss in the Texas portion alone by the year 2020.

FIGURE 1 (left)

Map of the Ogallala (High Plains) aquifer. After David E. Kromm and Stephen E. White, "Interstate Groundwater Management Preference Differences: The Ogallala Region," *Journal of Geography* 86, no. 1 (January–February 1987), p. 5. Adapted by permission.

FIGURE 2 (above)

Water levels in Floyd County, Texas, wells. After John E. Schefter, 1985, *Declining Ground-Water Levels and Pumping Costs: Floyd County, Texas,* U.S. Geological Survey Water Supply Paper No. 2275, Washington: Government Printing Office, p. 114.

about the same as that of the desalinated water. In addition, end-use patterns could be altered, such as upgrading traditional agricultural practices and adding greater efficiency to urban usage, to help reduce the demand for groundwater. The 1991 Persian Gulf War posed a grave threat to desalinization facilities in Saudi Arabia, and many in Iraq and Kuwait were damaged or destroyed.

Pollution of the Groundwater Resource

When surface water is polluted, groundwater also becomes contaminated because it is fed and recharged from surface-water supplies. Groundwater migrates very slowly compared to surface water.

Surface water flows rapidly and flushes pollution downstream, but sluggish groundwater, once contaminated, remains polluted virtually forever. Pollution can enter groundwater from industrial injection wells, septic tank outflows, seepage from hazardous-waste disposal sites, industrial toxic-waste dumps, residues of agricultural pesticides, herbicides, fertilizers, and residential and urban wastes in landfills. Thus, pollution can come either from a point source or from a large general area (a nonpoint source), and it can spread over a great distance, as illustrated in Figure 9–11.

Despite widespread publicity about groundwater pollution, few new institutional acts or laws have been established. Laws such as the Safe Drinking Water Act (1974) and the Resource Conservation and

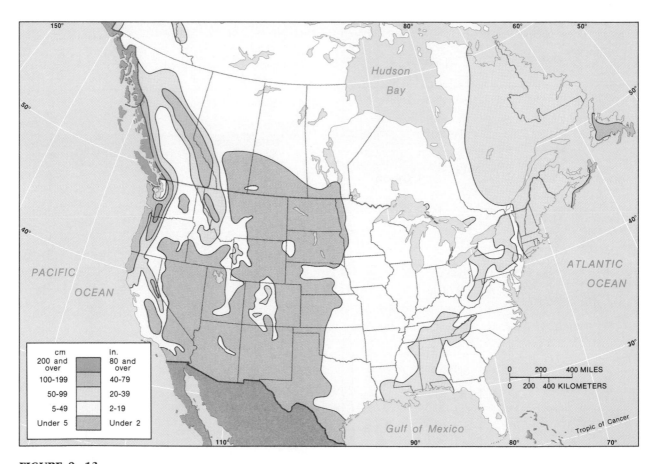

FIGURE 9–13
Distribution of runoff in the United States and Canada. Data courtesy of Water Resources Council for the United States and *Currents of Change* by Environment Canada.

Recovery Act (1976, 1984) either neglect or do not explicitly direct action for groundwater protection.

In 1988 the General Accounting Office (GAO) completed a nationwide survey of groundwater protection standards. It found that 38 states have nonspecific narrative standards, which are difficult to characterize because of their lack of uniformity, and that only 26 states have any numerical or empirical groundwater standards. In the face of government inaction, serious contamination of groundwater continues nationwide.

> One characteristic is the practical irreversibility of groundwater pollution, causing the cost of clean-up to be prohibitively high. . . . It is a questionable ethical practice to impose the potential risks associated with groundwater contamination on future generations when steps can be taken today to prevent further contamination.*

DISTRIBUTION OF STREAMS

Streams always have been important to human beings. They represent only 1250 km³ (300 mi³) of water, the smallest volume of any of the freshwater categories we have discussed. Yet, streams are the portion of the hydrologic cycle on which we most depend, representing four-fifths of all the water we use. Whether streams are perennial (constantly flowing) or intermittent, the total runoff that moves through them comes from surplus surface-water runoff, subsurface throughflow, and groundwater (Figure 9–13). Because of their rapid renewal, as compared to lake water or groundwater, streams are the best single measure of available water supply.

A stream's flow rate is termed its **discharge**, usually expressed as the volume of water per unit of time. In volume of runoff, the Western Hemisphere exceeds the Eastern Hemisphere, primarily because the Amazon River carries a far greater volume than any other, and because four of the world's longest river systems are in the Western Hemisphere. The Atlantic Ocean receives about 1.5 times more runoff than the Pacific Ocean. Table 9–4 lists the outflow locations, lengths, and discharge values for the world's largest rivers. (Chapter 14 examines more closely the hydrologic aspects of streams and their interaction with the landscape.)

In the United States, the conventional unit used by hydrologists to measure streamflow is *cubic feet per second (cfs)*; in Canada it is *cubic meters per second (cms)*, or billions of liters per day. In the eastern United States, or for large-scale assessments such as Figure 9–14, water managers speak in terms of *millions of gallons a day (MGD)* or *billions of gallons a day (BGD)*. In the western United States, where irrigated agriculture is so important, the measure frequently used is *acre-feet per year;* one acre-foot contains 325,872 gallons (43,560 ft³; 1234 m³; 1,233,429 liters).

Streamflows generally increase downstream because the area being drained increases. On the other hand, a stream that flows through an arid region usually decreases in discharge with distance because of high POTET rates. Such a stream is called an **exotic stream** and is exemplified by the Nile River (Table 9–4). In the United States the Colorado River flow also decreases with distance and in fact no longer actually reaches its mouth in the Gulf of California! The exotic Colorado River is depleted by upstream appropriation of water for agricultural and municipal uses, which is the subject of a Focus Study in Chapter 15. In other regions, stream drainage that does not flow into the ocean, but instead outflows through evaporation or subsurface gravitational flow, terminates in areas of **internal drainage**. Portions of Asia, Africa, Australia, Mexico, and the western United States have regions with such internal drainage patterns.

Estimates of available water supplies for the continents and a few selected nations are presented in Table 9–5. Population figures also are included to give an idea of the demand for the resource. Comparing the 182 mm (7.2 in.) mean annual runoff in the United States (48 states) to the 1300 mm (51.0 in.) in the Philippines indicates the runoff variability that occurs worldwide under varying climatic regimes.

OUR WATER SUPPLY

Water is critical for survival, yet the quality and quantity of our supply often is taken for granted. The

---

*James Tripp, 1984, "Groundwater Protection Strategies," *Groundwater Pollution, Environmental and Legal Problems.* Washington: American Association for the Advancement of Science, p. 137. Reprinted by permission.

TABLE 9–4

Largest rivers on Earth ranked by discharge volume.

Rank by Volume	Average Discharge at Mouth in Thousands of cms (cfs)	River (with Tributaries)	Outflow/Location	Length km (mi)	Rank by Length
1	212.5 (7500)	Amazon (Ucayali, Tambo, Ene, Apurímac)	Atlantic Ocean/Amapá-Pará, Brazil	6570 (4080)	2
2	79.3 (2800)	La Plata (Paraná)	Atlantic Ocean/Argentina	3945 (2450)	16
3	39.7 (1400)	Congo, also known as the Zaire (Lualaba)	Atlantic Ocean/Angola, Zaire	4630 (2880)	10
4	38.5 (1360)	Ganges (Brahmaputra)	Bay of Bengal/India	2898 (1800)	23
5	21.8 (770)	Yangtze (Chàng Chiang)	East China Sea/ Kiangsu, China	5980 (3720)	4
6	17.4 (614)	Yenisey (Angara, Selenga or Selenge, Ider)	Yenisey Gulf of Kara Sea/USSR	5870 (3650)	5
7	17.3 (611)	Mississippi (Missouri, Ohio, Tennessee, Jefferson, Beaverhead, Red Rock)	Gulf of Mexico/Louisiana	6020 (3740)	3
8	17.0 (600)	Orinoco	Atlantic Ocean/Venezuela	2737 (1700)	27
9	15.5 (547)	Lena	Laptev Sea/USSR	4400 (2730)	11
10	14.2 (500)	St. Lawrence	Gulf of St. Lawrence/Canada and U.S.	3060 (1900)	21
17	7.9 (280)	Mackenzie (Slave, Peace, Finlay)	Beaufort Sea, Northwest Territories	4240 (2630)	12
33	2.83 (100)	Nile (Kagera, Ruvuvu, Luvironza)	Mediterranean Sea/Egypt	6690 (4160)	1

U.S. water supply is derived from surface and groundwater sources that are fed by an average daily precipitation of 4244 BGD (Figure 9–14). That sum is based on an average annual precipitation value of 76.2 cm (30 in.) divided evenly among the 48 contiguous states (excluding Alaska and Hawaii). The Canadian resource greatly exceeds that of the United States.

Daily Water Budget

The 4244 BGD that makes up the U.S. water supply is not evenly distributed across the country or during the calendar year. Abundant water supplies in New England result in only about 1% of renewable available water being consumed each year, whereas in the dry Colorado River basin mentioned earlier, the discharge is completely consumed. In fact, by treaty and compact agreements, the Colorado actually is budgeted beyond its average discharge.

Figure 9–14 shows that the two principal routes of expenditure of the daily water supply are evapotranspiration from nonirrigated land and surplus. About 71% (3013 BGD) of the daily supply passes through nonirrigated land—farm crops and pastures; forest and browse (young leaves, twigs, and shoots); and noneconomic vegetation—and eventually returns to the atmosphere for another journey through the hydrologic cycle. Only 29%, or 1231 BGD, becomes surplus runoff, available for withdrawal and consumption. (This same surplus value appears in Table 9–5 under mean annual discharge for the 48 states.) In keeping with our earlier discussion, groundwater is linked to surface surpluses and thus is not presented as an independent water resource.

Consumptive uses are those that remove water from the budget at some point, without returning it farther downstream. **Withdrawal** refers to water that is removed from the supply, used for various purposes, and then returned to the supply. Withdrawn water represents opportunities to extend the resource through reuse and recycling. Often, however, withdrawn water is returned after being contaminated with chemical pollutants; or wastes from

industry, agriculture, and municipalities; or heat. Contaminated or not, because this water is returned to streams, it becomes a part of all water systems downstream. Thus, for example, the entire Missouri-Ohio-Mississippi River drainage is of immediate concern to residents of New Orleans, the last city to withdraw municipal water from the polluted river.

The estimated U.S. withdrawal of water for 1990 is 625 BGD, a sizeable increase from the 450 BGD withdrawn in 1980. Figure 9–14 illustrates the three main flows of this withdrawn water in the United States: irrigation (34%), industry (57%), and municipalities (9%). (In contrast, Canada uses only 7% of its withdrawn water for irrigation, and 84% for industry. Figure 9–15 compares a selection of other

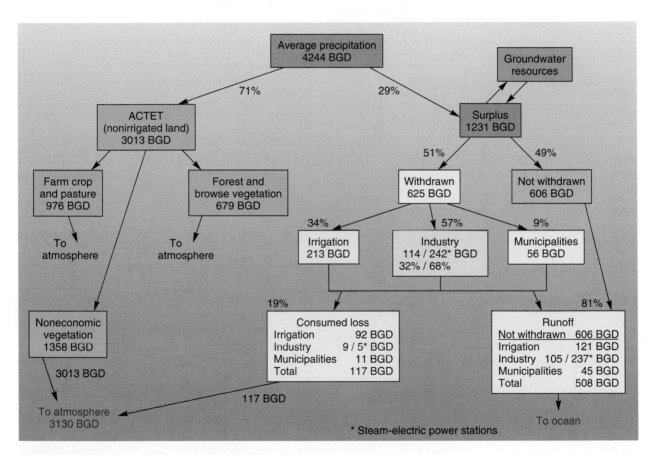

FIGURE 9–14
Daily water budget for the United States (lower 48 states) in billions of gallons a day (BGD). Adapted and updated from Abel Wolman, 1962, *Water Resources: A Report to the Committee on Natural Resources of the National Academy of Sciences,* Publication 1000–B, Washington: Natural Resources Council.

TABLE 9–5

Estimates of available global water supply.

	Land Area in Thousands of		Mean Annual Discharge (Water Supply) in		Mean Annual Runoff in		Population in Millions for	
	km²	(mi²)	km³ per yr	(BGD)	mm	(in.)	1990	2000
Africa	30,600	(11,800)	4220	(3060)	139	(5.5)	645	877
Asia	44,600	(17,200)	13,200	(9540)	296	(12.0)	3057	3544
China	9560	(3690)	2880	(2080)	300	(12.0)	1120	1256
India	3290	(1270)	1590	(1150)	485	(19.0)	832	962
Philippines	300	(116)	390	(282)	1300	(51.0)	61	75
USSR	22,400	(8650)	4350	(3150)	194	(7.6)	291	315
Australia-Oceania	8420	(3250)	1960	(1420)	245	(9.6)	27	30.4
Europe	9770	(3770)	3150	(2280)	323	(13.0)	500	513
North America (Canada, Mexico, U.S.)	22,100	(8510)	5960	(4310)	286	(11.0)	364	397
United States (48 states)	7820	(3020)	1620	(1231)	182	(7.1)	—	—
United States (50 states)	9360	(3620)	2340	(1700)	250	(9.9)	250	268
South America	17,800	(6880)	10,400	(7510)	583	(23.0)	299	360
Global (excluding Antarctica)	134,000	(51,600)	38,900	(28,100)	290	(11.0)	5249	6127

countries in their use of withdrawn water and graphically illustrates the differences between developed and developing countries.)

Returning to Figure 9–14, the 606 BGD that are not withdrawn flow as stream runoff, performing other important functions—transportation, waste removal and dilution, recreation, hydroelectric power generation, and provision of fish and wildlife habitats. Together with discharge water from irrigation, industry, and municipalities, this runoff returns to the ocean.

On an individual level, statistics suggest that urban dwellers directly use an average 180 gallons (680 liters) of water per person per day, whereas rural

FIGURE 9–15

Water withdrawal by sector for selected countries. After Gilbert R. White and others, 1982, *Resources and Needs: Assessment of the World Water Situation*, U.N. Water Conference, 1977.

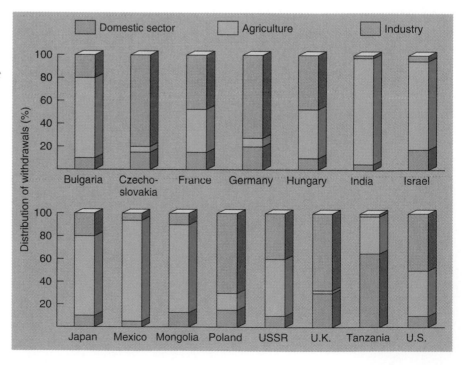

populations average only 70 gallons (265 liters) per person per day. However, indirectly each of us accounts for a much greater water demand. For the United States, dividing our total water withdrawal (625 BGD) by a population of 250 million people (1990 count) yields a per capita direct and indirect use of 2500 gallons (9500 liters) of water per day! Clearly, the average American lifestyle depends on enormous quantities of water and thus is vulnerable to shortfalls or quality problems.

Future Considerations

When the seemingly large supply of precipitation is budgeted as in Figure 9–14, the limits of the water resource become readily apparent, leading to obvious questions about how to satisfy the growing demands for water. Also, pollution limits the water-resource base in both the United States and Canada, so that even before quantity constraints are felt, quality problems may limit the health and growth of a region. Of course, the overall situation in Canada is quite different because of the large water supply and the lower overall demand generated by a smaller population of 27 million (1990 estimate).

Internationally, the world's water future appears even more complex. About one-fourth of the annual renewable water worldwide will be actively utilized by the year 2000. By that time, one-half billion people will be depending on the polluted Ganges River alone. Furthermore, of the 200 first-order river basins in the world (basins where rivers drain into an ocean, lake, or inland sea), 148 are shared by two nations, and 52 are shared by from three to ten nations. Clearly, water issues are international in scope, yet we continue toward a water crisis without a concept of a *world water economy* as a frame of reference.

SUMMARY

Using the water-balance approach, we can compare the supply of PRECIP to the natural demands of POTET for any area of land—a yard, farm, region, or for a global perspective. The uneven geographical distribution and seasonal variation of water supply and demand often requires intervention if civilization is to succeed in an area. Society interacts with surplus water supplies through its use of groundwater and surface runoff. Groundwater replenishment is tied to surface surpluses because groundwater does not exist as an independent water source.

In our present era, groundwater is being mined beyond natural recharge rates in many areas. The streams that drain the landscape often constitute our immediate water supply. Americans in the 48 contiguous states presently withdraw about one-half of the available surplus runoff for irrigation, industry, and municipal uses. Water-resource planning on a regional and global scale, using water-balance principles, is an important consideration for the future if societies are to have enough water of adequate quality.

KEY TERMS

Actual evapotranspiration
Aquiclude
Aquifer
Aquifer recharge area
Artesian water
Available water
Capillary water
Cone of depression
Confined aquifer
Consumptive uses
Deficit
Discharge
Drawdown

Evaporation
Evaporation pan
Evapotranspiration
Exotic stream
Field capacity
Gravitational water
Groundwater mining
Hygroscopic water
Internal drainage
Lysimeter
Overland flow
Permeability
Piezometric surface

Porosity	Transpiration
Potential evapotranspiration	Unconfined aquifer
Precipitation	Water detention
Rain gauge	Water table
Soil moisture storage	Wilting point
Specific yield	Withdrawal
Surplus	Zone of aeration
Total runoff	Zone of saturation

REVIEW QUESTIONS

1. What are the components that comprise the water-balance concept? How does the concept relate to the hydrologic cycle model?
2. Briefly describe the work and contribution of C. W. Thornthwaite in water-resource analysis.
3. Write out the water-balance equation, and work through it for one or two months, using specific data from Table 9–2. Try this "bookkeeping" method using data from Appendix A.
4. Explain how to derive actual evapotranspiration in the water-balance equation.
5. What is potential evapotranspiration? How do we go about measuring this potential rate? What method did Thornthwaite use to determine this value?
6. Explain the operation of soil moisture storage, soil moisture utilization, and soil moisture recharge. Include discussion of field capacity, capillary water, and wilting point concepts.
7. In the case of silt loam soil from Figure 9–6, roughly what is the available water capacity? How is this value derived?
8. What does the PRECIP–POTET bookkeeping procedure tell us about the individual months in the Kingsport, Tennessee, example?
9. Describe the different water balances in March and September in Kingsport.
10. In terms of water balance and water management, explain the logic behind the Snowy Mountains Scheme in southeastern Australia.
11. Are groundwater resources independent of surface supplies or interrelated with them? Explain your answer.
12. Make a simple sketch of the subsurface environment, labeling zones of aeration and saturation and the water table in an unconfined aquifer. Next, add a confined aquifer to the sketch.
13. At what point does groundwater utilization become groundwater mining? Use the Ogallala aquifer example to explain your answer.
14. What is the nature of groundwater pollution? Can contaminated groundwater be cleaned up easily? Explain.
15. Are there any differences in the freshwater inflow into the Atlantic Ocean and that into the Pacific Ocean? Why?
16. What are the five largest rivers on Earth in terms of discharge? Relate these to the weather patterns in each area. Can these flows be explained through an analysis of regional water budgets?
17. Describe the principal pathways involved in the water budget of the contiguous 48 states. What is the difference between consumptive use and withdrawal of water resources?
18. Characterize each of the sectors withdrawing water: irrigation, industry, and municipalities. What are the present usage trends in developed and developing nations?

SUGGESTED READINGS

Adler, Jerry, and William J. Cook, Stryker McGuire, Gerald C. Lubenow, Martin Kasindorf, Frank Maier, and Holly Morris. "Are We Running Out of Water? The Browning of America." *Newsweek* 23 (February 1981): 26–36.

Black, Peter E. *Thornthwaite's Mean Annual Water Balance*. Silviculture General Utility Library, Program GU–101. Syracuse, NY: State University College of Forestry, 1966.

Cairns, John, Jr., and Ruth Patrick, eds. *Managing Water Resources*. Environmental Regeneration Series. New York: Praeger, 1986.

Chelimsky, Eleanor. *The Nation's Water—Key Unanswered Questions About the Quality of Rivers and Streams*. Report to the Chairman, Subcommittee on Investigations and Oversight, Committee on Public Works and Transportation, House of Representatives, September 1986. Washington: General Accounting Office.

———. *Groundwater Quality—State Activities to Guard Against Contaminants*. Report to the Chairman, Subcommittee on Hazardous Wastes and Toxic Substances, Committee on Environment and Public Works, U.S. Senate, February 1988. Washington: General Accounting Office.

———. *Groundwater Standards—States Need More Information from EPA*. Report to the Chairman, Subcommittee on Hazardous Wastes and Toxic Substances, Committee on Environment and Public Works, U.S. Senate, March 1988. Washington: General Accounting Office.

———. "Fighting Groundwater Contamination." Testimony given before Subcommittee on Water Resources, Transportation, and Infrastructure, Committee on Environment and Public Works, U.S. Senate, May 1988. Washington: General Accounting Office.

Chow, Ven Te, ed. *Handbook of Applied Hydrology—A Compendium of Water-Resources Technology*. New York: McGraw-Hill, 1964.

Cobb, Charles E., Jr. "The Great Lakes' Troubled Waters." *National Geographic* 172, no. 1 (July 1987): 2–31.

Mather, John R., ed. *The Measurement of Potential Evapotranspiration*. Publications in Climatology, vol. 1. Seabrook, NJ: Johns Hopkins University, Laboratory of Climatology, 1954.

Postel, Sandra. *Water: Rethinking Management in an Age of Scarcity*. Worldwatch Paper 62. Washington: Worldwatch Institute, 1984.

Sidey, Hugh, Dan Goodgame, and David Brand. "The Big Dry," "Just Enough to Fight Over," and "Is the Earth Warming Up?" *Time* 132, no. 1 (4 July 1988): 12–18.

Thornthwaite, Charles W., and John R. Mather. *Instructions and Tables for Computing Potential Evapotranspiration and the Water Balance*. Publications in Climatology, vol. 3. Centerton, NJ: Drexel Institute of Technology, Laboratory of Climatology, 1957.

———. *The Water Balance*. Publications in Climatology, vol. 1. Centerton, NJ: Drexel Institute of Technology, Laboratory of Climatology, 1955.

Travis, Curtis, and Elizabeth L. Etnier, eds. *Groundwater Pollution—Environmental and Legal Problems*. American Association for the Advancement of Science Selected Symposia Series no. 95. Boulder, CO: Westview Press, 1984.

U.S. Geological Survey. *National Water Summary 1984*. Water-Supply Paper No. 2275. Washington: Government Printing Office, 1985.

———. *National Water Summary 1985*. Water-Supply Paper No. 2300. Washington: Government Printing Office, 1986.

Hawaiian landscape, Kauai, Hawaii. Photo by A. R. Christopherson.

10
Global Climate Systems

Earth experiences an almost infinite variety of weather. Even the same location may go through periods of changing weather. This variability, when considered along with the average conditions at a place over time, constitutes climate. Climates are so diverse that no two places on Earth's surface experience exactly the same climatic conditions, although general similarities permit grouping and classification.

In a traditional framework, early climatologists faced the challenge of identifying patterns as a basis for establishing climatic classifications. Currently at the forefront of scientific effort by climatologists is developing models that can simulate the vast interactions and causal relationships of the atmosphere and hydrosphere. This modern emphasis is explained by the fact that world climates are changing in complex ways. Possible prediction of such changes and their implications to society is an essential goal of contemporary climatology.

CLIMATE SYSTEM COMPONENTS

Our study of climate synthesizes the energy-atmosphere phenomena and water-weather phenomena discussed in the first nine chapters of this text. The condition of the atmosphere at any given place and time is called *weather*. The consistent behavior of weather over time, including its variability, represents a region's *climate*. Therefore, climate includes not just the averages but the extremes experienced in a place, for a climate may have temperatures that average above freezing every month, yet still threaten severe frost problems for agricultural activities.

Climatology, the study of climate, involves analysis of the patterns in time or space created by various physical factors in the environment. One type of climatic analysis involves discerning areas of similar weather statistics and grouping these into **climatic regions** that contain characteristic weather patterns. Climate classifications are an effort to formalize these patterns and determine their related implications to humans.

Weather components that combine to produce Earth's climates include insolation patterns, temperature, humidity, seasonal distribution and amounts of precipitation, atmospheric pressure and winds, types of weather disturbances, and cloud coverage and occurrence. Climates may be humid with distinct seasons, dry with consistent warmth, moist and cool—almost any combination is possible. There are places where it rains more than 20 cm (8 in.) each month and average monthly temperatures remain higher than 27°C (80°F) throughout the year. Other places may go without rain for 10 years at a time. (Probably none of you reading these words is in a climatic circumstance like either of these extreme examples.) Expected climatic patterns can be disrupted, as is the case with the recurring El Niño phenomenon, discussed in Focus Study 10–1.

Climates form an essential basis for *ecosystems,* the natural, self-regulating communities formed by plants and animals in their nonliving environment. On land, the basic climatic regions define large ecosystems, in regions called *biomes,* which include forest, grassland, savanna, tundra, and desert (Figure 10–1). Plant, soil, and animal communities are associated with these biomes. (In Chapters 19 and 20 we will discuss the various global ecosystems and principal biomes in detail.)

Because climate is continuously cycling through periodic change and is never really stable, these ecosystems should be thought of as being in a constant state of adaptation and response. The apparent global climatic warming trend is producing such changes in plant and animal distributions at this time. Climatologists speculate that changes in climate and natural vegetation in the next 50 years could exceed the total of all changes since the peak of the last ice age episodes, some 18,000 years ago. The possibility is that the operative factor behind these changes is more than likely human activity.

Radiation, Temperature, Pressure, and Air Masses

Before we synthesize the components of climate, let's briefly review key concepts from the first nine chapters. Uneven insolation over Earth's surface, varying with latitude, is the energy input for the climate system (Chapter 2 and Figure 2–12). Daylength and temperature patterns vary diurnally and seasonally. The principal controls of temperature are latitude, altitude, land-water heating differences, and the amount and duration of cloud cover. The pattern of world temperatures and annual temperature ranges are discussed in Chapter 5 and portrayed in Figures 5–7, 5–9, and 5–11.

Average temperatures and daylength are the basic factors that help us approximate POTET (potential evapotranspiration). These temperature variations are coupled with dynamic forces in the atmosphere forming Earth's patterns of atmospheric pressure and its global wind systems (see Figures 6–8 and 6–17). Important too are the locations and physical characteristics of air masses, those vast bodies of homogeneous air that form over oceanic and continental source regions.

Precipitation Input

The hydrologic cycle provides the basic means of transferring energy and mass through Earth's climate system. The moisture input to climate is precipitation in the forms of rain, sleet, snow, and hail. Figure 9–2 portrays the distribution of mean annual precipita-

tion in North America; Figure 10–2 shows the worldwide distribution of precipitation and identifies several patterns. In Figure 10–2, precipitation decreases from western Europe inland toward the heart of Asia, whereas high precipitation values dominate southern Asia where the mountains intercept moisture-laden monsoonal winds in the summer. Southern South America reflects a pattern of wet western (windward) slopes and dry eastern (leeward) slopes.

The lands dominated by subtropical high-pressure cells mark the location of Earth's deserts and regions of permanent drought, with bordering lands grading to grasslands and to forests as precipitation increases. The most consistently wet places on Earth straddle the equator in the Amazon region of South America, the Zaire region of Africa, and the Indonesian and southeast Asian area, all of which are influenced by equatorial low pressure and the intertropical convergence zone.

FIGURE 10–1
A northern coniferous forest biome, indicative of a cool, moist climate. Photo by author.

The El Niño Phenomenon

Climate is defined as the consistent behavior of weather over time. However, average weather conditions may include extremes that are departures from what is considered normal. Normally, as shown in Figure 6–21, the region off the west coast of South America is dominated by the northward-flowing Humboldt-Peru Current. These cold waters move toward the equator and join the westward movement of the south equatorial current. The Humboldt-Peru current is part of the overall counterclockwise circulation that normally guides the winds and surface ocean currents around the subtropical high-pressure cell dominating the eastern subtropical Pacific (visible on the world pressure maps in Figure 6–8). As a result, Guayaquil, Ecuador, normally receives 91.4 cm (36 in.) of precipitation per year with high pressures dominant, whereas islands in the western Pacific and the maritime landmasses of the Indonesian archipelago receive over 254 cm (100 in.), with low pressures dominant.

Occasionally, and for as-yet unexplained reasons, pressure patterns alter and shift from their usual locations, thus affecting surface ocean currents and weather on both sides of the Pacific. Unusually high pressure develops in the western Pacific and lower pressure in the eastern Pacific. This regional change is an indication of large-scale ocean-atmosphere interactions. Tradewinds normally moving from east to west weaken and can be replaced by an eastward (west-to-east) flow. Sea-surface temperatures off South America then rise above normal, sometimes becoming more than 8C° (14F°) warmer, replacing the normally cold, upwelling, nutrient-rich water along Peru's coastline. Such ocean-surface warming, the "warm pool," may extend to the International Date Line.

Fishermen working the waters off Peru have come to expect a slight warming of surface waters that lasts for a few weeks. Sometimes, however, the change in ocean temperature is more intense and is accompanied by strong changes of other environmental factors that may last for many months. The name El Niño ("the boy child") is used by Peruvian fishermen because these episodes periodically occur around the traditional December celebration time of Christ's birth (although they have occurred as early as spring and summer).

The shifting of atmospheric pressure and wind patterns across the Pacific is known as the Southern Oscillation and acts to initiate and support El Niño (Figure 1). Thus, the oceanic El Niño/Southern Oscillation (ENSO) designation is derived. National Oceanographic and Atmospheric

FIGURE 1

(a) Normal and (b) ENSO patterns across the Pacific Ocean. After Ramage, Colin S., "El Niño." Copyright © 1985 by Scientific American, Inc. All rights reserved.

Administration (NOAA) scientists speculate that ENSO events historically occurred in 1877–1878, 1884, 1891, 1899–1900, 1925, 1931, and 1941–1942, among other times. And, they are certain that such ENSO events occurred in 1953, 1957–1958, 1965, 1969–1970, 1972–1973, 1976–1977, and 1986–1987, with the most intense episode of this century during 1982–1983. The expected interval for occurrence is 3 to 5 years, but may range from 2 to 12 years.

The 1982–1983 El Niño, strengthened by an extremely strong Southern Oscillation, led scientists to further clarification of complex global interconnections among pressure patterns in the Pacific, ocean surface temperatures, occurrences of drought in some places, excessive rainfall in others, and the disruption of fisheries and wildlife. Estimates place the overall damage at more than $8 billion worldwide for that one event. It appears that the climate of one location is related to climates elsewhere—although it should be no surprise that Earth operates as a vast integrated system. "It is fascinating that what happens in one area can affect the whole world. As to why this happens, that's the question of the century. Scientists are trying to make order out of chaos."*

The anchovy catch off Peru dropped from 12 million tons in 1970 to a low of 0.5 million tons in 1983, following the warm water blockage of the cold, upwelling current by several ENSO events, principally in 1972–1973 and 1976–1977, culminating in the 1982–1983 episode. In addition, migratory birds did not arrive at expected places and times. Related effects occurred into the midlatitudes: droughts in South Africa, southern India, Australia, and the Philippines; strong hurricanes in the Pacific, including Tahiti and French Polynesia; and flooding in the southwestern United States and mountain states, Bolivia, Cuba, Ecuador, and Peru. The Colorado River flooding shown in the Focus Study in Chapter 15 was in large part attributable to the 1982–1983 El Niño.

Currently, remote-sensing capabilities of satellites and improved surface observations are permitting scientists to monitor ocean surface temperatures and to better predict the timing and intensity of each El Niño. Understanding will involve many principles of physical geography and will require a global-scale spatial perspective.

*Alan Strong of NOAA's National Environmental Satellite, Data, and Information Service, quoted by Barry Siegel in "El Niño: Global Climatic Patterns Become Chaotic," *Los Angeles Times*, August 17, 1983.

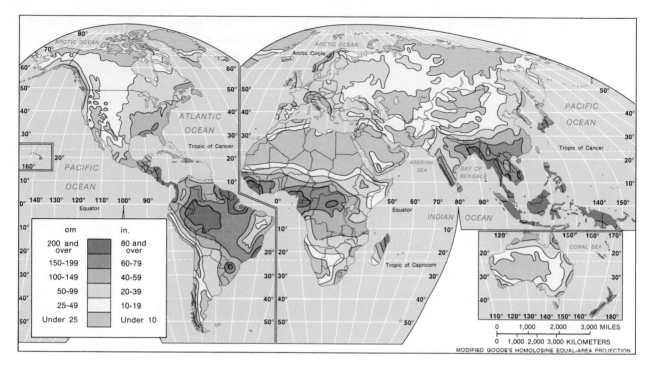

FIGURE 10–2
Worldwide average annual precipitation.

Simply comparing the two principal climatic components—temperature and precipitation—reveals important relationships indicative of the distribution of climate types (Figure 10–3). Synoptic analysis of weather elements—i.e., the characterization of weather over a region at a specific time—is like a snapshot of prevailing conditions. These simultaneous measurements and observations of weather elements are plotted on maps for analysis, important for the determination of climate types and

FIGURE 10–3
Temperature and precipitation schematic showing climatic relationships.

the development of computer models to simulate the world's climates.

Developing Climate Models

Imagine the tremendous complexity of building a mathematical model of all these interactive components over different time frames and at various scales! The most complex form of climate models, known as *general circulation models* or *GCMs*, are essentially based on mathematical models that were originally established for forecasting weather. At best they are only generalizations of reality, with the complex ocean-atmosphere interface, and the role of clouds, remaining quite elusive from climatologists. Adequate data availability is a problem.

Climatic models can be only as accurate as the data input and assumptions built into them. For instance, thermometers have been in use since the late 1600s, but a significant worldwide density of readings has been available only from around the turn of this century. Today, approximately 1000 stations worldwide balloon-launch automated instrument packages twice a day. These provide temperature, humidity, and pressure measurements at varying altitudes. New methods employing lasers (amplified light) to obtain these readings are gradually increasing the database.

To define manageable portions of Earth's climatic system for study, climatologists define statistical three-dimensional "boxes" that range from beneath the ocean to the tropopause and usually are squares roughly 300 nautical miles on a side (555 kilometers or 345 miles). Computer analysis must deal not only with the climatic components within each box, but also with the interaction among boxes. Even on a modern high-speed computer, a program can run for months and then only give general indications. An important goal of the ongoing multinational Global Change Program is to improve computer hardware and software and the database to create more accurate simulations and a better predictive capability of future climatic change.

At present, four complex GCMs (three-dimensional models) are in use in the United States: CCM—National Center for Atmospheric Research, GISS—Goddard Institute for Space Studies (NASA), GFDL—Geophysical Fluid Dynamics Laboratory (Princeton), and OSU—Oregon State University. These GCMs do not predict specific temperatures but rather offer various scenarios of global warming. Interestingly, they tend to agree on future global-scale climate changes but tend to vary in output for specific locales.

CLASSIFICATION OF CLIMATIC REGIONS

Consider this: climate cannot actually be observed and really does not exist at any particular moment. Climate is therefore a conceptual *statistical construction* from measured weather elements according to the assumptions of the classification scheme employed. The Greeks simplified their view of world climates into three zones: the torrid zone referred to warmer areas south of the Mediterranean, the frigid zone was to the north, and the area where they lived was labeled the temperate zone, which they considered the optimum climate.

Classification is the process of ordering or grouping data or phenomena in related classes. A classification based on causative factors—for example, the genesis of climate based on the interaction of air masses—is called a **genetic classification**. An **empirical classification** is based on statistics or other data used to determine general categories. Climate classifications based on temperature and precipitation data are empirical classifications. Generalizations are important organizational tools in science and are especially useful for the spatial analysis of climatic regions.

Just as there is no single universal climate classification system, neither is there a single set of empirical or genetic elements to which everyone agrees. Designing a comprehensive climate classification system is a most difficult task because there is such a great interactive complexity among weather elements. Any classification should be viewed as developmental, because it is always open to change and improvement. Such an evolutionary view is healthiest scientifically, because a dynamic classification can serve the similar function of a theory in generating further exploration and research.

Many climate classifications have been proposed over the years, based upon a variety of assumptions. Some are genetic and explanatory, based upon net radiation, thermal regimes, or air mass dominance over a region. Others are empirical and descriptive. One empirical classification system of particular interest was published by C. W. Thornthwaite in 1948. The essentials of Thornthwaite's system are the determination of moisture regions using aspects of the water-balance approach (discussed in Chapter 9) and vegetation types. Another empirical classification system, an alternative to Thornthwaite's and the one we describe here, is the Köppen classification system.

The Köppen Classification System

The Köppen classification system, widely used for its ease of comprehension, was designed by Wladimir Köppen (1846–1940), a German climatologist and botanist. First published in stages, his classification began with an article on heat zones in 1884. By 1900 he was considering plant communities in his selection of some temperature criteria, using a world vegetation map prepared by French plant physiologist A. de Candolle in 1855. Letter symbols then were added to designate climate types. Later he reduced the role played by plants in setting boundaries and moved his system strictly toward climatological empiricism. The first wall map showing world climates, coauthored with his student Rudolph Geiger, was introduced in 1928 and soon was widely used. Never completely satisfied with his system, Köppen continued to refine it until his death.

Classification Criteria. The basis of any empirical classification system is the choice of criteria used to draw lines between categories. Köppen chose mean monthly temperatures, mean monthly precipitation, and total annual precipitation to devise his spatial categories and boundaries. But we must remember that boundaries really are gray areas; they are transition zones of gradual change. The trends and overall patterns of boundary lines are more important than their precise placement, especially with the small scales generally used for world maps. The modified Köppen-Geiger system does not consider winds, temperature extremes, precipitation intensity, amount of sunshine, cloud cover, or net radiation. Yet the system is important because its correlations with the actual world are reasonable and the input data are standardized and readily available. For our

purposes of general understanding, the modified Köppen-Geiger classification is adequate.

Köppen's Climatic Designations. The Köppen system uses capital letters (A, B, C, D, E, plus H) to designate climatic categories from the equator to the poles. Five of the climate classifications are based upon thermal criteria:

A Tropical (equatorial regions)

C Mesothermal (Mediterranean, humid subtropical, marine west coast regions)

D Microthermal (humid continental, subarctic regions)

E Polar (polar regions)

H Highland (highlands at any latitude have lower temperatures—recall the normal lapse rate—and more efficient precipitation due to lower POTET demand, justifying separation of highlands from surrounding climates).

Only one is based on moisture as well:

B Dry, the deserts and steppes.

Additional lowercase letters are used to signify temperature and moisture conditions, as illustrated in Figure 10–4. Appendix A presents weather data for a selection of stations worldwide, grouped by Köppen climate type.

Figure 10–4 shows the criteria for each letter symbol within that climate type. For example, in an Af tropical rain forest climate, the A tells us that the average coolest month is above 18°C (64.4°F), and the f indicates that the weather station recording the measurement is constantly wet, with the driest month receiving at least 6 cm (2.4 in.) of precipitation. The Af climate dominates along the equator and equatorial rain forest.

In a Dfa climate, the D means that the average warmest month is above 10°C (50°F), with at least one month falling below 0°C (32°F); the f says that at least 3 cm (1.2 in.) of precipitation falls during every month; and the a indicates a warmest summer month averaging above 22°C (71.6°F). Thus, a Dfa climate is a humid-continental-hot-summer climate in the microthermal category. Using Figure 10–4 and the weather data for your city or town, see if you can determine your Köppen classification.

Tropical Climates

Consistently warm with all months averaging above 18°C (64.4°F); annual PRECIP exceeds POTET.

Af - Tropical rain forest:
All months receive PRECIP in excess of 6 cm (2.4 in.).

Am - Tropical monsoon:
A marked short dry season with 1 or more months receiving less than 6 cm (2.4 in.) PRECIP; an otherwise excessively wet rainy season. ITCZ 6-12 months dominant.

Aw - Tropical savanna:
Summer wet season, winter dry season; ITCZ dominant 6 months or less, winter water-balance deficits.

A World distribution of A climates.

Mesothermal Climates

Warmest month above 10°C (50°F); coldest month above 0°C (32°F) but below 18°C (64.4°F); seasonal climates.

Cfa, Cwa - Humid subtropical:
Hot summer; warmest month above 22°C (71.6°F) (a); year-round PRECIP (f), winter drought, summer wettest month 10 times more PRECIP than driest winter month (w).

Cfb, Cfc - Marine west coast:
Mild-to-cool summer; receives year-round PRECIP; warmest month below 22°C (71.6°F) with 4 months above 10°C (b); with 1 to 3 months above 10°C (c).

Csa, Csb - Mediterranean summer dry:
Pronounced summer drought with 70% of PRECIP in winter; hot summer with warmest month above 22°C (71.6°F) (a); mild summer; warmest month below 22°C.

C World distribution of C climates.

Microthermal Climates

Warmest month above 10°C (50°F); coldest month below 0°C (32°F); cool temperate-to-cold conditions; snow climates. In Southern Hemisphere, only in highland climates.

Dfa, Dwa - Humid continental:
Hot summer; warmest month above 22°C (71.6°F); year-round PRECIP (f), winter drought (w).

Dfb, Dwb - Humid continental:
Mild summer; warmest month below 22°C (71.6°F); year-round PRECIP (f), winter drought (w).

Dfc, Dwc, Dwd - Subarctic:
Cool summers, cold winters, year-round PRECIP (f), winter drought (w); with 1 to 4 months above 10°C (c); coldest month below -38°C (-36.4°F) (d), in Siberia only.

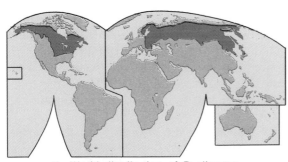

D World distribution of D climates.

FIGURE 10–4
Köppen-Geiger world climate classification system with locator maps for each of the principal climatic regions.

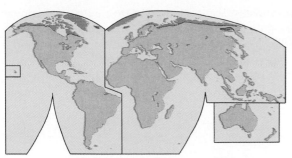

E World distribution of E climates.

Polar Climates

Warmest month below 10°C (50°F); always cold; ice climates.

ET - Tundra:
Warmest month between 0-10°C (32-50°F); PRECIP exceeds small POTET demand; snow cover 8-10 months.

EF - Ice cap:
Warmest month below 0°C (32°F); PRECIP exceeds very small POTET demand; the polar regions.

EM - Polar marine:
All months above -7°C (20°F), warmest month above 0°C; annual temperature range <17°C (30°F).

MOISTURE CLASSIFICATION

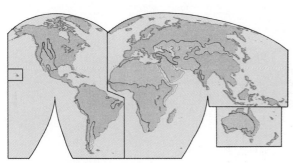

B World distribution of B climates.

Dry Arid and Semiarid Climates

POTET exceeds PRECIP in all **B** climates. Subdivisions based on PRECIP timing and amount and mean annual temperature. Boundaries determined by formulas and graphs.

BWh - Hot low-latitude desert

BWk - Cold midlatitude desert:
PRECIP less than $1/2$ POTET; mean annual temperature >18°C (64.4°F) (h); or <18°C (k). Earth's arid climates.

BSh - Hot low-latitude steppe

BSk - Cold midlatitude steppe:
PRECIP more than $1/2$ POTET but not equal to it; mean annual temperature >18°C (h); <18°C (k). Earth's semiarid climates.

B - Climate determinations (formulas in English units)

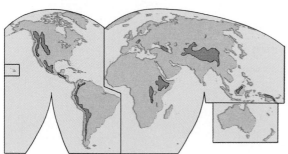

H World distribution of H climates.

Global Climate Patterns. The modified Köppen-Geiger classification system is presented on the world map in Figure 10–5. In the following sections, **climographs** are presented for cities that exemplify particular climates. These climographs show monthly temperature and precipitation, location coordinates, annual temperature range, total annual precipitation, annual hours of sunshine (as an indication of cloudiness), the local population, and a location map.

Discussions of soils, vegetation, and major terrestrial biomes, that fully integrate these global climate

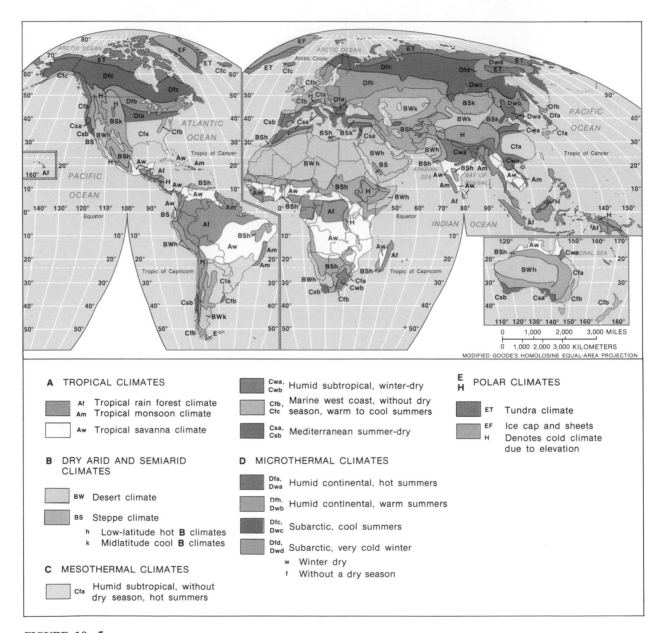

A TROPICAL CLIMATES

- Af Tropical rain forest climate
- Am Tropical monsoon climate
- Aw Tropical savanna climate

B DRY ARID AND SEMIARID CLIMATES

- BW Desert climate
- BS Steppe climate
    - h Low-latitude hot **B** climates
    - k Midlatitude cool **B** climates

C MESOTHERMAL CLIMATES

- Cfa Humid subtropical, without dry season, hot summers

- Cwa, Cwb Humid subtropical, winter-dry
- Cfb, Cfc Marine west coast, without dry season, warm to cool summers
- Csa, Csb Mediterranean summer-dry

D MICROTHERMAL CLIMATES

- Dfa, Dwa Humid continental, hot summers
- Dfb, Dwb Humid continental, warm summers
- Dfc, Dwc Subarctic, cool summers
- Dfd, Dwd Subarctic, very cold winter
    - w Winter dry
    - f Without a dry season

E **H** POLAR CLIMATES

- ET Tundra climate
- EF Ice cap and sheets
- H Denotes cold climate due to elevation

FIGURE 10–5
Map of world climates according to the Köppen-Geiger classification system.

patterns, are presented in Chapters 18, 19, and 20. Table 20–1 integrates all this information and will enhance your understanding of this chapter.

Tropical A Climates

In terms of total land and ocean area, tropical A climates are the most extensive, occupying about 36% of Earth's surface. The A climate classification extends along all equatorial latitudes, straddling the tropics from about 20° N to 20° S and stretching as far north as the tip of Florida and south-central Mexico, central India, and southeast Asia. The key temperature criterion for an A climate is that the coolest month must be warmer than 18°C (64.4°F), making these climates truly winterless. The consistent daylength and almost perpendicular Sun angle throughout the year generates this warmth. These A climates prevail in many developing countries, which are characterized by regions of poverty and struggling economies.

Tropical Rain Forest Climates (Af). Subdivisions of the A climates are based upon the distribution of precipitation during the year. Thus, in addition to consistent warmth, an Af climate is constantly moist, with no month recording less than 6 cm (2.4 in.) of precipitation. Indeed, most stations in Af climates receive in excess of 250 cm (100 in.) of rainfall a year. Not surprisingly, the water balances in these regions exhibit enormous water surpluses, creating the world's largest stream discharges in the Amazon and Congo (Zaire) Rivers.

The source of this abundant precipitation is convectional thunderstorms triggered by local heating, tending to peak each day from midafternoon to late evening inland and earlier in the day where marine influence is strong. This precipitation pattern coincides with the location of the intertropical convergence zone (ITCZ) discussed in Chapter 6. The ITCZ shifts with the high summer Sun throughout the year but influences Af regions during all 12 months.

The world climate map shows that the only interruption in the equatorial extent of Af climates occurs in the highlands of the Andes and in East Africa, where higher elevations produce lower temperatures. Mount Kilimanjaro, at 5895 m (19,340 ft), is within four degrees south of the equator but has permanent snow near its summit. Köppen placed such sites in the H (highland) climate designation.

The high rainfall of Af climates sustains lush ever-green broadleaf tree growth, producing Earth's equatorial and tropical rain forests. Their leaf canopy is so dense that light does not easily reach the forest floor, leaving the ground surface in low light and sparse plant cover. The main areas of dense surface vegetation occur along river banks, where light shines through to the surface. A major ecological issue facing the world society today is the rampant deforestation of these rain forests, discussed in more detail in Chapter 20.

The higher temperatures contribute to high bacterial action in the soil so that organic material is quickly lost, and certain minerals and nutrients are washed away by the heavy precipitation. The resulting soils are somewhat sterile and unable to support intensive agricultural activity without the addition of fertilizers.

Uaupés, Brazil (Figure 10–6), is characteristic of the Af classification. The month with the lowest precipitation receives nearly 15 cm (6 in.), and the annual range of temperature is barely 2C° (3.6F°). In all Af climates, the diurnal temperature range exceeds the annual range; in fact, day-night temperatures can range more than 11C° (20F°), which is more than five times the annual range. Other characteristic Af stations include Iquitos, Peru; Kiribati, Nigeria; Mbandaka, Zaire; Jakarta, Indonesia; and Singapore.

Tropical Monsoon Climates (Am). Tropical A climates that have a dry season—one month with rain of less than 6 cm (2.4 in.)—are candidates for this classification. The migration of the ITCZ affects these areas from 6 to 12 months of the year, with the dry season occurring when the convergence zone is not overhead. Am climates lie principally along coastal areas within the Af climatic realm and experience seasonal variation of winds and precipitation. Evergreen trees grade into thorn forests on the drier margins near the adjoining savanna climates. Typical Am stations include Freetown, Sierra Leone (west coast of Africa); Columbo, Sri Lanka; and Cairns, Australia.

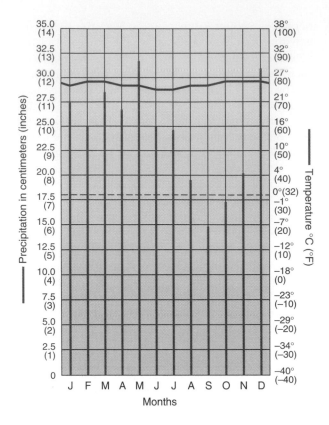

Station: Uaupés, Brazil **Af**   **Elevation:** 86 m (282.2 ft)
Lat/long: 0°08' S, 67°05' W   **Population:** 7500
Avg. Ann. Temp.: 25°C (77°F)   **Ann. Temp. Range:**
Total Annual Precipitation: 2C° (3.6F°)
            291.7 cm (114.8 In.)   **Ann. Hrs. of Sunshine:** 2018

FIGURE 10–6
Climograph for Uaupés, Brazil (Af climate).

Tropical Savanna Climates (Aw). The world climate map in Figure 10–5 shows the pattern of Aw climates poleward of the Af climates. Aw climates occur in Africa, South America, India, Southeast Asia, and the northern portion of Australia. The ITCZ dominates Aw climates for approximately six months or less of the year as it shifts with the high Sun. Summers are wetter than winters in Aw climates because of the influence of convectional rains that accompany the shifting ITCZ. The *w* designation signifies this winter-dry condition, when rains have left the area and higher pressures dominate. As a result, POTET rates exceed PRECIP in winter, and deficits in the water balance result.

Temperatures also fluctuate more than in Af climates and can each year feature two months of temperature maximums, because the Sun's direct rays are overhead twice during the year. Tropical savanna vegetation is associated with Aw climates, characterized by scattered trees throughout dominant grasslands. The trees, of course, must be drought resistant to cope with the highly variable precipitation (Figure 10–7).

Mérida, Mexico, on the northern edge of the Yucatán Peninsula, is a characteristic Aw station (Figure 10–8). Temperatures in Mérida are consistent with the pattern of tropical A climates, although the comparative winter dryness offers a few months of lower humidity. The summer rain, associated with thunderstorms, is sometimes augmented by tropical cyclones. In September of 1988 Hurricane Gilbert presented an extreme example of this moisture source, devastating most of Mérida with rain and strong winds.

Mesothermal C Climates

The mesothermal C climates occupy the second-largest percentage of Earth's surface, totaling about 27%. However, they rank only fourth when land area alone is considered. Together, A and C climates dominate over half of Earth's oceans and about one-third of its land area. The largest percentage of the world's population—approximately 55%—resides in the C climates.

FIGURE 10–7
Characteristic tropical savanna landscape in Kenya with plants able to cope with
seasonally dry water budgets. Photo by Gael Summer-Hebdon.

The word mesothermal suggests warm and temperate conditions, with the coldest month averaging below 18°C (64.4°F) but with all months averaging above 0°C (32°F). Originally, Köppen proposed that the coldest month be considered at the −3°C (26.6°F) isotherm as the boundary with colder climates, perhaps a more accurate criterion for Europe. However, for conditions in North America, the 0°C isotherm is more appropriate. The difference between the 0°C and −3°C isotherms covers an area about the width of the state of Ohio. A line denoting at least one month below freezing runs from New York City roughly along the Ohio River, trending westward until it meets the dry climates in the southeastern corner of Colorado. The line marking −3°C as the coldest month runs farther north along Lake Erie and the southern tip of Lake Michigan.

The C climates, and nearby portions of the D climates, are regions of great weather variability, for these are the latitudes of greatest air-mass conflict. The C climatic region marks the beginning of true seasonality, with contrasts in temperature as evidenced by vegetation, soils, and human lifestyle adaptations. Subdivisions of the C classification are based on precipitation variability as given in Figure 10–4.

Humid Subtropical Hot Summer Climates (Cfa, Cwa). The humid subtropical hot summer climates, with a warmest month above 22°C (71.6°F), form two principal types. The Cf type is moist all year, with the *f* indicating that all months receive precipitation above 3 cm (1.2 in.); the Cw type has a pronounced winter-dry period, designated by a *w*.

Station: Mérida, Mexico **Aw** **Elevation:** 22 m (72.2 ft)
Lat/long: 20°59' N, 89°39' W **Population:** 400,000
Avg. Ann. Temp.: 26°C (78.8°F) **Ann. Temp. Range:** 5C°(9F°)
Total Annual Precipitation: **Ann. Hrs. of Sunshine:** 2379
93 cm (36.6 in.)

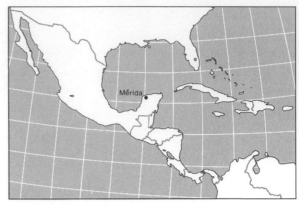

FIGURE 10–8
Climograph for Mérida, Mexico (Aw climate).

The Cfa climate, humid subtropical hot summer, is influenced during the summer by the maritime tropical air masses that are generated over warm coastal waters off eastern coasts. The warm, moist, unstable air forms convectional showers over land. In fall, winter, and spring, maritime tropical and continental polar air masses interact, generating frontal activity and frequent midlatitude cyclonic storms. Overall, precipitation averages 100 to 200 cm (40 to 80 in.) per year. Nagasaki, Japan, (Figure 10–9) is a characteristic station with no month receiving less than 7.5 cm (3.0 in.) of precipitation. Unlike Cfa stations in the United States, however, Nagasaki's summer precipitation is quite a bit higher than its winter precipitation because of the effects of the Asian monsoon.

Cfa climates are located in the eastern and east-central portions of the continents: areas of the United States, Argentina, and Brazil; southern European lowlands; eastern China; portions along the east coast of Australia; and a narrow strip along the far southeast African coast. Other Cfa climates include New Orleans; Nashville; Atlanta; Baltimore; Tokyo; Rosario, Argentina; and Sydney, Australia.

Cwa climates are related to the winter-dry, seasonal pulse of the monsoons and extend poleward from the Aw climates. Köppen identified the wettest Cwa summer month as receiving 10 times more precipitation than the driest winter month. Cherrapunji, India, mentioned in Chapter 8 as receiving the most precipitation in a single year, is an extreme example of this classification. In that location the contrast between the dry and wet monsoons is most severe, ranging from dry winds in the winter to torrential rains and floods in the summer. Downstream from the Assam Hills, such heavy rains produced floods in Bangladesh in 1988 and 1991, among other years. A representative station of this humid subtropical wet summer and dry winter regime is Ch'engtu, China. Figure 10–10 demonstrates the strong correlation between precipitation and the high Sun of summer. Other examples of Cwa climates include Allahabad, India; Lashio, Burma; Hong Kong; Tsinan, China; and Bandeirantes, Brazil.

Station: Nagasaki, Japan **Cfa** **Elevation:** 27 m (88.6 ft)
Lat/long: 32°44' N, 129°52' E **Population:** 449,000
Avg. Ann. Temp.: 16°C (60.8°F) **Ann. Temp. Range:**
Total Annual Precipitation: 21C° (37.8F°)
  195.7 cm (77 in.) **Ann. Hrs. of Sunshine:** 2131

FIGURE 10–9

Climograph for Nagasaki, Japan (Cfa climate), and the landscape of the Nagasaki region. Photo by Noboru Komine/Photo Researchers.

The habitability of the Cfa and Cwa climates and their ability to sustain populations are borne out by the concentrations in north-central India, the bulk of China's 1.1 billion people, and the many who live in climatically similar portions of the United States.

Marine West Coast Climates (Cfb, Cfc). Certainly the dominant climates of Europe and other middle-to-high-latitude west coasts are the marine west coast climates, featuring mild winters and cool summers. The *b* indicates that the warmest summer month is below 22°C (71.6°F), with at least four months averaging above 10°C (50°F); the *c* indicates that only one to three months are above 10°C; again, the *f*

denotes consistent moisture throughout the year. Chapter 5 discusses the moderating effects of a marine location on temperature patterns. Such climates demonstrate an unusual mildness for their latitude, extending all along the coastal margins of the Aleutians in the North Pacific, covering the southern third of Iceland in the North Atlantic, as well as Scandinavia and the British Isles. Unlike Europe, these climates in Canada, Alaska, Chile, and Australia are backed by mountains and remain restricted to coastal environs.

Representative marine west coast stations include Seattle; Vancouver, British Columbia; Valdivia, Chile; Hobart, Tasmania; Dublin, Ireland; Greenwich and

Station: Ch'engtu, China **Cwa** **Elevation:** 498 m (1633.9 ft)
Lat/long: 30°40' N, 104°04' E **Population:** 2,260,000
Avg. Ann. Temp.: 17°C (62.6°F) **Ann. Temp. Range:**
Total Annual Precipitation: 20C° (36F°)
            114.6 cm (45.1 in.) **Ann. Hrs. of Sunshine:** 1058

FIGURE 10–10
Climograph for Ch'engtu, China (Cwa climate).

London, England; Bordeaux, France; and Stuttgart, Germany. The cooler Cfc version of the marine west coast regime occurs at higher latitudes and elevations, such as Kodiak, Alaska; Reykjavik, Iceland; and Lerwick, Scotland. The climograph for Vancouver demonstrates the moderate temperature patterns and the annual temperature range for a Cfb station (Figure 10–11).

The dominant air mass in Cfb and Cfc climates is the maritime polar, which is cool, moist, and unstable. Weather systems forming along the polar front access these regions throughout the year, making weather quite unpredictable. Coastal fog, totaling a month or two of days, is a part of the moderating marine influence. Frosts are possible and tend to shorten the growing season.

An interesting anomaly relative to the marine west coast climate occurs in the eastern United States. Increased elevation in portions of the Appalachian highlands moderates summer temperatures in the Cfa classification, producing a Cfb designation. The climograph for Bluefield, West Virginia, (Figure 10–12) shows that its temperature and precipitation

patterns are a marine west coast type, despite its location in the east. Vegetation similarities between the Appalachians and the Pacific Northwest are quite noticeable, enticing many emigrants who relocate from the East to settle in these climatically familiar environments in the West.

Mediterranean Dry Summer Climates (Csa, Csb). Across the planet during summer months, shifting cells of subtropical high pressure block moisture-bearing winds from adjacent regions. As an example, in summer the continental tropical air mass over the Sahara in Africa shifts northward over the Mediterranean region and blocks maritime air masses and cyclonic systems. This shifting of stable, warm-to-hot, dry air over an area in summer and away from these regions in the winter creates a pronounced dry-summer and wet-winter pattern. The designation *s* specifies that at least 70% of annual precipitation occurs during the winter months.

Cool offshore currents (the California Current, Canary Current, Humboldt Current, Benguela Current, and West Australian Current) produce stability

Station: Vancouver, British Columbia   **Elevation:** sea level
Lat/long: 49°11' N, 123°10' W   **Population:** 431,000
Avg. Ann. Temp.: 10°C (50°F)   **Cfb**   **Ann. Temp. Range:**
Total Annual Precipitation: 16C°(28.8F°)
            104.8 cm (41.3 in.)   **Ann. Hrs. of Sunshine:**
                                        1723

FIGURE 10–11
Climograph of Vancouver, British Columbia. The natural vegetation of nearby
Victoria, on Vancouver Island, is representative of the marine west coast climate
(Cfb climate). Photo by author.

in overlying maritime tropical air masses along west
coasts, poleward of subtropical high pressure. The
world climate map shows Csa and Csb climates along
the western margins of North America, central Chile,
and the southwestern tip of Africa, as well as across
southern Australia and the Mediterranean basin—
the climate's namesake region.

Figure 10–13 compares the climographs of
Sevilla, Spain, (Csa) and San Francisco, California
(Csb). Coastal maritime effects moderate San Francisco's summer maximum warmest month to below
22°C (71.6°F). With greater refinement of the boundary, the *b* designation usually transitions into an *a*
designation no more than 24 to 32 km (15 to 20 mi)

inland. Along most west coasts, the warm, moist air
overlying cool ocean water produces frequent summer fog, discussed in Chapter 8. This type of fog is
not associated with the Csa climates of the Mediterranean region because water temperatures there are
higher during the summer months. Instead, the
Mediterranean regime features higher humidity values.

The Mediterranean dry-summer climate brings
water balance deficits in summer. Winter precipitation usually recharges soil moisture, but water
utilization usually exhausts soil moisture by late
spring. Thus, local water balances must be augmented by irrigation for large-scale agriculture,

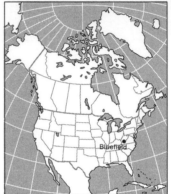

Station: Bluefield, West Virginia
Lat/long: 37°16' N, 81°13' W
Avg. Ann. Temp.:   **Cfb**
    12°C (53.6°F)
Total Annual Precipitation:
    101.9 cm (40.1 in.)

Elevation: 780 m (2559 ft)
Population: 16,000
Ann. Temp. Range:
    21C° (37.8F°)
Ann. Hrs. of Sunshine: not
    available

FIGURE 10–12
Climograph for Bluefield, West Virginia (Cfb climate); characteristic appearance of
the Appalachian highland region. Photo by author.

although some subtropical fruits, nuts, and vegetables are uniquely suited to this warm-to-hot, dry summer condition. Natural vegetation is a hard-leafed, drought-resistant variety known as *sclerophyllous vegetation;* in the western United States it is locally known as *chaparral.* Chapter 20 discusses the other types and local names for this type of vegetation in other parts of the world.

A few representative Csa stations include Sacramento; Rome; Adelaide and Perth, Australia; İzmir, Turkey; and Tunis, Tunisia. Coastal Csb stations include Santa Monica, California; Portland, Oregon; Valparaíso, Chile; and Lisbon, Portugal.

Microthermal D Climates

Humid microthermal climates experience a long winter season with some summer warmth. Microthermal signifies cool temperate-to-cold conditions, with at least one month averaging below 0°C (32°F) and at least one month above 10°C (50°F). Approximately 21% of Earth's land surface is influenced by D climates, equaling about 7% of Earth's total surface. These climates occur poleward of C climates and experience severe ranges of temperature related to continentality and air-mass conflicts. Because the Southern Hemisphere lacks substantial landmasses,

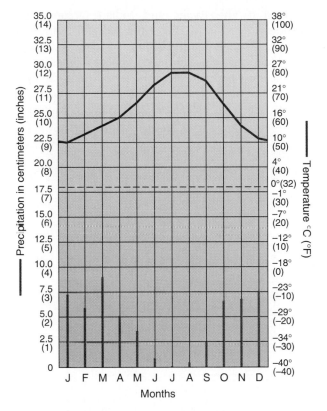

Station: Sevilla, Spain
Lat/long: 37°22' N, 6°00' W
Avg. Ann. Temp.:   **Csa**
    18°C (64.4°F)
Total Annual Precipitation:
    55.9 cm (22 in.)

Elevation: 13 m (42.6 ft)
Population: 651,000
Ann. Temp. Range:
    16C° (28.8F°)
Ann. Hrs. of Sunshine:
    2862

(a)

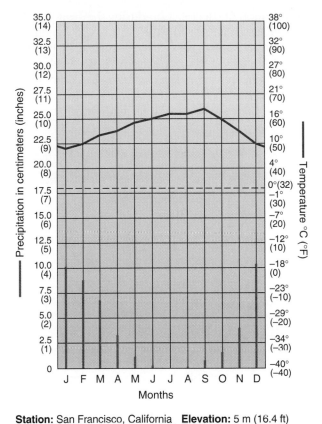

Station: San Francisco, California
Lat/long: 37°37' N, 122°23' W
Avg. Ann. Temp.:   **Csb**
    14°C (57.2°F)
Total Annual Precipitation:
    47.5 cm (18.7 in.)

Elevation: 5 m (16.4 ft)
Population: 750,000
Ann. Temp. Range:
    9C° (16.2F°)
Ann. Hrs. of Sunshine:
    2975

(b)

FIGURE 10–13
Climographs for (a) Sevilla, Spain (Csa climate), and (b) San Francisco (Csb climate).

D climates do not develop there except in highland areas. (Check Figure 10–5 and you can see that no D climates exist in the Southern Hemisphere.) The tertiary letters *a, b,* and *c* that are used with D climates carry the same meaning that they had with C climates. The D climates range from a Dfa in Chicago to the formidable extremes of a Dwd in the region of Verkhoyansk, Siberia.

Humid Continental Hot Summer Climates (Dfa, Dwa). The humid continental, hot-summer climates are distinguished from each other by the distribution of precipitation during the year. The dry winter associated with the vast Asian landmass, specifically Siberia, is exclusively assigned *w* because it is dominated by an extremely dry and frigid anticyclone in winter. The dry monsoons of southern and eastern Asia are produced in the winter months by this high-pressure system, with winds blowing outward to the Pacific and Indian Oceans. The intruding deep cold of continental polar air is a significant winter feature. Both Dfa and Dwa climates are influenced by maritime tropical air masses in the summer. In North America, frequent frontal activity is possible between conflicting maritime tropical and continental polar air masses, especially in winter.

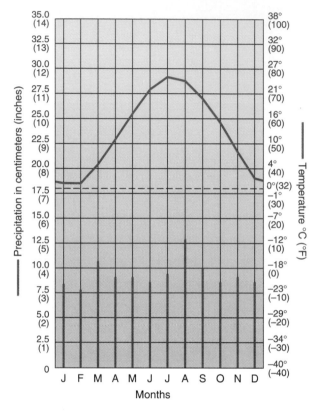

Station: New York, New York
Lat/long: 40°46' N, 74°01' W
Avg. Ann. Temp.:   **Dfa**
            13°C (55.4°F)
Total Annual Precipitation:
            112.3 cm (44.2 in.)

Elevation: 16 m (52.5 ft)
Population: 7,100,000
Ann. Temp. Range:
            24C° (43.2F°)
Ann. Hrs. of Sunshine:
            2564

(a)

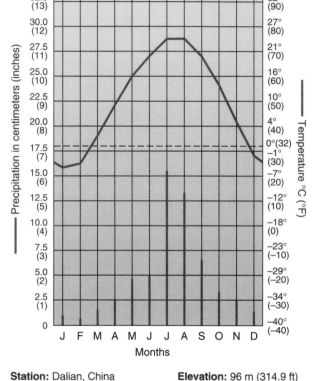

Station: Dalian, China
Lat/long: 38°54' N, 121°54' E
Avg. Ann. Temp.:   **Dwa**
            10°C (50°F)
Total Annual Precipitation:
            57.8 cm (22.8 in.)

Elevation: 96 m (314.9 ft)
Population: 4,270,000
Ann. Temp. Range:
            29C° (52.2F°)
Ann. Hrs. of Sunshine:
            2762

(b)

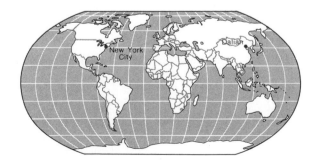

FIGURE 10–14
Climographs for (a) New York City (Dfa climate) and (b) Dalian, China (Dwa climate).

Station: Moscow, USSR   **Dfb**   **Elevation:** 156 m (511.8 ft)
Lat/long: 55°45' N, 37°34' E   **Population:** 9,900,000
Avg. Ann. Temp.: 4°C (39.2°F)   **Ann. Temp. Range:**
Total Annual Precipitation: 29C° (52.2F°)
     57.5 cm (22.6 in.)   **Ann. Hrs. of Sunshine:**
                                  1597

FIGURE 10–15
Climograph for Moscow, USSR (Dfb climate).

The climographs for New York City and Dalian, China, illustrate these two Da climates (Figure 10–14). Other examples of Dfa climates include Chicago; Omaha; Columbus, Ohio; Huron, South Dakota; and Varna, Bulgaria. Dwa climates are typified by Beijing and Tsingtao in China.

Originally, forests covered the Dfa climatic region of the United States as far west as the Indiana-Illinois border. Beyond that approximate line, the tall grass prairies extended westward to about the 98th meridian (98° W) and the approximate location of the 51 cm (20 in.) *isohyet* (line of equal precipitation), with the shortgrass prairies beyond to the west. The deep sod posed problems for the first settlers, as did the climate. However, native grasses soon were replaced with domesticated wheat and barley, and various eastern inventions helped open the region further (barbed wire, the self-scouring steel plow, well-drilling techniques, railroads, etc.). In the United States today, the Dfa climatic region is the location of corn, soybean, hog, and cattle production. The soybean was first domesticated in China but is now widely grown in this region, exceeded only by corn and wheat production.

Humid Continental Mild Summer Climates (Dfb, Dwb). Soils are thinner and less fertile in these cooler D climates, yet agricultural activity is important and includes dairy cattle, poultry, flax, sunflowers, sugar beets, wheat, and potatoes. Frost-free periods range from fewer than 90 days in the north to as long as 225 days in the south. Overall, precipitation is lower than it is in the hot summer regions to the south; however, heavier snowfall is notable and important to soil moisture recharge when it melts. Various snow-capturing strategies have been used, including fences and taller stubble left in fields to create snow drifts and thus more moisture retention on the ground.

The dry-winter aspect of this cool-summer climate occurs only in Asia, in a far-eastern area poleward of the winter-dry C climates. A representative Dwb station is Vladivostok, USSR, usually one of only two ice-free ports in the Soviet Union. Characteristic Dfb stations are Duluth, Minnesota, and Moscow and Saint Petersburg in the USSR. Figure 10–15 presents a climograph for Moscow, which is at 55° N, or about the same latitude as the southern shore of Hudson Bay in Canada.

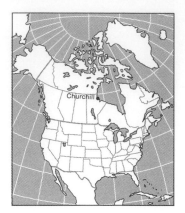

Station: Churchill, Manitoba   **Dfc**
Lat/long: 58°45' N, 94°04' W
Avg. Ann. Temp.: –7°C (19.4°F)
Total Annual Precipitation:
        44.3 cm (17.4 In.)

Elevation: 35 m (114.8 ft)
Population: 1400
Ann. Temp. Range:
        40C° (72F°)
Ann. Hrs. of Sunshine:
        1732

FIGURE 10–16
Climograph for Churchill, Manitoba (Dfc climate).

Subarctic Climates (Dfc, Dwc, Dwd). As we move farther poleward, seasonality becomes greater. The growing season, though short, is more intense during long summer days, with at least one to four months averaging above 10°C (50°F). The world climate map illustrates the land covered by these three cold climates: vast stretches of Alaska, Canada, northern Scandinavia, and the Soviet Union. Discoveries of minerals and petroleum reserves have led to new interest and settlement in extreme portions of these regions.

Those areas that receive 25 cm (10 in.) or more of precipitation per year on the northern continental margins are covered by the so-called "snow forest" of fir, spruce, larch, and birch—the boreal forests of Canada and the taiga of the Soviet Union. (*Taiga* is the more general term used in ecological studies.) These forests are a transition to the more northern open woodlands and to the tundra region of the far north. Köppen used the 10°C (50°F) temperature criterion for the warmest month in these climates because it is generally the lowest average tempera-

ture regime that can support a forest cover. Soils are thin in these lands once scoured by glaciers, and precipitation is low. However, POTET is also low, producing soils that are generally moist and either partially or totally frozen beneath the surface, a phenomenon known as *permafrost*.

The Churchill, Manitoba, climograph (Figure 10–16) shows average monthly temperatures below freezing for seven months of the year, during which time light snow cover and frozen ground persist. Churchill is representative of the Dfc subarctic climate, with an annual temperature range of 40C° (72F°) and a low precipitation total of 44.3 cm (17.4 in.), spread fairly evenly over the year but reaching a maximum in late summer. High pressure dominates Churchill during its cold winter—this is the source area for the continental polar air mass. Other Dfc climates are found at Fairbanks and Eagle, Alaska; Dawson, Yukon; Fort Vermillion, Alberta; and Trondheim, Norway. The Dwc and Dwd climates occur only within the USSR. Köppen selected the tertiary letter *d* for the intense cold of Siberia and north-

Station: Verkoyansk, USSR **Dwd** **Elevation:** 137m (449.5 ft)
Lat/long: 67°33' N, 135°23' E **Population:** 1400
Avg. Ann. Temp.: −15°C (5°F) **Ann. Temp. Range:**
Total Annual Precipitation: 63C° (113.4F°)
15.5 cm (6.1 in.) **Ann. Hrs. of Sunshine:**
not available

FIGURE 10–17
Climograph for Verkhoyansk, USSR (Dwd climate).

central and eastern Asia; it designates a coldest month with an average temperature lower than −38°C (−36.4°F).

A typical Dwd station is Verkoyansk, Siberia (Figure 10–17). For four months of the year average temperatures fall below −34°C (−29.2°F). Verkhoyansk frequently reaches minimum winter temperatures that are lower than −68°C (−90°F). However, as pointed out in Chapter 5 (Figure 5–8), higher summer temperatures in the same area produce the world's greatest annual temperature range from winter to summer, a remarkable 63C° (113.4F°). Winter lifestyles feature brittle metals and plastics, triple-thick window panes, and temperatures that render straight antifreeze a solid.

Polar Climates

The polar climates—ET, EF, and EM—cover about 19% of Earth's total surface, about 17% of its land area. These climates have no true summer similar to that of the lower latitudes. Poleward of the Arctic and

Antarctic circles, daylength increases in summer until it becomes continuous, yet average monthly temperatures never rise above 10°C (50°F). Daylength, which in part determines the amount of insolation received, and low Sun altitude are the principal climatic factors in these frozen and barren regions.

In an ET tundra climate, the temperature of the warmest month is between 0°C and 10°C (32°F and 50°F), and PRECIP is in excess of a very small POTET demand. The land is under continuous snow cover for 8 to 10 months, but when the snow melts and spring arrives, numerous plants appear on the landscape. Characteristic tundra vegetation is composed of stunted sedges, mosses, flowering plants, and lichens. Much of the area experiences permafrost conditions. The tundra is also the summer home of mosquitoes of legend and black gnats. Like the D climates, ET climates are strictly a Northern Hemisphere occurrence, except for elevated mountain locations in the Southern Hemisphere and a portion of the Antarctic Peninsula. A narrow band of

ET climates stretches along the northern extreme of Canada, Scandinavia, and Asia. As an example, at lower latitudes, the summit of Mount Washington in New Hampshire (1914 m or 6280 ft) partially qualifies as a highland ET climate of small scale.

Most all of Antarctica falls within the EF ice-cap climate as does the North Pole, with all months averaging below freezing. Both regions are dominated by dry, frigid air masses, with vast expanses that never warm above freezing. The area of the North Pole is actually a sea covered by ice, whereas Antarctica is a substantial continental landmass covered by Earth's greatest ice sheet. Winter minimums at the South Pole (July) frequently drop below the temperature of frozen CO_2 or "dry ice" ($-78°C$; $-109°F$).

Antarctica is constantly snow covered but receives less than 8 cm (3 in.) of precipitation each year. Because of consistent winds, it is difficult to discern what is new snow and what is old snow being blown

FIGURE 10–18
Ever-frozen Antarctic landscape, Earth's largest repository of freshwater. Photo by Jim Ferry.

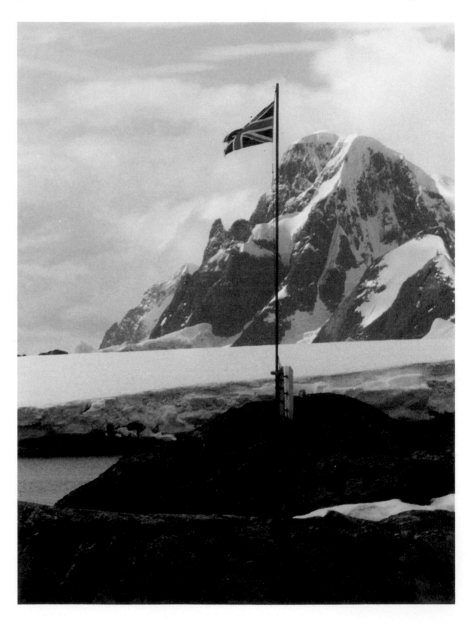

around. Antarctic ice is several kilometers thick and is the largest repository of freshwater on Earth. This ice represents a vast historical record of Earth's atmosphere, within which thousands of volcanic eruptions worldwide have deposited ash layers, and ancient combinations of atmospheric gases lie trapped in frozen bubbles (Figure 10–18).

The EM climate, or polar marine regime, was not proposed by Köppen, but by later researchers. EM stations are more moderate in winter, with no month below −7°C (20°F), yet they are not as warm as ET climates. Because of marine influences, annual temperature ranges do not exceed 17C° (30F°). EM stations occur along the Bering Sea, the tip of Greenland, northern Iceland, Norway, and in the Southern Hemisphere, generally over oceans between 50° and 60° S. Precipitation, which frequently falls as sleet, is greater in these regions than in continental E climates.

Dry Arid and Semiarid Climates

The B climates are the only ones that Köppen classified according to the amount and distribution of precipitation. These are the world's arid and semiarid regions—deserts and steppes possessing unique plants, animals, and physical features. The mountains, rock strata, long vistas, and the resilient struggle for life are all magnified by the dryness. Because of the sparseness of vegetation, the landscape is bare and exposed; POTET exceeds PRECIP in all parts of B climates, creating varying amounts of permanent water deficits that distinguish the different subdivisions within this climatic group. Vegetation is typically xerophytic, i.e., drought resistant, waxy, hard leaved, and adapted to aridity and low transpiration loss. Along water courses, plants called phreatophytes, or "water-well plants," penetrate to great depths for their water needs.

The B climates occupy more than 35% of Earth's land area, clearly the most extensive climate over land. The world climate map reveals the pattern of Earth's dry climates, which cover broad regions between 15° and 30° N and S. In these areas the subtropical high-pressure cells predominate, with subsiding, stable air and low relative humidity. Under generally cloudless skies, these subtropical deserts extend to western continental margins, where cool, stabilizing ocean currents operate offshore and summer advection fogs form. The Atacama Desert of Chile, the Namib Desert of Namibia, the Western Sahara of Morocco, and the Australian Desert each lie adjacent to a coastline. Extension of these dry regions into higher latitudes is associated with rain shadows, induced by orographic lifting over western mountains in North America and Argentina. Interior Asia, far distant from any moisture-bearing air masses, is also included within the B climates (Figure 10–19).

The major subdivisions are the BW deserts, where PRECIP is less than one-half of POTET, and the BS steppes, where PRECIP is more than one-half of POTET. In an effort to better approximate the dry climates, Köppen developed simple formulas to determine the usefulness of rainfall, based on the season in which it falls—whether it falls principally in the summer with a dry winter, or in the winter with a dry summer, or whether the rainfall is evenly distributed. Winter rains are the most effective because they fall at a time of lower POTET.

Köppen considered the summer dry if the wettest winter month was three times wetter than the driest summer month. The reverse reflects the role of POTET: winter is considered dry if the wettest summer month receives 10 times the precipitation of the driest winter month. Small graphs on Figure 10–4 help demonstrate these determinations. A tertiary lowercase letter is defined by the average annual temperature: temperatures above 18°C (64.4°F) are signified by an *h* for hot; those below 18°C are represented by a *k* for cold.

Hot Low-Latitude Desert Climates (BWh).

The BWh climates are Earth's true tropical and subtropical deserts. They are concentrated on the western side of continents, although Egypt, Somalia, and Saudi Arabia also fall within this classification. Rainfall is very unreliable in BWh climates, usually the result of local convectional showers during the summer season. Some regions receive near zero amounts, whereas others may receive up to 35 cm (14 in.). Specific annual and daily desert temperature regimes, including the highest record temperatures, are discussed in Chapter 5; surface energy budgets are covered in Chapter 4. Representative BWh stations include Yuma, Arizona; Aswan, Egypt; Lima,

Peru; Bahrain Island in the Persian Gulf; and Riyadh, Saudi Arabia (Figure 10–20).

Cold Midlatitude Desert Climates (BWk). On the world climate map, this classification covers only a small area: the southern Soviet Union, the Gobi Desert, and Mongolia in Asia; the central third of Nevada and areas of the American Southwest, particularly at high elevations; and Patagonia in Argentina. Because of lower temperatures and lower POTET values, rainfall must be low for stations to meet the BWk criteria; consequently, total average annual rainfall is only about 15 cm (6 in.). Albuquerque, New Mexico, with 20.7 cm (8.1 in.) of precipitation and an annual average temperature of 14°C (57.2°F) (which is 4C° or 7.2F° below the standard), is representative of this climate type (Figure 10–21). A comparison of the BWk and BWh stations shows an interesting similarity in the annual range of temperatures and a distinct difference in precipitation patterns.

Hot Low-Latitude Steppe Climates (BSh). These climates are generally located around the periphery of hot deserts, where shifting subtropical high-pressure cells create a distinct summer-dry and winter-wet precipitation pattern. This climate type is best identified around the periphery of the Sahara and in Iran, Afghanistan, and southern USSR. Along the southern margin of the Sahara lies the Sahel, a drought-tortured region of expanding desert conditions. Human populations in this region have suffered great hardship as desert conditions have gradually expanded over their countries. Average annual precipitation in BSh areas is usually below 60 cm (23.6 in.), less than the Köppen criterion for this classification of 86 cm (33.9 in.). A few sample stations include Santiago, Chile; Kayes, Mali; Las Palmas, Canary Islands; Daly Waters, Australia; and Poona, India.

Cold Midlatitude Steppe Climates (BSk). The BSk climates occur poleward of about 30° latitude

FIGURE 10–19
Desert plants are particularly adapted to the harsh environment. Photo by author.

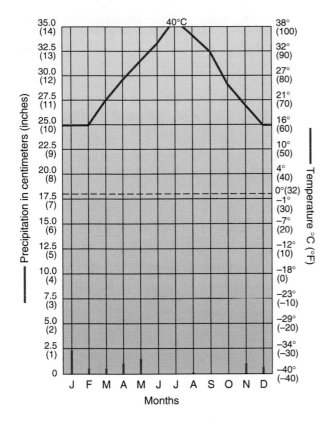

Station: Riyadh, Saudi Arabia **BWh**
Lat/long: 24°42' N, 46°43' E
Avg. Ann. Temp.: 26°C (78.8°F)
Total Annual Precipitation:
    8.2 cm (3.2 in.)

Elevation: 609 m (1998 ft)
Population: 1,308,000
Ann. Temp. Range:
    24C° (43.2F°)
Ann. Hrs. of Sunshine:
    not available

FIGURE 10–20
Climograph for Riyadh, Saudi Arabia (BWh climate).

Station: Albuquerque,   **BWk**
    New Mexico
Lat/long: 35°03' N, 106°37' W
Avg. Ann. Temp.: 14°C (57.2°F)
Total Annual Precipitation:
    20.7 cm (8.1 in.)

Elevation: 1620 m (5315 ft)
Population: 370,000
Ann. Temp. Range:
    24C° (43.2F°)
Ann. Hrs. of Sunshine:
    3420

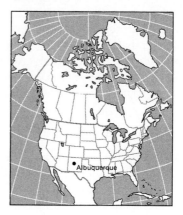

FIGURE 10–21
Climograph for Albuquerque, New Mexico (BWk climate).

Station: Semipalatinsk,   **BSk**
   USSR
Lat/long: 50°21' N, 80°15' E
Avg. Ann. Temp.: 3°C (37.4°F)
Total Annual Precipitation:
   26.4 cm (10.4 in.)

Elevation: 206 m (675.9 ft)
Population: 330,000
Ann. Temp. Range:
   39C° (70.2F°)
Ann. Hrs. of Sunshine:
   not available

FIGURE 10–22
Climograph for Semipalatinsk, USSR (BSk climate).

and the BWk climates. Such midlatitude steppes are not generally found in the Southern Hemisphere. As with other B climate classifications, BSk rainfall variability is large and undependable, ranging from 20 to 40 cm (7.9 to 15.7 in.). Not all rainfall is convectional, for cyclonic systems penetrate the continents; however, most produce little precipitation in this climate region. A sample of BSk stations includes Denver; Salt Lake City; Cheyenne, Wyoming; Lethbridge, Alberta; Ulaanbaatar, Mongolia; and Semipalatinsk, USSR (Figure 10–22).

FUTURE CLIMATE PATTERNS

The greenhouse effect and the contributive role of radiatively active gases such as carbon dioxide, CFCs, and methane are discussed in Chapters 3 and 5. Earlier in this chapter we mentioned the complexity of developing and running global circulation models to predict the climatic implications of increases in these gases, but there is little scientific doubt that air temperatures are the highest they have been since air-temperature recordings were begun in earnest

more than 100 years ago. In terms of **paleoclimatology**, the science that studies past climates (Chapter 17), Earth is within 1°C of equaling the highest average temperature of the past 125,000 years. Eight of the warmest years in this century were in the past decade (1990, 1988, and 1987 broke all records) and are thought to be attributable to a buildup of greenhouse gases. Richard Houghton and George Woodwell describe the present climatic condition:

> The world is warming. Climatic zones are shifting. Glaciers are melting. Sea level is rising. These are not hypothetical events from a science fiction movie; these changes and others are already taking place, and we expect them to accelerate over the next years as the amounts of carbon dioxide, methane, and other trace gases accumulating in the atmosphere through human activities increase.*

Scientists are attempting to determine the difference between forced fluctuations (human-caused) and unforced (natural) fluctuations as a key to

---

*Richard Houghton and George Woodwell, "Global Climate Change," *Scientific American,* April 1989, p. 36.

unraveling future climate trends. Because the gases generating any forced fluctuations are anthropogenic, various management strategies are possible to reduce their input. Several scenarios have been developed using computer-driven general circulation models (GCMs). These climate models produce results generally consistent with empirical evidence, although uncertainty remains as to actual overall global climate sensitivity, the role of oceans in heat absorption and transport, and the changeable role that clouds may or may not play.

Nevertheless, the confidence level is high for the current GCM projections of greenhouse warming substantiated by a database of billions of observations of the atmosphere. The 1990 report of the Intergovernmental Panel on Climate Change (IPCC), a broad-based group of over 200 scientists established in 1988 by the United Nations Environment Programme and representing the efforts of 25 countries, concluded that there is ". . . virtual unanimity among greenhouse experts that a warm-

ing is on the way and that the consequences will be serious." A minority of scientists persist as critics of the greenhouse warming model, and although no member of the IPCC agrees with these critics, all points of view should be considered.

Figure 10–23 graphs expected temperature changes up to the year 2050 for three scenarios. Scenario A assumes that activities continue with no real change in the behavior of society, which will produce a doubling of CO_2 by 2030 (assuming that no large volcanic eruptions inject into the atmosphere material that reflects or absorbs insolation). Scenario B represents a modification in human activities, leading to limited emissions frozen at current levels; this merely delays the doubling of CO_2 to the middle of the century. An unrealistic, yet most desirable, Scenario C involves drastic reductions in emissions and would produce significant delays in climatic warming.

Figure 10–24 presents three computer-generated predictions for July of 1990, 2000, and 2029, using the

FIGURE 10–23
Three predictions of temperature change influenced by trace gases. After J. Hansen, I. Fung, P. Store, *et al.,* "Prediction of Near-Term Climate Evolution," in *Proceedings of the First North American Conference on Preparing for Climate Changes,* Washington: Government Institutes, 1987, p. 38.

MEAN ΔT = +0.5°C   JULY 1990

MEAN ΔT = +0.6°C   JULY 2000

MEAN ΔT = +1.5°C   JULY 2029

ΔT(°C)

-3   -2   -1   0   1   2   3   5

FIGURE 10–24 (left)

July temperature projections for 1990, 2000, and 2029 for Scenario B. From Hansen, J., I. Fung, A. Lacis, S. Lebedeff, D. Rind, R. Ruedy, and G. Russ, "Global Climate Changes as Forecast by Goddard Institute for Space Studies Three-Dimensional Model," *Journal of Geophysical Research* 43, no. 8 (August 1988): plate 6. Published by the American Geophysical Union.

the IPCC, a warming of 2.5C° (4.5F°) is predicted for this timeframe (1990–2029), or on the low side of forecasts. With a carbon dioxide concentration of 550 ppm (expected by the year 2020), a temperature increase of 3C° (5.4F°) is forecast for the equatorial regions. This equatorial increase translates to a high-latitude and polar increase of about 10C° (18F°), notable on the maps.

Consistent with other computer-generated studies, unambiguous warming appears over low-latitude oceans and interior areas of Asia, especially China, with the greatest warming assumed to occur over Arctic and Antarctic regions of sea ice. In the United States, the East shows greater warming than the West as a result of maritime effects, pressure patterns, and winds—northerly winds are expected to increase in the West, whereas southerly winds are expected to increase in the East, because of a strengthening of the related subtropical high-pressure cells. These forecasts are valuable for the trends they show and the need for future research and appropriate public policy which they demand. Such climatic changes, if they occur, would have a major impact on the biosphere and challenging implications for society.

Consequences of Climatic Warming

The consequences of uncontrolled atmospheric warming are complex. Regional climate responses are expected as temperature, precipitation, soil moisture, and air-mass characteristics change. Although the ability to accurately forecast such regional changes is still evolving, some implications are proposed. For example, modern single-crop agriculture is more delicate and susceptible to temperature change, water demand and irrigation needs, and soil chemistry than traditional multicrop agriculture. Specifically, the southern and central grain-producing areas of North America are forecast to experience

assumptions of Scenario B. According to that projection, the year 2029 could experience a mean temperature 1.5C° higher than that of the 1950–1980 period. There is a wide range of uncertainty in the estimates, from 1.5 to 4.5C° (2.7–8.1F°); according to hot and dry weather by the middle of the next century as a result of higher temperatures. An increased probability of extreme heat waves is forecast for these U.S. grain regions. Available soil moisture is projected to be at least 10% less throughout the midlatitudes in the next 30 years.

Today, the United States and Canada sell wheat to China and the Soviet Union. If the Soviet grain areas become wetter and gain a longer frost-free period, grain sales could shift in the other direction. Specifically, as U.S. production decreases, the longer growing seasons, increased temperatures and thermal energy patterns, and possible increased precipitation at higher latitudes will benefit Canadian cropping patterns as well. The possibility exists that billions of dollars of agricultural losses in one region could be countered by billions of dollars of gain in another.

Scientists have postulated possible adaptations, such as changing crop varieties to late-maturing, heat-resistant types, and adjustments in soil-management practices, such as fertilizer applications and irrigation schedules. Studies completed at the Universities of Toronto and Wisconsin–Madison postulate, with an admitted degree of uncertainty, that greenhouse warming will increase the temperature of the Great Lakes, which in turn will be an advantage for various species of fish. On the other hand, a warming of small Canadian lakes might prove harmful to aquatic life. Also, land-based animals will have to adapt to changing patterns of available forage, for animals (mammalian fauna in particular) are habituated and adapted to the climate zones in which they live. Increased temperatures already are causing wide-ranging thermal stress to embryos as they reach their thermal limits. Warming of ocean water is endangering corals worldwide, and they appear to be deteriorating at record levels. This developing condition is discussed further in Chapter 16.

Polar and High-Latitude Warming Effects. Perhaps the most pervasive climatic effect of increased warming would be the rapid escalation of ice melt. The additional water, especially from continental ice masses that are grounded, would raise sea level worldwide. Satellite remote-sensing technology allows monitoring of global sea-ice and continental-ice distributions.

Scientists currently are studying the ice sheets of Greenland and Antarctica for possible changes in the operation of the hydrologic cycle, including snow-lines and the rate at which icebergs break off (calve) into the sea. The key area being watched is the **West Antarctic ice sheet**, where the Ross Ice Shelf holds back vast grounded ice masses. The late 1980s saw several of the largest-ever icebergs break away from Antarctica and the acceleration of several ice streams (channels of ice) flowing from the ice sheet. Several interpretations for these changes are possible.

The possible sea-level rise must be expressed as a range of values, based on a variety of assumptions that are under constant reassessment. Estimates of future rise in sea level range from a conservative 30–110 cm (1–4 ft) to a pessimistic 6 m (20 ft). These levels could be further augmented by melting of alpine and mountain glaciers, which have experienced about a 30% decrease in overall ice mass during this century.

A quick survey of world coastlines shows that even moderate changes could bring disaster of unparalleled proportions. At stake are the river deltas, lowland coastal farming valleys, and low-lying mainland areas. In the worst-possible scenario, a rise of 6 m (20 ft) would flood 20% of Florida; inundate the Mississippi floodplain as far inland as St. Louis; flood the Pampas of Argentina; inundate Venice; submerge the Bahamas, The Netherlands, the Maldives; and more. All surviving coastal areas would contend with high water and high tides—but again this is with a worst-possible scenario.

Solutions. The position is held by some government leaders that society has adapted to many problems in the past; capital can be raised and technology harnessed to block the flood and guard the coastlines and deltas. Although such adaptive strategies may be needed, no substantive long-range planning is going forward. Why are we waiting? Why not take action now? Whatever the actual impact—from major financial and cultural irritation to outright catastrophe—solutions exist to reduce the severity of the hazard. In addition, individual conservation and a demand for energy efficiency remain cheap ways to stretch present energy resources and reduce carbon dioxide emissions.

Participants at world conferences—such as those in The Hague, Ottawa, Geneva, and Cairo in 1989 and

1990—have agreed that governmental policies can profoundly alter the greenhouse effect. The 1989 Cairo Compact and the 1990 Intergovernmental Panel on Climate Change report to the United Nations World Climate Conference are steps in the right direction. The 1992 United Nations Conference on sustainable development is to set specific protocols for greenhouse gas emission limits and goals for emission reduction.

SUMMARY

Climate is a synthesis of weather phenomena at many scales, from planetary to local. Earth experiences a wide variety of climatic conditions. A modified empirical climate classification system developed by Köppen provides a basis for describing the various distinct environments on Earth: tropical A, mesothermal C, microthermal D, polar E, and arid B climates. Highland climates are assigned an H to distinguish them from the others and to denote the conditions created by elevation. The main climatic groups are divided into subgroups, based upon temperature and the seasonal timing of precipitation.

Various activities of present-day society possibly are producing climatic changes, particularly a global warming trend. The 1980s ended with a series of the highest average annual temperatures experienced since measurements began. There is little scientific doubt that this warming is related to the greenhouse effect and that various control options need to be activated in a concerted global effort, although some uncertainties remain to be resolved. Climatologists are constructing and refining general circulation models in an effort to better understand climate patterns in the next century.

KEY TERMS

Classification	Empirical classification
Climatic regions	Genetic classification
Climatology	Paleoclimatology
Climographs	West Antarctic ice sheet

REVIEW QUESTIONS

1. Define climate and compare it to weather. What is climatology?
2. What created the climatic disruption associated with El Niño? What were some of the changes and effects that occurred worldwide?
3. Evaluate the relationships among a climatic region, ecosystem, and biome.
4. How do radiation receipts, temperature, air pressure inputs, and precipitation patterns interrelate to produce climate types?
5. What are the differences between a genetic and an empirical classification system?
6. Describe Köppen's approach to climatic classification. What was the basis of the system he devised?
7. List and discuss each of the principal climate designations. In which one of these general types do you live? Which classification is the only type associated with the distribution and amount of precipitation?
8. What is a climograph, and how is it used to display climatic information?
9. Which of the major climate types occupies the most land and ocean area on Earth?
10. Characterize the tropical A climates in terms of temperature, moisture, and location.
11. Using Africa's tropical climates as an example, characterize the climates produced by the seasonal shifting of the ITCZ with the high Sun.
12. Mesothermal C climates occupy the second-largest portion of Earth's entire surface. Describe their temperature, moisture, and precipitation characteristics.

13. Explain the distribution of the Cfa and Csa climates at similar latitudes and the difference in precipitation patterns between the two types. Describe the difference in vegetation associated with these two climate types.
14. Which climates are characteristic of the Asian monsoon region? What do each of the letters in that Köppen classification indicate?
15. How can a marine west coast climate type (Cfb) occur in the Appalachian region of the eastern United States? Explain.
16. What role do offshore currents play in the distribution of the Csb climate designation? What type of fog is formed in these regions?
17. Discuss the climatic designation for the coldest places on Earth outside the poles. What do each of the letters in the Köppen classification indicate?
18. What is the largest single group of climates on Earth in terms of land area alone? Analyze the general pattern of this climate type.
19. In general terms, what are the differences among the four desert classifications? How are moisture and temperature distributions used to differentiate these subtypes?
20. Explain what is meant by the trace-gas scenarios shown in Figure 10–23. What are the implications of these trends?
21. Describe the potential climatic effects of global warming on polar and high-latitude regions. What are the implications of these climatic changes for persons living at lower latitudes? The effects on agricultural activity?

SUGGESTED READINGS

Barth, Michael C., and James G. Titus, eds. *Greenhouse Effect and Sea Level Rise*. New York: Van Nostrand Reinhold, 1984.

Griffiths, John F., and Dennis M. Driscoll. *Survey of Climatology*. Columbus: Macmillan Publishing Co., 1982.

Haurwitz, B., and J. M. Austin. *Climatology*. 2d ed. Englewood Cliffs, NJ: Prentice Hall, 1966.

Lydolph, Paul E. *The Climate of the Earth*. Totowa, NJ: Rowman & Allanheld, 1985.

Matthews, Samuel W. "Under the Sun—Is Our World Warming?" *National Geographic* 178, no. 4 (October 1990): 66–99.

Mintzer, Irving M. *A Matter of Degrees: The Potential for Controlling the Greenhouse Effect*. Research Report No. 5. Washington: World Resources Institute, 1987.

Oliver, John E., and Rhodes W. Fairbridge, eds. *The Encyclopedia of Climatology*. New York: Van Nostrand Reinhold Co., 1987.

Revkin, Andrew C. "Endless Summer: Living with the Greenhouse Effect." *Discover* 9, no. 10 (October 1988): 50–61.

Strahler, Arthur N., and Alan H. Strahler. *Modern Physical Geography*. 3d ed. New York: John Wiley & Sons, 1987.

Thomas, Robert. "Future Sea Level Rise and Its Early Detection by Satellite Remote Sensing." In James Titus, ed. *Effects of Changes in Stratospheric Ozone and Global Climate*. Vol. 4, *Sea Level Rise*. Washington: Environmental Protection Agency, 1986.

Thornthwaite, C. W. "An Approach Toward a Rational Classification of Climate." *Geographical Review* 38 (1948): 55–94.

Tooley, Michael J., and Ian Shennan, eds. *Sea-Level Changes*. Oxford: Basil Blackwell & the Institute of British Geographers, 1987.

Topping, John C., Jr., ed. "Coping with Climate Change." *Proceedings of the Second North American Conference On Preparing for Climate Change*. Washington: Climate Institute, 1989.

Trewartha, Glenn T., and Lyle H. Horn. *An Introduction to Climate*. 5th ed. New York: McGraw-Hill, 1980.

Wilcox, Arthur A. "Köppen After Fifty Years." *Annals of the Association of American Geographers* 58, no. 1 (March 1968): 12–28.

E arth is a dynamic planet whose surface is actively shaped by physical agents of change. Part Three is organized around two broad systems of these agents—endogenic and exogenic. The **endogenic system** (Chapters 11 and 12) encompasses internal processes that produce flows of heat and material from deep below the crust, powered by radioactive decay—the solid realm of Earth. Earth's surface responds by moving, warping, and breaking, sometimes in dramatic episodes of earthquakes and volcanic eruptions.

At the same time, the **exogenic system** (Chapters 13–17) involves external processes that set air, water, and ice into motion, powered by solar energy—the fluid realm of Earth's environment. These media are sculpting agents that carve, shape, and reduce the landscape. One such process—weathering—breaks up and dissolves the crust, making materials available for erosion, transport, and deposition by rivers, winds, wave action along coastlines, and flowing glaciers. Thus, Earth's surface is the interface between two systems, one that builds the landscape and one that reduces it.

PART THREE

The Earth-Atmosphere Interface

Grandview Point, near the junction of the Green and Colorado Rivers, central Utah. Photo by author.

Dramatic contrast in rock types near Window
Rock, Arizona. Photo by author.

11

The Dynamic Planet

The twentieth century is a time of great discovery about Earth's internal structure and dynamic crust, yet much remains undiscovered. This is a time of revolution in our understanding of how the present arrangement of continents and oceans evolved. One task of physical geography is to explain the spatial implications of all this new knowledge and its effect on Earth's surface and society.

Earth's interior is highly structured, with uneven heating generated by the radioactive decay of unstable elements. The results are irregular patterns of surface fractures, the occurrence of earthquakes and volcanic activity, and the formation of mountain ranges. The U.S. Geological Survey reports that, in an average year, continental margins and seafloors expand by 1.9 km^3 (0.46 mi^3). But, at the same time, 1.1 km^3 (0.26 mi^3) are consumed, resulting in a net addition of 0.8 km^3 (0.2 mi^3) to Earth's crust. This chapter describes Earth's endogenic system, for it is within Earth that we find the driving forces that effect the crust.

A new era of Earth-systems science is emerging, effectively combining various disciplines within the study of physical geography. Remember that geography is an approach—a spatial way of looking at things—the science that examines place, location, human-Earth relationships, the uniqueness of regions, and movement (Figure 1–1). The geographic essence of geology, geophysics, paleontology, seismology, and geomorphology are all integrated by geographers to produce an overall picture of Earth's surface environment and of specific physical phenomena that affect human society. Improved remote-sensing capabilities and the development of computer-based geographic information systems are enhancing these efforts.

THE PACE OF CHANGE

The **geologic time scale** (Figure 11–1) reflects currently accepted names and the relative and absolute time intervals that encompass Earth's history (eons, eras, periods, and epochs). The *sequence* in this scale is based upon the relative positions of rock strata above or below each other. An important general principle is that of *superposition,* which states that rock and sediment always are arranged with the youngest beds "superposed" near the top of a rock formation and the oldest at the base—if they have not been disturbed. The *absolute ages* on the scale, determined by scientific methods such as dating by radioactive isotopes, are also used to refine the time-scale sequence.

The guiding principle of Earth science is uniformitarianism, first proposed by James Hutton in his *Theory of the Earth* (1795) and later amplified by Charles Lyell in *Principles of Geology* (1830). **Uniformitarianism** assumes that *the same physical processes active in the environment today have been operating throughout geologic time*. The phrase "the present is the key to the past" is an expression coined to describe this principle. Uniformitarianism is a valuable generalization supported by unfolding evidence from modern exploration and from the landscape record of volcanic eruptions, earthquakes, and exogenic processes that shape the landscape.

In contrast, the philosophy of **catastrophism** attempts to fit the vastness of Earth's age and the complexity of its rocks into a shortened time span. Catastrophism holds that Earth is young and that mountains, canyons, and plains formed through catastrophic events that did not require eons of time. Because there is little physical evidence to support this idea, catastrophism is more appropriately considered a belief than a serious scientific hypothesis. Ancient landscapes, such as the Appalachians, represent a much broader time scale of events.

However, geologic time is punctuated by dramatic events, such as massive landslides, earthquakes, and volcanic episodes. Within the principle of uniformitarianism, these localized catastrophic events occur as small interruptions in the generally uniform processes that shape the slowly evolving landscape. The complexity of the biosphere is the product of a long evolutionary development period of past epochs dominated by different atmospheres, enormous land animals, mass extinctions, and the evolution of new species (Figure 11–2).

EARTH'S SURFACE RELIEF FEATURES

Earth's surface landscape is the place where we see its solid and fluid realms interact—the dynamic

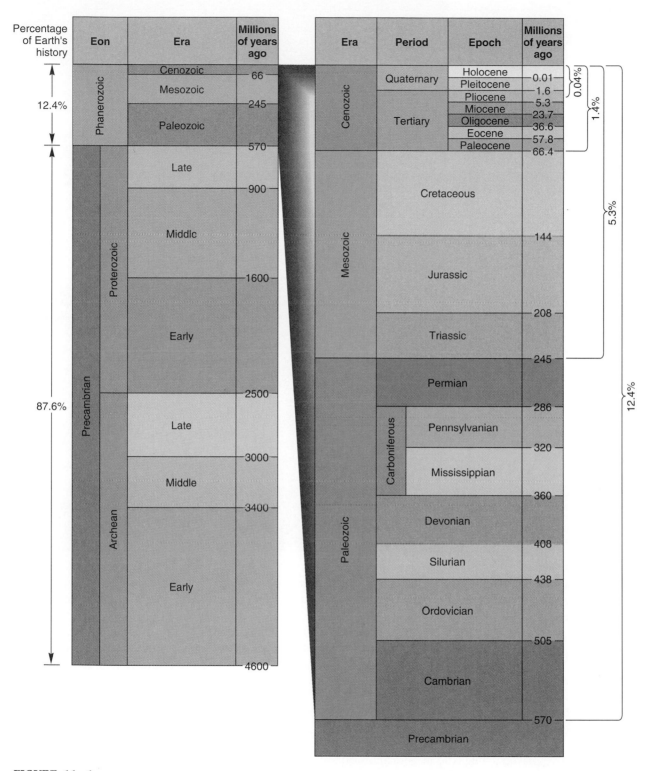

FIGURE 11–1

The geologic time scale was organized using relative-dating methods, with absolute dating determined through modern technological means. Note that over 87% of geologic time occurred during the Precambrian. Data from *1983 Geologic Time Scale,* Geological Society of America.

FIGURE 11-2
Ancient rock formations of the
Appalachians, which began
forming during the Alleghany
Orogeny over 250 million years
ago. Photo by John B. Yates,
Rainbow.

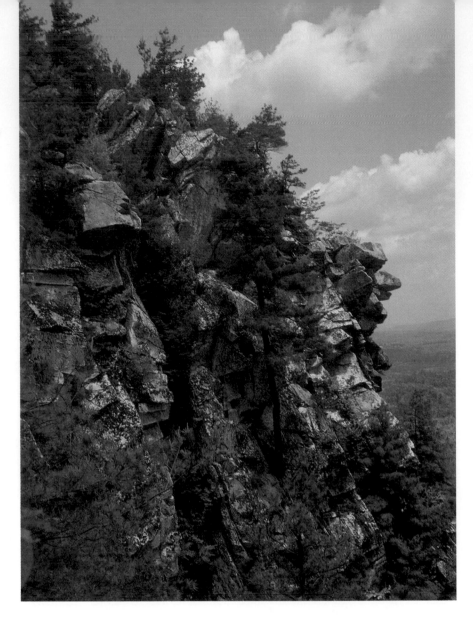

interplay of the endogenic and exogenic systems. This landscape is a composite of separate landforms composed of various rock structures.

Crustal Orders of Relief

Relief refers to vertical elevation differences in a local landscape. The undulating form that gives Earth's surface its character and general configuration is called **topography**, the feature portrayed so effectively on topographic maps. Earth's crustal landforms have played a vital role in human history: high mountain passes have both protected and isolated societies; ridges and valleys have dictated transportation routes; and vast plains have encouraged better methods of communication and travel.

For convenience, we group the landscape's topography into three "orders of relief." These classify landscapes from enormous ocean basins and continents down to local hills and valleys.

First Order of Relief. The broadest category of landforms includes huge **continental platforms** and **ocean basins**. Continental platforms are the masses of crust that reside above or near sea level, including the undersea continental shelves along the coastlines. The ocean basins are entirely below sea level. Approximately 71% of Earth's surface is covered by

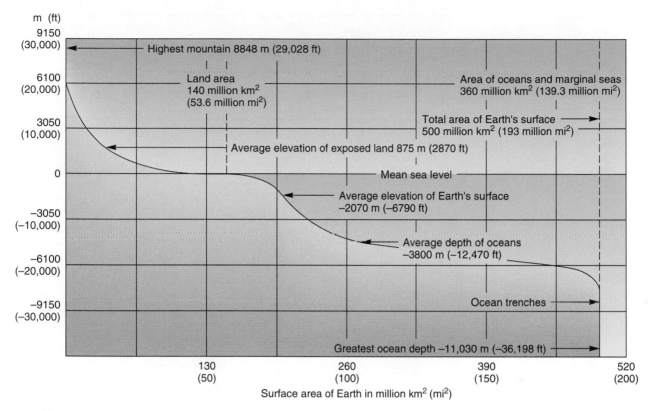

FIGURE 11–3
Hypsographic curve of Earth's surface, charting elevation as related to mean sea level. After Joseph E. Van Riper, *Man's Physical World,* p. 389, copyright © 1971 by McGraw-Hill. Adapted by permission.

water, with only 29% of its surface appearing as continents. The distribution of land and water in evidence today demonstrates a distinct water hemisphere and continental hemisphere, as shown in Chapter 7 (see Figure 7–4).

Second Order of Relief. Continental features that are classified in the second order of relief include continental masses, mountain masses, plains, and lowlands. A few examples are the Alps, Rocky Mountains (both Canadian and American), west Siberian lowland, and Tibetan Plateau. The great shields that form the heart of each continental mass are of this second order. In the ocean basins, second order of relief includes continental rises, slopes, abyssal plains, mid-ocean ridges, submarine canyons, and subduction trenches—all visible in the sea-floor illustration that opens Chapter 12.

Third Order of Relief. Individual mountains, cliffs, valleys, hills, and other landforms of smaller scale are included in the third order of relief. These features are identifiable as local landscapes.

Figure 11–3 is a *hypsographic curve,* a generalized curve showing the distribution of Earth's surface by elevation (the Greek *hypsos* means height). Relative to Earth's size, the surface generally is of low relief—in other words, Earth's topography is gentle when compared to the planet's diameter. The average elevation of Earth's overall surface is under water: −2070 m (−6790 ft) below mean sea level. The average elevation just for exposed land is +875 m (+2870 ft). For the ocean depths, the average elevation is −3800 m (−12,470 ft). From this you can see that, on the average, the oceans are much deeper than continental regions are high. Overall, the underwater ocean basins, ocean floor, and subma-

FIGURE 11–4

Earth's topographic regions. After Richard E. Murphy, "Landforms of the World," Map Supplement No. 9, *Annals of the Association of American Geographers* 58, no. 1 (March 1968). Adapted by permission.

rine mountain ranges represent the largest areal percentage of crust.

Earth's Topographical Regions

These landform orders can be further generalized into six topographic regions—plains, high tablelands, hills and low tablelands, mountains, widely spaced mountains, and depressions (Figure 11–4). Each type of topography is delineated by an arbitrary elevation or descriptive limit that is in common use (see legend in the figure). Four of the continents possess extensive plains, which are identified as areas with local relief of less than 100 m (325 ft) and slope angles of 5° or less. Some plains have high

elevations of over 600 m (2000 ft); in the United States the high plains achieve elevations above 1220 m (4000 ft). The Colorado Plateau and Antarctica are notable high tablelands, with elevations exceeding 1520 m (5000 ft). Africa is dominated by hills and low tablelands.

Mountain ranges are denoted by local relief exceeding 600 m (2000 ft) and appear in locations on each continent that probably are familiar to you.

EARTH'S STRUCTURE AND INTERNAL ENERGY

Along with the other planets and the Sun, Earth is thought to have condensed and congealed from a nebula of dust, gas, and icy comets about 4.6 billion years ago (Chapter 2). The oldest surface rock yet discovered on Earth lies in northwestern Canada and dates back 3.96 billion years. Scientists from Saint Louis University, the Canadian Geological Survey, and Australian National University certified the sample's age in November 1989. Previously, rocks from Greenland held the record at 3.8 billion years old. These finds tell us something very significant: Earth was forming continental crust nearly 4 billion years ago.

Throughout Earth's formation process, heat was accumulating. As the protoplanet compacted into a smaller volume, energy was transformed into heat through compression. In addition, significant heat was trapped in Earth from the decay of unstable isotopes of uranium, thorium, potassium, and other radioactive elements.

Earth in Cross Section

As Earth solidified, heavier elements slowly gravitated toward the center, and lighter elements slowly welled upward to the surface, concentrating in the crust. Consequently, Earth's interior is arranged roughly in concentric layers, each one distinct either chemically or thermally, with heat radiating outward from the center by conduction and then by physical convection at levels nearer the surface.

Our knowledge of Earth's *internal differentiation* into these rough concentric layers has been acquired entirely through indirect evidence, because we are unable to drill more than a few kilometers into Earth. There are several physical properties of Earth materials that enable us to approximate the nature of the interior. For example, when an earthquake or underground nuclear test sends shock waves through the planet, the cooler areas, which generally are more rigid, transmit these **seismic waves** at a higher velocity than do the hotter areas. Density also has an effect on seismic velocities, and fluid or plastic zones simply do not transmit some types of seismic waves (they absorb them). Thus, the distinctive ways in which seismic waves pass through Earth and the time they take to travel between two surface points help *seismologists* deduce the structure of Earth's interior (Figure 11–5). With an expanded network of scientific instruments, improvements in their sensitivity, and the increasing speed and capability of modern computers, our knowledge of Earth's interior is growing rapidly.

Figure 11–6 illustrates the dimensions of Earth's interior in comparison to surface distances in North America to give you a sense of size and scale.

Earth's Core. A third of Earth's entire mass, but only a sixth of its volume, lies in the core. The **core** is differentiated into two regions—inner core and outer core—divided by a transition zone of several hundred kilometers. The inner core is thought to be solid iron, with a density of 13.5 g per cm^3. (The metric relationship of grams per cubic centimeter is used to express such density. For comparison purposes, the density of water is 1.0 g/cm^3, and mercury is 13.0 g/cm^3.) The iron in the core is not pure, but probably is combined with silicon, and possibly oxygen and sulfur. The outer core is molten, metallic iron with lighter densities averaging 10.7 g/cm^3. Estimated core temperatures range from 3000°C (5400°F) to as high as 6650°C (12,000°F).

Earth's Magnetism. The fluid outer core generates at least 90% of Earth's magnetic field and the magnetosphere that surrounds and protects Earth from the solar wind. A present hypothesis by scientists from Cambridge University details spiraling circulation patterns in the outer core region that are influenced by Earth's rotation; this circulation generates electric currents, which in turn induce the magnetic field.

An intriguing feature of Earth's magnetic field is that it sometimes fades to zero and then returns to full strength with north and south magnetic poles reversed! In the process, the field does not blink on and off but instead oscillates slowly to nothing and then slowly regains its strength. This **magnetic reversal** has taken place nine times during the past 4 million years and hundreds of times over Earth's history. The average period of a magnetic reversal is 500,000 years, with occurrences as short as several thousand years possible. There appears to be a trend in recent geologic time toward more frequent reversals with shorter intervals. When Earth is without a magnetic field, a random pattern of magnetism results. The effects of these blanked-out episodes on life are still speculative, but without a magnetosphere, the surface environment is unprotected from cosmic radiation and solar wind.

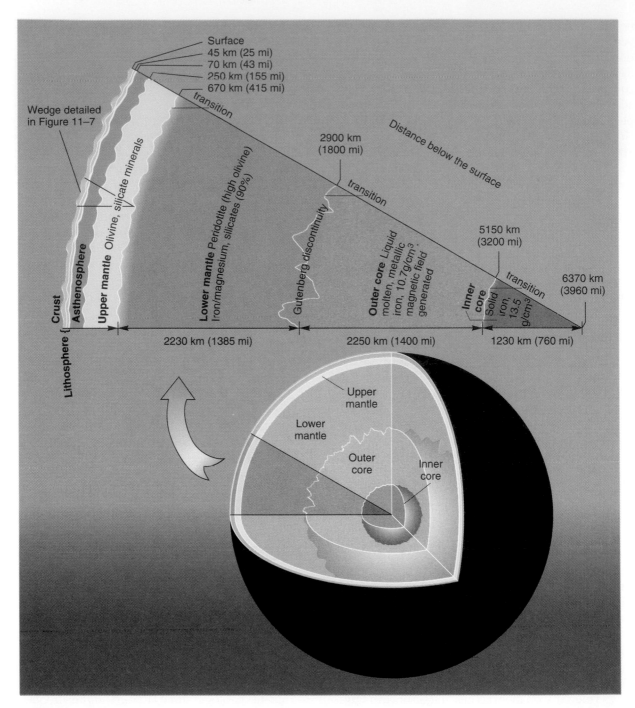

FIGURE 11-5
Earth's interior, with detailed cross-sectional wedge.

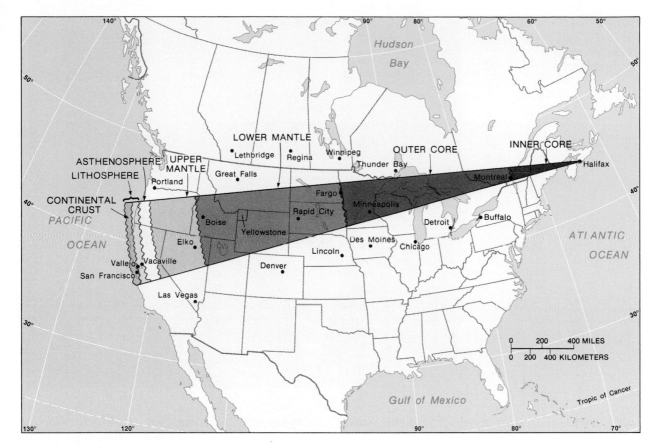

FIGURE 11–6
Surface map, using the distance from Halifax, Nova Scotia, to San Francisco to compare the distance from Earth's center to the surface. The continental crust thickness is the same as the distance between Vallejo (in the eastern portion of San Francisco Bay) and the city of San Francisco.

The reasons for these magnetic reversals are unknown; however, the spatial patterns they create at Earth's surface represent an important diagnostic tool in understanding the evolution of landmasses and the movements of the continents. When new iron-bearing rocks solidify from molten material at Earth's surface, the small magnetic particles in the rocks align according to the orientation of the magnetic poles at that time. This pattern is then locked in place as the rock solidifies. All across Earth, rocks bear this identical record of reversals as measurable magnetic stripes. In addition, they record the migration over time of Earth's magnetic poles, which do not coincide with its geographic poles (axis). (The difference in location of the magnetic and geographic poles is shown in Figure 17–22 for both the north and south polar regions.) Later in this chapter we shall show the importance of this migration and these magnetic reversals.

Earth's Mantle. A transition zone of several hundred kilometers marks the top of the outer core and the beginning of the mantle. Scientists at the California Institute of Technology analyzed the behavior of more than 25,000 earthquakes and determined that this transition area is bumpy and uneven, with ragged peak-and-valleylike formations. Some of the motions in the mantle may be created by this rough texture at what is called the **Gutenberg discontinuity**. A discontinuity is a place where a change in physical properties occurs between two regions deep in Earth's interior.

FIGURE 11–7
The structure of the crust and
its relation to the
asthenosphere.

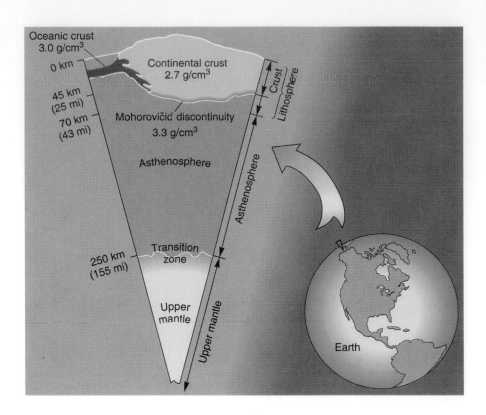

The **mantle** represents about 80% of Earth's total volume, with average densities of 4.5 g/cm³. The mantle is rich in oxides and silicates of iron and magnesium (FeO, MgO, and SiO_2) and is more dense and tightly packed in the lower region, grading to lesser densities in the upper region. Of the mantle's volume, 50% is in the lower mantle, which is comprised of rock that probably has never been at Earth's surface. The upper mantle is separated from the lower mantle by a broad transition zone, 400–500 km (250–310 mi) thick and centered about 670 km (415 mi) below the surface. The entire mantle experiences a gradual temperature increase with depth.

The extreme upper mantle, just below the crust, is known as the **asthenosphere**, or plastic layer. It contains pockets of increased heat from radioactive decay and is susceptible to convective currents in these hotter (and therefore less dense) materials. The depths affected by these convection currents are the subject of much scientific speculation. Because of this dynamic condition, the asthenosphere is the least rigid region of the mantle, with densities averaging 3.3 g/cm³. About 10% of the asthenosphere is molten in asymmetrical patterns and hot spots. The resulting slow movement in this zone immediately below the crust causes movement of the crust and creates tectonic activity—the folding, faulting, and general deformation of surface rocks.

Lithosphere and Crust. The lithosphere includes the entire **crust**, which represents only 0.01% of Earth's mass, and the upper mantle down to a depth of about 70 km (43 mi), with imprecise transition marking the asthenosphere boundary. The boundary between the crust and the rest of the lithospheric upper mantle is another discontinuity called the **Mohorovičić discontinuity**, or *Moho* for short, named for the Yugoslavian seismologist who determined that seismic waves change at this depth, owing to sharp contrasts of materials and densities.

Figure 11–7 illustrates the relationship of the crust to the rest of the lithosphere and the asthenosphere. Crustal areas below mountain masses extend farther downward, perhaps to 50 or 60 km (31 to 37 mi), whereas the crust beneath continental interiors averages about 30 km (19 mi) in thickness and only 5 km (3 mi) under the oceanic crust. The composition and texture of the two types of crust are quite different. Continental crust is basically **granite**; it is crystalline, and high in silica, aluminum, potassium, calcium, sodium, and other constituents. (Sometimes continental crust is called *sial,* shorthand for silica-alumina.) Oceanic crust is **basalt**; it is granular, and high in silica, magnesium, and iron. (Sometimes oceanic crust is called *sima,* shorthand for silica-magnesia.)

The principle of buoyancy (that something less dense, like wood, floats in denser things like water) and the principle of balance were further developed in the 1800s into the important principle of **isostasy** to explain certain movements of Earth's crust. Think of Earth's outer crust as floating on the denser layers

beneath, much as a boat floats on water. With a greater load (e.g., ice, sediment, mountains), the crust tends to ride lower in the asthenosphere. Without that load (e.g., when ice melts), the crust rides higher, in a recovery uplift known as an *isostatic rebound*. Thus, the entire crust is in a constant state of compensating adjustment, or isostasy, slowly rising and sinking in response to its own weight, and pushed and dragged about by currents in the asthenosphere (Figure 11–8).

GEOLOGIC CYCLE

Earth's crust is in an ongoing state of change, being formed, deformed, moved, and broken down by

FIGURE 11–8

Isostatic adjustment of the crust. Reprinted by permission of Merrill, an imprint of Macmillan Publishing Company, from *The Earth,* 2d ed., by Edward J. Tarbuck and Frederick K. Lutgens, p. 463. Copyright © 1987 by Merrill Publishing.

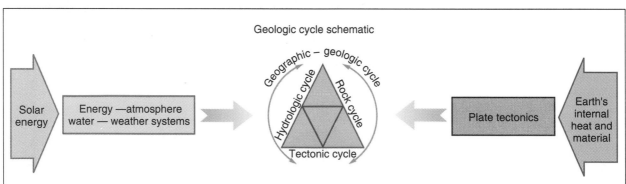

FIGURE 11–9
Geologic cycle, showing the interactive relationship of the rock cycle, tectonic cycle, and hydrologic cycle.

physical, chemical, and biological processes. The endogenic system is at work building landforms, while the exogenic system works to wear them down. This vast give-and-take at the Earth-atmosphere interface is called the **geologic cycle**. It is fueled by Earth's internal heat and by solar energy, influenced by the leveling force of Earth's gravity. Figure 11–9 illustrates the geologic cycle, combining many of the elements discussed in this text.

As you can see, the geologic cycle itself is composed of three principal cycles—hydrologic, rock, and tectonic. The hydrologic cycle, detailed in Chapter 7, transports and deposits Earth's varied materials. The rock cycle produces the three basic rock types found in the crust. The tectonic cycle brings heat energy and new material to the surface, creating movement and deformation of the crust.

Rock Cycle

Of Earth's crust, 99% is composed of only eight natural elements, and only two of these—oxygen and silicon—account for 74% (Table 11–1). Oxygen is the most reactive gas in the lower atmosphere, readily combining with other elements. For this reason, the percentage of oxygen is higher in the crust than in the atmosphere. The relatively large percentages of lightweight elements such as silicon and aluminum in the crust are explained by the internal differentiation process in which less dense elements migrated toward the surface, as discussed earlier.

A **mineral** is an element or combination of elements that forms an inorganic natural compound. A mineral can be described with a specific symbol or formula and possesses specific qualities. Silicon (Si) readily combines with other elements to produce the *silicate* mineral family, which includes quartz, feldspar, amphibole, and clay minerals, among others. Another important mineral family is the *carbonate* group, which features carbon in combination with oxygen and other elements such as calcium, magnesium, and potassium. Of the nearly 3000 minerals, only 20 are common, with just 10 of those making up 90% of the minerals in the crust.

A *rock* is an assemblage of minerals bound together. Literally thousands of rocks have been identified, all the result of three rock-forming processes: igneous, sedimentary, and metamorphic.

TABLE 11–1
Common elements in Earth's crust.

Element	Percent of Earth's Crust by Weight
Oxygen (O)	46.60
Silicon (Si)	27.72
Aluminum (Al)	8.13
Iron (Fe)	5.00
Calcium (Ca)	3.63
Sodium (Na)	2.83
Potassium (K)	2.70
Magnesium (Mg)	2.09
Total	98.70

Figure 11–10 illustrates the interrelationships among these three processes in the **rock cycle**. The figure also identifies these processes as they relate to both endogenic and exogenic systems.

Igneous Processes. Rocks that solidify and crystallize from a molten state are called **igneous rocks**. Most rocks in the crust are igneous. They form from **magma**, which is molten rock beneath the surface (hence the name *igneous,* which means "fire-formed" in Latin). Magma is fluid, highly gaseous, and under tremendous pressure. It is either *intruded* into preexisting crustal rocks, known as *country rock,* or *extruded* onto the surface as **lava**. The cooling history of the rock—how fast it cooled, and how steadily the temperature dropped—determines its texture and degree of crystallization. These range from coarse-grained (slower cooling, with more time for larger crystals to form) to fine-grained or glassy (faster cooling). Figure 11–11 illustrates various igneous forms.

Intrusive igneous rock that cools slowly in the crust forms a **pluton**, a general term for any intrusive igneous rock body, regardless of size or shape. The largest pluton form is a **batholith** (Figure 11–11), an irregular-shaped mass with a surface exposure greater than 100 km^2 (40 mi^2) that has invaded layers of crustal rocks. Batholiths form the mass of many large mountain ranges, such as the Sierra Nevada batholith, the Idaho batholith, and the Coast Range batholith of Canada and extreme northern Washington State—Figure 11–12(a).

Smaller plutons include the magma conduits of ancient volcanoes, which have cooled and hardened

FIGURE 11-10

A rock-cycle schematic,
demonstrating the relationships
among igneous, sedimentary,
and metamorphic processes and
the endogenic and exogenic
systems. Adapted by permission
of Merrill, an imprint of
Macmillan Publishing Company,
from *Physical Geology,* 4th ed.,
by Robert J. Foster, p. 43.
Copyright © 1983 by Merrill
Publishing.

FIGURE 11-11

Basic igneous forms and structures. Adapted by permission of Merrill, an imprint of
Macmillan Publishing Company, from *The Earth,* 2d ed., by Edward J. Tarbuck and
Frederick K. Lutgens, p. 63. Copyright © 1987 by Merrill Publishing.

(a)

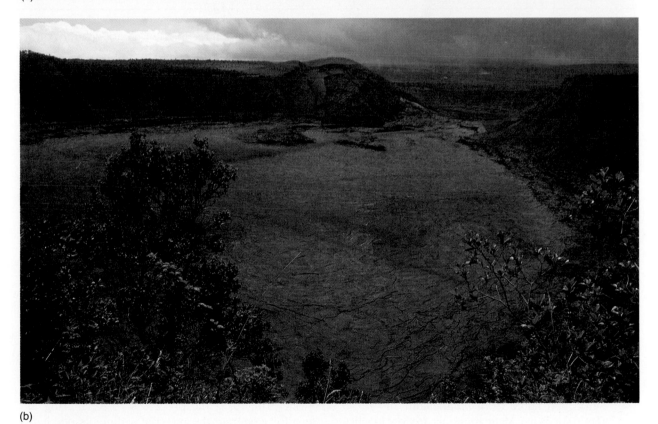

(b)

FIGURE 11–12

(a) Exposed Sierra Nevada batholith, an intrusive igneous rock; (b) basaltic lava flows at the Halemaumau crater in Hawaii, an extrusive igneous rock. (a) Photo by author; (b) photo by A. R. Christopherson.

TABLE 11–2
Igneous rock minerals.

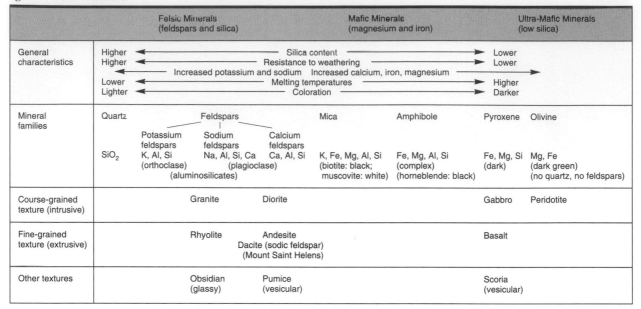

	Felsic Minerals (feldspars and silica)			Mafic Minerals (magnesium and iron)		Ultra-Mafic Minerals (low silica)	
General characteristics	Higher ⟵ Silica content ⟶ Lower Higher ⟵ Resistance to weathering ⟶ Lower ⟵ Increased potassium and sodium Increased calcium, iron, magnesium ⟶ Lower ⟵ Melting temperatures ⟶ Higher Lighter ⟵ Coloration ⟶ Darker						
Mineral families	Quartz SiO₂	Feldspars Potassium feldspars K, Al, Si (orthoclase)	Sodium feldspars Na, Al, Si, Ca (plagioclase) (aluminosilicates)	Calcium feldspars Ca, Al, Si	Mica K, Fe, Mg, Al, Si (biotite: black; muscovite: white)	Amphibole Fe, Mg, Al, Si (complex) (hornblende: black)	Pyroxene Fe, Mg, Si (dark)	Olivine Mg, Fe (dark green) (no quartz, no feldspars)
Course-grained texture (intrusive)	Granite	Diorite					Gabbro	Peridotite
Fine-grained texture (extrusive)	Rhyolite	Andesite Dacite (sodic feldspar) (Mount Saint Helens)					Basalt	
Other textures	Obsidian (glassy)	Pumice (vesicular)					Scoria (vesicular)	

in the crust. Those that form parallel to layers of sedimentary rock are termed *sills*; those that cross layers of invaded country rock are called *dikes*. Magma also can bulge between rock strata and produce a lens-shaped body called a *laccolith*, a form of sill. In addition, magma conduits themselves may solidify in roughly cylindrical forms that stand starkly above the landscape when finally exposed by weathering and erosion. Ship Rock, New Mexico, is such a feature, as suggested in Figure 11–11. All of these intrusive forms can be exposed by exogenic actions.

The categorization of igneous rocks is usually done by mineral composition and texture (Table 11–2). The two broad categories are:

1. *Felsic* igneous rocks—derived both in composition and name from *fel*dspar and *sil*ica (SiO₂). Felsic minerals are generally high in silica, aluminum, potassium, and sodium, with low melting points. Rocks formed from felsic minerals generally are lighter in color and density than mafic mineral rocks.
2. *Mafic* igneous rocks—derived both in composition and name from *ma*gnesium and *f*erric (Latin for iron). Mafic minerals are low in silica and high in magnesium and iron, with high melting points. Rocks formed from mafic min-

erals are darker in color and of greater density than felsic mineral rocks.

The table shows that the same magma which produces coarse-grained granite when slowly cooled beneath the surface also forms a fine-grained rhyolite as its rapidly cooled volcanic counterpart. If it cools rapidly, magma having a silica content comparable to granite and rhyolite may form the dark, smoky, glassy-textured rock called *obsidian* or volcanic glass. Another glassy rock called *pumice* forms when escaping gases bubble a frothy texture into the lava. Pumice is full of small openings, is light in weight, and low enough in density to float in water.

On the mafic side, basalt is the most common fine-grained extrusive igneous rock. It comprises the bulk of the ocean floor and appears in the lava flows in Figure 11–12(b). Its intrusive counterpart, formed by slow cooling of the parent magma, is *gabbro*.

Sedimentary Processes. Most sedimentary rocks are derived from preexisting rocks, or from organic materials (such as bone and shell) that form limestone. The exogenic processes of weathering and erosion generate the material sediments needed to form these rocks. Bits and pieces of former rocks—principally quartz, feldspar, and clay minerals—are eroded and then mechanically transported (by water, ice, wind, and gravity) to other sites where they are deposited. In addition, some minerals are dissolved into solution and form sedimentary deposits by precipitating from those solutions; this is an important process in the oceanic environment.

Various sedimentation forms are created under different environmental regimes—deserts, glaciers, beaches, tropics, and so on (Figure 11–13). Characteristically, sedimentary rocks are laid down by wind, water, and ice in horizontally layered beds. These layered strata are important records of past ages, and **stratigraphy** is the study of their sequence (superposition), spacing, and spatial distribution for clues to the age and origin of the rocks.

The cementation, compaction, and hardening of sediments into **sedimentary rocks** is called **lithification**. Various cements fuse rock particles together; lime ($CaCO_3$, or calcium carbonate) is the most common, followed by iron oxides (Fe_2O_3), and silica (SiO_2). Particles also can be united by drying (dehydration), heating, or chemical reactions. The two primary sources of sedimentary rocks—the mechanically transported bits and pieces of former rock and the dissolved minerals in solution—are known as *clastic sediments* and *chemical sediments,* respectively. Clastic sedimentary rocks are derived from weathered and fragmented rocks that are further worn in transport. Table 11–3 lists the range of sizes—everything from boulders to microscopic clay particles—and the form they take as lithified rock.

Chemical sedimentary rocks are not formed from physical pieces of broken rock, but instead are dissolved, transported in solution, and chemically precipitated out of solution. The most common chemical sedimentary rock is **limestone**, which is lithified calcium carbonate ($CaCO_3$). A similar form is dolomite, which is lithified calcium-magnesium carbonate, $CaMg(CO_3)_2$. About 90% of all limestone comes from organic sources—deposits of shell and bone that have been lithified. Once formed, these

FIGURE 11–13
Differential erosion on sedimentary strata. Photo by author.

TABLE 11–3
Clastic sediment sizes and related rock types.

Unconsolidated Sediment	Grain Size	Lithified Rock
Boulders, cobbles	>80 mm	Conglomerate (breccia, if pieces are angular)
Pebbles, gravel	>2 mm	
Coarse sand	0.5 to 2.0 mm	Sandstone
Medium-to-fine sand	0.02 to 0.05 mm	Sandstone
Silt	0.002 to 0.02 mm	Siltstone (mudstone)
Clay	<0.002 mm	Shale

SOURCE: U.S. Department of Agriculture, Soil Conservation Service.

rocks are susceptible to being dissolved in solution and assume unique characteristic forms, as discussed in the weathering section of Chapter 13.

Another sedimentary rock of organic origin is coal. Coal is an important nonrenewable fossil-fuel resource and is the topic of Focus Study 11–1.

Chemical sediments also form from inorganic sources when water evaporates and leaves behind a residue of salts. These **evaporites** may exist as common salts, such as gypsum or sodium chloride (table salt), to name only two, and often appear as flat, layered deposits across a dry landscape. This process is dramatically demonstrated in the pair of photographs in Figure 11–14, taken in Death Valley National Monument one day and one month after a record 2.57 cm (1.01 in.) rainfall.

Chemical deposition also occurs in the water of natural hot springs from chemical reactions between minerals and oxygen; this is a sedimentary process related to **hydrothermal activity**. Deposits of travertine, a form of calcium carbonate, at Mammoth Hot Springs in Yellowstone National Park are an example.

Another occurs near the Nevada-California border in Mono Lake, an alkaline lake high in carbonates and sulfates that presently has an overall salinity of 95 ‰ (the ocean average is 35 ‰). Carbonates in Mono Lake interact with calcium-rich hot springs to produce a rock called *tufa*. The tufa deposits grew tall in Mono Lake as long as the springs were beneath the surface of the lake. However, since the 1940s, tributary streams have been diverted out of the region, lowering the lake's surface more than 20 m (65 ft), reducing its surface area by more than half, exposing these chemical sedimentary deposits and alkaline shorelines, and damaging the lake's natural ecology and habitats for wildlife and birds. Some tufa towers now rise over 10 m (30 ft) above the shoreline, as shown in the foreground of Figure 7–19(g). Interestingly, tributary flows were restored by court order in April 1991 and the lake level is again rising.

Metamorphic Processes. Any rock, either igneous or sedimentary, may be transformed into a **metamorphic rock**, by going through profound physical and/or chemical changes under increased pressure and temperature. (The name metamorphic comes from the Greek, meaning to "change form.") Metamorphic rocks generally are more compact than the original rock and therefore are harder and more resistant to weathering and erosion.

Several conditions can cause metamorphism. In the tectonic cycle, the major causes are high temperatures and high compressional stresses occurring in subsurface rock over millions of years. Igneous rocks are compressed during collisions of portions of Earth's crust (described under "Plate Tectonics" later in this chapter). Or, igneous rocks may be sheared and stressed along earthquake fault zones. Sometimes they simply are crushed under great weight when a crustal area is thrust beneath other crust.

Another metamorphic condition occurs when sediments collect in broad depressions in Earth's crust and, because of their own weight, create enough pressure in the bottommost layers to transform the sediments into metamorphic rock. Also, molten magma rising within the crust may initiate metamorphic processes in adjacent rock, which is "cooked" by the intruding magma, a process called *contact metamorphism*. Figure 11–15 shows one of the best-known examples of metamorphic rock—marble.

Metamorphic rocks may be changed both physically and chemically from the original rocks. If the

(a)

(b)

FIGURE 11-14

A Death Valley landscape (a) one day after a record rainfall event, and (b) one month after. Photos by author.

mineral structure demonstrates a particular alignment after metamorphism, the rock is considered **foliated**; some minerals may appear as wavy striations (streaks or lines) in the new rock. On the other hand, parent rock with a more homogeneous makeup may produce a *nonfoliated* rock. Table 11-4 lists some metamorphic rock types, their parent rocks, and resultant textures.

The ancient roots of mountains are composed predominantly of metamorphic rocks. For example, at the bottom of the inner gorge of the Grand Canyon in Arizona, a Precambrian metamorphic rock called the Vishnu Schist is over 2 billion years in age. Despite the fact that this schist is harder than steel, the Colorado River has cut down into the uplifted Colorado Plateau, exposing and eroding these old rocks. Thus, the canyon itself is much younger than the schist. Within the overall 1.6 km-deep (1.0 mi)

canyon, this river-cut inner gorge represents the bottom 455 m (1500 ft).

You have seen how the rock-forming processes yield the igneous, sedimentary, and metamorphic materials of Earth's crust. Next we will look at how vast **tectonic processes** press, push, and drag portions of the crust in large-scale movements, causing the continents to "drift." The facts that the continents drift and the young seafloor spreads from rifts are revolutionary discoveries that came to light only in this century.

CONTINENTAL DRIFT

Have you ever looked at a world map and noticed that a few of the continental landmasses appear to fit

FIGURE 11-15

An example of metamorphic rock—a marble formation in Antarctica. Photo courtesy of U.S. Geological Survey.

Coal: An Important Sedimentary Rock

Coal is an important sedimentary rock that develops from living organisms; it is really a biochemical rock, very rich in carbon. Photosynthesis in plants is the process that fixes organic carbon into the rock cycle. During the Carboniferous period (360–286 million years ago in Figure 11–1) lush, plant-rich swamps and forests lived and died. In addition, photosynthesizing organisms flourished in tropical waters, as well as in bogs and forests in more temperate latitudes. As these carbon-fixing lifeforms perished, organic debris collected in water-soaked bogs. The water was important, for this low-oxygen environment was necessary to limit natural decomposition in the accumulating layers. Bacterial action freed various elements and thereby concentrated the carbon in the debris.

Coal forms in these bogs through progressive stages. First, pieces of former plants can be easily recognized in *peat,* a brownish, compacted fibrous organic sediment. Peat has been burned as a low-grade fuel for centuries, producing a low-heat energy output and a great deal of air pollution. Following more compaction and hardening, the soft, coal-like substance known as *lignite* forms. Further lithification leads to *bituminous coal,* the first true sedimentary rock in this sequence. Burial of coal deposits under sedimentary overburden results in the formation of coal seams, usually between beds of shale or sandstone. Finally, where conditions are right, metamorphism of bituminous coal results in *anthracite coal,* a metamorphic rock that is extremely hard and therefore difficult to mine.

Sedimentary and metamorphic coal deposits are found from the Arctic to Antarctica and on every continent, yet another confirmation of continental drift. About 30–35% of the world's reserves are in the United States and Canada (Figure 1), 25% in the Soviet Union, 15% in China, and lesser amounts in Australia, Great Britain, India, and Poland, among others. The largest consumer of coal in the United States is the electric power industry, which uses 70% of mined coal for the production of steam in power plant boilers.

Estimates of how long these known reserves of coal will last are variable and are based on assumptions regarding the annual rate of increase in production. There are at least 500 years of coal remaining in the United States and Canada if extraction and consumption rates stay about where they were in the 1980s (about 1 billion tons a year), with bituminous coal comprising 65% of the production. However, if these reserves are subjected to an annual production growth of 3 to 5%, the lifetime of the reserves will be reduced to 100 years or less. Coal is a valuable source of chemicals for many important uses, including medicines, so its depletion for the

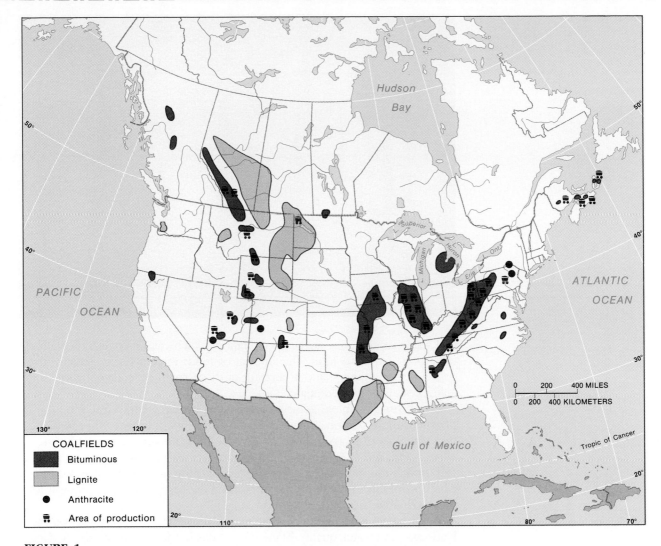

FIGURE 1
Types of coal deposits in North America. Adapted by permission of Merrill, an imprint of Macmillan Publishing Company, from *Geography and Development,* 3rd ed., by James S. Fisher, p. 123. Copyright © 1989 by Merrill Publishing.

production of electricity—a high-grade source of energy used generally for low-grade energy applications—would be a tragedy for future generations.

Another concern related to coal is an impurity commonly found within it, sulfur. Sulfur emissions are a major air pollutant when released to the atmosphere through power-plant and industrial smokestacks. In addition, coal is baked to produce coke, a purified carbon that is a basic raw material for steelmaking; in the coking process, large amounts of sulfur oxides are released into the atmosphere. The combination of sulfur oxides

FIGURE 2
Strip mine spoil banks in southern West Virginia. Photo by author.

and water vapor forms acidic precipitation, which causes environmentally damaging acid deposition. The amount of sulfur in coal varies spatially, even over a few kilometers, and as a result of these environmental concerns low-sulfur coal is in great demand.

A further problem associated with coal is that, when burned, it produces more carbon dioxide per unit consumed than any other fossil fuel. The role of increased carbon dioxide in stimulating the greenhouse effect is detailed in Chapters 5 and 10. Strategies to reduce atmospheric carbon dioxide must include energy conservation and reduced coal use, which would in turn extend the lifetime of the resource—preserving it for more valuable uses and applications.

One final environmental aspect of coal extraction is its impact on the landscape and water quality. Many coal seams lie deep beneath the surface, roughly 100 m (300 ft) or deeper, and are mined by drilling shafts and tunnels. These weaken the overlying rock strata, which crack, causing subsidence of the land surface and disrupting natural drainage. Other coal seams lie near the surface and are surface mined (strip mined) by removing the overburden, extracting the coal, and replacing the overburden, as prescribed by law. However, in many cases the disturbed area is not returned to its original topography, and the landscape remains marred by deposits of mining debris called "spoil banks." Another problem with both types of mining is that water and air combine with sulfur in the mine debris to form bright orange "acid mine drainage" that pollutes streams. Major regional controversies revolve around these mining scars and related reclamation issues (Figure 2).

TABLE 11–4
Metamorphic rocks.

Parent Rock	Metamorphic Equivalent	Texture
Shale (clay minerals)	Slate	Foliated
Granite, slate, shale	Gneiss	Foliated
Basalt, shale, periodotite	Schist	Foliated
Limestone, dolomite	Marble	Nonfoliated
Sandstone	Quartzite	Nonfoliated

together like pieces of a jigsaw puzzle—particularly South America and Africa? The incredible reality is that the continental pieces indeed were once fitted together! Continental landmasses not only migrated to their present locations but continue to move at annual speeds up to 6 cm (2.4 in.) per year. We say that the continents are "adrift" because convection currents in the asthenosphere are dragging them around. The key point is that the arrangement of continents and oceans we see today is not permanent but is in a continuing state of change.

A Brief History

In 1620, the English philosopher Sir Francis Bacon noted gross similarities between the shapes of Africa and South America (although he did not suggest that they had drifted apart). Others wrote—unscientifically—about such apparent relationships, but it was not until much later that a valid explanation was proposed. In 1912, German geophysicist and meteorologist Alfred Wegener publicly presented in a lecture his idea that Earth's landmasses migrate. His book, *Origin of the Continents and Oceans,* appeared in 1915. Wegener today is regarded as the father of this concept, called **continental drift**. But scientists at the time, knowing little of Earth's interior structure and bound by an inertial mindset as to how continents and mountains were formed, were unreceptive to Wegener's revolutionary proposal and, in a nonscientific spirit, rejected it outright.

Wegener postulated that all landmasses were united in one supercontinent approximately 225 million years ago, during the Triassic period—see Figure 11–19(b). (As you read this, it may be helpful to refer to the geologic time scale in Figure 11–1.) This one landmass he called **Pangaea**, meaning "all Earth." Although his initial model kept the landmasses together far too long and his idea about the driving mechanism was incorrect, Wegener's overall configuration of Pangaea was right. To come up with

his Pangaea fit, he began studying the research of others, specifically the geologic record, the fossil record, and the climatic record for the continents. He concluded that South America and Africa were related in many complex ways. He also concluded that the large midlatitude coal deposits, which stretch from North America to Europe to China, and date to the Permian and Carboniferous periods (245–360 million years ago), existed because these regions once were more equatorial in location and therefore covered by lush vegetation that became coal.

As modern scientific capabilities built the case for continental drift, the 1950s and 1960s saw a revival of interest in Wegener's concepts, and finally confirmation. Table 11–5 highlights developments from the time of Wegener's proposal to the present. Aided by an avalanche of discoveries, the theory today is nearly universally accepted as an accurate model of the way Earth's surface evolves, and virtually all Earth scientists accept the fact that continental masses move about.

Sea-Floor Spreading

The key to establishing the fact of continental drift was a better understanding of the seafloor. The seafloor has a remarkable feature: an interconnected worldwide mountain chain (ridge) some 65,000 km (40,000 mi) in extent and averaging more than 1000 km (600 mi) in width. How did it get there? In the early 1960s, geophysicists Harry H. Hess and Robert S. Dietz proposed **sea-floor spreading** as the mechanism that builds this mountain chain and drives continental movement.

Hess said that these submarine mountain ranges, called the **mid-ocean ridges**, were the direct result of upwelling flows of magma from hot areas in the upper mantle and asthenosphere. When mantle convection brings magma up to the crust, the crust is fractured and new seafloor is formed, building the ridges and spreading laterally. Figure 11–16 suggests

THE DYNAMIC PLANET • **315**

TABLE 11–5
Stages in the development of a revolution.

Date	Development
1912–1915	Continental drift proposed by Wegener
1915–1930	Great debate
1930–1950	Stalemate (a lost cause in the United States; continuing debate elsewhere)
1950–1960	Revival of interest Exploration of ocean floor Fossil magnetism and polar wandering
1960–1962	Sea-floor spreading
1963	Oceanic magnetic anomalies associated with sea-floor spreading; the magnetic "tape recorder" Hot spots
1963–1966	Polarity reversals of Earth's magnetic field; fossil magnetism and accurate rock dating for lavas on land and for deep-sea sediment cores
1966–1967	Revolution proclaimed by scientists
1967–1968	Spreading and drift incorporated in plate tectonics Earthquake synthesis
1968–1970	Deep-sea drilling by *Glomar Challenger*
1970	"Geopoetry becomes geofact," *Time* magazine (and other more scientific sources)
Currently	Recent geological features are being related to present plates and plate boundaries; ancient geological features are being analyzed to reconstruct plate history

SOURCE: Peter J. Wyllie, *The Way the Earth Works*. Copyright © 1976 by John Wiley & Sons, Inc.

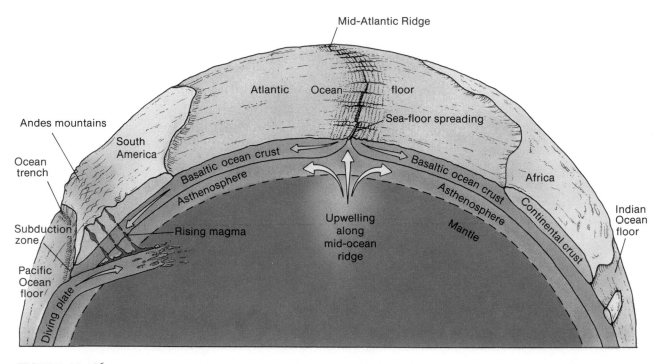

FIGURE 11–16
Sea-floor spreading, upwelling currents, subduction, and plate movements, shown in cross section. After Peter J. Wyllie, *The Way The Earth Works*. Copyright © 1976, by John Wiley & Sons, Inc. Adapted by permission of John Wiley & Sons, Inc.

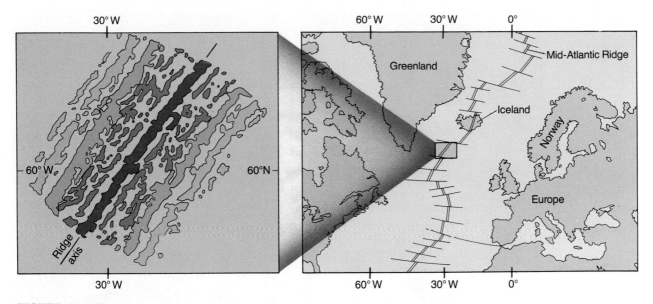

FIGURE 11–17
Magnetic reversals recorded in the seafloor south of Iceland along the Mid-Atlantic Ridge. Magnetic reversals reprinted from *Deep-Sea Research* 13, J. R. Heirtzler, S. Le Pichon, and J. G. Baron, Copyright 1966, Pergamon Press, p. 247.

that the ocean floor is rifted and scarred along a mid-ocean ridge, with arrows indicating the direction in which the floor spreads.

As new crust is generated and the seafloor spreads, the alignment of the magnetic field in force at the time dictates the alignment of magnetic particles in the cooling rock, creating a kind of magnetic tape recording in the seafloor. Figure 11–17 illustrates the mirror images that develop on either side of the sea-floor rift as a result of these magnetic reversals recorded in the rock. If you measure off the same distance from the rift to either side, you find the same orientation of magnetic content in the rock. Thus, the periodic reversals of Earth's magnetic field have proven a valuable clue to understanding sea-floor spreading.

As the age of the seafloor was determined from these sea-floor recordings of Earth's magnetic-field reversals and other measurements, the complex harmony of the two concepts of continental drift and sea-floor spreading became clearer. The youngest crust anywhere on Earth is at the spreading centers of the mid-ocean ridges (Figure 11–18). With increasing distance from these centers, Earth's surface gets increasingly older. The oldest seafloor is in the western Pacific near Japan, an area called the

Pigafetta Basin that dates to 170 million years ago (during the Jurassic period).

Overall, the seafloor is relatively young—nowhere does it exceed 200 million years in age, remarkable when you remember that Earth's age is 4.6 billion years. The reason is that oceanic crust is being consumed by subduction beneath continental crust along Earth's deep oceanic trenches. The discovery that the seafloor is so young demolished earlier geologic thinking that the oldest rocks would be found there.

Subduction

In contrast to the upwelling zones along the mid-ocean ridges are the areas of descending crust elsewhere. We must remember that the basaltic ocean crust has a density of 3.0 g/cm^3, whereas the continental crust averages a significantly lighter 2.7 g/cm^3. As a result, when continental crust and oceanic crust collide, the heavier ocean floor will dive beneath the lighter continent, thus forming a descending **subduction zone** (Figure 11–16). The world's **oceanic trenches** coincide with these subduction zones and are the deepest features on Earth's surface. (The deepest is the Mariana Trench near

FIGURE 11–18
Relative age of the oceanic crust. Adapted with the permission of W. H. Freeman and Company from *The Bedrock Geology of the World* by R. L. Larson, et al. Copyright © 1985.

KEY

0–2 m.y.	58–66 m.y.
2–5 m.y.	66–84 m.y.
5–24 m.y.	84–117 m.y.
24–37 m.y.	117–144 m.y.
37–58 m.y.	144–208 m.y.

Guam, which descends below sea level to −11,033 m or −36,198 ft).

The subducted portion of crust is dragged down into the asthenosphere where it remelts, is recycled, and eventually migrates back toward the surface through deep fissures and cracks in basement rock. Volcanic mountains form inland from subduction zones—for example, the Andes in South America and the Cascade Range from northern California to the Canadian border—providing an important clue for the study of continental drift. The fact that spreading ridges and subduction zones are areas of earthquake and volcanic activity provides further evidence for *plate tectonics*, which by 1968 had become the all-encompassing term for these crustal processes.

The Formation and Breakup of Pangaea

The supercontinent of Pangaea and its subsequent breakup into today's continents represents only the last 225 million years of Earth's 4.6 billion years, or only the most recent 1/23 of Earth's existence. During the other 22/23 of geologic time, other things were happening. For example, Figure 11–19(a) depicts the movement of the *plates*—the slabs of Earth's crust—with a map of the suspected pre-Pangaea arrangement approximately 465 million years ago. The landmasses as we know them are barely recognizable.

Pre-Pangaea. Figure 11–19(a) shows the pre-Pangaea arrangement of 465 million years ago (during the Middle Ordovician period). The outlines of South America, Africa, India, Australia, and southern Europe appear within the continent called Gondwana in the Southern Hemisphere. What is to become North America and Greenland is the inverted continent of Laurentia, which approximately straddles the equator near the location of the present-day Fiji and Samoa island groups. The other land groups are identified.

Pangaea. Figure 11–19(b) shows an updated version of Wegener's Pangaea, 225 to 200 million years ago (Triassic-Jurassic periods). Areas of North America, North Africa, the Middle East, and Eurasia were near the equator and therefore were covered by plentiful vegetation. The landmasses focused near the South Pole and attached to the sides of Antarctica were covered by ice. Today, those portions of South America, southern Africa, India, Australia, and Antarctica bear the imprint of this shared glacial blanket.

The India plate (the slab of Earth's crust that included modern India) appears larger on the map than the same area today because the Tibetan Plateau comprised the northern portion of the plate. We also can see that Africa shared a common connection with both North and South America. Today, the Appalachian Mountains in the eastern United States and the Atlas Mountains of northwestern Africa reflect this common ancestry; they are, in fact, portions of the same mountain range. At the time of Figure 11–19(b), the Atlantic Ocean did not exist. The only oceans were Panthalassa ("all seas"), which would become the Pacific, and the Tethys Sea (partly enclosed by the African and the Eurasian plates), which formed the Mediterranean Sea, with trapped portions becoming the Caspian Sea.

Pangaea Breaks Up. Figure 11–19(c) shows the movement of plates that had occurred by 135 million years ago (the beginning of the Cretaceous period). New seafloor that had formed is highlighted. The active spreading center had rifted North America away from landmasses to the east, shaping the coast of Labrador. India was further along in its long journey, with a spreading center to the south, and a subduction zone to the north where the leading edge of the India plate was diving beneath Eurasia. At that time a rift also began in what was to be the South Atlantic. And, Antarctica and Australia were pulled into isolation from the other masses.

Modern Continents Take Shape. Figure 11–19(d) shows the arrangement 65 million years ago (beginning of the Tertiary period). Sea-floor spreading along the Mid-Atlantic Ridge had grown some 3000 km (almost 1900 mi) in 70 million years. Africa had moved northward about 10° in latitude, leaving Madagascar split from the mainland and opening up the Gulf of Aden. The rifting along what would be the Red Sea had begun. The India plate had moved three-fourths of the way to Asia, as Asia continued to rotate clockwise. Of all the major plates, India traveled the farthest. From this time until the present, according to Dietz and Holden, more than half of the ocean floor was renewed.

The Continents Today. Figure 11–19(e) shows modern geologic time (the late Cenozoic Era). The northern reaches of the India plate have underthrust the southern mass of Asia through subduction, forming the Himalayas in the upheaval created by the collision. Plate motions continue to this day and will proceed into the future.

PLATE TECTONICS

Plate tectonics is the inclusive conceptual model for Wegener's continental drift proposal, sea-floor spreading, and other related aspects of this twentieth-century revolution in Earth sciences. Tectonic refers to the deformation of Earth's crust as a result of internal forces, which can form various structures in the lithosphere. Tectonic processes include folds, faults, warps, fractures, earthquakes, and volcanic activity.

Earth's present crust is divided into at least 12 plates, of which about half are major and half are

(a) 465 M.Y. AGO

(b) 225 M.Y. AGO

(c) 135 M.Y. AGO

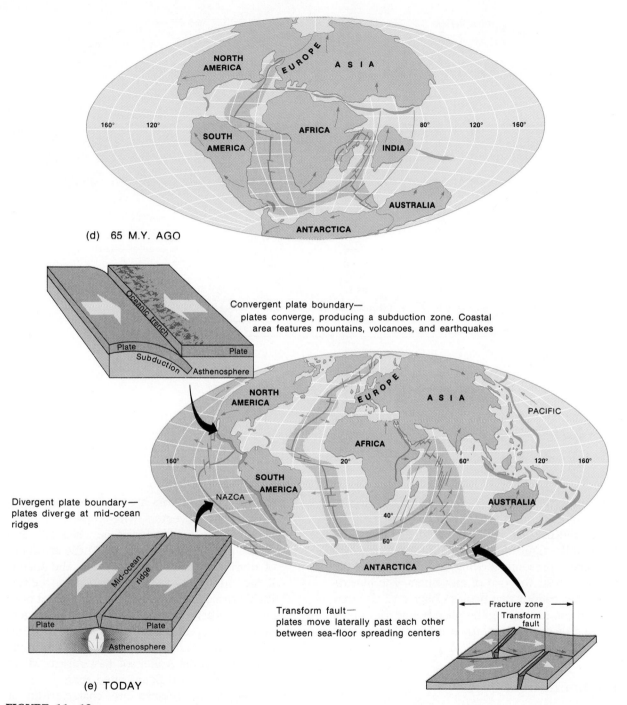

(d) 65 M.Y. AGO

Convergent plate boundary—
plates converge, producing a subduction zone. Coastal
area features mountains, volcanoes, and earthquakes

Divergent plate boundary—
plates diverge at mid-ocean
ridges

Transform fault—
plates move laterally past each other
between sea-floor spreading centers

(e) TODAY

FIGURE 11–19

The formation and breakup of Pangaea and the types of motions occurring at plate boundaries. (a) from R. K. Bambach, Scotese, and Ziegler, "Before Pangaea: The Geography of the Paleozoic World," *American Scientist* 68 (1980): 26–38, reprinted by permission; (b) through (e) from Robert S. Dietz and John C. Holden, *Journal of Geophysical Research* 75, no. 26 (10 September 1970): 4939–56, copyright by the American Geophysical Union.

FIGURE 11–20

Earth's 12 lithospheric plates and their motions. Each arrow represents 20 million years of movement with its length, indicating that the Pacific plate is moving more rapidly than the Atlantic plates. Adapted from U. S. Geodynamics Committee and Sheldon Judson, Kenneth S. Deffeyes, and Robert B. Hargraves, *Physical Geology,* © 1976, p. 28. Reprinted by permission of Prentice Hall, Inc., Englewood Cliffs, NJ.

minor, in terms of size. These broad plates (Figure 11–20) are composed of literally hundreds of smaller pieces that have migrated together. The arrows in the figure indicate the direction in which each plate is presently moving, and the length of the arrows suggests the rate of movement. The illustration of the seafloor that begins Chapter 12 also may be helpful in identifying the plate boundaries in the following discussion. Try to correlate this map with that illustration.

Plate Boundaries

The boundaries where plates meet are clearly dynamic places. The block diagrams in Figure 11–19(e) show the three general types of motion and interaction that occur in those areas:

- *Divergent boundaries* are characteristic of sea-floor spreading centers, where upwelling material from the mantle forms new seafloor, and crustal plates are spread apart. These are zones of tension. Examples include the Mid-Atlantic Ridge and the divergent boundary along the East Pacific rise, which gives birth to the Nazca plate and the Pacific plate. While most divergent boundaries occur at mid-ocean ridges, there are a few within continents themselves. An example is the Great Rift Valley of East Africa, where continental crust is being rifted apart.

- *Convergent boundaries* are characteristic of collision zones, where areas of continental and/or oceanic crust collide. These are zones of compression. Examples include the subduction zone off the west coast of South and Central

America and the area along the Japan and Aleutian trenches. Along the western edge of South America, the Nazca plate collides and is subducted beneath the South American plate, creating the Andes Mountains chain and related volcanoes in the process. The collision of India and Asia mentioned earlier is another example of a convergent boundary.

- *Transform boundaries* occur where plates slide laterally past one another at right angles to a sea-floor spreading center, neither diverging nor converging, and usually with no volcanic eruptions. Examples include the right-angle fractures stretching across the Mid-Atlantic ridge, the East Pacific rise, and the southeast Indian ridge.

Transform Faults. In 1965, another piece of the tectonic puzzle fell into place when University of Toronto geophysicist Tuzo Wilson first described the nature of transform boundaries and their relationship to earthquake activity. All the spreading centers on Earth's crust feature these lateral scars that stretch out on both sides of spreading centers (Figure 11–21). Some are a few hundred kilometers long; others, such as those along the East Pacific rise, stretch out 1000 km or more (over 600 mi). Across the entire ocean floor these boundaries are the location of **transform faults**, always parallel to the direction in which the plate is moving. How do they occur?

You can see in Figure 11–21 that mid-ocean ridges are not simple straight lines. When a mid-ocean rift begins, it opens at points of weakness in the crust. The fractures you see in the figure began as a series of offset breaks in the crust in each portion of the spreading center. As new material rises to the surface, building the mid-ocean ridges and spreading the plates, these offset areas slide past each other, in horizontal faulting motions. The resulting long *fracture zone,* which follows these breaks in Earth's crust, is active only along the fault section between spreading centers, as shown in Figure 11–21.

Along transform faults the motion is purely one of horizontal displacement—no new crust is formed or old crust subducted. The fracture zone is marked by parallel rifts (valleys) and ridges. Beyond the spreading centers, the two sides of the fracture zones are joined and inactive. In fact, the plate pieces on either side of the fracture zone are moving in the same direction, away from the spreading center. The name "transform" was assigned because of this apparent transformation in the direction of the fault movement. The famous San Andreas fault system in California, where continental crust has overridden a transform system, is related to this type of motion and is discussed in the next chapter.

Earthquake and Volcanic Activity. Plate boundaries are the primary location of Earth's earthquake and volcanic activity, and the correlation of these phenomena is an important aspect of plate tectonics. Earthquakes and volcanic activity are discussed in more detail in the next chapter, but their general relationship to the tectonic plates is important to point out here. Earthquake zones and volcanic sites are identified in Figure 11–22. The "ring of fire" surrounding the Pacific basin, named for the frequent incidence of volcanoes, is most evident. The subducting edge of the Pacific plate thrusts deep into the crust and asthenosphere, producing molten material that makes its way back toward the surface, causing active volcanoes. Such processes occur similarly at plate boundaries throughout the world.

Hot Spots

A dramatic aspect of plate tectonics is the estimated 50 to 100 **hot spots** across Earth's surface. These are individual sites of upwelling material from the asthenosphere, noted on Figure 11–22. Hot spots occur beneath both oceanic and continental crust and appear to be deeply anchored in the mantle, tending to remain fixed relative to migrating plates. Thus, the area of a plate that is above a hot spot is locally heated for the brief geologic time it is there (a few hundred thousand or million years).

An example of an isolated hot spot is the one that has formed the Hawaiian-Emperor Islands chain (Figure 11–23). The Pacific plate has moved across this hot, upward-erupting plume for almost 80 million years, with the resulting string of volcanic

FIGURE 11–21
Transform faults. Appearing along fracture zones, a transform fault is only that
section between spreading centers where adjacent plates move in opposite
directions. Adapted from B. Isacks, J. Oliver, and L. R. Sykes, *Journal of Geophysical
Research* 73, (1968): 5855–99, copyright by the American Geophysical Union.

islands moving northwestward away from the hot
spot. Thus, the age of each island or seamount in the
chain increases northwestward from the island of
Hawaii, as you can see from the ages marked in the
figure. The oldest island in the Hawaiian part of the
chain is Kauai, approximately 5 million years old; it is
weathered, eroded, and deeply etched with canyons
and valleys.

The big island of Hawaii actually took less that 1
million years to build to its present stature. The
youngest island in the chain is still a seamount, a
submarine mountain that does not reach the surface.
It rises 3350 m (11,000 ft) from the ocean floor but is
still 975 m (3200 ft) beneath the ocean surface. Even
though this new island will not see the tropical Sun
for about 10,000 years, it is already named Loihi.

Farther away from the hot spot near Hawaii, the island of Midway rises as a part of the same system. From there the Emperor Seamounts stretch northwestward until they reach about 40 million years of age. At that point, this linear island chain shifts direction northward, evidently reflecting a change in the movement of the Pacific plate at that earlier time. At the northernmost extreme, the mounts that formed about 80 million years ago are now approaching the Aleutian Trench, where they eventually will be subducted beneath the Eurasian plate.

Most linear volcanic chains have a hot spot at one end, where mantle material works its way up through the passing plate. The track left by these eruptions matches the direction of the plate. Yellowstone National Park, in northwestern Wyoming, is above such a hot spot, and because the North American plate is moving westward, the hot spot track has been cast through Idaho, Oregon, and Washington. If the plate continues drifting westward, the phenomenon of Yellowstone, including thermal springs, lava flows, and earthquake and volcanic activity, eventually will appear in the Dakotas.

Iceland is an example of an active hot spot sitting astride a mid-ocean ridge; it also is the best example of a segment of mid-ocean ridge rising above sea level. This hot spot has generated enough material to form Iceland and continues to cause eruptions from

FIGURE 11–22

Earthquake and volcanic activity in relation to major tectonic plate boundaries, and principal hot spots. Earthquake/volcano data from *Earthquakes* by Bruce A. Bolt. Copyright © 1978, 1988 W. H. Freeman and Company; reprinted with permission. Hot spot locations from "Hot Spots on the Earth's Surface," Kevin C. Burke and J. Tuzo Wilson, Copyright © 1976 by Scientific American, Inc. All rights reserved.

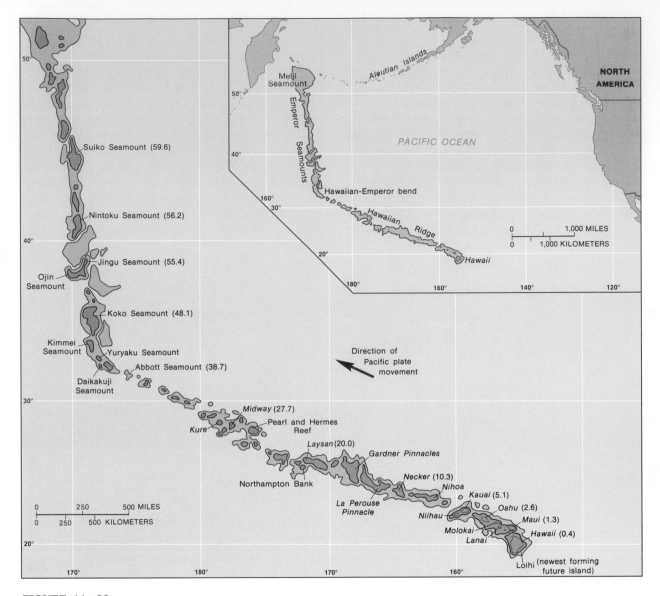

FIGURE 11–23
Hawaii and the linear volcanic chain of islands known as the Emperor Seamounts. Ages of islands and seamounts in the chain are shown in millions of years. After David A. Clague, "Petrology and K-Ar [Potassium-Argon] Ages of Dredged Volcanic Rocks from the Western Hawaiian Ridge and the Southern Emperor Seamount Chain," *Geological Society of America Bulletin* 86 (1975): 991.

deep in the asthenosphere. As a result, Iceland is still growing in area and volume. The youngest rocks are near the center of Iceland, with rock age increasing toward the eastern and western coasts. Iceland is a living example of our changing world on a most dynamic planet.

SUMMARY

The Earth-atmosphere interface is where the endogenic system, powered by heat energy from within the planet, and the exogenic system, powered by insolation and influenced by gravity, work together to produce Earth's surface. The geologic time scale is an effective device for organizing the great span of Earth's history. Earth's surface relief features can be conveniently categorized into three topographic orders of relief. The geologic cycle is a model of endogenic-exogenic process interactions that comprises three other cycles: the hydrologic cycle, tectonic cycle, and rock cycle. The rock cycle characterizes the three rock-forming processes—igneous, sedimentary, and metamorphic.

A knowledge of Earth's interior is important in understanding its magnetic field and the processes of crustal motion: upwelling, sea-floor spreading, and subduction. Twelve major plates of Earth's crust migrate about the surface in direct response to endogenic forces beneath the crust. The theory of plate tectonics has brought a twentieth-century revolution in Earth sciences. The present configuration of the oceans and the continents bears the imprint of these vast tectonic forces.

KEY TERMS

Asthenosphere	Magma
Basalt	Magnetic reversal
Batholith	Mantle
Catastrophism	Metamorphic rock
Continental drift	Mid-ocean ridges
Continental platforms	Mineral
Core	Mohorovičić discontinuity
Crust	Ocean basins
Endogenic system	Oceanic trenches
Evaporites	Pangaea
Exogenic system	Plate tectonics
Foliated	Pluton
Geologic cycle	Relief
Geologic time scale	Rock cycle
Granite	Sea-floor spreading
Gutenberg discontinuity	Sedimentary rocks
Hot spots	Seismic waves
Hydrothermal activity	Stratigraphy
Igneous rocks	Subduction zone
Isostasy	Tectonic processes
Lava	Topography
Limestone	Transform faults
Lithification	Uniformitarianism

1. To what extent is Earth's crust actively building at this time in its history?
2. Define the endogenic and the exogenic systems. Describe their energy sources.
3. How is geologic time organized? What is the basis for the time scale? What era, period, and epoch are we living in?
4. Contrast uniformitarianism and catastrophism.
5. Define topography and relief relative to landforms.
6. What is meant by orders of relief? Describe each of the three principal orders of relief in the world. List examples.
7. What order of relief is the Appalachian Mountains? the Bonneville salt flats in western Utah? the Canadian prairies?
8. Make a simple sketch of Earth's interior, label each layer, and list its physical characteristics, temperature, composition, and range of size on your drawing.
9. What is the present thinking on how Earth generates its magnetic field? Is this field constant, or does it change? Explain the implications of your answer.
10. Describe the asthenosphere. Why and under what circumstances is it mobile? What are the consequences of this movement?
11. What is a discontinuity? Describe the two principal discontinuities within Earth.
12. Define isostasy and isostatic rebound, and explain the crustal equilibrium concept.
13. Diagram the upper mantle and crust. Label the density of the layers and include generalizations as to composition.
14. Illustrate the geologic cycle and define each component: rock cycle, tectonic cycle, and hydrologic cycle.
15. What is a mineral? A mineral family? Name the most common minerals on Earth.
16. Describe igneous processes. What is the difference between intrusive and extrusive forms?
17. Characterize felsic and mafic minerals. Give examples of both coarse- and fine-grained textures.
18. From what sources and particle sizes are the materials for sedimentary processes derived? Briefly describe sedimentary processes and lithification.
19. What is metamorphism? Name some original parent rocks and their metamorphic equivalents.
20. Briefly review the history of continental drift, sea-floor spreading, and the all-inclusive plate tectonics theory. What was Alfred Wegener's role?
21. Define upwelling and describe related features associated with such action on the ocean floor. Define subduction and explain the process.
22. What was Pangaea? What happened to it during the past 225 million years?
23. Characterize the three types of plate boundaries and the probable conditions associated with each type.
24. Why is there a relationship between plate boundaries and volcanic and earthquake activity?
25. Discuss the relationship between the hot spot beneath Hawaii and the Pacific plate on which the islands have formed.

SUGGESTED READINGS

Alexander, Tom. "A Revolution Called Plate Tectonics Has Given Us a Whole New Earth." *Smithsonian Magazine* 5, no. 10 (January 1975): 30–40.

———. "Plate Tectonics Has a Lot to Tell Us About the Present and Future Earth." *Smithsonian Magazine* 5, no. 11 (February 1975): 38–47.

Ballard, Robert D. *Exploring Our Living Planet*. Washington: National Geographic Society, 1988.

———. "Window on Earth's Interior." *National Geographic* 150, no. 2 (August 1976): 228–49.

Bloxham, Jeremy, and David Gubbins. "The Evolution of the Earth's Magnetic Field." *Scientific American* 261, no. 6 (December 1989): 68–75.

Bolt, Bruce A. *Inside the Earth—Evidence from Earthquakes*. San Francisco: W. H. Freeman, 1982.

Continents Adrift and Continents Aground—Readings from Scientific American. San Francisco: W. H. Freeman, 1976. "The Dynamic Earth"; "The Earth's Core"; "The Earth's Mantle"; "The Continental Crust." *Scientific American* 249, no.3 (September 1983): entire issue.

Gore, Rick. "Our Restless Planet Earth." *National Geographic* 168 (August 1985): 142–81.

Hurley, Patrick M. "The Confirmation of Continental Drift." *Scientific American* 218, no. 4 (April 1968): 52–64.

Miller, Russel, and the editors of Time-Life Books. *Continents in Collision*. Planet Earth Series. Alexandria, VA: Time-Life Books, 1983.

Overbye, Dennis. "The Shape of Tomorrow—How Earth Will Look in 250 Million Years." *Discover* 3, no. 11 (November 1982): 20–25.

Powell, Corey S. "Trends in Geophysics—Peering Inward." *Scientific American* 254, no. 6 (June 1991): 100–11.

Press, Frank, and Raymond Siever. *Earth*. 4th ed. New York: W. H. Freeman, 1986.

Tarbuck, Edward J., and Frederick K. Lutgens. *The Earth—An Introduction to Physical Geology*. 3d ed. Columbus: Macmillan, 1990.

Uyeda, Seiya. *The New View of the Earth—Moving Continents and Moving Oceans*. Translated by Masako Ohnuki. San Francisco: W. H. Freeman, 1978.

Vink, Gregory E., W. Jason Morgan, and Peter R. Vogt. "The Earth's Hot Spots." *Scientific American* 252, no. 4 (April 1985): 50–57.

Wyllie, Peter J. *The Way the Earth Works: An Introduction to the New Global Geology and Its Revolutionary Development*. New York: John Wiley & Sons, 1976.

Floor of the Oceans, 1975, by Bruce C. Heezen
and Marie Tharp. Copyright © 1980 by Marie
Tharp. Reproduced by permission of Marie Tharp,
1 Washington Ave., South Nyack, NY 10960.

12
Tectonics and Volcanism

Tectonic forces were brought home to hundreds of millions of television viewers as they watched the 1989 baseball World Series from Candlestick Park near San Francisco. Less than half an hour before the call to "play ball," a powerful earthquake rocked the region and turned sportscasters into newscasters and sports fans into disaster witnesses. Such tectonic activity has repeatedly deformed, recycled, and reshaped Earth's crust during its 4.6-billion-year existence. The principal tectonic and volcanic zones lie along plate boundaries or in areas where Earth's crustal plates are affected by processes in the asthenosphere. The arrangement of continents and oceans, the origin of mountain ranges, and the locations of earthquake and volcanic activity are all the result of these dynamic endogenic processes.

THE OCEAN FLOOR

The illustration that opens this chapter is a striking representation of Earth with its blanket of water removed. The ocean floor is revealed to us through decades of direct and indirect observation. Careful examination of this portrayal provides a helpful review of concepts learned in the previous chapter, laying the foundation for this and subsequent chapters. The scarred ocean floor is clearly visible, its sea-floor spreading centers marked by over 64,000 km (40,000 mi) of oceanic ridges, its subduction zones indicated by deep oceanic trenches, and its transform faults stretching at angles between portions of oceanic ridges.

On the floor of the Indian Ocean you can locate the track along which the India plate traveled to its collision with the Eurasian plate. You also can spot the vast sediment patterns in the Indian Ocean, south of the Indus and Ganges rivers on either side of India. Sediments derived from the Himalayan Range blanket the floor of the Bay of Bengal (south of Bangladesh) to a depth of 20 km (12.4 mi). The principle of isostasy can be examined in central and west-central Greenland where the weight of ice has pressed portions of the land far below sea level.

In the area of the Hawaiian Islands, the hot spot track is marked by a chain of islands and seamounts which you can follow along the Pacific plate from Hawaii to the Aleutians. You can find Iceland's position on the Mid-Atlantic Ridge. Subduction zones south and east of Alaska and Japan as well as along the western coast of South and Central America are quite visible. In addition, you can follow the East Pacific Rise northward as it trends beneath the west coast of the North American plate, disappearing under earthquake-prone California. The continents, submerged continental shelves and slopes, and the expanse of the sediment-covered abyssal plain are all identifiable on this illustration. (See if you can correlate this sea-floor illustration with the map of crustal plates and plate boundaries shown in Figures 11–20 and 11–22.)

CRUSTAL FORMATION PROCESSES

Tectonic activity is generally slow when viewed in human terms, but when it is abrupt, the resulting change is often dramatic. Endogenic processes result in uplift and new landforms, with major mountain building occurring along plate boundaries. These uplifted crustal regions are quite varied, depending on composition, structure, and operative processes. Nonetheless, they generally can be thought of in three categories, all of which are discussed in this chapter: (1) residual mountains and shields, formed from inactive remnants of older rock—the ancient heart of continental crust; (2) tectonic mountains and landforms, produced by active folding and faulting movements that deform the crust; and (3) volcanic features, formed by the surface accumulation of molten rock from eruptions of subsurface materials. Thus, several distinct processes operate in concert to produce continental crust.

Continental Crust

Continental crust is less dense and more complex in both content and structure than the relatively homogeneous oceanic crust. Materials from the mantle rise at the spreading centers, flow outward as ocean floor, and collide with the continents, only to plunge back into the mantle, remelt, and work back up to form more continent as intrusive igneous rock. The upwelling material along the mid-ocean ridges originating in the asthenosphere is less than 50% silica and is rich in iron and magnesium. However, a subducting oceanic plate works its way under a

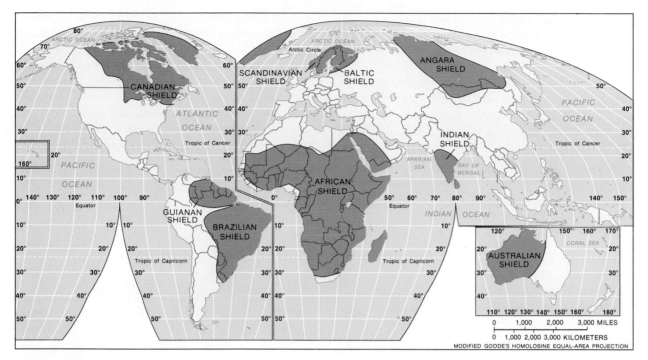

FIGURE 12–1
Major continental shields. After Richard E. Murphy, "Map Supplement No. 9—Landforms of the World," *Annals of the Association of American Geographers* 58, no. 1 (March 1968).

continental plate, melting and taking with it sediment and trapped water, plus various elements which are melted from the crust and drawn into the mixture. As a result, the molten magma that migrates upward from a subducted plate contains 50–75% silica, is high in aluminum, and has a viscous (thick) texture. Bodies of viscous magma may reach the surface in explosive volcanic eruptions, or they may stop short and become subsurface intrusive bodies in the crust, cooling slowly to form crystalline plutons.

All continents have a nucleus of old crystalline rock on which the continent grows. *Cratons* are the cores, or heartland regions, of the continental crust. They generally are low in elevation and old (Precambrian, more than 570 million years in age). Those regions where various cratons and ancient mountains are exposed at the surface are called **continental shields**. Figure 12–1 shows the principal areas of shield exposure.

Terranes. Each of Earth's major plates actually is a collage of many crustal pieces acquired from a variety of sources. Accretion, or accumulation, has occurred as crustal fragments of ocean floor, curved chains (or arcs) of volcanic islands, and other pieces of continental crust have been swept aboard the edges of continental shields. These migrating crustal pieces, which have become attached to the plates, are called **terranes** (not to be confused with "terrain," which refers to the topography of a tract of land). Displaced terranes have histories different from those of the continents that accept them, are usually framed by fractures, and are distinguished in rock composition and structure from their new continental homes.

These processes of continental construction are well captured in James Michener's novel *Alaska:*

> But the immediate task is to understand how this trivial ancestral [continental] nucleus could aggregate to itself the many additional segments of rocky land which would ultimately unite to comprise the Alaska we know. Like a spider waiting to grab any passing fly, the nucleus remained passive but did accept any passing terranes—those unified

agglomerations of rock considerable in size and adventurous in motion—that wandered within reach . . . . And the great plates of Earth's crust never rest . . . .*

In the region surrounding the Pacific, accreted terranes are particularly prevalent. At least 25% of the

*James A. Michener, *Alaska* (New York: Random House, 1988), p. 5 and 9. Reprinted by permission.

growth of western North America can be attributed to the accretion of terranes since the early Jurassic period (190 million years ago). A good example is the Wrangell Mountains, which lie just east of Prince William Sound and the city of Valdez, Alaska. The **Wrangellia terranes**—a former volcanic island arc and associated marine sediments from near the equator—migrated approximately 10,000 km (6200 mi) to form the Wrangell Mountains and three other

FIGURE 12–2
Wrangellia terranes, highlighted among the other terranes along the western margin of North America. After "The Growth of Western North America" by David L. Jones, Allan Cox, Peter Coney, and Myrl Beck, illustration by Andrew Tomko. Copyright © 1982 by *Scientific American,* Inc. All rights reserved.

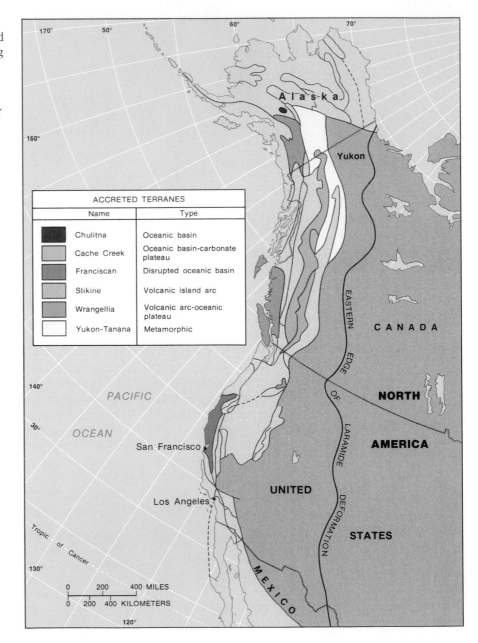

ACCRETED TERRANES	
Name	Type
Chulitna	Oceanic basin
Cache Creek	Oceanic basin-carbonate plateau
Franciscan	Disrupted oceanic basin
Stikine	Volcanic island arc
Wrangellia	Volcanic arc-oceanic plateau
Yukon-Tanana	Metamorphic

FIGURE 12–3
General types of stress and resulting strain.

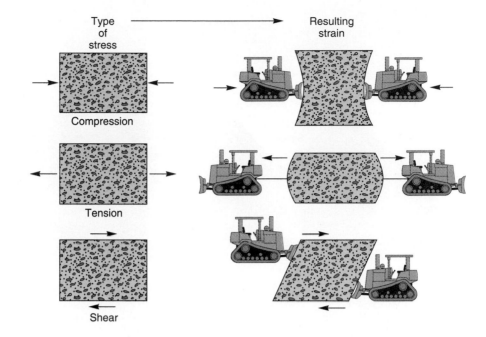

Type of stress → Resulting strain

Compression

Tension

Shear

distinct areas along the western margin of the continent (Figure 12–2). These discoveries, barely a decade old, demonstrate how continents possibly are assembled.

Deformation Processes

Earth's crust bends and breaks along zones of strain, which are stress-induced zones in deformed rock. The strain is caused by *compression* (shortening or folding), *tension* (stretching or faulting), and *shearing* (tangential or transform stress when two pieces slide past each other), as shown in Figure 12–3. The patterns created by these processes are evident in the landforms we see today.

Folding. When layered flat strata are subjected to compressional forces, the original structure is bent and deformed. Convergent plate boundaries worldwide create intensely deformed rock in a process known as **folding**. If we take pieces of thick cloth, stack them flat on a table, and then slowly push the sides of the stack together, the cloth layers will bend and form similar folds (Figure 12–4). If we then draw a line down the center axis of a resulting ridge and trough, we are able to see how the names of the folds are assigned. Along the ridge of a fold, layers seem to *slope downward away from the axis,*

resulting in an **anticline**. In the trough of a fold, however, layers *slope downward toward the axis,* producing a **syncline**.

If the axis of either type of fold is not horizontal (parallel to Earth's surface), the layers then plunge (dip down) at an angle. Knowledge of how folds are angled to Earth's surface and where individual structures are located is important for resource recovery. Petroleum, for example, collects in the upper portions of folds in permeable rock layers such as sandstone.

Figure 12–4 further illustrates different kinds of folds that have been weathered and reduced by exogenic processes. Within a syncline, a residual ridge of strata may form, caused by differential resistance of the rocks to exogenic processes. Compressional forces often push folds far enough that they actually overturn on their own strata, as you can see in the figure. Further stress eventually fractures the rock strata along distinct lines, and some overturned folds are thrust up, causing a considerable shortening of the original strata. The Canadian Rocky Mountains and the Appalachian Mountains illustrate well the complexity of the resulting folded landscape.

Satellites let us view many of these structures from an orbital perspective, as in Figure 12–5. Northwest of the Strait of Hormuz, just north of the Persian Gulf,

FIGURE 12-4
Folded landscape and the basic types of fold structures.

are the Zagros Mountains of Iran. Originally, this area was a dispersed terrane that had separated from the Eurasian plate. However, the collision produced by the northward push of the Arabian block is now shoving this terrane back into Eurasia and forming an active margin known as the Zagros crush zone, a zone of continued collision more than 400 km (250 mi) wide. In the figure, anticlines form the parallel ridges; active weathering and erosion processes expose underlying strata.

Earth's continental crust also is subjected to broad warping forces, which produce an up-and-down bending of strata that is too great in extent to be considered folding. Such forces can be the result of mantle convection, isostatic adjustments, and/or swelling associated with hot spot activity. Warping features range from small, individual, foldlike structures called *basins* and *domes* up to regional features the size of the Ozark Mountain complex in Arkansas and Missouri, or the Black Hills of South Dakota. The Richat dome structure in the West African nation of Mauritania, shown in Figure 12-6, is an example of this type of deformation.

Faulting. When rock strata are strained beyond their ability to remain a solid unit, they fracture, and one side is displaced relative to the other side in a

process known as **faulting**. Thus, fault zones are areas of crustal movement. At the moment the fault line shifts, a sharp release of energy occurs, called an **earthquake** or quake (Figure 12-7).

The surface along which the two sides of a fault move is called the *fault plane*. A cliff formed by faulting is commonly called a *fault scarp*. The tilt and orientation of the fault plane provides the basis for naming the three basic types of faults. A *normal fault,* or tension fault, moves vertically along an inclined fault plane—Figure 12-7(a). The overlying block is called the hanging wall, from old mining terminology when such fault zones were the location of mining activity; it shifts downward relative to the footwall block. The exposed fault plane sometimes is evident along the base of faulted mountains, where individual ridges are truncated by the movements of the fault and appear as triangular facets at the ends of the ridges.

Compressional forces associated with converging plates produce blocks that move upward relative to the footwall block so that the hanging-wall block overlies it. This is called a *reverse fault,* or compression fault, and appears on the surface similar in form to a normal fault, although more collapse and landslides may occur from the hanging wall component—Figure 12-7(b). If the fault plane forms a low

FIGURE 12–5
Southern Zagros Mountains in the Zagros crush zone between the Arabian and Eurasian plates. Photo courtesy of National Aeronautics and Space Administration.

FIGURE 12–6

(a) An upwarped dome and a structural basin; (b) the Richat dome structure of Mauritania. (a) After Joseph E. Van Riper, *Man's Physical World,* p. 436, copyright 1971 by McGraw-Hill; (b) courtesy of National Aeronautics and Space Administration.

(a)   Dome   Basin

(b)

angle relative to the horizontal, the fault is termed a *thrust fault,* or overthrust fault, indicating that the overlying block has shifted far over the underlying block. In the Alps, three to five such overthrusts are in evidence from the compressional forces of the ongoing African plate collision with the Eurasian plate. Beneath the Los Angeles basin, such overthrust faults produce a high risk of earthquakes.

If movement along a fault line is horizontal, it is called a **strike-slip fault**, or transcurrent fault—Figure 12–7(c). The movement is described as *right-lateral* or *left-lateral,* depending on the relative motion perceived when you observe movement on one side of the fault relative to the other side. Although strike-slip faults do not produce cliffs (scarps), they do create linear *rift valleys,* as is the

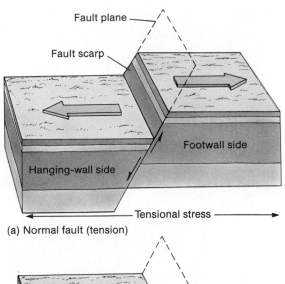

Fault plane

Fault scarp

Footwall side

Hanging-wall side

Tensional stress

(a) Normal fault (tension)

Hanging-wall side

Footwall side

Compressional stress

(b) Thrust or reverse fault (compression)

Right-lateral

Left-lateral

(c) Strike-slip fault (lateral)

FIGURE 12–7

Basic types of faults: (a) normal fault (tension); (b) reverse fault (compression); and (c) strike-slip fault (transcurrent or horizontal). After Joseph E. Van Riper, *Man's Physical World,* p. 437. Copyright 1971 by McGraw-Hill.

North American plate

Pacific plate

FIGURE 12–8

The San Andreas fault in the eastern margin of the central Coast Ranges of California. This view is toward the south, so the Pacific plate is on the right in the picture (the west side). The Pacific plate is moving northward relative to the North American plate on the left (the east side). Photo courtesy of U.S. Geological Survey.

case with the San Andreas fault system of California (Figure 12–8). The rift is clearly visible on the photograph where the edges of the North American and Pacific plates are tearing past one another as a result of transform faults associated with a former sea-floor spreading center. In its westward drift, the North American plate rode over portions of this center, as shown in Figure 12–13. Consequently, the San Andreas system is described as a series of

FIGURE 12–9
Example of a tilted fault block: the Teton Range in Wyoming. Photo by author.

transform (associated with former spreading center), strike-slip (horizontal motion), right-lateral (one side moving to the right hand of the other) faults.

In the interior western United States, the Basin and Range Province experienced tensional forces caused by uplifting and thinning of the crust, which cracked the surface to form aligned pairs of normal faults and a distinctive landscape (please refer to Figure 15–18). The term **horst** is applied to upward-faulted blocks; **graben** refers to downward-faulted blocks. Examples of horst and graben landscapes include the Great Rift Valley of East Africa (associated with crustal spreading), which extends northward to the Red Sea, and the Rhine graben through which the Rhine River flows in Europe.

The Sierra Nevada of California and the Grand Tetons of Wyoming are examples of **tilted fault block** mountain ranges, in which a normal fault on one side

of the range has produced a tilted landscape of dramatic relief (Figure 12–9). Slowly cooling magma intruded into those blocks and formed granitic cores of coarsely crystalline rock. After tremendous tectonic uplift and the removal of overlying material through weathering, erosion, and transport, those granitic masses are now exposed. In some areas, the overlying material originally covered these batholiths by more than 7500 m (25,000 ft).

Orogenesis

Orogenesis literally is the birth of mountains (*oros* comes from the Greek for "mountain"). An *orogeny* is a mountain-building episode that thickens continental crust. It can occur through large-scale deformation and uplift of the crust. It also may include the capture of migrating terranes and cementation of them to the continental margins, and the intrusion of granitic magmas to form plutons. These granite masses often are exposed following uplift and removal of overlying materials. Uplift is the final act of the orogenic cycle. Earth's major chains of folded and faulted mountains, called *orogens,* bear a remarkable correlation to the plate tectonics model.

No orogeny is a simple event; many involve previous developmental stages dating back millions of years, and the processes are ongoing today. Major orogens include the Rocky Mountains, produced during the Laramide orogeny (40–80 million years ago); the Sierra Nevada in the Sierra Nevadan orogeny (35 million years ago, with older batholithic intrusions dating back 130–160 million years); the Appalachians and the Valley and Ridge Province formed by the Alleghany orogeny (250–300 million years ago, preceded by at least two earlier orogenies); and the Alps of Europe in the Alpine orogeny (20–120 million years ago and continuing to the present, with many earlier episodes).

Examination of a world map reveals two large mountain chains. The relatively young mountains along the western margins of the North and South American plates stretch from the tip of Tierra del Fuego to the massive peaks of Alaska, comprising the **Cordilleran system**. The mountains of southern Asia, China, and northern India continue in a belt through the upper Middle East to Europe and the European Alps, constituting the **Eurasian-Himalayan system** (see Figure 12–12).

Types of Orogenies. Figure 12–10 illustrates convergent plate-collision patterns associated with orogenesis. Shown in (a) is the *oceanic plate–continental plate collision* type of orogenesis. This occurred along the Pacific coast of the Americas and has formed the Andes, the Sierra of Central America, the Rockies, and other western mountains. Their buildup was augmented by accretion of displaced terranes, which were cemented during their collision with the continental mass. Also, note the associated volcanic activity inland from the subduction zone.

Shown in (b) is the *oceanic plate–oceanic plate collision,* where two portions of oceanic crust collide. This has formed the chains of island arcs and volcanoes that continue from the southwestern Pacific to the western Pacific, the Philippines, the Kurils, and through portions of the Aleutians. Both of these collision types are active around the Pacific Rim. Both are thermal in nature, because the diving plate melts and works its way back toward the surface as molten rock. This region around the Pacific is known as the **circum-Pacific belt** or, more popularly, the **ring of fire**.

Shown in (c) is the *continental plate–continental plate collision,* which occurs when two large continental masses collide. Here the orogenesis is quite mechanical; large masses of continental crust are subjected to intense folding, overthrusting, faulting, and uplifting. As mentioned earlier, the collision of India with the Eurasian landmass produced the Himalayan Mountains. That collision is estimated to have shortened the overall continental crust by as much as 1000 km (620 mi), and to have produced sequences of thrust faults at depths of 40 km (25 mi). The disruption created by that collision has reached far under China, and frequent earthquakes there signal the continuation of this collision. The Himalayas feature the tallest above-sea-level mountains on Earth, including Mount Everest at 8848 m (29,028 ft) elevation.

Appalachians. The old, eroded fold-and-thrust belt of the eastern United States contrasts with the younger mountains of the western portions of North America. As noted, the **Alleghany orogeny** followed at least two earlier orogenic cycles of uplift and the accretion of several captured terranes. (In Europe this is called the Hercynian orogeny.) The original material for the Appalachians and Valley and Ridge Province resulted from the collisions that produced Pangaea. In fact, the Atlas Mountains of northwest Africa were originally connected to the Appalachians, but have rafted apart, embedded in their own tectonic plates. Similarly, folded and faulted rock structures in the British Isles and Greenland demonstrate a past relationship with the Appalachians.

The linear folds of the Appalachian system are well displayed on both satellite images and surface maps (Figure 12–11). These dissected ridges are cut through by rivers, forming *water gaps* that greatly influenced the migration and settlement patterns during the early days of the United States. The initial flow of people, goods, and ideas was directed by this topography.

FIGURE 12–10

Types of plate convergence:
(a) oceanic-continental,
(b) oceanic-oceanic,
(c) continental-continental.
Adapted by permission of
Merrill, an imprint of Macmillan
Publishing Company, from *The
Earth,* 2d ed., by Edward J.
Tarbuck and Frederick K.
Lutgens, p. 412. Copyright ©
1987 by Merrill Publishing.

(a)

(b)

(c)

FIGURE 12–11

Appalachian Mountain region of central Pennsylvania, shown on a *Landsat* image and a U.S. Geological Survey topographic quadrangle map (1:250,000 scale). Both illustrate the highly folded nature of this region. Courtesy of U.S. Geological Survey and National Aeronautics and Space Administration.

World Structural Regions

A brief examination of the distribution of world structural regions helps summarize the information presented about the three rock-forming processes (igneous, sedimentary, metamorphic), plate tectonics, landform construction, and orogenesis. Figure 12–12 defines seven essential structural regions, highlights Earth's major mountain systems, and allows interesting comparison with the chapter-opening illustration of the ocean floor.

EARTHQUAKES

Crustal plates do not move smoothly past one another. Instead, stress causes strain in the rocks along the plate boundaries until the sides release and lurch into new positions. In the Liaoning Province of northeastern China, ominous indications of possible tectonic activity began in 1970. Foreboding symptoms included land uplift and tilting, increased numbers of minor tremors, and changes in the region's magnetic field—all of this after almost 120 years of quiet.

These precursors of coming tectonic events continued for almost five years before Chinese scientists took the bold step of forecasting an earthquake. Finally, on 4 February 1975 at 2:00 P.M., some 3 million people were evacuated in what turned out to be a timely manner; the quake struck at 7:36 P.M., within the predicted time frame. Ninety percent of the buildings in the city of Haicheng were destroyed, but thousands of lives were saved, and success was proclaimed—an earthquake had been forecasted and preparatory action taken, for the first time in history.

Only 17 months later, at Tangshan in the northeastern province of Hebei (Hopei), an earthquake occurred on 28 July 1976 without any warning precursors from which a forecast could be prepared. Consequently, this quake destroyed 95% of the buildings and 80% of the industrial structures, severely damaged more than half the bridges and highways, and killed a quarter of a million people! The jolt was strong enough to throw people against the ceilings of their homes. An old, undetected fault had ruptured and offset 1.5 m (5 ft) along an 8 km (5 mi) stretch through Tangshan, devastating large areas just 145 km (90 mi) southeast of Beijing, China's capital city. How is it possible that these two tectonic events produced such different human consequences?

Earthquake Essentials

Earthquakes associated with faulting are referred to as tectonic earthquakes. Their vibrations are transmitted as waves of energy throughout Earth's interior and are detected with a **seismograph**, an instrument that records vibrations in the crust. Earthquakes are rated on two different scales, an intensity scale and a magnitude scale.

Intensity scales are useful in classifying and describing terrain, construction, and local damage conditions following an earthquake. Earthquake intensity is rated on the arbitrary Mercalli scale, a Roman numeral scale from I to XII representing "barely felt" to "catastrophic total destruction." It was designed in 1902 and modified in 1931 to be more applicable to conditions in North America.

Earthquake *magnitude* is estimated according to a system designed by Charles Richter in 1935. First, the amplitude of seismic waves is recorded on a seismograph located at least 100 km (62 mi) from the epicenter of the quake. That measure is then charted on the **Richter scale**, which is open ended and logarithmic; that is, each whole number on the scale represents a 10-fold increase in the measured wave amplitude. Translated into energy, each whole number demonstrates a 31.5-fold increase in the amount of energy released. Thus, a 3.0 on the Richter scale represents 31.5 times more energy than a 2.0 and 992 times more energy than a 1.0. It is difficult to imagine the power released by the Tangshan quake, which was rated a 7.8 on the Richter scale.

Table 12–1 presents the characteristics and frequency of earthquakes that are expected annually worldwide. Table 12–2 lists a sampling of significant earthquakes. Note each earthquake's location, magnitude, and intensity rating (if available), and total loss of life (if known).

The subsurface area along a fault plane, where the motion of seismic waves is initiated, is called the *focus,* or hypocenter (see Figure 12–14). The area at the surface directly above this subsurface location is the *epicenter.* Shock waves produced by an earthquake radiate outward from both the focus and

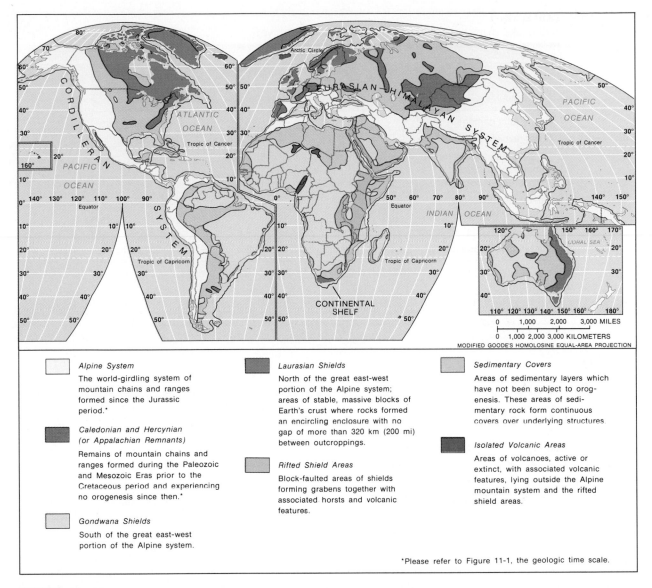

FIGURE 12–12
World structural regions and major mountain systems. Because each structural region in this figure includes related landforms adjacent to the central feature of the region, some of the regions appear wider than the features themselves. After Richard E. Murphy, "Map Supplement No. 9—Landforms of the World," *Annals of the Association of American Geographers* 58, no. 1 (March 1968). Adapted by permission.

epicenter areas. An *aftershock* may occur after the main shock, sharing the same general area of the epicenter. A *foreshock* also is possible, preceding the main shock. The greater the distance from an epicenter, the less severe the shock that is experienced. At a distance of 40 km (25 mi), only 1/10 of the full effect of an earthquake normally is felt. However, if the ground is unstable, distant effects can be magnified, as they were in Mexico in 1985 and in San Francisco in 1989.

TABLE 12–1

Earthquake characteristics and frequency expected each year.

Characteristic Effects in Populated Areas	Approximate Intensity (modified Mercalli scale)	Approximate Magnitude (Richter scale)	Number per Year
Nearly total damage	XII	>8.0	1 every few years
Great damage	X–XI	7–7.9	18
Considerable-to-serious damage to buildings; railroad tracks bent	VIII–IX	6–6.9	120
Felt-by-all, with slight damage to buildings	V–VII	5–5.9	800
Felt-by-some to felt-by-many	III–IV	4–4.9	6200
Not felt, but recorded	I–II	2–3.9	500,000

SOURCE: Charles F. Richter, *Elementary Seismology*, W. H. Freeman, 1958, and others.

TABLE 12–2

A sampling of significant earthquakes.

Year	Date	Location	Deaths	Intensity	Magnitude
1556	Jan. 23	Shensi, China	830,000	—	—
1737	Oct. 11	Calcutta, India	300,000	—	—
1811	Dec. 16	New Madrid, Missouri	—	—	—
1812	Jan. 23	New Madrid, Missouri	—	—	—
1812	Feb. 7	New Madrid, Missouri	Several	XI–XII	—
1857	Jan. 9	Fort Tejon, California	—	X–XI	—
1865	Oct. 8	San Francisco, California	—	IX	—
1870	Oct. 21	Montreal to Quebec, Canada	—	IX	—
1906	Apr. 18	San Francisco, California	3000	XI	8.25
1923	Sept. 1	Kwanto, Japan	143,000	XII	8.2
1939	Dec. 27	Erzincan, Turkey	40,000	XII	8.0
1960	May 21–30	Southern Chile	5700	—	8.5
1964	Mar. 28	Southern Alaska	131	X–XII	8.6
1970	May 31	Northern Peru	66,000	—	7.8
1971	Feb. 9	San Fernando, California	65	VII–IX	6.5
1972	Dec. 23	Managua, Nicaragua	5000	X–XII	6.2
1975	Feb. 4	Liaoning Province, China	Few	X	7.4
1976	Jul. 27	Tangshan, China	250,000	XI–XII	7.6
1985	Sept. 19	Mexico City, Mexico	7000	IX–XII	8.1
1988	Dec. 7	Armenian U.S.S.R.–Turkey border	30,000	XII	6.9
1989	Oct. 17	Loma Prieta (near Santa Cruz, California)	67	VII–IX	7.1

Mexico City, currently the world's second-largest city, is positioned on the soft, moist sediments of an ancient lake bed. On 19 September 1985, during rush-hour traffic, two major earthquakes (8.1 and 7.6 on the Richter scale) struck 400 km (250 mi) southwest of Mexico City. The epicenter was on the seafloor off Mexico and Central America, and the old lake bed beneath Mexico City magnified the shock waves by more than 500%. As a result, 250 buildings collapsed, some 8000 structures experienced damage, and about 7000 people perished.

The Nature of Faulting. We earlier described specific types of faults and faulting motions. How a fault actually breaks is still under investigation, but the basic process is described by the **elastic-rebound theory**. Generally, two sides along a fault appear to be locked by friction, resisting any movement despite the powerful motions of adjoining pieces of crust. This stress continues to build strain along the fault surfaces, storing elastic energy like a wound-up spring. When movement finally does occur as the strain build-up exceeds the frictional lock, energy is released abruptly, returning both sides of the fault to a condition of less strain. Think of the fault plane as a surface with irregularities that act as sticking points, preventing movement, similar to two pieces of wood held together by drops of glue rather than an even coating of glue. Research scientists at the USGS and the University of California have identified these small areas of high strain as **asperities**. They are the points that break and release the sides of the fault.

If the fracture along the fault line is isolated to a small asperity break, the quake will be small in magnitude. However, if the break involves the release of strain along several asperities, the quake will be greater in extent and will involve the shifting of massive amounts of crust. Clearly, as some asperities break (perhaps recorded as small foreshocks), the strain is increased on surrounding asperities. Thus, small earthquakes in an area may be precursors to a major quake.

The San Francisco Earthquakes. In 1906 an earthquake devastated San Francisco, a city of 400,000 at the time, and three-quarters of a million today. Faulting was evident over 435 km (270 mi), and this prompted intensive research to discover the nature of faulting. (Realize that this occurred six

years before Wegener's continental drift hypothesis was proposed.) The elastic-rebound theory developed as a result, and today quakes such as the one near San Francisco in 1989 can be interpreted.

The fault system involved was the San Andreas, which provides a good example of the evolution of a spreading center overridden by an advancing continental plate (Figure 12–13). In the figure, the East Pacific Rise developed as a spreading center with associated transform faults (a), while the North American plate was progressing westward after the breakup of Pangaea. Forces then shifted the transform faults toward a northwest-southeast alignment along a weaving axis (b). Finally, the western margin of North America overrode those shifting transform faults (c). In relative terms, the motion along the fault is right lateral, whereas in absolute terms the North American plate is still moving westward.

The earthquake mentioned at the outset of this chapter involved a portion of the San Andreas fault approximately 16 km (9.9 mi) east of Santa Cruz and 95 km (59 mi) south of San Francisco. The quake occurred at 5:04 P.M. Pacific Standard Time and registered 7.1 on the Richter scale. A fault had ruptured at a focus unusually deep for the San Andreas system—more than 18 km (11.5 mi) below the surface.

Unlike previous earthquakes—such as the one in 1906, when the plates shifted a maximum of 6.4 m (21 ft) relative to each other—there was no evidence of plate movement at the surface. Instead, the fault plane solution suggested in Figure 12–14(a) shows the two plates moving approximately 2 m (6 ft) past each other deep below the surface, with the Pacific plate thrusting 1.3 m (4.3 ft) upward. This is an unusual motion for the San Andreas fault, especially when compared to the apparent motion in 1906, and perhaps is a clue that this fault is more complex than previously thought.

The damage from this quake was extreme in the region of the Loma Prieta epicenter and in the San Francisco Bay Area, where houses and freeways are built on fill and bay mud. Like the situation in Mexico City, unstable soils magnified the effects of a distant quake. In only 15 seconds more than 2 km (1.2 mi) of freeway overpass and a section of the San Francisco–Oakland Bay Bridge collapsed, $8 billion in damages was generated, 14,000 people were displaced from their homes, 4000 were injured, and 67 people were killed—Figure 12–14(b). (Your

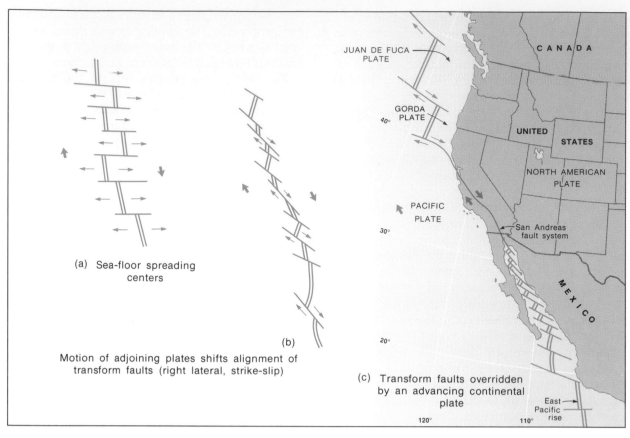

FIGURE 12–13
Formation of the San Andreas fault system as a series of transform faults.

author and his wife remember it well; as we worked on this book in our living room, with the TV on for the game, the whole house wavered and lights swung as the quake struck about 200 km (125 mi) from our home near Sacramento.)

One potentially positive discovery from this disaster was that Stanford University scientists reported unusually great changes in Earth's magnetic field—about 30 times more than normal—three hours before the main shock. These measurements were made 7 km from the epicenter, raising hopes that short-term warnings might be possible in the future.

Paleoseismology and Earthquake Forecasting

Principal zones of earthquake activity occur near plate boundaries, as the shaded areas in Figure 11–22 demonstrate. The correlation is such that

earthquakes have provided key diagnostic evidence in understanding plate tectonics, and in turn, that correlation offers a long-term forecast of earthquake sites. The challenge is to discover how to predict the *specific time and place* for a quake in the short term.

One approach is to examine the history of each plate boundary and determine the frequency of earthquakes in the past, a study called **paleoseismology**. Paleoseismologists then construct maps that provide an estimate of expected earthquake activity. Areas that are quiet and overdue for an earthquake are termed **seismic gaps**, that is, an area that forms a gap in the earthquake occurrence record and therefore a place that possesses accumulated strain. The area along the Aleutian Trench subduction zone had three such gaps until the great 1964 Alaskan earthquake (8.6 on the Richter scale) filled one of them (Figure 12–15).

The areas around San Francisco and northeast of Los Angeles represent other such gaps where the

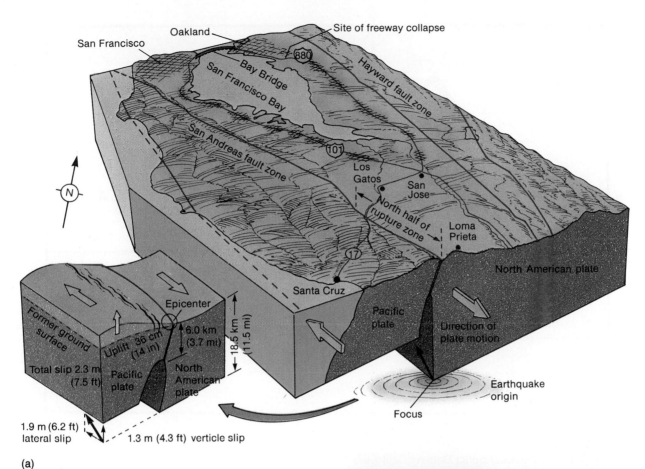

Oakland

San Francisco

Site of freeway collapse

880

Bay Bridge

San Francisco Bay

Hayward fault zone

San Andreas fault zone

101

Los Gatos

San Jose

North half of rupture zone

Loma Prieta

North American plate

Santa Cruz

17

Pacific plate

Direction of plate motion

Former ground surface

Epicenter

Uplift 36 cm (14 in)

6.0 km (3.7 mi)

18.5 km (11.5 mi)

Total slip 2.3 m (7.5 ft)

Pacific plate

North American plate

Earthquake origin

Focus

1.9 m (6.2 ft) lateral slip

1.3 m (4.3 ft) verticle slip

(a)

FIGURE 12–14

The 1989 Loma Prieta, California, earthquake: (a) fault-plane diagram of the lateral and thrust movement at depth for this earthquake; (b) Route 880 Cypress Freeway collapse. (a) After "The Loma Prieta Earthquake of October 17, 1989," U.S. Geological Survey pamphlet by Peter J. Ward and Robert A. Page. Washington: November 1989, p. 1. (b) Photo courtesy of California Department of Transportation.

(b)

fault system appears to be locked by friction and accumulating strain. The 1989 Loma Prieta earthquake was predicted in 1988 by the U.S. Geological Survey as having a 30% chance of occurring with a 6.5 magnitude, within 30 years. The actual quake filled a portion of the seismic gap in that region.

In southern California, along a presently locked portion of the fault system in the Mojave Desert, the San Andreas fault experienced a major earthquake in 1857. The seismic history of the area shows a recurrence interval for severe quakes of 140 to 150 years, so the area is approaching the time for a major earthquake (a 7.5 Richter rating or higher). Such a quake could have devastating consequences on the Los Angeles metropolitan area. Because strain accumulates at about 3.5 cm/year (1.4 in./year), more than

FIGURE 12–15
Seismic gap map. Data from Lamont-Doherty Geological Observatory, Columbia
University. Originally published in "Seismic Gaps and Plate Tectonics: Seismic
Potential for Major Boundaries," *Pageoph,* p. 117, 1979. Reprinted by permission.

4.7 m (15.4 ft) of strain is estimated to exist along the
plate boundary in southern California.

Except for isolated successes in China, Japan, and
the United States, earthquake prediction and forecasting has proven to be a much more difficult task
than scientists originally imagined. Prediction is
thought to be a matter of assessing earthquake
precursors. Scientists still are trying to assemble an
accurate model of the microevents that precede an
earthquake, including the detection of small asperity
breaks. *Dilitancy* refers to the slight increase in bulk
volume produced by small cracks during deformation in stressed rock as it accumulates strain. These
prequake physical changes are thought to be detectable through various methods, although their complete meaning remains elusive.

The affected region may tilt and swell in response
to accumulated strain. An indicator of dilitancy is
thought to be the increased amount of radon gas that
appears dissolved in groundwater, liberated from
the newly formed cracks in the rock. Also, the ability
of the rock to transmit an electrical current and
compressional waves decreases in measurable
amounts. In addition, the latest discovery is that
certain electromagnetic emissions occur prior to a
quake. The complexity of the interactions along a
fault continue to challenge science.

Earthquake Preparedness and Planning

Actual implementation of an action plan to reduce
deaths, injuries, and property damage from earthquakes is very difficult. The political environment
adds additional complexity; sadly, the impact of an
accurate earthquake prediction would be viewed as a
negative threat to a region's economy (Figure

12–16). If we examine the potential socioeconomic impact of earthquake prediction on an urban community, it is difficult to imagine a chamber of commerce, bank, real estate agent, tax assessor, or politician who would privately welcome such a prediction. Long-range planning is a complex subject.

A valid and applicable generalization seems to be that humans are unable or unwilling to perceive hazards in a familiar environment. Such an axiom of human behavior certainly helps explain why large populations continue to live and work in earthquake-prone settings in developed countries. Similar questions also can be raised about populations in areas vulnerable to other disasters.

VOLCANISM

A **volcano** forms at the end of a central vent or pipe that rises from the asthenosphere through the crust into the volcanic mountain, usually forming a **crater**, or circular surface depression at the summit. Magma rises and collects in a magma chamber deep below the volcano until conditions are right for an eruption. This subsurface magma produces tremendous heat and in some areas it boils groundwater, as seen in the thermal features of Yellowstone National Park. The steam given off is a potential energy source if it is accessible. Such **geothermal energy** has provided heating and electricity for more than 90 years in Iceland, New Zealand, and Italy. North of San Francisco, some 1300 megawatts of electrical capacity are generated from steam brought up through geothermal wells and fed through turbines.

The magma that actually issues from the volcano is termed **lava** (molten rock). It can occur in many different textures and forms, which accounts for the varied behavior of volcanoes and the different landforms they build. In this section we will look at

FIGURE 12–16
Socioeconomic impacts and adjustment to an earthquake prediction. After J. E. Haas and D. S. Mileti, 1976, *Soicoeconomic Impact of Earthquake Prediction on Government, Business, and Community,* Boulder: University of Colorado, Institute of Behavioral Sciences. Used by permission.

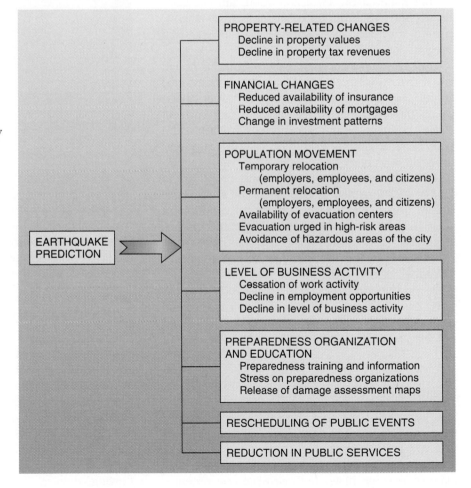

EARTHQUAKE PREDICTION

PROPERTY-RELATED CHANGES
Decline in property values
Decline in property tax revenues

FINANCIAL CHANGES
Reduced availability of insurance
Reduced availability of mortgages
Change in investment patterns

POPULATION MOVEMENT
Temporary relocation
(employers, employees, and citizens)
Permanent relocation
(employers, employees, and citizens)
Availability of evacuation centers
Evacuation urged in high-risk areas
Avoidance of hazardous areas of the city

LEVEL OF BUSINESS ACTIVITY
Cessation of work activity
Decline in employment opportunities
Decline in level of business activity

PREPAREDNESS ORGANIZATION AND EDUCATION
Preparedness training and information
Stress on preparedness organizations
Release of damage assessment maps

RESCHEDULING OF PUBLIC EVENTS

REDUCTION IN PUBLIC SERVICES

FIGURE 12–17

Principal mechanisms for each type of volcanic activity. After U.S. Geological Survey, *The Dynamic Planet,* 1989.

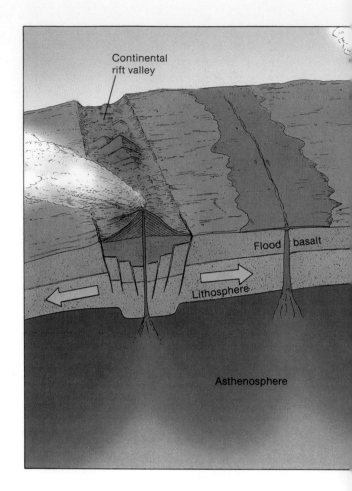

five related landforms and their origins: cinder cones, calderas, composite volcanoes, shield volcanoes, and flood basalt plateaus.

Lava, gases, and **tephra** (pulverized rock and clastic materials ejected violently during an eruption) pass through the vent to the surface and build a volcanic landform. One such landform is a **cinder cone**, a small cone-shaped hill usually less than 450 m (1500 ft) high, with a truncated top formed from cinders that accumulate during moderately explosive eruptions. Cinder cones are made of tephra and scoria (a vesicular, cindery rock), as exemplified by Paricutín in southwestern Mexico. Another distinctive landform is a large basin-shaped depression called a **caldera** (Spanish for kettle). It forms when summit material on a volcanic mountain collapses inward after an eruption or other loss of magma, forming a caldera such as Crater Lake in southern Oregon.

Over 1300 volcanoes exist on Earth, although fewer than 600 are *active* (have had at least one eruption in recorded history). North America has about 70 (mostly inactive) along the western margin of the continent, Mount Saint Helens in Washington State being a famous example. In an average year about 50 volcanoes erupt worldwide, varying from modest activity to major explosions. Eruptions in remote locations and at depths on the seafloor go largely unnoticed, but the occasional eruption of great magnitude near a population center makes headlines. More than 200,000 people have been killed from volcanic eruptions in the past 400 years. Volcanoes also produce some benefits. These include production of materials that produce fertile soils, as in Hawaii; conditions that can provide geothermal energy; and even new real estate added to Iceland, Japan, Hawaii, and elsewhere.

Locations of Volcanic Activity

The location of volcanic mountains on Earth is a function of plate tectonics and hot spot activity. Volcanic activity occurs in three areas: (1) along subduction boundaries at continental plate–oceanic plate or oceanic plate–oceanic plate convergence; (2) along sea-floor spreading centers on the ocean floor and areas of rifting on continental plates; and (3) at hot spots, where individual plumes of magma rise through the crust. Figure 12–17 illustrates these three types of locations, which you can compare to the active volcano sites and plate boundaries shown in Figure 11–22.

Types of Volcanic Activity

The variety of forms among volcanoes makes them hard to classify; most fall in transition areas between one type and another. Even during a single eruption, a volcano may behave in several different ways. The primary factors in determining an eruption type are (1) the chemistry of its magma related to its source and (2) its viscosity. Viscosity is magma's resistance to flow, or degree of fluidity; it ranges from low viscosity (very fluid) to high viscosity (thick and flowing slowly). We will consider two types of

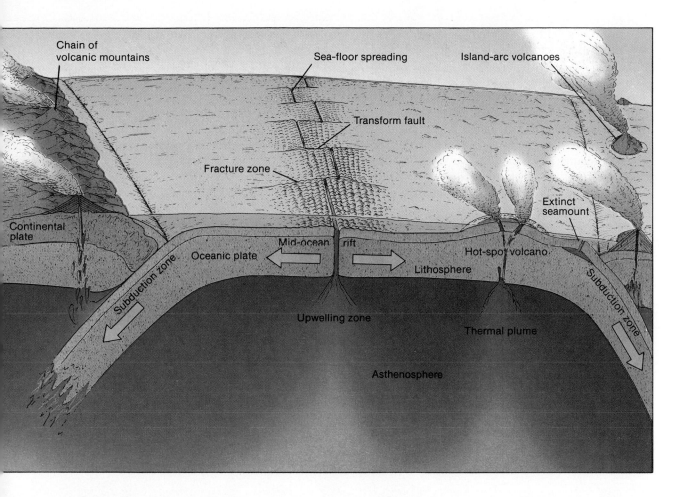

Chain of volcanic mountains

Sea-floor spreading

Island-arc volcanoes

Transform fault

Fracture zone

Extinct seamount

Continental plate

Mid-ocean rift

Oceanic plate

Subduction zone

Hot-spot volcano

Lithosphere

Subduction zone

Upwelling zone

Thermal plume

Asthenosphere

eruptions—effusive and explosive—and the characteristic landforms they build.

Effusive Eruptions. These are the relatively gentle eruptions that produce enormous volumes of lava annually on the seafloor and in places like Hawaii. Direct eruptions from the asthenosphere produce a low-viscosity magma which is very fluid and yields a dark, basaltic rock (less than 50% silica and rich in iron and magnesium). Gases readily escape from this magma because of its texture, causing an **effusive eruption** that pours out on the surface, with relatively small explosions and little tephra. Dramatic fountains of basaltic lava sometimes shoot upward, powered by jets of rapidly expanding gases (Figure 12–18).

An effusive eruption may come from a single vent or from a flank eruption through side vents in surrounding slopes. If such vents form a linear opening, they are called *fissures* and sometimes form a dramatic curtain of fire during eruptions. In Iceland, active fissures are spread throughout the plateau landscape. In Hawaii, rift zones capable of erupting tend to converge on the central crater, or vent. The interior of such a crater is often a sunken caldera, which may fill with low-viscosity magma during an eruption, forming a molten lake, which may then overflow lava downslope to the sea.

On the island of Hawaii the continuing Kilauea eruption represents the longest eruptive episode in Hawaiian history, active since 1823! During 1989–1990, lava flows from Kilauea actually consumed several visitor buildings in the Hawaii Volcanoes National Park and homes in Kalapana, a nearby town. Kilauea has produced more lava than any other vent on Earth.

A typical mountain landform built from effusive eruptions is gently sloped, gradually rising from the surrounding landscape to a summit crater, similar in outline to a shield of armor lying face up on the ground, and therefore called a **shield volcano**. The shield shape and size of Mauna Loa in Hawaii is distinctive when compared to Mount Rainier in Washington, which is a different type of volcano (we will explain shortly) and the largest in the Cascade Range (Figure 12–19). The height of the shield is the result of successive eruptions, flowing one on top of another. Mauna Loa is one of five shield volcanoes that comprise the island of Hawaii, and has taken at least 1 million years to accumulate its mass. Mauna Kea is slightly taller, but Mauna Loa is the most massive single mountain on Earth.

In rifting areas, effusive eruptions send material out through elongated fissures, forming extensive sheets on the surface. The Columbia Plateau of the northwestern United States represents the eruption of **flood basalts** (Figure 12–20). These extensive sheet eruptions also episodically flow from the mid-ocean ridges, forming new seafloor in the process. Worldwide, the volume of material produced in effusive fissure eruptions along spreading centers is far greater than that of the continental eruptions. Iceland continues to form in this effusive manner, with upwelling basaltic flows creating new Icelandic real estate.

Explosive Eruptions. Volcanic activity along subduction zones produces the well-known explosive volcanoes. Magma produced by the melting of subducted oceanic plate and other materials is thicker (more viscous) than magma from effusive volcanoes; it is 50–75% silica and high in aluminum. Consequently, it tends to block the magma conduit inside the volcano, allowing pressure to build, leading to an **explosive eruption**. This magma forms a lighter dacitic rock at the surface (see *dacite* in Table 11–2), as illustrated in the comparison in Figure 12–21.

The term **composite volcano** is used to describe explosively formed mountains. (They are sometimes called stratovolcanoes, but because shield volcanoes also can exhibit a stratified structure, composite is the preferred term.) Composite volcanoes tend to

FIGURE 12–18
Kilauea fountain eruption in Hawaii. Photograph by J. D. Griggs/U.S. Geological Survey.

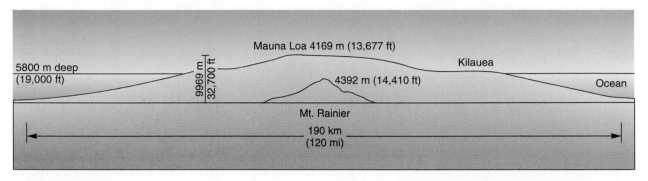

FIGURE 12–19
Comparison of Mauna Loa and Mount Rainier. After U.S. Geological Survey, 1986, *Eruption of Hawaiian Volcanoes,* Washington: Government Printing Office.

have steep sides and are more conical than shield volcanoes and therefore are also known as *composite cones.* If a single summit vent remains stationary, a remarkable symmetry may develop as the mountain grows in size, as demonstrated by Mount Orizaba in Mexico, Mount Fuji in Japan, the pre-1980-eruption shape of Mount Saint Helens in Washington, and the Mayon Volcano in the Philippines (Figure 12–22).

As the magma in a composite volcano forms plugs near the surface, those blocked-off passages build tremendous pressure, keeping the trapped gases compressed and liquified. When the blockage can no longer hold back this pressurized inferno, explosions equivalent to megatons of TNT blast the tops and sides off these mountains, producing much less lava but larger amounts of tephra, which includes volcanic ash (<2 mm in diameter), dust, cinders, lapilli (up to 32 mm in diameter), scoria (volcanic slag), pumice, and **pyroclastics** (explosively ejected rock pieces). Table 12–3 presents some of the notable composite volcano eruptions that have oc-

FIGURE 12–20
Flood basalts characteristic of the Columbia Plateau. Photo by author.

Basalt   Dacite

FIGURE 12–21
Comparison of basalt from Hawaii and dacite from Mount Saint Helens. Photograph by J. D. Griggs/U.S. Geological Survey.

curred. Focus Study 12–1 details the much-publicized 1980 eruption of Mount Saint Helens.

Unlike Hawaii Volcanoes National Park, where tourists gather at observation platforms to watch the relatively calm effusive eruptions, composite volcanoes do not invite close inspection and can explode with little warning:

- In A.D. 79, Mount Vesuvius buried the city of Pompeii, Italy, in three days of tephra and pyroclastic eruptions, even though the volcano had been dormant for centuries.
- In 1902 on the Caribbean island of Martinique, the beautiful port city of Saint-Pierre was devastated in a few minutes by an eruption of Mount Pelée. That eruption featured a lateral blast of incandescent ash and super-hot gases, known as a "glowing cloud" or **nuée ardente** in French. Despite months of precursory rumbling, ash, hot mudflows, and minor bursts from the

FIGURE 12–22
A typical composite volcano with its cone-shaped form.

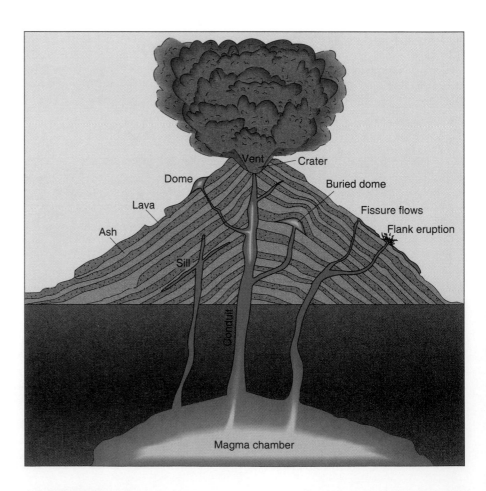

The 1980 Eruption of Mount Saint Helens

Probably the most studied and photographed composite volcano on Earth is Mount Saint Helens, located 70 km (45 mi) northeast of Portland, Oregon, and 130 km (80 mi) south of the Tacoma-Seattle area of Washington. Mount Saint Helens is the youngest and most active of the Cascade Range of volcanoes, which form a line from Mount Lassen in California to Mount Baker in Washington. The Cascade Range is the product of the Juan de Fuca sea-floor spreading center off the coast of northern California, Oregon, Washington, and British Columbia, and the plate subduction that occurs offshore.

Associated events began on 20 March 1980 with a sharp earthquake at 3:48 P.M., registering 4.1 on the Richter scale. The first eruptive outburst occurred one week later, beginning with a 4.5 Richter quake and continuing with a thick black plume of ash and the development of an initial crater about the size of a football field. April 1 marked the first harmonic tremor, or volcanic earthquake, from which it was inferred that magma was on the move within the mountain. *Harmonic tremors* are slow, steady vibrations, unlike the sharp releases of energy associated with earthquakes and faulting. A growing bulge on the north side of the mountain indicated the direction of the magma flow within the volcano. A bulge represents the greatest risk from a composite volcano, for it could mean a potential lateral burst across the landscape.

On Sunday, May 18, at 8:27 A.M., the area north of the mountain was rocked by a 5.0 Richter quake, the strongest to that date. The mountain, with its 245 m (800 ft) bulge, was shaken, but nothing else happened. Then a second quake (5.1) hit five minutes later, loosened the bulge, and launched the eruption. David Johnson, a volcanologist with the U.S. Geological Survey, was only 8 km (5 mi) from the mountain, servicing monitoring instruments when he saw it begin. He radioed headquarters in Vancouver, Washington, saying, "Vancouver, Vancouver, this is it!" but he perished in the eruption that followed.

The slumping north face of the mountain produced the greatest landslide witnessed in recorded history; about 2.75 km³ (0.67 mi³) of rock, ice, and trapped air, all fluidized with steam, tumbled down at speeds approaching 250 kmph (155 mph). Landslide materials did not come to rest until they reached 21 km (13 mi) into the valley, blanketing the forest, covering a lake, and filling the rivers below. A series of photographs, taken at 10-second intervals from the east looking west, records this sequence (Figure 1).

FIGURE 1
The Mount Saint Helens eruption sequence and corresponding schematics. Photo
sequence by Keith Ronnholm.

As the contents of the mountain exploded, a surge of hot gas (about 300°C or 570°F), steam-filled ash, pyroclastics, and a nuée ardente moved northward, hugging the ground and traveling up to 400 kmph (250 mph) for a distance of 28 km (17 mi). The eruption continued with intensity for nine hours, first clearing out old rock from the throat of the volcano and then blasting new material (Figure 2).

As destructive as such eruptions are, they also are constructive, for this is the way in which a volcano builds height. Before the eruption, Mount Saint Helens was 2950 m (9677 ft) tall; the eruption blew away 418 m (1370 ft). But today, Mount Saint Helens is growing by building a lava dome within the crater. The thick lava rapidly and repeatedly plugs and breaks in a series of dome eruptions that may continue for several decades. The dome already is over 300 m (1000 ft) high, so that a new mountain is being born from the eruption of the old. An ever-resilient ecology is recovering as plants and animals alike reclaim the devastation.

FIGURE 2
Scorched earth and tree blowdown area, covering some 38,950 hectares (95,000 acres). Photo by author.

TABLE 12−3
Composite volcano eruptions.

Date	Name or Location	Number of Deaths	Amount Extruded (mostly pyroclastics, in km³ (mi³)
Prehistoric	Yellowstone, Wyoming	Unknown	2400(576)
4600 B.C.	Mount Mazama (Crater Lake, Oregon)	Unknown	50−70(12−17)
1900 B.C.	Mount Saint Helens	Unknown	4(0.95)
A.D. 79	Mount Vesuvius, Italy	20,000	3(0.7)
1500	Mount Saint Helens	Unknown	1(0.24)
1815	Tambora, Indonesia	66,000	80−100(19−24)
1883	Krakatoa, Indonesia	36,000	18(4.3)
1902	Mount Pelée, Martinique	29,000	Unknown
1912	Mount Katmai, Alaska	Unknown	12(2.9)
1943−1952	Paricutín, Mexico	0	1.3(0.3)
1980	Mount Saint Helens	54	4(0.95)
1985	Nevado del Ruiz, Colombia	23,000	1(0.24)
1991	Mount Pinatubo, Philippines	400	Current

FIGURE 12−23
Crater Lake caldera, Oregon, formerly Mount Mazama. Photo by author.

mountain, politicians discouraged citizens from fleeing, and all but two of 30,000 people in the town perished.

- In 1883 in Indonesia, an entire group of islands centered on the volcano Krakatoa (between the big islands of Java and Sumatra) was obliterated in two days of explosions. The ash cloud reached into the mesosphere, and the sounds of that August blast were heard in central Australia, the Philippines, and even 4820 km (3000 mi) away in the Indian Ocean! (The sound took some 4 hours to reach the listener, who noted the noise in his ship's log.) What was not blown into the sky fell back into the collapsed remains of a large caldera 275 m (900 ft) under the sea.

- Around 4500 B.C. in southern Oregon, a large caldera was left after Mount Mazama's violent eruption. However, that caldera was fed by snowmelt, rainwater, and subsurface water to form a lake 9.7 km (6 mi) in diameter and 610 m (2000 ft) deep, with a later cone forming a small island in the middle—the beautiful Crater Lake (Figure 12–23).

SUMMARY

The continents are formed as a result of plate tectonics, with old cores of the continents being enlarged through accretion of dispersed terranes. The crust is deformed by folding, broad warping, and faulting, which produce characteristic landforms. Orogenesis, the process of mountain building, has produced Earth's major mountain belts—the Cordilleran system of the Americas and the Eurasian-Himalayan system that stretches from the Alps in Europe to the Himalayas in southern Asia.

Earthquakes generally occur along plate boundaries and affect segments of the world's population. More is being learned all the time about the nature of faulting: stress and the buildup of strain, irregularities along fault plane surfaces, and the way faults rupture. Earthquakes have claimed many lives and caused a great deal of damage through the centuries, so earthquake prediction and improved planning are active concerns of seismology, the study of earthquake waves and Earth's interior.

Volcanoes offer direct evidence of the makeup of the asthenosphere. Material from hot spots upwells to the surface, or plate subduction generates magma that rises beneath the surface, collects in chambers, and erupts through surface vents. Volcanoes are of two general types, based on the chemistry and viscosity of the magma involved—effusive and explosive. Volcanic activity has created some destructive moments in history, but also has replenished the land and constantly creates new seafloor and real estate.

KEY TERMS		
	Alleghany orogeny	Eurasian-Himalayan system
	Anticline	Explosive eruption
	Asperities	Faulting
	Caldera	Flood basalts
	Cinder cone	Folding
	Circum-Pacific belt	Geothermal energy
	Composite volcano	Graben
	Continental shields	Horst
	Cordilleran system	Lava
	Crater	Nuée ardente
	Earthquake	Orogenesis
	Effusive eruption	Paleoseismology
	Elastic-rebound theory	Pyroclastics

Richter scale   Syncline

Ring of fire   Tephra

Seismic gaps   Terranes

Seismograph   Tilted fault block

Shield volcano   Volcano

Strike-slip fault   Wrangellia terranes

REVIEW QUESTIONS

1. How does the ocean floor map (chapter-opening illustration) bear the imprint of the principles of plate tectonics? Describe and explain.
2. Define "tectonic" and use the concept in two or three sentences.
3. What is a craton? Relate this structure to the continental shields, and describe where shield regions exist in North America.
4. What are the migrating terranes that came together to form continental masses? Give an example of such a terrane.
5. Diagram a folded landscape in cross section, and identify the features of the folding.
6. Define the four basic types of faults. How do these relate to earthquakes and seismic activity?
7. How did the Basin and Range Province evolve in the western United States? What other examples exist of this type of landscape?
8. Define orogenesis. What is meant by the birth of mountain chains?
9. Name some major orogenies.
10. Identify and detail Earth's two large mountain chains.
11. How are plate boundaries related to episodes of mountain building? Do different types of plate boundaries produce differing orogenic episodes?
12. Relate tectonic processes to the formation of the Appalachians and the Alleghany orogeny.
13. Describe the differences between the two earthquakes that occurred in China in 1975 and 1976.
14. Differentiate between the Mercalli and the Richter scales. How are these used to describe an earthquake?
15. What is the relationship between an epicenter and the focus of an earthquake?
16. What local conditions in Mexico City and San Francisco severely magnified the energy felt from each quake?
17. How do the elastic-rebound theory and asperities help explain the nature of faulting?
18. Describe the San Andreas fault and its relationship to sea-floor spreading movements.
19. How does the seismic gap concept relate to expected earthquake occurrences? Have any gaps correlated with earthquake events in the recent past? Explain.
20. What do you see as the biggest barrier to effective earthquake prediction?
21. What is a volcano?
22. Where do you expect to find volcanic activity in the world? Why?
23. Compare effusive and explosive eruptions. Why are they different? What types of landforms are produced by each type?

SUGGESTED READINGS

Anderson, Don L. "The San Andreas Fault." *Scientific American* 225 (November 1971): 52–66.

Berger, A. R., ed. *Episodes (International Geoscience Magazine)* 10, no. 4 (December 1987): 234–313. Entire issue reviews plate tectonics, geology, and U.S. mineral resources.

Bolt, Bruce A. *Earthquakes—A Primer*. San Francisco: W. H. Freeman, 1978.

Bullard, Fred M. *Volcanoes of the Earth*. Austin: University of Texas Press, 1976.

Decker, Robert W., Thomas L. Wright, and Peter H. Stauffer. *Volcanism in Hawaii*. 2 vols. U.S. Geological Survey Professional Paper 1350. Washington: Government Printing Office, 1987.

Editors of Time-Life Books. *Volcano*. Planet Earth Series. Alexandria, VA: Time-Life Books, 1982.

Findley, Rowe. "Eruption of Mount St. Helens." *National Geographic* 159, no. 1 (January 1981): 3–65.

Francis, Peter, and Stephen Self. "Collapsing Volcanoes." *Scientific American* 256 (June 1987): 90–97.

———. "The Eruption of Krakatoa." *Scientific American* 249 (November 1983): 172–87.

Hoblitt, Richard P. *Observations of the Eruptions of July 22 and August 7, 1980, at Mount St. Helens, Washington*. U.S. Geological Survey Professional Paper 1335. Washington: Government Printing Office, 1986.

Iacopi, Robert. *Earthquake Country*. Menlo Park, CA: Lane Books, 1971.

Keller, Edward A. *Environmental Geology*. 5th ed. Columbus: Macmillan, 1988.

Macdonald, G. A., A. T. Abbott, and D. C. Cox. *Volcanoes in Hawaii*. Honolulu: University of Hawaii Press, 1983.

Maranto, Gina. "Inferno in Paradise—Double Eruption of Hawaii's Volcanoes." *Discover* 5, no. 6 (June 1984): 88–93.

Mayberry, Robert H. *Violent Forces of Nature*. Mt. Airy, MD: Lomond Publications, 1986.

McDowell, Bart. "When the Earth Moves." *National Geographic* 169, no. 5 (May 1986): 638–75.

Murphy, Richard E. "Map Supplement No. 9, Landforms of the World." *Annals of the Association of American Geographers* 58, no. 1 (March 1968).

Newell, Reginald E., and Adarsh Deepak, eds. *Mount St. Helens Eruptions of 1980*. Washington: National Aeronautics and Space Administration, 1982.

Richter, Charles F. *Elementary Seismology*. San Francisco: W. H. Freeman, 1958.

U.S. Geological Survey Staff. "The Loma Prieta, California, Earthquake: An Anticipated Event." *Science* 247 (January 19, 1990): 286–93.

Walker, Bryce. *Earthquake*. Planet Earth Series. Alexandria, VA: Time-Life Books, 1982.

Ward, Peter L., and Robert A. Page. *The Loma Prieta Earthquake of October 17, 1989*. Washington: Government Printing Office, 1990.

Alabama Hills and Inyo Mountains of eastern
California. Photo by author.

13
Weathering and Mass Movement

We have discussed the internal processes of our planet and how the endogenic landforms of the crust are produced. However, as the landscape is formed, a variety of exogenic processes simultaneously operate to wear it down. Perhaps you have noticed highways in mountainous and cold climates that appear rough and broken. Roads that experience freezing weather seem to pop up in chunks each winter. Or, maybe you have seen older marble structures such as tombstones, etched and dissolved by rainwater. Similar physical and chemical weathering processes to these are important to the overall reduction of the landscape and the release of essential minerals from bedrock. In addition, mass movement of surface materials rearranges land-

forms, providing often-dramatic reminders of the power of nature.

One of the benefits you receive from a physical geography course is an enhanced appreciation of the scenery. Whether you go on foot, or by car, train, or plane, travel is an opportunity to experience Earth's varied landscapes and witness the active processes that produce them. We now begin a five-chapter examination of various denudation processes that make up the exogenic system.

LANDMASS DENUDATION

Geomorphology is a science that analyzes and describes the origin, evolution, form, and spatial

FIGURE 13–1
John Wesley Powell developed his base level concept during his extensive exploration of the west. He first ventured down the Colorado River in 1869 in heavy oak boats, drifting by Dead Horse Point and the bend in the river shown in the photo. Photo by author.

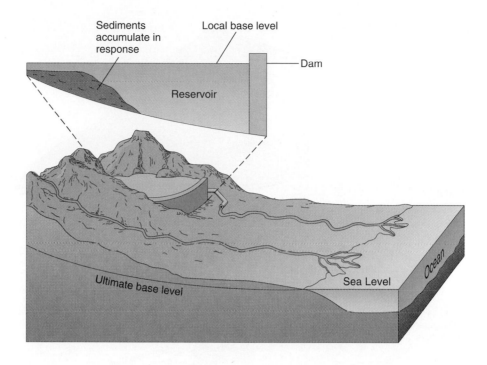

FIGURE 13–2
Ultimate and local base level concepts.

distribution of landforms. It is an important aspect of the study of physical geography. **Denudation** is a general term referring to all processes that cause reduction or rearrangement of landforms. The principal denudation processes affecting surface materials include weathering, mass movement, erosion, transportation, and deposition.

Early Theories

American geologist and ethnologist John Wesley Powell (1834–1902) was an early director of the U.S. Geological Survey, explorer of the Colorado River, and a pioneer in understanding the landscape. In 1875 he put forward the idea of **base level**, or a level below which a stream cannot erode its valley further (Figure 13–1). The hypothetical absolute or ultimate base level is sea level (which is the average level between high and low tides). You can imagine base level as a surface extending inland from sea level, inclined gently upward, under the continents. Ideally, this is the lowest practical level for all denudation process. Although base level is a very useful concept, no satisfactory working definition has yet been agreed upon.

Powell also recognized that not every landscape degraded down to sea level, for other base levels

seemed to be in operation. A local, or temporary, base level may control a regional landscape and the lower limit of local streams. That local base level might be a river, a lake, a hard and resistant rock structure, or a human-made dam. In arid landscapes, with their intermittent precipitation, local control is provided by valleys, plains, or other low points (Figure 13–2).

Interactions between the structural elements of the land and denudation processes are complex and represent a kind of struggle between the internal and external processes. The endogenic processes build and create *initial* landscapes, whereas the exogenic processes work toward low relief, an ultimate condition of little change, and the stability of *sequential* landscapes. Various theories have been proposed to model these denudation processes and account for the appearance of the landscape at different developmental stages, not only for humid temperate climates, but also for coastal, arid desert, and equatorial landscapes.

William Morris Davis (1850–1934), a famed American geographer and geomorphologist, proposed the geographic cycle of landscape development. He later named it the erosion cycle, or the **geomorphic cycle model**. Davis theorized that a landscape goes through an initial uplift that is accompanied by little

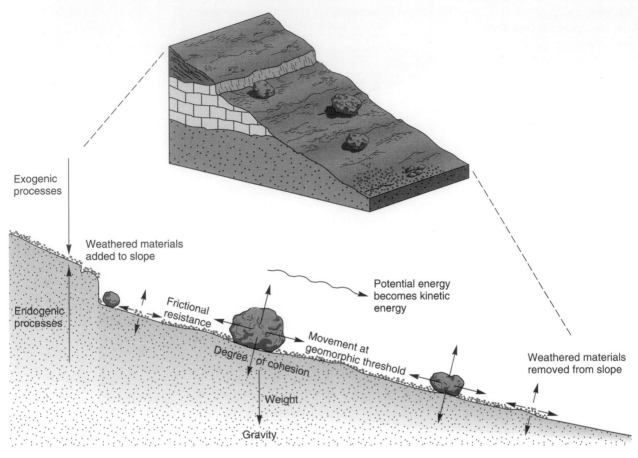

FIGURE 13–3
Characterization of forces acting on materials on an inclined slope.

erosion or removal of materials. The landscape then enters a prolonged period of stability, an idea he later modified. The raised elevation of the landscape is such that streams begin flowing more rapidly downhill, cutting both headward (upstream) and downward. According to this cyclic model, the landscape eventually evolves into an old erosional surface (detailed further in Chapter 14).

But Davis's theory, which helped launch the science of geomorphology and was innovative at the time, was too simple and did not account for the processes being observed as systems theory entered geomorphology. Academic support for his cyclic and evolutionary development model declined by 1960. Although not generally accepted today, his thinking about the evolution of landscapes is still influential.

Dynamic Equilibrium Approach to Landforms

In reality, a landscape behaves as an open system, with highly variable inputs of energy and materials: uplift creates the *potential energy of position* above base level and therefore a disequilibrium of relief and energy. The Sun provides *radiant energy* that converts into heat energy. The hydrologic cycle imparts *kinetic energy* through mechanical motion. *Chemical energy* is made available from the atmosphere and various reactions within the crust. In response to this input of energy and materials, landforms constantly adjust toward a condition of equilibrium. As physical factors fluctuate in an area, the surface constantly responds in search of an equilibrium condition, with every change producing compensating actions and reactions. The balancing act between tectonic uplift and the reduction rates of weathering and erosion, between the resistance of crust materials and the ceaseless attack of denudation processes, is summarized in the **dynamic equilibrium model**.

A dynamic equilibrium is different from one in which an equilibrium fluctuates around an average. Instead, according to current thinking, landscapes show ongoing adaptations to the ever-changing conditions of rock structure, climate, local relief, and elevation. Various endogenic episodes, such as faulting, or exogenic episodes, such as a heavy rainfall, may episodically provide new sets of relationships for the landscape. (See Figure 1–5 for a graphic distinction between a steady-state equilibrium and a dynamic equilibrium). The system, then, arrives at a **geomorphic threshold**—the point at which the system breaks through to a new set of equilibrium relationships—and rapidly realigns landscape materials accordingly.

Slopes. Material loosened by weathering is susceptible to erosion and transportation. However, if material is to move downslope, the forces of erosion must overcome the force of friction, the cohesion of particles to each other, and the inertial resistance to movement (Figure 13–3). If the angle is such that the pull of gravity overcomes frictional forces, or if the impact of raindrops dislodges material, or if the force of water flowing downhill lifts and carries material, then erosion of particles and transport downslope can occur.

Slopes, as parts of landscapes, are open systems and seek an angle of equilibrium among the forces described here. Conflicting forces work together on slopes to establish an optimum compromise incline that balances these forces. A geomorphic threshold (change point) is reached when any of the conditions in the balance is altered. All the forces on the slope then compensate by adjusting to a new dynamic equilibrium. The recently disturbed hillslope in Figure 13–4 is in the midst of compensating adjustment. The disequilibrium was created by the failure of saturated slopes, which a day earlier were underwater. (A dam break downstream rapidly emptied the reservoir and exposed these slopes, leading to their ultimate failure.)

Why hillslopes are shaped in certain ways, how slope elements evolve or age, and how hillslopes behave during rapid, moderate, or slow uplift are topics of active study and research. Now, with the concepts of base level and landmass denudation in

FIGURE 13–4
A slope in disequilibrium. The unstable, saturated soils gave way, leaving the roadbed destroyed and the guardrail stretching through the air. Photo by author.

FIGURE 13–5
(a) A typical cross section of a hillside; (b) sand dunes and the parent material in the background. Photo by author.

(a)

(b)

mind, let's examine specific exogenic processes that operate to reduce landforms.

WEATHERING PROCESSES

Rocks at or near Earth's surface are exposed to both physical and chemical weathering processes. **Weathering** encompasses a group of processes by which surface and subsurface rock disintegrates into mineral particles or dissolves into minerals in solution. Weathering does not transport the weathered materials; it simply generates these raw materials for transport by the agents of wind, water, ice, and gravity. In most areas, the upper surface of **bedrock** is partially weathered to broken-up rock called **regolith**. In some areas regolith may be missing or undeveloped, thus exposing an outcrop of unweathered bedrock. Loose surface material comes from further weathering of regolith and from transported and deposited regolith—Figure 13–5(a).

Bedrock is the "parent rock" of weathered regolith and any soils that might develop from it (traceable through similarities in composition while the soils are relatively youthful). For example, the sand and soils in Figure 13–5(b) derive their color and character from the bedrock in the background. This unconsolidated **sediment** and weathered rock form the **parent material** from which soil evolves. Weathering is greatly influenced by the character of the bedrock: hard or soft, soluble or insoluble, broken or unbroken. Also important in determining weathering rates are climatic elements, including the amount of precipitation, overall temperature patterns, and any freeze-thaw cycles. Other significant factors are the presence and position of the water table and the topographic orientation of exposed slopes (which controls exposure to Sun, wind, and precipitation).

Figure 13–6 shows the relationship between annual precipitation and temperature and the physical (mechanical) or chemical weathering processes in an area. You can see that physical weathering dominates in drier, cooler climates, whereas chemi-

cal weathering dominates in wetter, warmer climates. Extreme dryness reduces most weathering to a minimum, as is experienced in desert climates (BW). (Refer to Chapter 10 for the key to these climate designations.) In the hot, wet tropical and equatorial rain-forest climates (Af), most rocks weather rapidly, and the weathering tends to be deep below the surface.

Another form of weathering is the freeze-thaw action that alternately expands the water in rocks (freezing) and then contracts it (thawing), creating forces great enough to mechanically split the rocks. Freezing actions are important in humid microthermal climates (Df) and in subarctic and polar regimes (Dfc, Dw, E).

Vegetation is also a factor in weathering. Although vegetative cover can protect rock, it also can provide organic acids from the partial decay of organic matter and thus hasten chemical weathering. Plants roots can enter crevices and break up a rock, exerting

FIGURE 13–6

Relationship among temperature and rainfall and various types of weathering. (Note that both horizontal and vertical axes are ordered in descending values.) From L. C. Peltier, *Annals of the Association of American Geographers* 40, (1950) 214–36. Adapted by permission.

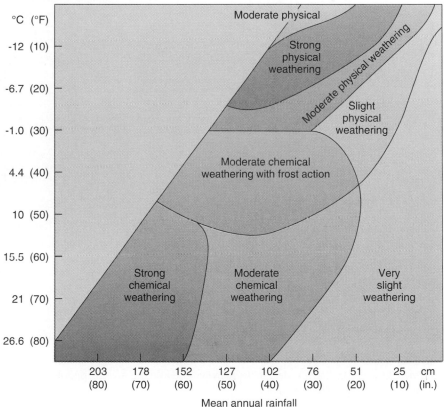

enough pressure to drive rock segments apart, thereby exposing greater surface area to other weathering processes. (You may have observed how tree roots can heave the sections of a sidewalk or driveway sufficiently to raise and crack the cement.) The differing resistances of rock, coupled with these variations in the intensity of weathering, result in **differential weathering**.

Physical Weathering Processes

When rock is broken and disintegrated without any chemical alteration, the process is called **physical weathering** or mechanical weathering. By breaking up rock, physical weathering greatly increases the surface area on which chemical weathering may operate. In the complexity of nature, both physical and chemical weathering probably operate together in a combination of processes. We separate them here for the convenience of discussion.

Crystallization. Especially in arid climates, where dry weather draws moisture to the surface of rock formations, some minerals develop crystals. As the water evaporates, salt crystals form from the dissolved mineral salts. As this process continues over time and the crystals grow and enlarge, they exert a force great enough to spread apart individual grains and begin breaking up the rock. Such crystallization is a form of physical weathering. In the Colorado Plateau of the Southwest, deep indentations continue to develop in sandstone cliffs, especially above impervious layers, where salty water is forced out of the rock strata. Crystallization then loosens the sand grains and subsequent erosion and transportation by water and wind complete the sculpturing process. Over 1000 years ago, Native Americans built entire villages in these weathered niches at Mesa Verde in Colorado and Canyon de Chelly in Arizona (pronounced "canyon duh shay"), among other places (Figure 13–7).

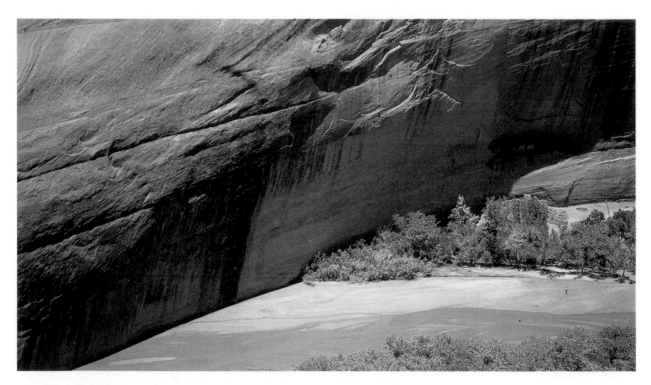

FIGURE 13–7
Cliff dwelling site in Canyon de Chelly, Arizona, used by the Anasazi people until about 900 years ago. Photo by author.

FIGURE 13–8

Freeze-Thaw Action. Water expands as much as 9% of its volume as it freezes (see Chapter 7). This expansion creates a powerful mechanical force that can exceed the tensional strength of rock. In a weathering action called *frost-wedging,* ice crystals grow in preexisting cracks in rock and push the sides apart along joints or fractures. Figure 13–8 shows this action on blocks of rock, causing joint-block separation, although cracking and breaking can be in any shape. The work of ice probably begins in small openings, gradually expanding until rocks are cleaved (portions split off). Softer supporting rock underneath the slabs already has weathered physically—an example of differential weathering.

Several cultures have used this principle of frost-wedging to quarry rock. Pioneers in the early American West drilled holes in rock, poured water in them, and then plugged them. During the cold winter months, expanding ice broke off large blocks along lines determined by the hole patterns. In spring the blocks were hauled for construction. Because nature directs its own frost-wedging program, spring can be a risky time to venture into mountainous terrain. As ice melts with the rising temperatures, newly fractured rock pieces fall without warning and may even start rock slides. The falling rock pieces may *shatter* on impact—another form of physical weathering (Figure 13–9).

Pressure-Release Jointing. To review from Chapter 11, rising magma that is deeply buried and subjected to high pressures forms intrusive igneous rocks called plutons, which cool slowly and produce coarse-grained, crystalline granitic rocks. As the landscape is subjected to uplift, the weathering regolith that covers such a pluton can be eroded and transported away, eventually exposing the pluton as a mountainous batholith. As the tremendous weight of the overburden is removed from the granite, the pressure of deep burial is relieved. The granite responds with a slow but enormous heave, and in a process known as *pressure-release jointing,* layer after layer of rock peels off in curved slabs, thinner toward the top of the formation and thicker toward the sides. As these slabs weather they slip off in a process called **spalling**, forming an **exfoliation dome.** This exfoliation process creates arch-shaped and dome-shaped features on the exposed landscape (Figure 13–10). Such domes probably are the largest single weathering features on Earth (in areal extent).

On a smaller scale, **spheroidal weathering** of rock is more a chemical weathering process than a physical process, although both are involved. The sharp edges and corners of boulders and rocks are weathered by water that penetrates joints and fractures, creating a rounded, spheroidal appearance similar to that of exfoliation, although not the result

FIGURE 13–9
Shattered rock debris, the
product of a large rockfall in
Yosemite National Park. Photo
by author.

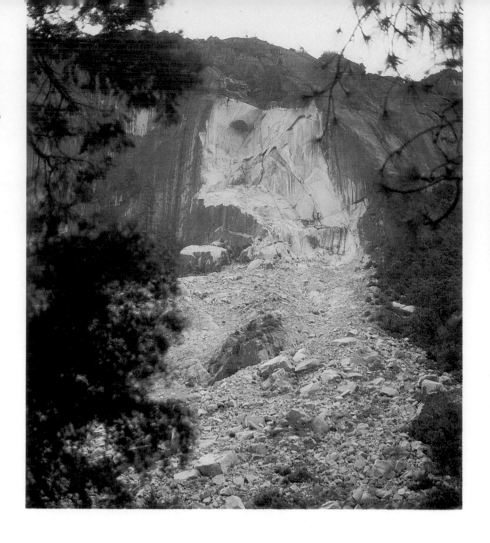

FIGURE 13–10
Characteristic dome structures,
formed by exfoliation processes
on Half Dome in Yosemite
National Park, California. Photo
by author.

of pressure-release jointing. A boulder can be attacked from all sides, shedding spherical shells of decayed rock like the layers of an onion. This type of weathering is visible in the photograph that opens this chapter.

Frozen Ground Phenomena

Periglacial regions occupy about 20% of Earth's land surface at high latitude or high altitude. These regions typically are in cold subarctic Dfc, Dfd, and polar E climates (especially ET climate), with seasonally snow-free ground next to land that is covered by permanent ice. Various processes occur in these periglacial regions that relate to physical weathering as well as to mass movement, climates (Chapter 10), and soils (Chapter 18). Much periglacial land is permanently frozen below the surface, a condition known as **permafrost**, or frozen ground. The zone between the permafrost layer and the surface is subjected to consistent daily and seasonal freeze-thaw cycles, perhaps to a meter in depth. As the water-ice volume changes, coarser particles in this active layer of the soil are moved toward the surface.

The periglacial environment features several unique landforms that are characteristic of frost-action weathering processes and low temperatures. Water working into a crack gradually expands the width and depth of the crack and forms an ice wedge which can reach several meters in depth. In addition, the freezing and thawing of the ground may create a unique **patterned ground** of polygonal forms. The patterned ground may include arranged and sized rocks at the surface that coalesce into stone *polygon nets* (Figure 13–11). Also, large areas of frozen ground (soil-covered ice) may develop a heaved-up, circular, ice-cored mound called a *pingo* which rises above a flat landscape, sometimes over 60–70 m high (195–230 ft). Pingos rise as freezing water and ice expand, sometimes as a result of freezing artesian groundwater.

Populations in areas that experience frozen ground phenomena encounter various difficulties. Because thawed ground above the permafrost zone frequently shifts in periglacial environments, the maintenance of road beds and railroad tracks is a particular problem. In addition, any building placed directly on frozen ground will begin to melt itself into the defrosting soil. Thus, the melting of perma-

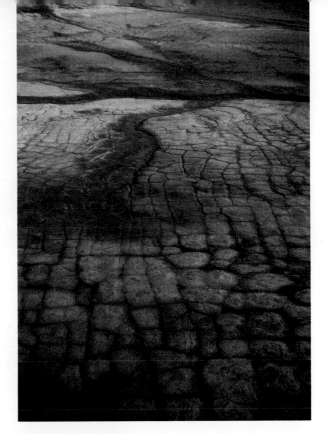

FIGURE 13–11
An example of polygonal nets formed by patterned ground in Alaska, a specified frozen ground phenomenon. Photo courtesy of U.S. Geological Survey.

frost can create subsidence in structures and complete failure of building integrity. Proper construction in periglacial regions dictates placing structures above the ground to allow air circulation beneath. This airflow allows the ground to cycle through its normal annual temperature pattern and not melt from conduction of building heat (Figure 13–12). Likewise, the Alaskan oil pipeline was constructed above the ground on racks to avoid melting the frozen ground, causing shifting that could rupture the line, disrupt oil delivery to the port, and cause severe environmental damage.

One type of frozen ground phenomenon involves the actual flowing of soil and regolith, a process called **gelifluction**. This happens because permafrost prevents the downward movement of subsurface water. In this icebound type of soil flow, movement up to 5 cm per year (2.0 in.) can occur on slopes as gentle as a degree or two. Gentle downslope movement of saturated surface material can occur in various climatic regimes and is generally referred to

FIGURE 13–12
Building failure due to the melting of permafrost south of Fairbanks, Alaska. Adapted from U.S. Geological Survey. Copyright © 1987 by Merrill Publishing. Photo courtesy of O. J. Ferrians, Jr./U.S. Geological Survey.

as **solifluction** in environments that are above freezing, although soil flow is more apparent in periglacial regions with solifluction used as a somewhat generic term. The cumulative effect of this landflow can be an overall flattening of a rolling landscape, with identifiable sagging surfaces and scalloped patterns in the downslope soil movements, somewhat characteristic of mass movement processes discussed later.

Chemical Weathering Processes

Chemical weathering refers to actual decomposition and decay of the constituent minerals in rock due to chemical alteration of those minerals. A familiar example of chemical weathering is the eating away of cathedral facades and etching of tombstones caused by increasingly acid precipitation. Water is essential to chemical weathering and, as shown in Figure 13–6, the chemical breakdown becomes more intense as both temperature and precipitation increase. Although individual minerals vary in susceptibility, no rock-forming minerals are completely unresponsive to chemical weathering.

An example of chemical weathering occurs when the oxygen dissolved in water oxidizes certain metallic elements to form oxides. Perhaps most familiar is the "rusting" of iron in a rock or soil that produces a reddish-brown stain of iron oxide. These oxides are physically weaker than the original iron-bearing mineral and more susceptible to further weathering.

Hydrolysis and Hydration. When minerals chemically combine with water, the process is called **hydrolysis**. Hydrolysis is a decomposition process that breaks down silicate minerals in rocks. Water is not simply absorbed in hydrolysis but actively participates in chemical reactions to produce different compounds and minerals.

In Chapter 11, Table 11–2 shows the mineral ingredients in igneous rocks and their varying resistance to such weathering. On the mafic end of the table, olivine and peridotite are most susceptible to chemical weathering, with stability gradually increasing toward the felsic minerals. Among the feldspars, calcium feldspar has the least resistance, followed by sodium feldspar; potassium feldspar is the strongest of the feldspar family. On the far left side of the table, quartz is very resistant to chemical weathering. You can see that, because of the nature of the constituent minerals, basalt weathers faster chemically than does granite.

When weaker constituent minerals in rock dissolve by hydrolysis, the interlocking crystal network in the rock fails, and granular disintegration takes place. Such disintegrating granite may appear etched, corroded, and softened. By-products of this chemical weathering of granite include clay, a soluble potassium bicarbonate salt, and silica. As clay is formed from the minerals in granite (especially feldspars), quartz particles are left behind. The resistant quartz may wash downstream, eventually becoming sand on some distant beach.

Another process involving water, but no chemical change, is **hydration**. It actually is a physical weathering process, in which water is added to a mineral, initiating swelling and stress within the rock, mechanically forcing grains apart as the constituents expand. This can lead to granular disintegration and further susceptibility of the rock to chemical weathering, especially by oxidation and carbonation.

Carbonation and Solution. The simplest form of chemical weathering occurs when a mineral dissolves into **solution**—for example, when sodium chloride (common table salt) dissolves in water. Water is called the *universal solvent* because it is capable of dissolving at least 57 of the natural elements and many of their compounds. It readily dissolves carbon dioxide, thereby yielding precipitation containing carbonic acid. This acid is strong enough to react with many minerals, especially limestone, in a process called *carbonation*. Such chemical weathering transforms minerals containing calcium, magnesium, potassium, and sodium into *carbonates*.

When rainwater attacks formations of limestone (which is calcium carbonate), the constituent minerals are dissolved and the reaction forms a calcium bicarbonate solution. Walk through an old cemetery and you can observe the carbonation of marble, a metamorphic form of limestone. Weathered limestone and marble formations appear pitted and weathered wherever adequate water is available for carbonation. In this era of human-induced increases of acid deposition by precipitation, carbonation processes are greatly enhanced.

Karst Topography and Landscapes Dominated by Solution

Limestone is so abundant on Earth that many landscapes are composed of it. These areas consequently are quite susceptible to chemical weathering. Such weathering creates a specific landscape of pitted, bumpy surface topography, poor surface drainage, and well-developed solution channels underground. Remarkable labyrinths of underworld caverns also may develop. These are the hallmarks of **karst topography**, originally named for the Krš Plateau in Yugoslavia where these processes were first studied. Approximately 15% of Earth's land area has some developed karst, with outstanding examples found in southern China, Japan, Puerto Rico, Cuba, the Yucatán of Mexico, Kentucky, Indiana, New Mexico, and Florida (Figure 13–13).

For a limestone landscape to develop into karst topography, several important prerequisites must exist. The limestone formation must contain 80% or more calcium carbonate for solution processes to proceed effectively. Complex patterns of joints in the otherwise impermeable limestone are needed for water to concentrate its solution activity and to form routes to subsurface drainage channels. Local relief also is important to the relationship between the ground surface and the water table—the subsurface *vadose zone,* or zone of aeration, above the water table. Finally, vegetation cover can determine the amount of organically derived acids that enhance solution processes. The role of climate in providing optimum environmental conditions for karst processes remains the subject of debate.

As with all exogenic processes, time is a factor. Earlier in this century, karst landscapes were thought to progress through evolutionary stages of develop-

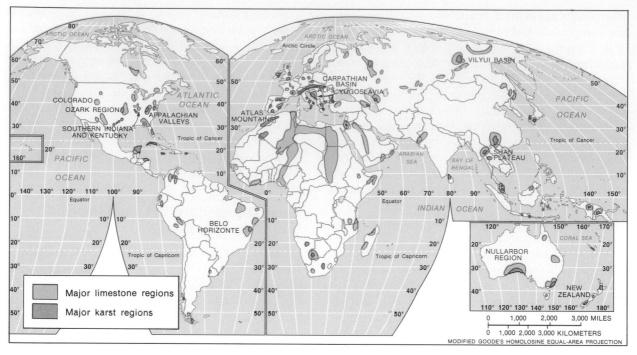

FIGURE 13–13

Distribution of karst landscapes worldwide. From R. E. Snead, *Atlas of World Physical Features,* p. 76. Copyright © 1972 by John Wiley & Sons. Adapted by permission of John Wiley & Sons, Inc.

ment, as if they were aging. Today they are thought to be locally unique, customized to specific conditions, with little evidence that different regions evolve sequentially along similar lines. Nonetheless, mature karst landscapes do display certain characteristic forms (Figure 13–14).

The weathering of limestone landscapes creates many **sinkholes**, which form in circular depressions. If these collapse through the roof of an underground cavern, the term *doline* is also used, and the landscape is referred to as *doline karst*. A gently rolling limestone plain might be pockmarked with sinkholes and dolines 2–100 m deep (7–330 ft) and 10–1000 m wide (33–3300 ft). The area southwest of Orleans, Indiana, has over 1000 sinkholes and dolines in just 2.6 km² (1 mi²). The Lost River in southern Indiana, a "disappearing stream," flows over 12.9 km (8 mi) underground before it resurfaces; its dry bed can be seen on the lower left of Figure 13–14(b).

Through solution and collapse, dolines may coalesce to form *uvalas,* which are depressions up to several kilometers in diameter as in a karst valley. In

Florida, several sinkholes have made the news because lowered water tables have caused their collapse into underground solution caves, taking with them homes, businesses, and even new cars from an auto dealership. A complex doline karst landscape in which dolines intersect is called a *cockpit karst*. The dolines can be symmetrically shaped in certain circumstances; one at Arecibo, Puerto Rico, is perfectly shaped for a radio telescope installation (Figure 13–15).

Another type of karst topography forms in the wet tropics, where deeply jointed, thick limestone beds are weathered into gorges and isolated resistant blocks. These resistant cones and towers are most remarkable in several areas of China, where an otherwise lower-level plain is interrupted by *tower karst* up to 200 m (660 ft) high (Figure 13–16).

Caves and Caverns. Most caves in the world are composed of limestone rock, because it is so easily dissolved by carbonation. The largest limestone

(a) Idealized features of a portion of karst topography in southern Indiana; (b) map of 1022 sinkholes occurring in a 1 mi² area southwest of Orleans, Indiana. Both adapted from William D. Thornbury, *Principles of Geomorphology.* (a) Illustration by William J. Wayne, p. 326; (b) after C. A. Malott, p. 321. Copyright © 1954 by John Wiley & Sons. Adapted by permission of John Wiley & Sons, Inc.

Labels in figure (a): Sinkholes · Uvala · Disappearing streams · Deeply entrenched permanent stream · Karst valley

Labels in figure (b): 1.6 km (1.0 mi) · 1.6 km (1.0 mi) · Indiana · Orleans · 40°N · 86°W · Dry bed · River

Legend:
- Dolines 30.5 m (100 ft) or more across and 3-9.1 m (10-30 ft) deep
- Sinkhole ponds
- Small sinkholes with visible openings
- Small sinkholes without visible openings

FIGURE 13–14

caverns in the United States are Mammoth Cave in Kentucky, Carlsbad Caverns in New Mexico, and Lehman Cave in Nevada. Limestone formations in which the Carlsbad Caverns formed were themselves deposited 200 million years ago when the region was covered by shallow seas. Regional uplifts associated with the Rockies (the Laramide orogeny 40 to 80 million years ago) then brought the region above sea level, leading to active cave formation (Figure 13–17).

FIGURE 13–15

Cockpit karst topography near Arecibo, Puerto Rico, used for the dish antenna of a radio telescope. The Arecibo Observatory is part of the National Astronomy and Ionosphere Center, which is operated by Cornell University under contract with the National Science Foundation. Photo courtesy of Cornell University.

Caves generally form just beneath the water table, where a later lowering of the water level exposes them to further development. The *dripstones* that form under such conditions are produced as water containing dissolved minerals slowly drips from the cave ceiling. Calcium carbonate precipitates out of the evaporating solution, literally a molecular layer at a time, and accumulates at a point below on the cave floor. The depositional features called *stalactites* grow down from the ceiling and *stalagmites* build up from the floor, sometimes growing until they connect and form a continuous *column* (Figure 13–18). A dramatic subterranean world is thus created.

MASS MOVEMENT PROCESSES

On 13 November 1985, at 11 P.M., after a year of earthquakes, 10 months of harmonic tremors, a growing bulge on its northeast flank, and 2 months of small summit eruptions, Nevado del Ruiz in central Colombia, South America, violently erupted in a lateral explosion. The volcano, northernmost of two dozen dormant peaks in the Cordilleran Central of Colombia, had erupted six times during the past 3000 years, killing 1000 people during its last eruption in 1845.

But on this night, the familiar tephra, lava, and the blast itself were not the problem. The hot eruption melted about 10% of the ice on the mountain's snowy peak, liquefying mud and volcanic ash, sending a hot mudflow downslope. Such a flow is called a *lahar,* an Indonesian word referring to flows of volcanic origin. This lahar moved rapidly down the Lagunilla River toward the villages below. The wall of mud was at least 40 m (130 ft) high as it approached Armero, a regional center with a population of

FIGURE 13–16
Tower karst of the Kwangsi Province, Yangzhou, China. Photo by Stephen Cunha.

25,000. The city slept as the lahar buried its homes and citizens: 23,000 people were killed there and in other afflicted river valleys, thousands were injured, and 60,000 were left homeless. The debris flow from the volcano is today a permanent grave for its victims.

Although less devastating to human life, the mudflow generated by the eruption of Mount Saint Helens was also a lahar. Not all mass movements are this destructive, but such processes are important in the overall denudation of the landscape.

Mass Movement Mechanics

Physical and chemical weathering processes create an overall weakening of surface rock, which makes it more susceptible to the pull of gravity. The term **mass movement** applies to all unit movements of materials propelled and controlled by gravity, such as the lahar just described. These movements can range from dry to wet, slow to fast, small to large, and from free-falling to gradual or intermittent in motion. The term mass movement is sometimes used interchangeably with *mass wasting,* although the latter also is applied to gravitational movement of nonunified material. To combine the concepts, we can say that mass movement of material works to waste slopes and provide raw material for erosion, transportation, and deposition.

The Role of Slopes. All mass movements occur on slopes. If we try to pile dry sand on the beach, the grains will flow downward until an equilibrium is achieved. The steepness of a slope that determines where loose material comes to rest varies, depending on the size and texture of the grains; this is called the *angle of repose.* It represents a balance of driving and resisting forces.

FIGURE 13–17
Carlsbad Caverns in New Mexico, where a series of underground caverns include rooms over 1200 m (4000 ft) long and 190 m (625 ft) wide. Although a national park since 1930, unexplored portions remain. Photo by author.

The driving force in mass movements is gravity, working in conjunction with the weight, size, and shape of the grains or surface material, the degree to which the slope is oversteepened, and the amount and form of moisture available—whether frozen or fluid. The greater the slope angle, the more susceptible the surface material is to mass movement. The resisting force is the shearing strength of slope material, that is, its cohesiveness and internal friction working against mass movement. To reduce shearing strength is to increase shearing stress, which eventually reaches the point at which gravity overcomes friction.

If the rock strata in a slope are underlain by clays, shales, and mudstones, all of which are highly susceptible to hydration, the rock strata will tend toward mass movement because less driving force energy is required. However, if the rock strata are such that material is held back from slipping, then more driving force energy may be required, such as that generated by an earthquake (see Figure 13–3).

In the Madison River Canyon near West Yellowstone, Montana, on the Wyoming border, a blockade of dolomite (a magnesium-rich rock in the carbonate group) had held back a deeply weathered and oversteepened slope (with a 40°–60° slope angle) for untold centuries. Then, shortly after midnight on 17 August 1959, an earthquake measuring 7.5 on the Richter scale broke the dolomite structure along the foot of the slope and released 32 million m³ (1.13

billion ft³) of mass, which moved downslope at 95 kmph (60 mph), creating gale force winds through the canyon. The material continued more than 120 m (390 ft) up the opposite canyon slope, entombing campers beneath about 80 m (260 ft) of rock and debris.

The mass of material also effectively dammed the Madison River and created a new lake, dubbed Quake Lake. The landslide debris dam established a new temporary base level for the canyon. A channel was quickly excavated by the U.S. Army Corps of Engineers to prevent a disaster below the dam, for if Quake Lake overflowed the landslide dam, the water would quickly erode a channel and thereby release the entire contents of the new lake onto farmland downstream. These events convey a dramatic exam-

ple of the role of slopes in creating mass land movements (Figure 13–19).

Classes of Mass Movements

For convenience, four basic classifications of mass movement are used: fall, slide, flow, and creep. Each involves the pull of gravity working on a mass until the critical shearing strength is reduced to the point that the mass falls, slides, flows, or creeps downward. The Madison River Canyon event was a type of slide, whereas the Nevado del Ruiz lahar mentioned earlier was a flow.

Fall. A rockfall is simply a quantity of rock that falls through the air and hits a surface. During a rockfall

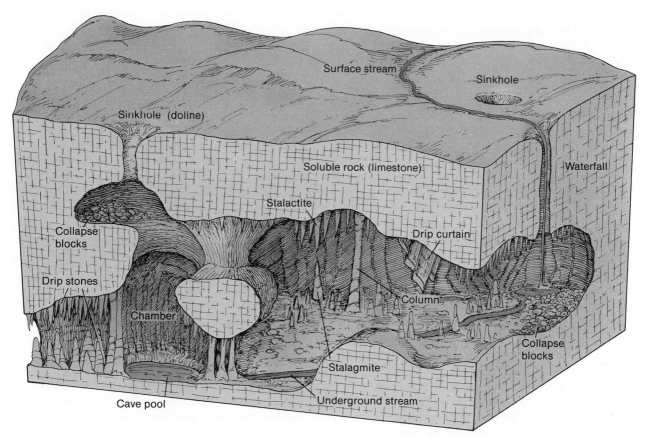

FIGURE 13–18
Formation of underground caves in limestone.

FIGURE 13-19
Cross section of the geologic structure of the Madison River Canyon at the location
of the landslide triggered by the 1959 earthquake. After Jarvis B. Hadley, "Landslides
and Related Phenomena Accompanying the Hebgen Lake Earthquake of 17 August
1959," U.S. Geological Survey Professional Paper 435–K, p. 115.

individual pieces fall independently, and characteristically form a pile of irregular broken rocks called
a *talus cone* at the base of a steep slope (Figure
13–20).

A **debris avalanche** is a mass of falling and
tumbling rock, debris, and soil. It is differentiated
from a debris slide by the tremendous velocity
achieved by the onrushing materials. In the Peruvian
Andes in 1970, at Yungay-Ranrahirca, Peru, a powerful earthquake knocked loose a debris avalanche that
attained speeds of 300 kmph (185 mph) in a
downward plunge of 3660 m (12,000 ft), devastating
towns and claiming lives.

Slide. A sudden rapid movement of a cohesive mass
of regolith and/or bedrock that is not saturated with
moisture is a **landslide**. It is literally a large amount of
material failing simultaneously. Surprise is what
creates the danger, for the downward pull of gravity
wins the struggle for equilibrium in an instant of
time. Slides occur in one of two basic forms—
translational or rotational (Figure 13–21). *Translational slides* involve movement along a planar (flat)

surface roughly parallel to the angle of the slope. The
Madison Canyon landslide described earlier was a
translational slide. Flow and creep patterns also are
considered translational in nature.

Rotational slides occur when surface material
moves along a concave surface. Frequently, underlying clay presents an impervious surface to percolating water. As a result, water flows along the clay
surface, undermining the overlying block. The simplest form of rotational slide is a rotational slump, in
which a small block of land shifts downward. The
upper surface of the slide appears to rotate backward
and often remains intact. The surface may rotate as a
single unit, or it may present a stepped appearance.

On 27 March 1964, Good Friday, the Pacific plate
plunged a bit further beneath the North American
plate near Anchorage, Alaska. More than 80 mass
movement events were triggered by the resulting
earthquake. One particular housing subdivision in
Anchorage experienced a sequence of translational
movements accompanied by rotational slumping.
Figure 13–22 shows one view of the ground failure
throughout the housing area.

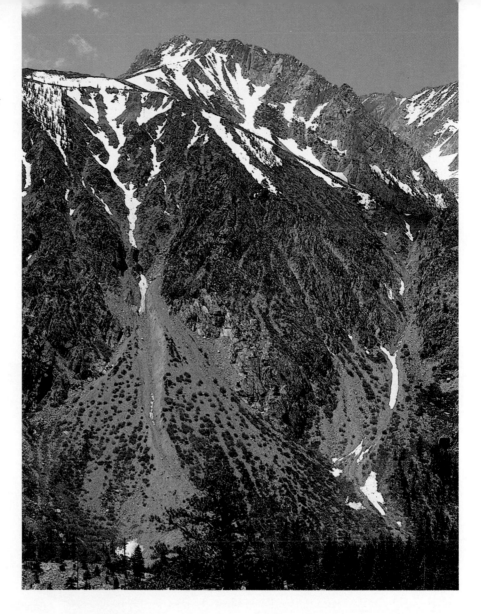

FIGURE 13–20
Rockfall and talus cone at the base of a steep slope. Photo by author.

Flow. When the moisture content of moving material is high, the suffix *-flow* is used, as in *earthflow* and *mudflow*. Heavy rains frequently saturate barren mountain slopes and set them moving, as was the case east of Jackson Hole, Wyoming, in the spring of 1925. A slope above the Gros Ventre River (pronounced "grow vaunt") broke loose and slid downward as a unit. The slide occurred because sandstone formations rested on weak shale and siltstone, which became moistened and soft, offering little resistance to the overlying strata.

Because of melted snow and rain, the water content of the Gros Ventre landslide was high enough to classify it as an earthflow. About 37 million m³ (1.3 billion ft³) of wet soil and rock moved down one side of the canyon and surged 30 m (100 ft) up the other side. The earthflow dammed the river and formed a lake, as did the landslide across the Madison River in the Hebgen Lake area in 1959. However, in 1925 equipment was not available to excavate a channel, so the new lake filled. Two years later, the lake water broke through the temporary base-level dam, transporting a tremendous quantity of debris over the region downstream.

Creep. A persistent mass movement of surface soil is called *soil creep*. Individual soil particles are lifted and disturbed by the expansion of soil moisture as it

FIGURE 13–21
Principal types of slides. Adapted by permission of Merrill, an imprint of Macmillan Publishing Company, from *The Earth,* 2d ed., by Edward J. Tarbuck and Frederick K. Lutgens, p. 184. Copyright © 1987 by Merrill Publishing.

Translational Slide

Rotational Slide

Scarp

Surface of slide

Rotational slump block

Earthflow

freezes, by cycles of moistness and dryness, by diurnal temperature variations, or even by grazing livestock or digging animals. In the freeze-thaw cycle, particles are lifted at right angles to the slope by freezing soil moisture, as shown in Figure 13–23. However, when the ice melts, the particles are pulled straight downward by the force of gravity. As the process is repeated, the soil cover gradually creeps its way downslope. The overall wasting of a slope may cover a wide area and may cause fence posts, utility poles, and even trees to lean downslope. Various strategies are used to arrest the mass

movement of slope material—grading the terrain, building terraces and retaining walls, planting ground cover—but the persistence of creep usually wins.

Human-Induced Mass Movements

Every highway roadcut or surface-mining activity that exposes a slope can hasten mass wasting because the oversteepened surfaces are thrown into a new search for equilibrium. Large open-pit strip mines—such as the Bingham Copper Mine west of Salt Lake City, the

FIGURE 13–22
Anchorage, Alaska, landslide, showing mass movement patterns throughout a housing subdivision. Courtesy of U.S. Geological Survey.

FIGURE 13–23
Soil creep and its effects.

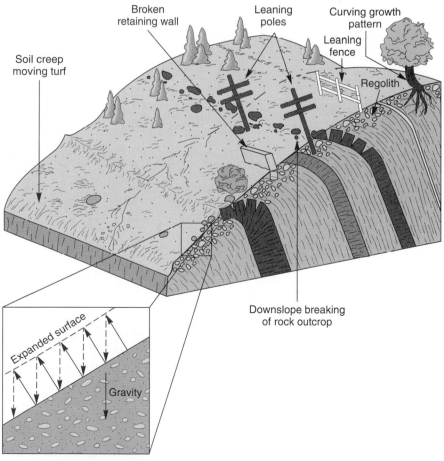

Berkeley Pit in Butte, Montana, and the extensive strip mining for coal in the East and West—are examples of human-induced mass movements, generally called **scarification**. At the Bingham Copper Mine, a mountain literally was removed. The disposal of *tailings* (mined ore of little value) and waste material is a significant problem with such large excavations because the tailing piles prove unstable and susceptible to further weathering, mass wasting, or wind dispersal. Wind dispersal is a particular problem with uranium tailings in the West due to their radioactivity.

SUMMARY

The exogenic system tears down the landscape through processes of landmass denudation. Various models to characterize landmass denudation and slope reduction have been attempted. Both physical and chemical weathering derive raw materials from parent bedrock that are essential for soil formation. Physical weathering refers to the actual breakup of rock into smaller pieces, whereas chemical weathering is the chemical altering of rock. The physical action of water when it freezes (expands) and thaws (contracts) is a powerful agent in shaping the landscape. At the same time, the chemical action of carbon dioxide dissolved in rainwater attacks susceptible rock, such as limestone. Karst topography refers to distinctively pitted and weathered limestone landscapes. The formation of massive caverns is related to these processes. Mass movement of Earth's surface produces some dramatic moments, including rockfalls, landslides, mudflows, and soil creep, all of which are mass movements of soil and debris and are important aspects of overall landmass denudation processes. In addition, human mining activity has created massive scarification of landscapes.

REVIEW QUESTIONS

1. Define geomorphology and describe its relationship with physical geography.
2. How do the base level and the local base level control how far a stream can erode its channel?
3. Define landmass denudation. What processes are included in the concept?
4. How does the construction of a reservoir alter the relationship between a stream and base level?
5. Give a brief overview of Davis's geomorphic cycle model. What was his basic thinking in setting up this cyclic model? What was Davis's principal contribution to the models of landmass denudation?
6. Describe how a landscape operates in response to constantly adjusting conditions. Do you think a landscape ever reaches a stable, old-age condition? Explain.
7. When a hillslope system reaches a geomorphic threshold, what occurs?
8. Relative to slopes, what is meant by an "angle of equilibrium"? Can you apply this to the photograph in Figure 13–4?
9. Describe the processes of physical and chemical weathering on an open expanse of bedrock. How does regolith develop? How is sediment derived?
10. Describe the relationship between climatic conditions and rates of weathering activities.
11. What is the interplay between the resistance of rock structures and weathering variabilities?
12. Why is freezing water such an effective physical weathering agent?
13. What are the weathering processes that produce a granite dome? Describe the sequence of events that takes place.
14. What is the principal factor in producing the unique landforms associated with frozen ground phenomena? How is the actual flowing of the soil surface possible?

15. What is hydrolysis? How does it affect rocks?
16. Describe the development of limestone topography. What is the name applied to such landscapes? From what was this name derived?
17. Differentiate among sinkholes, dolines, uvalas, and cockpit karst. Within which form is the radio telescope at Arecibo, Puerto Rico?
18. Describe the type of mudflow associated with volcanic eruptions.
19. Define the role of slopes in mass movements—angle of repose, driving force, resisting force, and so on.
20. What events occurred in the Madison River Canyon in 1959?
21. What are the classes of mass movement? Describe each briefly and differentiate among these classes.
22. What is scarification, and why is it considered a type of mass movement? Give an example of scarification.

SUGGESTED READINGS

Alden, William C. "Landslide and Flood at Gros Ventre, Wyoming." *Transactions of the American Institute of Mining Engineers* 76 (February 1928): 347–61.

Bloom, Arthur L. *Geomorphology*. Englewood Cliffs, NJ: Prentice-Hall, 1978.

Davis, William Morris. "The Geographical Cycle." *Journal of Geography* 14, no. 5 (November 1899): 481–504.

Fairbridge, Rhodes W., ed. *The Encyclopedia of Geomorphology*. Encyclopedia of Earth Sciences Series, vol. 3. New York: Reinhold Book Company, 1968.

Jackson, Donald Dale, and the editors of Time-Life Books. *Underground Worlds*. Planet Earth Series. Alexandria, VA: Time-Life Books, 1982.

Kerr, Richard A. "Landslides from Volcanoes Seen as Common." *Science* 224 (20 April 1984): 275–76.

Krantz, William B., Kevin J. Gleason, and Nelson Caine. "Patterned Ground." *Scientific American* (December 1988): 68–76.

McDowell, Bart. "Eruption in Colombia." *National Geographic* 165, no. 5 (May 1986): 640–53.

Peterson, Cass. "Killing Mountains for Coal—Federal Strip Mining Laws Have Not Healed Kentucky's Scars." *Washington Post National Weekly Edition* (29 June 1987): 6–7.

Price, Larry W. *The Periglacial Environment, Permafrost, and Man*. Commission on College Geography, Resource Paper no. 14. Washington: Association of American Geographers, 1972.

Rice, R. J. *Fundamentals of Geomorphology*. 2d ed. White Plains, NY: Longman, 1988.

Schumm, Stanley A. "Geomorphic Thresholds and Complex Response of Drainage Systems." *Proceedings of the 4th Annual Geomorphology Symposium*. Binghamton, NY, 1973, 299–310.

Spencer, Edgar Winston. *The Dynamics of the Earth*. New York: Thomas Y. Crowell, 1972.

Thornbury, William D. *Principles of Geomorphology*. New York: John Wiley & Sons, 1965.

U.S. Geological Survey, National Park Service, Coast and Geodetic Survey, and Forest Service. *The Hebgen Lake, Montana Earthquake of August 17, 1959*. U.S. Geological Survey Professional Paper 435. Washington: Government Printing Office, 1964.

Photo by author.

14

River Systems and Landforms

FLUVIAL PROCESSES AND LANDSCAPES

The Drainage Basin System
Streamflow Characteristics
Stream Gradient
Stream Deposition

FLOODS AND RIVER MANAGEMENT

Streamflow Measurement
Additional Flood Considerations

SUMMARY

Key Terms
Review Questions
Suggested Readings

Earth's rivers and waterways form vast arterial networks that both shape and drain the continents, transporting the by-products of weathering, mass movement, and erosion. To call them Earth's lifeblood is not an exaggerated metaphor, inasmuch as rivers redistribute mineral nutrients important for soil formation and plant growth and generally serve society in many ways. Not only do rivers provide us with essential water supplies, but they also receive, dilute, and transport wastes and provide critical cooling water for industry. Rivers have been of fundamental importance throughout human history. This chapter discusses the dynamics of river systems and their landforms.

At any one moment approximately 1250 km^3 (300 mi^3) of water flows through Earth's waterways. Even though this volume represents only 0.003% of all freshwater and 0.0001% of all water, the work performed by this dynamic flow makes it the dominant agent of landmass denudation. Figure 9–13 portrays North America's surface runoff, our basic water supply. Of the world's rivers, the Amazon and the Paraná of South America, the Zaire (Congo) of Africa, the Ganges of India, and the Chang Chiang (Yangtze) of Asia have the greatest flow volume per unit of time, or *discharge* (see Table 9–4). In North America, the Missouri-Ohio-Mississippi, Saint Lawrence, and Mackenzie systems produce the largest discharge.

FLUVIAL PROCESSES AND LANDSCAPES

Stream-related processes are termed **fluvial** (from the Latin *fluvius,* meaning river). Insolation is the driving force of fluvial systems, operating through the hydrologic cycle and working under the influence of gravity. Individual streams vary greatly,

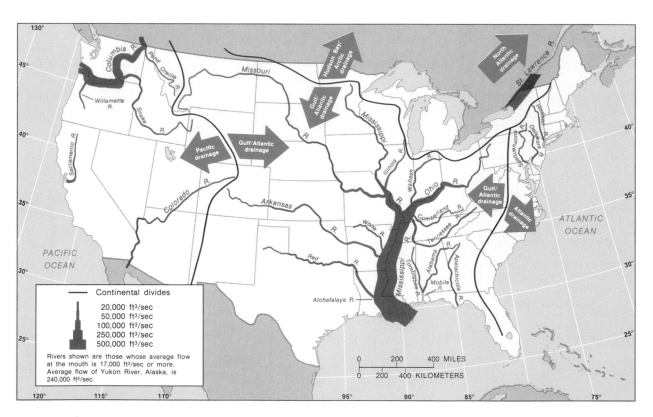

FIGURE 14–1
Continental divides separate the major drainage basins that empty into the Pacific, Atlantic, Gulf of Mexico, and to the north through Canada into Hudson Bay and the Arctic Ocean. After U.S. Geological Survey.

depending on the climate in which they operate, the variety of surface composition and topography over which they flow, and the length of time they have been operating in a specific setting.

Denudation by wind, water, and ice that dislodges, dissolves, or removes surface material is called **erosion**. Thus, streams produce *fluvial erosion,* which supplies weathered and wasted sediments for **transport** to new locations, where they are laid down in a process known as **deposition**. **Alluvium** is the general term for the clay, silt, and sand transported by running water.

The work of streams modifies the landscape in many ways. Landforms are produced by the erosive action of flowing water and the deposition of stream-transported materials. Rivers create floodplains during episodic flood events, which humans attempt to moderate and control. Let's begin our study by examining a basic fluvial unit—the drainage basin.

The Drainage Basin System

Streams are organized into areas or regions called **drainage basins**. A drainage basin is the spatial geomorphic unit occupied by a river system. A drainage basin is defined by ridges that form *drainage divides,* i.e., the ridges are the dividing lines that control into which basin precipitation drains. Drainage divides define **watersheds**, the catchment areas of the drainage basin. The United States and Canada are divided by several **continental divides**; these are extensive mountain and highland regions that separate drainage basins, sending flows either to the Pacific, or to the Gulf of Mexico and the Atlantic, or to Hudson Bay and the Arctic Ocean (Figure 14–1). The principal drainage basins in the United States and Canada are mapped in Figure 14–2. These form water-resource regions and provide a spatial framework for water studies and water planning.

An example of a specific drainage basin is the Delaware River basin, which has its headwaters in the Catskill Mountains of New York. This basin encompasses 33,060 km^2 (12,890 mi^2) and includes parts of five states in the stream's length, 595 km (370 mi) from headwaters to the mouth. Delaware Bay, which eventually enters the Atlantic Ocean, marks the end of the river system (Figure 14–3).

A major drainage basin system, such as the one created by the Mississippi-Missouri-Ohio river system, is made up of many smaller drainage basins, which in turn are comprised of even smaller basins, each divided by specific watersheds. Each drainage basin gathers and delivers its precipitation and sediment to a larger basin, concentrating the volume in the main stream. For example, rainfall in north-central Pennsylvania generates the streams that flow into the Allegheny River. The Allegheny River then joins with the Monongahela River at Pittsburgh to form the Ohio River. The Ohio flows southwest and connects with the Mississippi River at Cairo, Illinois, and eventually flows on past New Orleans to the Gulf of Mexico (Figure 14–1). Each contributing tributary adds its discharge and sediment load to the larger river. In our example, sediment weathered and eroded in north-central Pennsylvania is transported thousands of kilometers and accumulates as the Mississippi delta on the floor of the Gulf of Mexico.

Drainage basins are open systems whose inputs include precipitation, the minerals and rocks of the regional geology, and both the uplift and subsidence provided by tectonic activities. System outputs of water and sediment leave through the mouth of the river. Change that occurs in any portion of a drainage basin can affect the entire system as the stream adjusts to carry the appropriate load relative to discharge and velocity. If a region is uplifted or brought to a geomorphic threshold by other factors, the relationships within the drainage basin system are put into a disequilibrium, producing a transition period.

Drainage Density and Patterns. One measure of the overall efficiency of a drainage basin is expressed as the **drainage density**, which is determined by dividing the length of the channel by the area of its drainage basin. The number and length of channels in a given area are an expression of the landscape's regional texture and surface appearance. Initially, overland flows concentrate in *rills,* or small-scale indentations, which may develop further into *gullies* and *stream courses* with specific drainage patterns. The resultant **drainage pattern** is an arrangement of channels determined by slope, differing rock resistance, climatic and hydrologic variability, and structural controls imposed by the landscape.

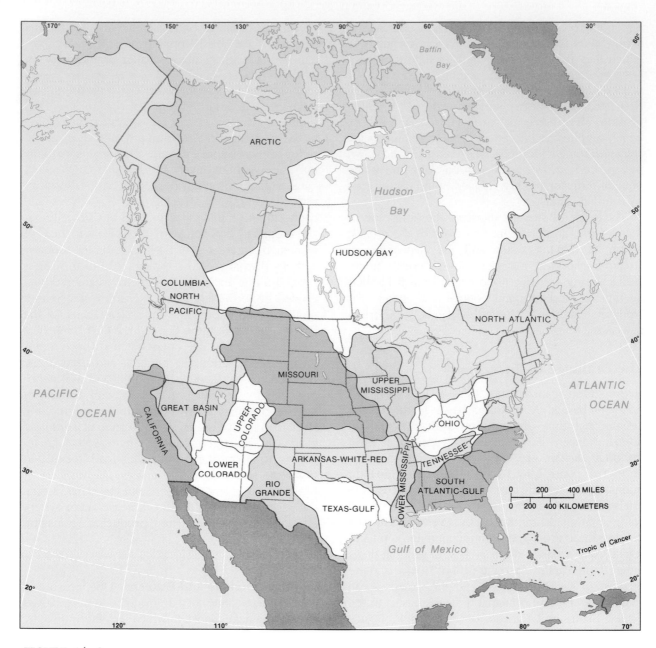

FIGURE 14–2

Major drainage basins in the United States and Canada. Data from Environment Canada, *Currents of Change* and from U.S. Water Resources Council, *The Nation's Water Resources.*

The seven most common drainage patterns encountered in nature are represented in Figure 14–4. The most familiar pattern is *dendritic* (a); this treelike pattern is similar to that of many natural systems, such as capillaries in the human circulatory system or the vein patterns in leaves. Energy expended by this drainage system is efficient because the overall length of the branches is minimized. The *Landsat* image in Figure 14–5 is of the Ohio River area near the junction of the West Virginia, Ohio, and

FIGURE 14–3
The Delaware River drainage basin. After U.S. Geological
Survey, 1986, "Hydrologic Events and Surface Water
Resources," *National Water Summary 1985,* Water
Supply Paper 2300. Washington: Government Printing
Office, p. 30.

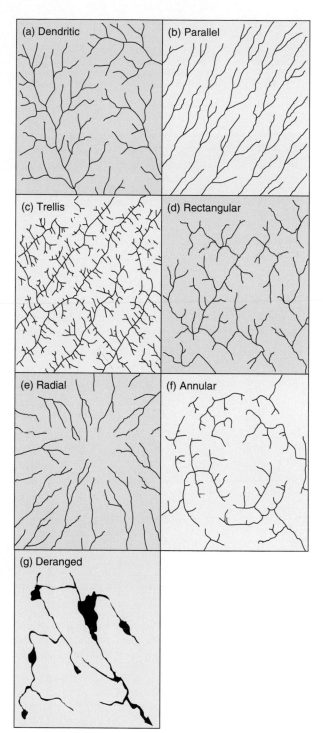

FIGURE 14–4
Seven drainage patterns. After A. D. Howard, "Drainage
Analysis in Geological Interpretation: A Summation,"
*Bulletin of American Association of Petroleum Geologists
51,* 1967, p. 2248. Adapted by permission.

FIGURE 14–5
Drainage pattern of a highly dissected topography near the junction of the West Virginia, Ohio, and Kentucky borders. *Landsat* image 2386–15280 and 15282 courtesy of National Aeronautics and Space Administration.

Kentucky borders. The high-density drainage pattern and intricate dissection of the land occur because of the region's generally level sandstone, siltstone, and shale strata, which are easily eroded, given the humid mesothermal climate. Fluvial action, along with other denudation processes, is responsible for this dissected topography.

The *trellis* drainage pattern (c) is characteristic of dipping or folded topography, which exists in the nearly parallel Valley and Ridge Province of the East, where drainage patterns are influenced by rock structures of variable resistance and folded strata. Figure 12–11 presents an orbital image of this region that shows this distinctive drainage pattern. The principal streams are directed by the parallel folded structures, whereas smaller streams are at work on nearby slopes, joining the main streams at right angles. The simple sketch in Figure 14–4 suggests that the headward-eroding part of one stream could break through a drainage divide and *capture* the headwaters of another stream in the next valley, and indeed this does happen. The sharp bends in two of the streams in the illustration are called *elbows of capture* and are evidence that one stream has breached a drainage divide.

The remaining drainage patterns in Figure 14–4 are caused by other specific structural conditions. *Parallel* drainage (b) is associated with steep slopes and some relief. A faulted and jointed landscape forms a *rectangular* pattern, directing stream courses in patterns of right-angle turns (d). A *radial* drainage pattern (e) results from streams flowing off a central peak or dome, such as occurs on a volcanic mountain. *Annular* patterns (f) are produced by structural domes, with concentric patterns of rock strata guiding stream courses. In areas having disrupted surface patterns, such as the glaciated shield regions of Canada and northern Europe, a deranged pattern (g) is in evidence, with no clear geometry in the drainage and no true stream valley pattern.

Stream Orders. Another important way to understand the linear aspect of stream channels in a drainage basin is the concept of **stream orders**, which are convenient hierarchies of stream size and relationships in a given basin. In 1945 the American hydrologist Robert E. Horton introduced this quantitative methodology, which was modified by the geomorphologist Arthur Strahler in the 1950s, and subsequently by others.

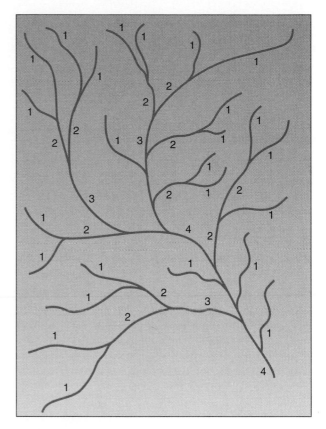

FIGURE 14–6
Stream orders of a characteristic fourth-order stream.

Think of a nesting of streams, from smallest to largest, as they join to form a main stream (Figure 14–6). A *first-order stream* has no tributaries and therefore is without smaller branches; it is the smallest stream in the system and has the smallest drainage basin. A *second-order* stream receives at least two first-order tributaries. A *third-order stream* is formed by the joining of two second-order streams; a *fourth-order stream* is formed by the joining of two third-order streams; and so forth. (However, note that the addition of a first-order stream to a second-order stream does not constitute a third order in the system depicted here.) As you move upstream, the number of tributaries increases, resulting in a first-order stream count that is usually four or five times greater than the second-order count.

Stream order designations proceed numerically until the main stream is reached; that order designation then applies all the way to the sea. The complexity of the Mississippi River system, the largest in the United States, makes it a 10th-order main stream. The Amazon River is identified as a 12th or 13th-order system. Higher designations generally result from increased length and drainage area. Figure 14–6 illustrates one version of this hierarchical system.

Streamflow Characteristics

Stream channels vary in width and depth. The streams that flow in them vary in velocity and in the sediment load they carry. All of these factors may increase with increasing discharge. Discharge is calculated by multiplying the velocity of the stream by its width and depth for a specific cross section of the channel. Discharge is expressed either in cubic meters per second (cms) or cubic feet per second (cfs). Figure 14–7 illustrates the relationship of discharge to each of these factors. It shows graphically that velocity actually increases with greater discharge downstream, despite the common misconception that streamflow becomes more sluggish. (The increased velocity downstream often is masked by reduced turbulence in the water.)

The cross-section profile of a stream varies over time, especially during heavy floods. Figure 14–8 shows changes in the San Juan River channel in Utah that occurred during a flood. The increase in discharge increases the velocity and therefore the carrying capacity of the river as the flood progresses. As a result, the river's ability to scour materials from its bed is enhanced. Such scouring represents a powerful clearing action, especially in the excavation of alluvium from the stream bed by a stream in flood.

You can see in the figure that the San Juan River's channel was deepest on 14 October, when floodwaters were highest. Then, as the discharge returned to normal, the kinetic energy of the river was reduced, and the bed again filled as sediments were redeposited. The process depicted in Figure 14–8 moved a depth of about 3 m (10 ft) of sediment from this cross section of the stream channel. Such channel adjustments occur as the stream system continuously works toward an equilibrium to balance discharge, velocity, and sediment load.

Stream Erosion. The erosional work of a stream carves and shapes the landscape through which it flows. Several types of erosional processes are

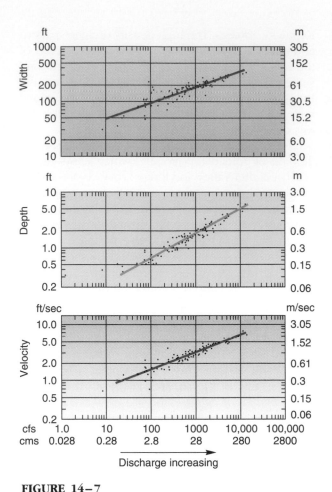

FIGURE 14–7

Relation of stream velocity, depth, and width to stream discharge at Powder River, Locate, Montana. Discharge is shown in cubic meters per second (cms) and cubic feet per second (cfs). After Luna Leopold and Thomas Maddock, Jr., 1953, "The Hydraulic Geometry of Stream Channels and Some Physiographic Implications," *U.S. Geological Survey Professional Paper 252,* Washington: Government Printing Office, p. 7.

operative. **Hydraulic action** is the work of *turbulence* in the water—the eddies of motion. Running water causes friction in the joints of the rocks in a stream channel. A hydraulic squeeze-and-release action works to loosen and lift rocks. As this debris moves along, it mechanically erodes the streambed further, through a process of **abrasion**, with rock particles grinding and carving the streambed.

The upstream tributaries in a drainage basin usually have small and irregular discharges, with most of the stream energy expended in turbulent eddies. As a result, hydraulic action in these upstream sections is at maximum, even though the coarse-textured load of such a stream is small. The downstream portions of a river, however, move much larger volumes of water past a given point and carry larger suspended loads of sediment. Thus, both volume and velocity are important determinants of the amount of energy expended in the erosion and transportation of sediment (Figure 14–9).

FIGURE 14–8
Stream channel cross sections showing the progress of a 1941 flood on the San Juan River near Bluff, Utah. After Luna Leopold and Thomas Maddock, Jr., 1953, "The Hydraulic Geometry of Stream Channels and Some Physiographic Implications," *U.S. Geological Survey Professional Paper 252,* Washington: Government Printing Office, p. 32.

Figure 14–10 presents an important graph showing the relationships of velocity and erosion/deposition. It illustrates that sediment particles of various grain sizes are deposited at velocities below the lower line on the graph, whereas they are eroded at velocities above the upper line. Interestingly, finer clays have such a cohesiveness among particles that a stream can actually erode and transport coarser sands, as well as noncohesive silts and clays, more easily than the fine clays.

Stream Transport. You may have watched a river or creek after a heavy rainfall, the water colored brown by the heavy sediment load being transported. Eroded materials are transported by four processes: solution, suspension, saltation, and traction (Figure 14–11). *Competence,* which is a stream's ability to move particles of specific size, is a function of stream velocity. The total possible load that a stream can transport is its *capacity.*

Solution refers to the dissolved load of a stream, especially the chemical solution derived from minerals such as limestone or dolomite or from soluble salts. The undesirable salt content that hinders human use of some rivers comes from dissolved rock formations and from springs in the stream channel. The **suspended load** consists of fine particles physically held aloft in the stream, with the finest particles not deposited until the stream velocity slows to near zero. The **bed load** refers to those coarser materials that are dragged along the bed of the stream by **traction** or are rolled and bounced along by **saltation** (from the Latin *saltim,* which means "by leaps or jumps"). Particles transported by saltation are too large to remain in suspension, a determination directly related to stream velocity and its ability to retain particles in suspension. Saltation is also a process in the transportation of materials by wind (see Chapter 15).

The first Spanish explorers to visit the Grand Canyon reported in their journals that they were kept awake at night by the thundering sound of boulders tumbling along the Colorado River bed. Such sounds today are substantially lessened because of the reduced velocity and discharge of the Colorado resulting from the many dams and control facilities that now trap sediments and reduce bed load capacity.

(a)

(b)

FIGURE 14–9
The relationship between volume and velocity in (a) a turbulent mountain stream and (b) the downstream portion of the Mississippi River near Natchez. Photos by author.

FIGURE 14–10
Relationship of velocity and sediment size to erosion, transportation, and deposition. After Åke Sundborg (after F. Hjulström), "The River Klarälven: A Study of Fluvial Processes," *Geografiska Annaler* 38, no. 187 (1956). Adapted by permission.

FIGURE 14–11
Fluvial transportation of eroded materials through saltation, traction, suspension, and solution.

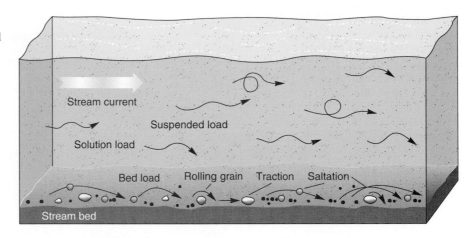

Figure 14–12 shows the annual water discharge and suspended sediment load for the Colorado River at Yuma, Arizona, from 1905 to 1964. The completion of Hoover Dam in the 1930s was the principal cause in the reduction of suspended sediment. The later construction of Glen Canyon Dam upstream from Hoover Dam further reduced stream competence and capacity. In fact, Lake Powell, which formed upstream behind Glen Canyon Dam, is forecast to fill with sediment over the next 100 years. River flows

FIGURE 14–12
Water and sediment discharge at Yuma, Arizona, 1905–1964. After U.S. Geological Survey, 1985, "Hydrologic Events, Selected Water-Quality Trends, and Ground Water Resources," *National Water Summary 1984,* Geological Survey Water Supply Paper 2275, Washington: Government Printing Office, p. 55.

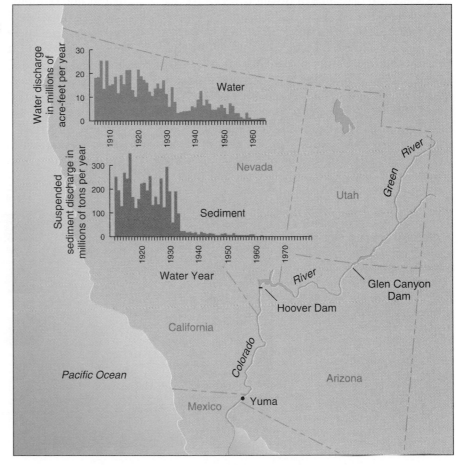

FIGURE 14–13

Laminar and turbulent stream-flow patterns. Adapted by permission of Merrill, an imprint of Macmillan Publishing Company, from *The Earth,* 2d ed., by Edward J. Tarbuck and Frederick K. Lutgens, p. 198. Copyright © 1987 by Merrill Publishing.

Laminar flow

Turbulent flow

through the Grand Canyon today frequently are blue-green rather than the muddy red of the past and can fluctuate up to 4.3 m (14 ft) at a time as hydroelectric production at Glen Canyon adjusts to daily electrical demands in distant cities.

Flow Characteristics and Channel Patterns.
Flow patterns in a stream are generally of two types, laminar and turbulent. A low-velocity flow, perhaps over a smooth streambed, can produce a *laminar flow* in which individual clay and other fine particles move along evenly in generally parallel flows. *Stream lines,* an imaginary set of lines connecting water molecules or fluid particles in a stream, remain distinct and in the direction of flow. Also, deeper

water away from the friction of the channel sides tends to move in a laminar flow (Figure 14–13).

In streams with increased velocity—shallower streams, or in places where the channel is rough, as in a section of rapids—the flow becomes *turbulent*. Complex turbulent flows propel sand, pebbles, and even boulders, increasing the action of suspension and saltation. And as bed load material is rolled, lifted, and dragged along, channel abrasion is further enhanced. The *laminar velocity* of a stream is the flow rate that marks a transition between these two flow types.

Viewed in cross section, the most rapid velocities in a stream are at the center and near the surface, corresponding with the deepest parts of the stream

FIGURE 14–14

Areal view and cross sections of a meandering stream, showing the location of maximum flow velocity. Adapted by permission of Merrill, an imprint of Macmillan Publishing Company, from *Earth Science,* 5th ed., by Edward J. Tarbuck and Frederick K. Lutgens, p. 76. Copyright © 1988 by Merrill Publishing.

Maximum velocity

Maximum velocity

Maximum velocity

Erosion of cut bank

Deposition of point bar

FIGURE 14–15
Braided stream pattern in Rio de Hauynamoto, Mayarit, Mexico. Photo courtesy of U.S. Geological Survey.

channel (Figure 14–14). Velocities decrease closer to the sides and bottom of the channel because of the frictional drag on the water flow. The maximum velocity line in a stream migrates from side to side along the channel, deflected by the stream's curves.

A **meandering** channel pattern is common for a stream that slopes gradually, producing a sinuous (snakelike) form weaving across the landscape. The outer portion of each meandering curve is subject to the greatest scouring erosive action and can be the site of a steep bank called a **cut bank** (Figure 14–14). On the other hand, the inner portion of a meander receives sediment fill and forms a deposit called a **point bar**. As meanders develop, these scour-and-fill features gradually work their way downstream.

If the load in a stream exceeds the capacity of the stream, sediments accumulate as an **aggradation** in the stream channel (the opposite of degradation) as the channel builds up through deposition. With excess sediment, a stream becomes a maze of interconnected channels laced with sediments that form a *braided* pattern (Figure 14–15). Braiding often occurs when there is some reduction of discharge, such as diurnal or seasonal conditions,

FIGURE 14–16
Ideal longitudinal profile of a graded stream.

that affects a stream's transportation ability. Locally, braiding may also occur with the addition of a new sediment load which is frequently associated with glacial meltwaters.

Stream Gradient

Every stream has a degree of inclination or **gradient**, which is the rate of decline in elevation from its headwaters to its mouth, generally forming a concave-shaped slope (Figure 14–16). Theoretically, a stream gradient becomes **graded** when the load carried by the stream and the landscape through which it flows become mutually adjusted, forming a state of dynamic equilibrium among erosion, transported load, deposition, and the stream's capacity. Attainment of a graded condition does *not* mean that the stream is at its lowest gradient, but rather that it represents a balancing among erosion, transportation, and deposition over time along a specific portion of the stream.

Characteristically, the *longitudinal profile* of a graded stream features a steeper slope upstream and a more gradual slope downstream. This constantly descending curve assumes a concave shape for complex reasons related to the stream's ability to do just enough work to accomplish the transport of the load it receives (Figure 14–16).

One problem with applying the graded stream concept in an absolute sense, however, is that an individual stream can have both graded and ungraded portions and may have graded sections without having an overall graded slope. In fact, variations and interruptions in a graded *profile of equilibrium* occur as a rule rather than the exception, making a universally acceptable definition difficult.

Nickpoints. When the longitudinal profile of a stream shows an abrupt change in gradient, such as at a waterfall or an area of rapids and cascades, the point of interruption is termed a **nickpoint** (also spelled knickpoint). Figure 14–17 shows a stream with such an interruption. Nickpoints can result from a stream flowing across a zone of hard, resistant rock, or from various tectonic uplift episodes, such as might occur along a faultline. Temporary blockage in a channel, caused by a landslide or a log jam, could also be considered a nickpoint; when the log jam breaks, the stream quickly readjusts its channel to its former grade. Earlier assumptions that all nickpoints are the result of changes in some distant base level, such as sea level, are in question today.

One of the more interesting and beautiful features associated with a gradient break is a waterfall. At its edge a stream becomes free-falling, moving at high velocity and causing increased abrasion on the channel below. This action generally undercuts the waterfall, and eventually the rock ledge at the lip of the fall collapses, causing the waterfall to shift a bit farther upstream (Figure 14–17). The height of the waterfall is gradually reduced, however, as debris accumulates at its base.

At Niagara Falls on the Ontario–New York border, glaciers advanced and receded over the region, exposing resistant rock strata underlain by less resistant shales. As the less resistant material continues to weather away, the overlying rock strata collapse, allowing the falls to erode farther upstream toward Lake Erie. In fact, the falls have retreated more than 11 km (6.8 mi) from the steep face of the Niagara escarpment (long cliff) during the past 12,000 years. Flows over the American Falls at Niagara have even been reduced to a trickle from time to time by upstream flow controls so that the escarpment could be examined for possible reinforcement to save the waterfall (Figure 14–18). As this example demonstrates, a nickpoint should be thought of as a relatively temporary feature on the landscape.

The Davis Geomorphic Cycle. Whether or not streams follow some cyclic pattern to an ideal grade was an issue addressed earlier in this century by William Morris Davis, whose evolutionary concepts of denudation (Chapter 13) also included fluvial processes. Drawing from G. K. Gilbert, an important geomorphologist of the era, Davis incorporated the graded-stream concept into his model, identifying denudation stages of *youth, maturity,* and *old age* (Figure 14–19).

Following uplift and a related drop in base level, Davis thought that a stream would quickly downcut narrow **V**-shaped canyons in its youth, leaving broad interfluvial uplands—Figure 14–19(a). Features such as waterfalls, rapids, and lakes are eliminated early as the stream evolves to a more graded profile, that is, a smoothed gradient more appropriate to existing discharge and transportation requirements.

In Davis's mature stage, the parent stream achieves a nearly graded profile and begins to widen its valley by cutting laterally into banks and the surrounding upland areas. A floodplain wider than the stream channel forms as the stream scours from side to side—Figure 14–19(b). Late maturity is achieved when the floodplain broadens and the low stream gradient produces a wide, meandering flow pattern and floodplain—Figure 14–19(c).

A most controversial aspect of Davis's model, and one largely rejected by later geomorphic thinking, is

FIGURE 14–17
Longitudinal stream profile showing nickpoints produced by resistant rock strata.

FIGURE 14–18
Niagara Falls, with the American Falls portion almost completely shut off; Horseshoe
Falls in the background is still flowing. Photo courtesy of the New York Power
Authority.

his final stage known as a **peneplain,** or an old
erosional surface. It is doubtful that a peneplain has
ever been formed, nor has significant evidence been
produced to demonstrate the existence of one in the
past, although many have been proposed and de-
bated. The problem is that the endogenic processes
do not remain stable or inactive for the long periods
of time that his evolutionary model requires.

Davis eventually allowed for some alteration in his
ideal cycle, such as renewed uplift. In such a case, the
new base level relationship imposed on the system
brings back youthful textures and processes, so that
the region is said to be *rejuvenated*. Mature features,
such as river meanders, actively return to downcutting and become **entrenched meanders** in the
landscape. Davis finally viewed the resulting landscape as one of multicyclic geomorphology, or a

composite topography. Most landscapes are, in reality, similar to this multicyclic condition, owing to the
aggressiveness of tectonic activity. As one example,
the Colorado Plateau was uplifted during the past 10
million years, and many rejuvenated meandering
streams again sought an equilibrium by actively
cutting their channels deeper (Figure 14–20).

A complete model has yet to replace the cyclic
model of Davis, now abandoned, although landscapes are generally viewed as operating in a state of
dynamic equilibrium. However, as suggested by
Schumm and Lichty, two modern geomorphologists,
the validity of cyclic or functional landscape models
may depend on the time frame selected. Over the
long span of geologic time, cyclic models of evolutionary development might explain, for example,
aspects of the disappearance of entire mountain

FIGURE 14–19
Davis's geomorphic model showing (a) youth, (b) maturity, and (c) late maturity. After Ward's Natural Science Establishment, Inc., Rochester, NY, and reprinted by permission of Merrill, an imprint of Macmillan Publishing Company, from *Earth Science,* 5th ed., by Edward J. Tarbuck and Frederick K. Lutgens, p. 79. Copyright © 1988 by Merrill Publishing Company.

ranges through denudation. But these generalizations would not be applicable to "steady time," the adjustments ongoing in a portion of a drainage basin. In between these two time frames lies the realm of "graded time," or the conditions of dynamic equilibrium.

FIGURE 14–20
Rejuvenated landscape of entrenched meanders of the San Juan River near Mexican Hat, Utah. Photo by Randall Christopherson.

Stream Deposition

Deposition is the next logical event after weathering, mass movement, erosion, and transportation. In deposition, a stream deposits alluvium, or unconsolidated sediments, thereby creating specific depositional landforms.

As discussed earlier, stream meanders tend to migrate downstream through the landscape. As a result, the landscape near such a river bears meander scars of residual deposits from the previous river channels. Former point-bar deposits leave low-lying ridges, creating a bar-and-swale relief (ridges and slight depressions). The photograph of the Itkillik River in Alaska, Figure 14–21(a), clearly shows both meanders and meander scars.

Streams often are used as natural political boundaries, but it is easy to see how disagreements might arise when boundaries are based on river channels that shift around. For example, the Ohio and Mississippi rivers meander and shift their positions and therefore the boundaries based upon them, and can do so rapidly during times of flood. Instead, boundaries should always be fixed by surveys independent of river locations. Such surveys have been completed along the Rio Grande near El Paso, Texas, and along the Colorado River, between Arizona and California, permanently establishing political boundaries separate from changing river locations.

When a meandering stream erodes its outside bank as the curve migrates downstream, the neck of land created by the looping meander eventually erodes through and forms a *cutoff*—Figure 14–21(b). When the former meander becomes isolated from the rest of the river, the resulting **oxbow lake** may gradually fill in with silt or may again become part of the river when it floods. The Mississippi River is many miles shorter today than it was in the 1830s because of artificial cutoffs that were dredged across these necks of land to improve navigation and safety.

Floodplains. The low-lying area near a stream channel that is subjected to recurrent flooding is a **floodplain**. It is formed when the river leaves its channel during times of high flow. Thus, when the river channel changes course or when floods occur, the floodplain is inundated with water. When the

water recedes, alluvial deposits generally mask the underlying rock. Figure 14–22 illustrates a characteristic floodplain, with the present river channel embedded in the plain's alluvial deposits.

On either bank of the river are **natural levees**, which are by-products of flooding. When flood waters arrive, the river overflows its banks, loses velocity as it spreads out and drops a portion of its sediment load to form the levees. Larger sand-sized particles drop out first, forming the principal component of the levees, with finer silts and clays deposited farther from the river. Successive floods

(a)

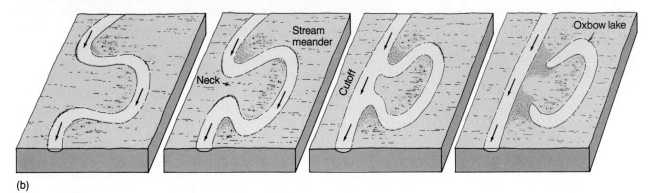

(b)

FIGURE 14–21
(a) Itkillik River in Alaska; (b) development of river meanders and oxbow lakes. [a] Photo courtesy of U.S. Geological Survey; (b) reprinted by permission of Merrill, an imprint of Macmillan Publishing Company, from *Earth Science,* 5th ed., by Edward J. Tarbuck and Frederick K. Lutgens, p. 79. Copyright © 1988 by Merrill Publishing Company.

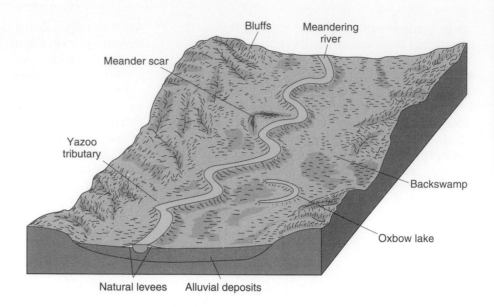

FIGURE 14–22
Typical floodplain landscape and related features.

Bluffs

Meandering river

Meander scar

Yazoo tributary

Backswamp

Oxbow lake

Natural levees   Alluvial deposits

increase the height of the levees and may even raise the overall elevation of the channel bed so that it is *perched* above the surrounding floodplain.

Notice on Figure 14–22 an area labeled **backswamp** and a stream called a **yazoo tributary**. The natural levees and elevated channel of the river prevent this tributary from joining the main channel, so it flows parallel to the river and through the backswamp area. (The terminology was derived from the Yazoo River in the southern part of the Mississippi floodplain.)

Society often chooses to build cities and conduct activities on vulnerable floodplains, encouraged by government assurances of artificial flood protection. This fact is difficult to rationalize because floodplains are certain to be flooded periodically. To increase their sense of security, people living on a floodplain may seek government assistance in building artificial levees on top of natural levees. Artificial levees not only increase the capacity in the channel; they also lead to even greater floods when they are topped or when they fail. Perhaps the best use of some floodplains is for agriculture, because inundation generally delivers nutrients to the land with each new alluvial deposit. However, some floodplains that are covered with sand and gravel may not be as suitable for agriculture (Figure 14–23). Are there any river floodplains in the vicinity where you live? If so, how would you assess their status relative to hazard perception, planning, and zoning?

Stream Terraces. Several factors may renew stream energy and landscape relationships so that a stream again scours downward with increased erosion. The resulting entrenchment of the river into its own floodplain produces **alluvial terraces** on either side of the valley, which look like topographic steps above the river. Alluvial terraces generally appear paired at similar elevations on either side of the valley (Figure 14–24). If there appear to be more than one set of paired terraces, the assumption is that the valley has gone through more than one episode of adjustment.

If the terraces on either side of the valley do not match in elevation, then entrenchment actions must have been continuous as the river meandered from side to side, with each meander cutting a terrace slightly lower in elevation in the downstream direction. Thus, alluvial terraces represent an original depositional feature, a floodplain, which is subsequently eroded by a stream that has experienced changes in stream load and capacity.

River Deltas. The mouth of a river marks the point where the river reaches a base level. Its forward velocity rapidly decelerates as it enters a larger body of standing water, with the reduced velocity causing its transported load to be in excess of its capacity. Coarse sediments drop out first, with finer clays carried to the extreme end of the deposit. This depositional plain formed at the mouth of a river is

FIGURE 14–23
Levee break and flooding along the Sacramento River during the February 1986
floods. Photo by California Department of Water Resources.

called a **delta**, named after the triangular shape of the Greek letter *delta:* Δ, which was perceived by Herodotus in ancient times to be similar to the shape of the Nile River delta.

Each flood stage deposits a new layer of alluvium over the surface of the delta so that it grows outward.

At the same time river channels divide into smaller channels known as *distributaries,* which appear as a reverse of the dendritic drainage pattern discussed earlier. The Danube River delta is located in Romania, where the river enters the Black Sea, and is an excellent example of an *arcuate delta* (arc-shaped),

FIGURE 14–24
Alluvial terraces are formed as a stream cuts into alluvial fill. After W. M. Davis, *Geographical Essays,* New York: Dover, 1964 (1909), p. 515.

Alluvial terraces (paired)

FIGURE 14–25
Danube River delta in Romania, shown in a *Landsat* mosaic and outline map. Image courtesy of the National Aeronautics and Space Administration.

similar to those of the Nile, Ganges, and Indus rivers. The *Landsat* image in Figure 14–25 shows the Danube's many meander scars, an oxbow lake, and its distributaries. Other deltaic formations include those of the Mississippi River, which has an elongate form called a *bird-foot delta,* and the Tiber River, with its *estuarine delta,* or one that is in the process of filling an **estuary**, which is the seaward mouth of a river, where the river's freshwater encounters seawater.

The Mississippi River delta is divided among many distributaries, with sediments carried beyond the delta into the Gulf of Mexico. The resulting shape of the delta features a long channel with distributaries near the end. Over the past 120 million years the Mississippi has deposited sediments downstream all the way from southern Illinois. During the past 5000 years, seven distinct deltaic complexes have formed along the Louisiana coast. The seventh and current subdelta has been building for at least the last 500 years. Each lobe reflects distinct course changes in the Mississippi River, probably where the river broke through its natural levees. In 1966, Kolb and Lopik, two engineering geologists, prepared a map of this recent deltaic history, which shows the relatively smaller size and difference in configuration of the present delta—Figure 14–26(a).

The tremendous weight of the sediments in the Mississippi River is creating isostatic adjustments in the crust. The entire region of the delta is subsiding, thereby placing ever-increasing stress on natural and artificial levees and other structures along the lower Mississippi. Severe problems are a certainty for existing and planned settlements unless further intervention or relocation efforts take place. Past protection and reclamation efforts by the U.S. Army Corps of Engineers apparently have only worsened the flood peril.

An additional problem for the lower Mississippi Valley is the possibility, in a worst-case flood, of the river's breaking from its existing channel and assuming a new route to the Gulf of Mexico. The Atchafalaya River, now blocked off from the Mississippi at some 480 km (300 mi) from its mouth, would be less than one-half the distance to the Gulf. The principal problem that is emerging is not just water, but sediment deposition. Recent reports generally indicate that it is only a matter of time and that it would be prudent to prepare for a channel change.

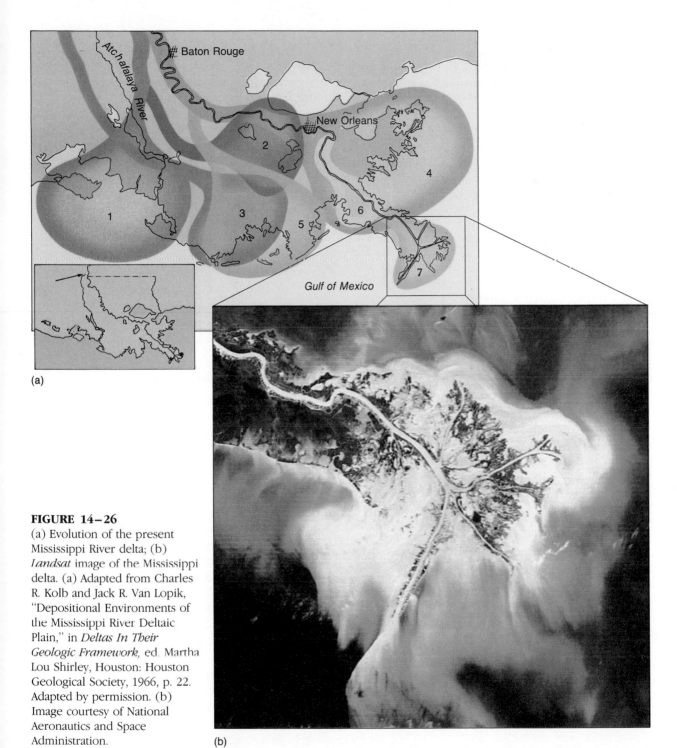

FIGURE 14–26
(a) Evolution of the present Mississippi River delta; (b) *Landsat* image of the Mississippi delta. (a) Adapted from Charles R. Kolb and Jack R. Van Lopik, "Depositional Environments of the Mississippi River Deltaic Plain," in *Deltas In Their Geologic Framework,* ed. Martha Lou Shirley, Houston: Houston Geological Society, 1966, p. 22. Adapted by permission. (b) Image courtesy of National Aeronautics and Space Administration.

The Amazon River, which exceeds 175,000 cms discharge (6.2 million cfs) and carries sediments far into the Atlantic, lacks a true delta. Its mouth, 160 km (100 mi) wide, has formed a subaqueous (underwater) deltaic plain deposited on a sloping continental shelf. As a result, the mouth is braided into a broad maze of islands and channels. Other rivers that lack significant sediment or whose discharge into the ocean faces strong erosive currents also lack deltaic formations. The Columbia River of the U.S. northwest lacks a delta because of offshore currents.

FLOODS AND RIVER MANAGEMENT

Throughout history, civilizations have settled floodplains and deltas, especially since the agricultural revolution that occurred some 8000 years B.P.* when the fertility of floodplain soils was discovered. Early villages generally were built away from the area of flooding, or on stream terraces, because the floodplain was the location of intense farming. However, as the era of commerce grew, sites near rivers became important for transportation: port and dock facilities and river bridges to related settlements. Also, because water is a basic industrial raw material used for cooling and for diluting and removing wastes, waterside industrial sites became desirable. However, human activities on vulnerable flood-prone lands require planning to reduce or avoid disaster.

Catastrophic floods continue to be a threat. In Bangladesh, intense monsoonal rains and tropical cyclones in 1988 and 1991 created devastating floods over the country's vast alluvial plain (130,000 km² or 50,000 mi²). One of the most densely populated countries on Earth, Bangladesh was more than three-fourths covered by floodwaters. Excessive forest harvesting in the upstream portions of the Ganges-Brahmaputra River watersheds increased runoff and added to the severity of the flooding. Over time the increased load carried by the river was deposited in the Bay of Bengal, creating new islands. These islands, barely above sea level, became sites of new settlements and farming villages. When the recent floodwaters finally did recede, the lack of freshwater—coupled with crop failures, disease, and pestilence—led to famine and the death of tens of

*years before the present.

thousands. About 30 million people were left homeless and many of the alluvial-formed islands were gone.

Floods and floodplains are rated statistically in terms of expected intervals between flooding. A *10-year flood* is one that is expected to occur once every 10 years (i.e., it has a 10% probability of occurring in any one year). Such a frequency would cause the floodplain to be labeled as one of moderate threat. A 50-year or 100-year flood is of greater and perhaps catastrophic consequence, but is also less likely to occur in a given year. These statistical estimates are probabilities that events of varying significance will occur randomly during any single year of the multiyear frame of reference. Of course, two decades might pass without a 10-year flood, or 10-year flood volumes could occur three years in a row. This concept is a purely statistical way of looking at the flood threat for planning and management purposes.

Streamflow Measurement

A flood is a high water level that overflows the natural (or artificial) banks along any portion of a stream. Understanding flood patterns for a drainage basin is as complex as understanding the weather, for floods and weather are equally variable, and both include a level of unpredictability. However, to develop the best possible management of flooding, the behavior of each large watershed and stream is measured and analyzed. Unfortunately, such data often are not available for small basins or for the changing landscapes of urban areas.

The key is to measure *streamflow*—the height and discharge of a stream. A **staff gauge**, a pole placed in a stream bank and marked with water heights, is used to measure stream level. With a fully measured cross section, stream level can be used to determine discharge. A **stilling well** is sited on the stream bank and a gauge is mounted in it to measure stream level. A movable current meter can be used to sample velocity at various locations. Approximately 11,000 stream gauge stations are used in the United States (an average of over 200 per state). Of these, 7000 have continuous stage and discharge records operated by the U.S. Geological Survey. Many of these stations automatically telemeter data to satellites, from which information is retransmitted to

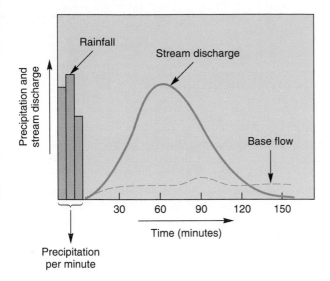

FIGURE 14–27
Typical hydrograph for a stream. After *Water: A Primer* by Luna B. Leopold. Copyright © 1974 by W. H. Freeman and Company. Adapted by permission.

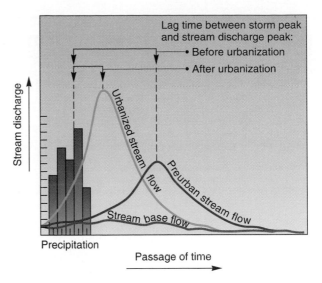

FIGURE 14–28
Effect of urbanization on a typical stream hydrograph.

regional centers. Environment Canada's Water Survey of Canada maintains more than 3000 gauging stations.

Hydrographs. If we study streamflow measurements, we can understand channel characteristics as conditions vary. A graph of stream discharge over a time period for a specific place is called a **hydrograph**. The hydrograph in Figure 14–27 shows the relationship between stream discharge and precipitation input. During dry periods, at low water stages, the flow is described as *base flow* and is largely maintained by contributions from the local water table. When rainfall occurs in some portion of the watershed, the runoff collects and is concentrated in streams and tributaries. The amount, location, and duration of the rainfall episode determine the *peak flow*. Also important is the nature of the surface in a watershed; for example, a hydrograph for a specific portion of a stream changes after a forest fire or urbanization of the watershed.

Human activities have enormous impact on water flow in a basin. The effects of urbanization are quite dramatic, both increasing and hastening peak flow (Figure 14–28). In fact, urban areas produce runoff patterns quite similar to those of deserts. The sealed surfaces of the city drastically reduce infiltration and

soil moisture recharge, behaving much like the hard, nearly barren surfaces of the desert.

A useful engineering tool is the **unit hydrograph**, which depicts a unit depth, such as a cm or in., of effective rainfall spread uniformly over a given drainage basin and received during a specific period (1 to 24 hours, depending on basin size). The unit hydrograph demonstrates how a basin behaves under various rainfall patterns and is used to design water-management facilities and to assess potential impact on cities downstream.

Additional Flood Considerations

Hurricane Camille, discussed in Chapter 8, is of interest here as well. In 1969 Camille's remnants combined with an existing low-pressure area and moved west-to-east over the entire James River basin in Virginia. Rainfall amounts unofficially reached 78.7 cm (31 in.) within the three-day storm period. The basin's narrow, steep valleys quickly concentrated runoff into a flood surge that peaked at Richmond, Virginia, as a 1000-year flood occurrence. That flood on the James River was the greatest since record-keeping began in 1771.

> Most of the residents of the mountain hollows, hamlets, and town were asleep when the storm began. Little warning was possible . . . . Rapidly

Floodplain Management Strategies

Detailed measurements of streamflows and floods have been rigorously kept in the United States only for about 100 years, in particular since the 1940s. At any selected location along any given stream, the probable maximum flood (PMF) is a hypothetical flood of such a magnitude that there is virtually no possibility it will be exceeded. Because floods are produced by the collection and concentration of rainfall, hydrologists speak of a corollary, the probable maximum precipitation (PMP) for a given drainage basin, which is an amount of rainfall so great that it will never be exceeded.

These probable parameters for a flood in a specific location along a stream are used by hydrologic engineers to establish a "design flood" against which to take protective measures. For urban areas near creeks, planning maps often include survey lines for a 50-year or a 100-year floodplain; such maps have been completed for most U.S. urban areas. The designation of a design flood usually is used to enforce planning restrictions and special insurance requirements. Unfortunately, the scenario all too often goes like this: (1) minimal zoning precautions are not carefully supervised; (2) a flooding disaster occurs; (3) the public is outraged at being caught off guard; (4) businesses and homeowners are surprisingly resistant to stricter laws and enforcement; and (5) eventually another flood refreshes the memory and promotes more knee-jerk planning. As strange as it seems, there is little indication that our risk perception improves as the risk increases.

A strategy in some larger river systems is to develop artificial floodplains by constructing bypass channels to accept seasonal or occasional floods. When not flooded, the bypass channel can serve as farmland, often benefiting from the occasional inundations. When the river reaches flood stage, large gates called *weirs* are opened, allowing the water to enter the bypass channel. This alternate route relieves the main channel of the burden of carrying the entire discharge.

rising streams and landslides caused by the unprecedented rainfall destroyed homes as the occupants slept. . . . the raging floods passed out of the small headwater streams and consolidated in the normally placid James River . . .*

Ironically, the remnants of Hurricane Agnes in 1972—only three years later—produced severe floods in that same region which also had a magnitude reaching the 1000-year probability of occurrence. So much for feeling safe with probability estimates for potential disasters!

The benefit of any levee, bypass, or other project intended to prevent flood destruction is measured in avoided damage and is used to justify the cost of the protection facility. Thus, ever-increasing damage leads to the justification of ever-increasing flood control structures. All such strategies are subjected to cost-benefit analysis, but bias is a serious drawback because such an analysis usually is prepared by an

*H.J. Thompson, "The James River Flood of August 1969," *Weatherwise* 22, no. 5 (October 1969): 183.

(a)   (b)

FIGURE 1
A reservoir during (a) dry and (b) wet weather conditions. Photos by author.

Dams and reservoirs are common streamflow control methods within a watershed. For conservation purposes, a dam holds back seasonal peak flows for distribution during low-water periods. In this way streamflows are regulated to assure year-round water supplies. Dams also are constructed for flood control, to hold back excess flows for later release at more moderate discharge levels. Adding hydroelectric power production to these functions of conservation and flood control can define a modern multipurpose reclamation project.

The function of a reservoir is to provide flexible storage capacity within a watershed to regulate river flows, especially in a region with variable precipitation. Figure 1 shows one reservoir during drought conditions and during wetter weather six years later.

Unfortunately, reservoir construction also involves negative considerations. The area upstream from a dam becomes permanently drowned. In mountainous regions, this may mean loss of white water rapids and recreational sections of a river. In agricultural areas the ironic end result may be that a hectare of farmland is inundated upstream in order to preserve a hectare of farmland downstream. Furthermore, dams built in warm and arid climates lose substantial water to evaporation, compared to

agency or bureau with a vested interest in building more flood-control projects.

As suggested in an article, "Settlement Control Beats Flood Control,"* there are other ways to protect populations than with enormous, expensive, sometimes environmentally disruptive projects. Strictly zoning the floodplain is one approach, but flat, easily developed floodplains near pleasant rivers

*Walter Kollmorgen, *Economic Geography* 29, no. 3 (July 1953): 215.

might be perceived as desirable for housing, and thus weaken political resolve. This strategy would set aside the floodplain for farming or passive recreation, such as a riverine park, golf course, or plant and wildlife sanctuary, or for other uses that are not hurt by natural floods. Kollmorgen's study concludes that "urban and industrial losses would be largely obviated by set-back levees and zoning and thus cancel the biggest share of the assessed benefits which justify big dams." Focus Study 14–1 takes a closer look at floodplain management.

FIGURE 2
The failed remains of Teton Dam in Idaho. Photo by author.

the free-flowing streams they replace. Reservoirs in the southwestern United States can lose 3 to 4 meters (10–13 feet) of water a year. Also, sedimentation can reduce the effective capacity of a reservoir and can shorten a dam's life span, as mentioned earlier regarding Glen Canyon Dam.

One final concern is the need for great care in geologic assessment of the dam site, for dam failures are more common than you might realize. For example, in 1972 two dams failed, one near Rapid City, South Dakota, and another at Buffalo Creek, West Virginia. The General Accounting Office estimates that about 1900 unsafe dams exist near urban areas. The Teton Dam, near Rexburg, Idaho, collapsed 5 June 1976, releasing more than 303 billion liters (80 billion gallons) of water, destroying 41,000 hectares (100,000 acres) of farmland, killing 16,000 head of livestock, and causing more than $1 billion in property damage (Figure 2). Congressional testimony at the time disclosed: "The principal human cause of failure of the Teton Dam was very poor site selection."

SUMMARY

River systems, fluvial processes and landscapes, floodplains, and river control strategies are important topics as populations inhabit affected areas and as demands for limited water resources increase. The basic fluvial system is a drainage basin. Within it, streams are ordered as tributaries. They gather overland flows and concentrate them into the main stream. Streams establish a graded profile over distance, with interruptions triggering adjustments.

Various landforms are associated with the action of flowing water: terraces, levees, deltas, and flood-plains. Floodplains have been an important site of human activity throughout history. Rich soils, bathed in fresh nutrients by floodwaters, attract agricultural activity and urbanization. Illogically, floodplains are settled, despite our knowledge that they are apt to be devastated by floods. Collective efforts through government agencies have been undertaken to reduce flood probabilities. Such management attempts include the construction of artificial levees, bypasses, straightened channels, diversions, dams, and reservoirs. Society is still learning how to live in a sustainable way with river systems.

KEY TERMS

Abrasion	Gradient
Aggradation	Hydraulic action
Alluvial terraces	Hydrograph
Alluvium	Meandering
Backswamp	Natural levees
Bed load	Nickpoint
Continental divides	Oxbow lake
Cut bank	Peneplain
Delta	Point bar
Deposition	Saltation
Drainage basins	Staff gauge
Drainage density	Stilling well
Drainage pattern	Stream orders
Entrenched meanders	Suspended load
Erosion	Traction
Estuary	Transport
Floodplain	Unit hydrograph
Fluvial	Watersheds
Graded	Yazoo tributary

REVIEW QUESTIONS

1. What role is played by rivers in the hydrologic cycle?
2. Define the term fluvial. What is a fluvial process?
3. What is the basic spatial geomorphic unit of an individual river system? How is this unit delimited? Define the key terms used.
4. Follow the river system from the Allegheny River to the Gulf of Mexico, analyze the pattern of tributaries, and describe the channel.

5. Describe a drainage basin. Define the various patterns that commonly appear in nature.
6. What is a stream order designation? How does the concept apply to major river systems such as the Mississippi or the Amazon?
7. What was the impact of flood discharge on the channel of the San Juan River near Bluff, Utah? Why did these changes take place?
8. How does stream discharge do its erosive work? What are the processes at work on the channel?
9. Differentiate between stream competence and stream capacity.
10. How does stream transport of sediments occur? What processes are at work?
11. Describe the flow characteristics of a meandering stream. What is the nature of the flow in the channel, the erosional and depositional features, and the typical landforms created?
12. Explain these statements: (a) all streams have a gradient but not all streams are graded, and (b) graded streams may have ungraded segments.
13. How is Niagara Falls an example of a nickpoint? If left unattended, what do you think would eventually take place at Niagara Falls?
14. What did William Morris Davis intend to illustrate with his geomorphic cycle model? How did he construct the stages of landmass denudation?
15. What are the arguments against the formation of an old erosional surface such as a peneplain?
16. What is meant by the idea that the validity of cyclic or equilibrium models depends on the time frame being considered? Explain and discuss.
17. Describe the evolution of a floodplain. How are natural levees, oxbow lakes, backswamps, and yazoo tributaries formed?
18. What is a river delta? What are the various delta forms? Give some examples.
19. Has the Mississippi delta been stable for long? Explain.
20. What do you see as the major consideration regarding floodplain management? How would you describe the general attitude of society toward natural hazards and disasters?
21. Specifically, what is a flood? How are such flows measured and tracked?
22. Differentiate between a hydrograph from a natural terrain and one from an urbanized area.
23. In the article "Settlement Control Beats Flood Control," what do you think the author meant by the title? Explain your answer, using information presented in the chapter.

SUGGESTED READINGS

Burton, Ian, and Robert W. Kates. "The Perception of Natural Hazards in Resource Management." *Natural Resources Journal* 3, no. 3 (January 1964): 412–41.

Chorley, Richard J., Stanley A. Schumm, and David E. Sugden. *Geomorphology*. New York: Methuen, 1985.

Chow, Ven Te, ed. *Applied Hydrology—A Compendium of Water-Resources Technology*. New York: McGraw-Hill, 1964.

Clark, Champ, and the editors of Time-Life Books. *Flood*. Planet Earth Series. Alexandria, VA: Time-Life Books, 1982.

Davis, William Morris, *Geographical Essays*. Edited by Douglas Wilson Johnson. New York: Dover Publications, 1964 (reprint of 1909 edition). (Note especially p. 249–50, and 381–412.)

Deutsch, Morris, Donald R. Wiesnet, and Albert Rango, eds. "Satellite Hydrology." *Proceedings of the Fifth Annual William T. Pecora Memorial Symposium on Remote Sensing*. Minneapolis: American Water Resources Association, 1981.

Kazman, Raphael G. *Modern Hydrology*. New York: Harper & Row, 1965.

Leopold, Luna B. *Hydrology for Urban Planning—A Guidebook on the Hydrologic Effects of Urban Land Use*. Circular 554. Washington: U.S. Geological Survey, 1968.

————. *Water—A Primer*. San Francisco: W. H. Freeman, 1974.

————. and Thomas Maddock, Jr. "The Hydraulic Geometry of Stream Channels and Some Physiographic Implications". *Geological Survey Professional Paper 252*. Washington: Government Printing Office, 1953.

Pearse, P. H., F. Bertrand, and J. W. MacLaren. *Currents of Change—Final Report, Inquiry on Federal Water Policy*. Ottawa: Environment Canada, 1985.

Pringle, Laurence, and the editors of Time-Life Books. *Rivers and Lakes*. Planet Earth series. Alexandria, VA: Time-Life Books, 1985.

Rudloe, Jack, and Anne Rudloe. "Trouble in Bayou Country: Louisiana's Atchafalaya." *National Geographic* (September 1979): 377–97.

Schumm, Stanley A. *The Fluvial System*. New York: John Wiley & Sons, 1977

Shirley, Martha Lou, and James A. Ragsdale. *Deltas in their Geologic Framework,* 2 vols. Houston: Houston Geological Society, 1966.

U.S. Geological Survey. "Hydrologic Events, Selected Water-Quality Trends, and Ground Water Resources." *National Water Summary 1984*. Geological Survey Water Supply Paper 2275. Washington: Government Printing Office, 1985.

————. "Hydrologic Events and Surface Water Resources," *National Water Summary 1985*. Water Supply Paper 2300. Washington: Government Printing Office, 1986.

Monument Valley, Navajo Tribal Park,
Utah-Arizona border. Photo by author.

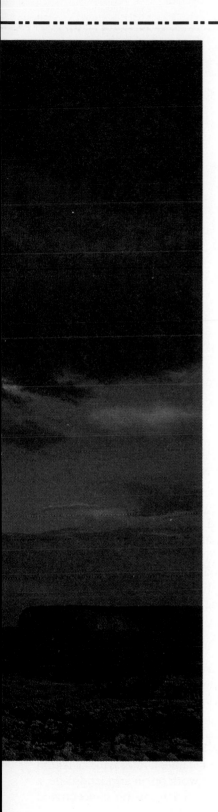

15
Eolian Processes and Arid Landscapes

A rid landscapes display unique landforms and lifeforms: ". . . instead of finding chaos and disorder the observer never fails to be amazed at a simplicity of form, an exactitude of repetition and a geometric order. . . ."*

Wind is an agent of erosion, transportation, and deposition. Its effectiveness has been the subject of much debate; in fact, wind at times was thought to produce major landforms. Presently, wind is regarded as a relatively minor exogenic agent, but it is significant enough to deserve our attention.

In this chapter we examine the work of wind, associated processes, and resulting landforms. In addition, we'll look at desert landscapes, where water remains the major erosional force but where an overall lack of moisture and stabilizing vegetation allows wind processes to operate. Evidence of this fact are extensive sand seas and sand dunes of infinite variety. Beach and coastal dunes, which form in many climates, also are influenced by wind and are discussed in the following chapter.

THE WORK OF WIND

Wind-eroded, wind-transported, and wind-deposited materials are called **eolian** (also spelled aeolian; named for Aeolus, the ruler of the winds in Greek mythology). Much eolian research was accomplished by a British major, Ralph Bagnold, who was stationed in Egypt in 1925. Bagnold was an engineering officer who spent much of his time in the deserts west of the Nile, where he measured, sketched, and developed hypotheses about the wind and desert forms. His often-cited work, *The Physics of Blown Sand and Desert Dunes,* was published in 1941 following the completion of wind-tunnel simulations in London.

The actual ability of wind to move materials is small compared to that of other transporting agents such as water and ice, because air is so much less dense than these other media. Bagnold studied the ability of wind to transport sand over the surface of a dune. Figure 15–1 shows that a wind of 50 kmph (30 mph) can move approximately one-half ton of sand per day over a one-meter-wide section of dune. The graph also demonstrates how rapidly the amount of transported sand increases with wind speed.

*From Ralph A. Bagnold, *The Physics of Blown Sand and Desert Dunes,* Methuen, 1941.

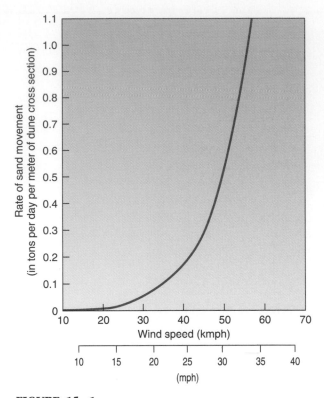

FIGURE 15–1
Sand movement relative to wind velocity, as measured over a meter-wide (39 in.) strip of ground surface. After Ralph A. Bagnold, 1941, *The Physics of Blown Sand and Desert Dunes,* Methuen & Co. Adapted by permission.

The highest-velocity winds are needed to move both the largest and the smallest particles, with intermediate-sized grains moved most easily. Small particles are difficult to move because they exhibit a mutual cohesiveness and usually present a smooth surface to the wind. Both wind velocity and turbulence increase with height.

Eolian Erosion

Two principal wind-erosion processes are **deflation,** the removal and lifting of individual loose particles, and *abrasion,* the grinding of rock surfaces with a "sandblasting" action by particles captured in the air.

Deflation. Deflation literally blows away unconsolidated or noncohesive sediment. After wind deflation and sheetwash do their work on an arid landscape, a **desert pavement** is formed from the concentrated pebbles and gravels left behind. That pavement, which resembles a cobblestone street, protects underlying sediment from further deflation (Figure 15–2). Desert pavements are so common that many provincial names have been used for them—for example, *gibber plain* in Australia, *gobi* in China, and in Africa, *lag gravels* or *serir* (or *reg* desert if some fine particles remain). Water also should be considered in the formation of lag desert plains that wind deflation has modified.

One of the environmental impacts of the 1991 Persian Gulf War involves the disruption of desert pavement. Thousands of square kilometers of stable desert pavement was shattered by bombardment with tens of thousands of tons of explosives and disrupted by movement of heavy equipment. The resulting loosened sand and silt is now available for deflation, and thus threatens cities and farms with increased dust and sand accumulations.

Wherever wind encounters loose sediments, deflation may also form basins. Called **blowout depressions**, these range in size from small indentations of less than a meter up to areas hundreds of meters wide and many meters deep. Chemical weathering is important in the formation of a blowout, for it removes the cementing materials that give particles their cohesiveness. Large depressions occurring in the Sahara are at least partially formed by deflation. The enormous Qattara Depression just inland from the Mediterranean Sea in the Western Desert of Egypt, which covers 18,000 km² (6950 mi²), is now about 130 m (427 ft) below sea level at its lowest point.

Abrasion. Sandblasting is commonly used to clean stone surfaces on buildings or to remove unwanted markings from streets. Abrasion by wind-blown particles is nature's version of sandblasting and is especially effective at polishing exposed rocks when the abrading particles are hard and angular. Variables that affect the rate of abrasion include the hardness of vulnerable surface rocks and the wind velocity. Abrasive action is restricted to the area immediately above the ground, usually no more than a meter or two in height. Rocks exposed to eolian abrasion appear pitted, grooved, or polished and usually are aerodynamically shaped in a specific direction, according to the flow of airborne particles. Rocks that bear such evidence of eolian erosion are called **ventifacts**.

On a larger scale, deflation and abrasion are capable of streamlining rock structures that are aligned parallel to the most effective wind direction, leaving behind distinctive, elongated ridges called **yardangs**. These can range from meters to kilometers in length and up to many meters in height. Abrasion is concentrated on the windward end of each yardang, with deflation operating on the leeward portions. The Sphinx in Egypt perhaps formed as a yardang, suggesting a head and body to the ancients, which some think lead them to complete the sculpture artificially.

Many other outstanding examples of yardangs occur worldwide: the Ica Valley of southern Peru has yardangs reaching 100 m (330 ft) in height and several kilometers in length, and yardangs in the Lūt

FIGURE 15–2
(a) How desert pavement is formed from larger rocks and fragments left after deflation and sheetwash; (b) a typical desert pavement. Photo by author.

Desert of Iran attain 150 m (490 ft) in height. Some of these larger formations are detectable on satellite imagery. In fact, remote sensing by orbiting spacecraft has disclosed curious features on Mars that suggest yardangs.

Eolian Transportation

As mentioned in Chapter 6, atmospheric circulation is capable of transporting fine material such as volcanic ash worldwide within a number of days. But the distance that wind transports particles varies greatly with their size. Only the finest dust particles travel significant distances, and consequently the finer material suspended in a *dust* storm is lifted much higher than the coarser particles of a *sand* storm, which may be lifted only about 2 m (6.5 ft). People living in areas of frequent dust storms are faced with very fine particles infiltrating their homes and businesses through even the smallest cracks. People living in desert regions, where frequent sand storms occur, contend with the sandblasting of painted surfaces and etched windows.

The term *saltation* was used in Chapter 14 to describe movement of particles along stream beds by water. The term saltation also is used in eolian processes to describe the wind transport of grains along the ground, grains usually larger than 0.2 mm

(0.008 in.). About 80% of wind transport of particles is accomplished by this skipping and bouncing action (Figure 15–3). In comparison with fluvial transport, in which saltation is accomplished by hydraulic lift, eolian saltation is executed by aerodynamic lift, elastic bounce, and impact (compare Figures 15–3 and 14–11).

Saltating particles crash into other particles, knocking them both loose and forward. This causes **surface creep**, which slides and rolls particles too large for saltation and represents about 20% of the material being transported. In a desert or along a beach, you can hear the myriad saltating grains of sand produce a slight hissing sound, almost like steam escaping, as they bounce along and collide with surface particles.

Wind exerts a drag or frictional pull on surface particles. Bagnold studied the relationship between wind velocity and grain size, determining the *fluid threshold* (minimum wind speed) required for initial movement of grains of various sizes (Figure 15–4). A slightly lower wind velocity suffices if the particle already has been set into motion by the impact of a saltating grain. Bagnold termed this lesser velocity the *impact threshold*. Once in motion, particles continue to be transported by lower wind velocities. As you can see from the graph, the impact threshold is less than the fluid threshold.

FIGURE 15–3
Eolian saltation and surface creep—forms of sediment transportation.

FIGURE 15–4
Relationship between wind velocity and grain size for winds measured at a height of 2 m (6.6 ft) above the surface. After Ralph A. Bagnold, 1941, *The Physics of Blown Sand and Desert,* London: Methuen & Co. Adapted by permission.

Through processes of weathering, erosion, and transportation, mineral grains are removed from parent rock and redistributed elsewhere. In the photograph that opens this chapter, and especially in the landscape shown in Figure 13–5(b), the relation-ship between the composition and color of the sandstone and the derived sandy surface is visible. Wind action does not significantly participate in the weathering process that frees individual grains of sand from the parent rock, but it is active in relocating the weathered grains.

Eolian Depositional Landforms

The smallest features shaped by individual saltating grains are **ripples** (Figure 15–5). Ripples form in crests and troughs positioned transversely (at a right angle) to the direction of the wind. Their formation is influenced by the length of time particles are airborne. Eolian ripples are different from fluvial ripples because the impact of saltating grains is very slight in water.

A common assumption is that most deserts are covered by sand. Instead, desert pavements predominate across most subtropical arid landscapes; only about 10% of desert areas are covered with sand. Sand grains generally are deposited as transient ridges or hills called dunes. A **dune** is a wind-sculpted accumulation of sand. An extensive area of dunes, such as that found in North Africa, is characteristic of an **erg desert**, which means **sand sea** (Figure 15–6). Several of the Saharan sand seas, such as the Grand Erg Oriental in the central Sahara, exceed 1200 m (4000 ft) in thickness and cover 192,000 km² (75,000 mi²). This sand sea has been active for over 1.3 million years and has average dune heights of 120 m (400 ft).

FIGURE 15–5
Sand ripple patterns, which may later become fixed patterns in rock during lithification. Photo by author.

FIGURE 15–6
Grand Erg sand sea in the
Sahara of North Africa. *Landsat*
imagery courtesy of National
Aeronautics and Space
Administration.

Dune Movement and Form. Dune fields tend to migrate in the direction of effective, sand-transporting winds. In this regard, stronger seasonal winds or winds from a passing storm may prove more effective than average prevailing winds. Dunes that move actively are called *freedunes* and reflect most dynamically the interaction between fluid atmospheric winds and moving sand. However, because the sand is moving close to the ground, it may encounter an obstruction such as stabilizing vegetation or a rock outcrop, resulting in a *tied dune,* or one that is fixed in place.

Study the dune in Figure 15–7 and you can see that winds characteristically create a gently sloping *windward side* (*stoss* side), with one or more steeply sloped *slipfaces* on the *leeward side*. A dune usually is asymmetrical in one or more directions. The angle of a slipface is the angle at which loose material is stable—its *angle of repose;* thus, the constant flow of new material makes a slipface a type of *avalanche slope*. As sand moves over the crest of the dune to the brink, it builds up and avalanches as the slipface continually adjusts, seeking its angle of repose (usually 30° to 34°). In this way, a dune migrates in the direction of the effective wind, as suggested by the successive dune profiles in Figure 15–7.

. . . I see hills and hollows of sand like rising and falling waves. Now at midmorning, they appear paper white. At dawn they were fog gray. This evening they will be eggshell brown.*

The ever-changing form of these eolian deposits is part of their beauty, but their many wind-shaped styles have made classification difficult. Although many terms exist, we can simplify dune forms into three classes—*crescentic, linear,* and *star* dunes (Figure 15–8).

Crescentic dunes are crescent-shaped mounds or ridges of sand that form in response to a fairly unidirectional wind pattern. The crescentic group is the most common class, with related forms including *barchan* dunes (limited sand), *transverse* dunes (abundant sand), *parabolic* dunes (vegetation controlled), and *barchanoid* ridges (rows of coalesced barchans).

Linear dunes generally form sets of long parallel ridges separated by sheets of sand or bare ground, although they may occur as isolated ridges. Linear dunes characteristically are much longer than they are wide, with some exceeding 100 km (60 mi) in length. The effective winds which produce these

*Janice E. Bowers, *Seasons of the Wind,* Flagstaff, AZ: Northland Press, 1985, p. 1.

dunes are principally bidirectional, so the slipface alternates from side to side. Related forms include *longitudinal dunes* (common name) and the *seif* (from the Arabic word for "sword"), which is a sharper, narrower version of a linear dune with a more sinuous crest.

Star dunes are the mountain giants of the sandy desert. They form in response to complicated, changing wind patterns and have multiple slipfaces. They are rosette-shaped or pinwheel-shaped, with several radiating arms rising and joining to form a common central peak or crest. The best examples of star dunes are in the Sahara, where they approach 200 m (650 ft) in height (Figure 15–9).

Figure 15–10 illustrates the correlation of active sand regions with the locations of tropical, continental interior, and coastal deserts. It is interesting to note the limited extent of desert areas covered by active sand dunes—only about 10% of all continental land between 30° N and 30° S.

Loess Deposits

Pleistocene glaciers advanced and retreated in many parts of the world, leaving behind large glacial outwash deposits of fine-grained clays and silts (<0.06 mm or 0.0023 in.). These materials were blown great distances by the wind and redeposited in unstratified, homogeneous deposits named **loess** (pronounced "luss") by peasants working along the Rhine River valley in Germany where such deposits were first encountered. No specific landforms were

created; instead, loess appears to cover existing landforms with a thick blanket of material that assumes the texture of the preexisting landscape. Because of its own binding strength (coherent structure), loess weathers and erodes into steep bluffs, or vertical faces. When a bank is cut into a loess deposit, it generally will stand vertically (Figure 15–11), although it can fail if saturated.

Figure 15–12 shows the worldwide distribution of loess deposits. Significant accumulations of loess throughout the Mississippi and Missouri valleys form continuous deposits 15–30 m thick (50–100 ft). Loess deposits also occur in eastern Washington State and Idaho. This silt explains the fertility of the soils in these regions, for loess deposits are well drained, deep, and characterized by excellent moisture retention. Loess deposits also cover much of the Soviet Ukraine, central Europe, China, the Pampas-Patagonia regions of Argentina, and lowland New Zealand.

In Europe and North America, loess is thought to be derived mainly from glacial and periglacial sources. The vast deposits of loess in China, covering more than 300,000 km² (115,800 mi²), are thought to be derived from desert rather than glacial sources. Accumulations in the Loess Plateau of China exceed 300 m (984 ft) thickness, forming some complex weathered badlands and some good agricultural land. These wind-blown deposits are interwoven with much of Chinese history and society. New research also has discovered that plumes of wind-blown dust from African deserts have moved across the Atlantic Ocean to enrich soils of the Amazon rain

FIGURE 15–7
Dune cross section, showing the pattern of successive slipfaces as the dune migrates in the direction of the effective wind.

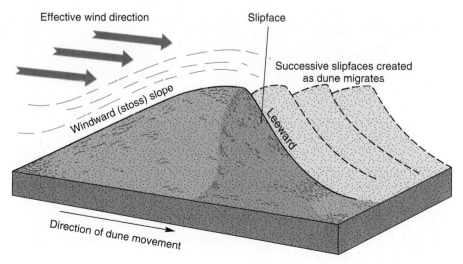

Class	Type	Description
Crescentic	Barchan	Crescent-shaped dune with horns pointed downwind. Winds are constant with little directional variability. Limited sand available. Only one slipface. Can be scattered over bare rock or desert pavement or commonly in dune fields.
	Transverse	Asymmetrical ridge, transverse to wind direction (right angle). Only one slipface. Results from relatively ineffective wind and abundant sand supply.
	Parabolic	Role of anchoring vegetation important. Open end faces upwind with U-shaped "blow-out" and arms anchored by vegetation. Multiple slipfaces, partially stabilized.
	Barchanoid Ridge	A wavy, asymmetrical dune ridge aligned transverse to effective winds. Formed from coalesced barchans; look like connected crescents in rows with open areas between them.

Barchan

Transverse

Parabolic

Barchanoid ridge

FIGURE 15–8

Major classes of dune forms. Adapted from Edwin D. McKee, 1979, *A Study of Global Sand Seas,* U.S. Geological Survey Professional Paper 1052, Washington: Government Printing Office.

Class	Type	Description
Linear	Longitudinal	Long, slightly sinuous, ridge-shaped dune, aligned parallel with the wind direction; two slipfaces. Can be 100 m high and 100 km long. The "draas" at the extreme is up to 400 m high. Results from strong effective winds varying in one direction.
	Seif	After Arabic word for "sword"; a more sinuous crest and shorter than longitudinal dunes. Rounded towards upwind direction and pointed downwind.
Star Dune		The giant of dunes. Pyramidal or star-shaped with 3 or more sinuous radiating arms extending outward from a central peak. Slipfaces in multiple directions. Results from effective winds shifting in all directions. Tend to form isolated mounds in high effective winds and connected sinuous arms in low effective winds.
Other	Dome	Circular or elliptical mound with no slipface. Can be modified into barchanoid forms.
	Reversing	Asymmetrical ridge form intermediate between star dune and transverse dune. Wind variability can alter shape between forms.

FIGURE 15–8
(continued)

FIGURE 15–9
Star dune in the Namib Desert of Namibia in southwest Africa. Photo by Comstock.

forest of South America, the southeastern United States, and the Caribbean islands.

When subjected to overgrazing or to dryness during episodes of drought, these fine sediments can be lifted by winds to form severe dust storms. This is what occurred in the Dust Bowl of the American Great Plains during the 1930s. Such episodes can have devastating consequences that disrupt economic activities, cause loss of topsoil, and even bury farmsteads.

OVERVIEW OF DESERT LANDSCAPES

Dry climates occupy about 26% of Earth's land surface and, if all semiarid climates are considered, perhaps as much as 35% of all land, constituting the largest single climatic region on Earth. The spatial distribution of these dry lands is related to several factors: subtropical high-pressure cells between 15° and 35° N and S, rain shadows to the lee of mountain ranges, and great distance from moisture-bearing air masses. Figure 15–13 portrays this distribution according to the modified Köppen climate classification used in this text. These areas possess unique landscapes created by the interaction of intermittent water, weathering processes, and wind. Rugged desert landscapes contrast sharply with the vegetation-covered, rounded slopes characteristic of humid regions.

The daily surface energy balance for El Mirage, California, presented in Figure 4–16, highlights the high sensible heat conditions and intense ground heating in the desert. Such areas receive a high input of insolation through clear daylight skies and experience high radiative heat losses at night. A typical desert water balance is given in Figure 9–8 for Phoenix, Arizona, showing high potential evapo-

FIGURE 15–10
Worldwide distribution of active and stable sand regions and areas. After R. E. Snead, *Atlas of World Physical Features,* p. 134, copyright © 1972 by John Wiley & Sons. Adapted by permission of John Wiley & Sons, Inc.

transpiration demand, low precipitation supply, and prolonged summer deficits. Fluvial processes in the desert generally are dominated by intermittent run-ning water, with hard, poorly vegetated desert pavement yielding high runoff during rainstorms.

Desert Fluvial Processes

Precipitation events in a desert may be rare indeed, a year or two apart, but when they do occur, a dry stream bed fills with a torrent called a **flash flood**. Such channels may fill in a few minutes and surge briefly during and after a storm. Depending on the region, such a dry stream bed is known as a **wash**, an **arroyo** (Spanish), or a **wadi** (Arabic). A desert highway that crosses a wash usually is posted to warn drivers not to proceed if rain is in the vicinity, for a flash flood can suddenly and without warning sweep away anything in its path.

When washes fill with surging flash flood waters, a unique set of ecological relationships develops. Crashing rocks and boulders break open seeds that respond to the timely moisture and germinate. Other plants and animals also spring into brief life cycles as the water irrigates their limited habitats.

FIGURE 15–11
Loess deposit near Rockford, Illinois. Photo by Kenneth R. Olson.

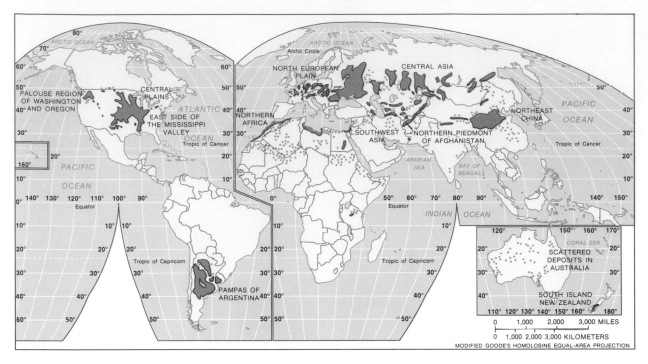

FIGURE 15–12
Worldwide loess deposits. Small dots represent small scattered loess formations.
After R. E. Snead, *Atlas of World Physical Features,* p. 138, copyright © 1972 by John
Wiley & Sons. Adapted by permission of John Wiley & Sons, Inc.

At times of intense rainfall, remarkable scenes fill the desert. Figure 15–14 shows two photographs taken just one month apart in a large sand dune field in Death Valley, California. The rainfall event that intervened produced 2.57 cm (1.01 in.) of precipitation in one day, in a place that receives only 4.6 cm (1.83 in.) in an average year. The river in the photograph continued to run for hours and then collected in low spots on the hard, underlying clay surfaces. The water was quickly consumed by the high evaporation demand so that, in just a month, these short-lived watercourses were dry and covered with accumulations of alluvial materials.

As runoff water evaporates, salt crusts may be left behind on the desert floor at low elevations. This intermittently wet-and-dry low area in a region of closed drainage is called a **playa**. Accompanying our earlier discussion of evaporites, Figure 11–14 shows such a playa in Death Valley, covered with salt precipitate just one month after this record rainfall event.

Permanent lakes and continuously flowing rivers are uncommon features in the desert, although the Nile River and the Colorado River are notable exceptions. Both these rivers are *exotic streams* with their headwaters in a wetter region. The plight of the Colorado River is described in detail in Focus Study 15–1.

Alluvial Fans. In arid climates, with their intermittent water flow, a particularly noticeable fluvial landform is the **alluvial fan**, or alluvial cone, which occurs at the mouth of a canyon where it exits into a valley. The fan is produced by flowing water that loses velocity as it leaves the constricted channel of the canyon and therefore drops layer upon layer of sediment along the base of the mountain block. Water then flows over the surface of the fan and produces a braided drainage pattern, shifting from channel to channel with each moisture event (Figure 15–15). A continuous apron, or **bajada** (Spanish for "slope"), may form if individual alluvial fans coalesce

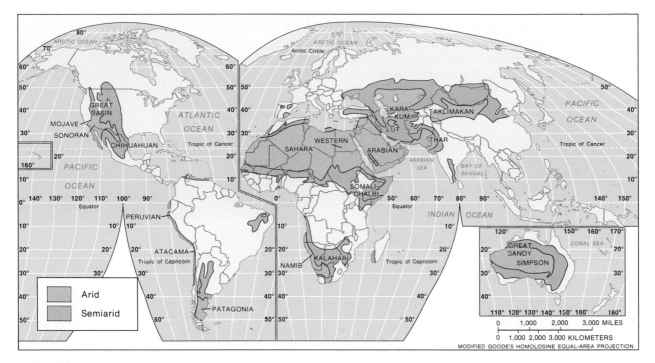

FIGURE 15–13
Worldwide distribution of arid and semiarid lands, based on the Köppen climatic
classification system.

(a)   (b)

FIGURE 15–14
(a) The day following a 2.57 cm (1.01 in.) rainfall in the Stovepipe Wells dune field
of Death Valley; (b) one month later in the same location. Photos by author.

FIGURE 15-15

Alluvial fan in a desert landscape (right); topographic map of Cedar Creek alluvial fan (below). Topographic map is the Ennis Quadrangle, 15-minute series, scale 1:62,500, contour interval = 40 ft; latitude/longitude coordinates for mouth of canyon are 45°22′ N, 111°35′ W. Photo by author; map courtesy of U.S. Geological Survey.

The Colorado River: A System Out of Balance

An exotic stream is one that headwaters in a humid region of water surpluses but then flows out of that environment and mostly through arid lands. The East African mountains and plateaus and the Colorado Rockies represent such source areas for the Nile River and the Colorado River, respectively. The Colorado rises on the high slopes of Mount Richthofen (3962 m or 13,000 ft) in Rocky Mountain National Park and flows almost 2317 km (1440 mi) to where a trickle of water disappears in the sand, kilometers short of its former mouth in the Gulf of California (Figure 1).

Orographic precipitation totaling over 102 cm (40 in.) per year (mostly snow) falls in the Rockies, feeding the Colorado headwaters. But at Yuma, Arizona, near the river's end, annual precipitation is a scant 8.9 cm (3.5 in.), an extremely small amount when compared to the annual potential evapotranspiration demand of 140 cm (55 in.).

From its source region, the Colorado River quickly leaves the humid Rockies and spills out into the arid desert of western Colorado and eastern Utah. At Grand Junction, Colorado, on the Utah border, annual precipitation is only 20 cm (8 in.). After carving its way through the intricate labyrinth of canyonlands in Utah, the river enters Lake Powell, 945 m (3100 ft) lower in elevation than the river's point of origin in the Rockies some 982 km (610 mi) upstream. The Colorado then flows through the Grand Canyon chasm formed by its own erosive handiwork.

West of the Grand Canyon, the river turns south, tracing its final 644 km (400 mi) as the Arizona-California border. Along this stretch sits Hoover Dam, just east of Las Vegas; Davis Dam, built to control the releases from Hoover; Parker Dam for the water needs of Los Angeles; three more dams for irrigation water (Palo Verde, Laguna, and Imperial); and finally Morelos Dam at the Mexican border. Mexico owns the end of the river and whatever water is left. Overall, the drainage basin encompasses 641,025 km^2 (247,500 mi^2) of mountain, basin and range, plateau, canyon, and desert landscapes, in parts of seven states and two countries.

A Brief History

John Wesley Powell (1834–1902), the first to successfully navigate the Colorado River through the Grand Canyon, was named the first director of the U.S. Bureau of Ethnology and later director of the U.S. Geological Survey (1881–1892). Powell perceived that the challenge of the West was too great for individual efforts and believed that solutions to problems such

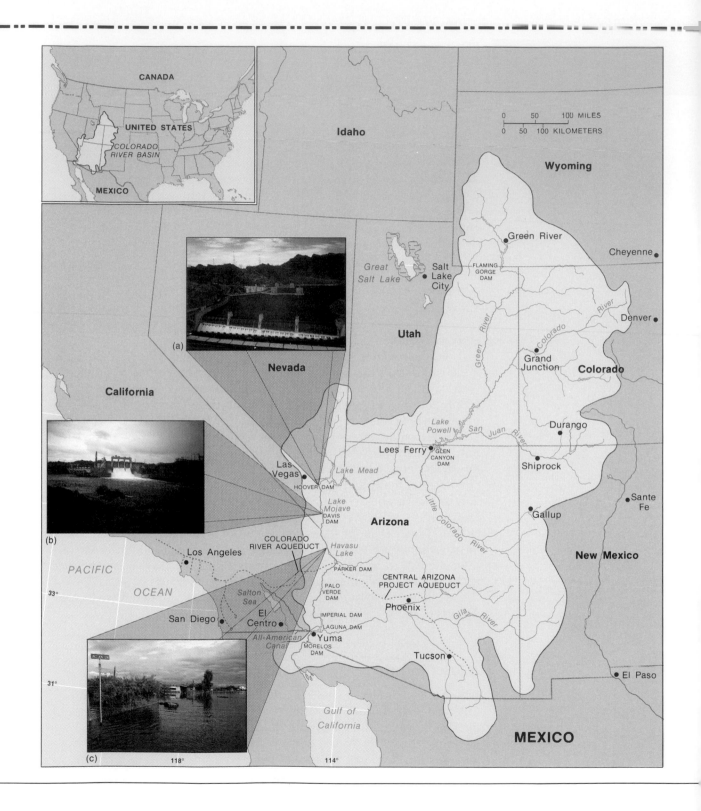

as water availability could be met only through private cooperative efforts. His 1878 study (reprinted 1962), *Report of the Lands of the Arid Region of the United States,* is a conservation landmark.

Today, Powell probably would be skeptical of the intervention by government agencies in building large-scale reclamation projects. An anecdote in Wallace Stegner's *Beyond The Hundredth Meridian* relates that, at an 1893 international irrigation conference held in Los Angeles, Powell observed development-minded delegates bragging that the entire West could be conquered and reclaimed from nature. Powell spoke against that sentiment: "I tell you, gentlemen, you are piling up a heritage of conflict and litigation over water rights, for there is not sufficient water to supply the land." He was booed from the hall. But history has shown Powell to be correct.

The Colorado River Compact was signed by six of the seven basin states in 1923. (The seventh, Arizona, signed in 1944, the same year as the Mexican Water Treaty). With this compact the Colorado River basin was divided into an upper basin and a lower basin, arbitrarily separated for administrative purposes at Lees Ferry. Congress adopted the Boulder Canyon Act in 1928, authorizing Hoover Dam as the first major reclamation project on the river. Also authorized was the All-American Canal into the Imperial Valley, which required an additional dam.

Los Angeles then began its project to bring Colorado River water 390 km (240 mi) from still another dam and reservoir on the river to their city. Shortly after Hoover Dam was finished and downstream enterprises were thus offered flood protection, the other projects were quickly completed. There are now eight major dams on the river and many irrigation works. The latest effort to redistribute Colorado River water is the Central Arizona Project, which carries water to the Phoenix area.

Flow Characteristics

Exotic stream flows are highly variable, and the Colorado is no exception. For 1917, the discharge measured at Lees Ferry totaled 24 million acre-feet (maf), whereas in 1934 it reached only 5.03 maf. In 1977, the discharge dropped to 5.02 but rose to an all-time high of 24.5 maf in 1984. In addition to this variability, approximately 70% of the year-to-year discharge occurs between April and July.

FIGURE 1 (left)
The Colorado River basin showing division of the upper and lower basins near Lees Ferry in northern Arizona. Photos portray (a) Hoover Dam spillways in operation; (b) Davis Dam, and (c) flooded Topock Estates subdivision near Needles, California, in 1983. Photos by author.

TABLE 1
Estimated Colorado River budget.

Water Demand	Quantity (maf)
In-basin consumptive uses (75% agricultural)*	11.5
Central Arizona Project (rising to 2.8 maf)	1.0
Mexican allotment (1944 Treaty)	1.5
Evaporation from reservoirs	1.5
Bank storage at Lake Powell	0.5
Phreatophytic losses (water-demanding plants)	0.5
Budgeted total demand	16.5
(1930–1980 average flow of the river	13.0)

*7.0 maf in lower basin; includes 5.5 pumped out-of-basin, 5.1 of which goes to California.

The average flows between 1906 and 1930 were almost 18 maf a year with averages dropping to 13 maf during the last 50 years. As a planning basis for the Colorado River Compact, the government used average river discharges from 1914 up to the treaty signing in 1923, an exceptionally high 18.8 maf. That amount was perceived as more than enough for the upper and lower basins each to receive 7.5 maf and, later, enough for Mexico to receive 1.5 maf in the 1944 Mexican Water Treaty.

We might question whether proper long-range planning should rely on the providence of high variability. Tree-ring analyses of past climates have disclosed that the only other time Colorado discharges were at the 1914–1923 level was between A.D. 1606 and 1625! However, our thinking about Colorado River flows seems never to have accepted this reality, for the dependable flows of the river have been consistently overestimated. This predicament is shown in an estimated budget for the river (Table 1): clearly the situation is out of balance, for there is not enough discharge to meet budgeted demands.

Water Losses at Glen Canyon Dam

Glen Canyon Dam was completed and began water impoundment in 1963, 27.4 km (17 mi) north of the basin division point at Lees Ferry. The deep, fluted, inner gorges of Glen, Navajo, Labyrinth, Cathedral, and myriad other canyons slowly were flooded by advancing water. The Bureau of Reclamation stated that the dam's purpose was to regulate flows between the basins, although many thought that Lake Mead, behind Hoover Dam, could have served this administrative-regulatory function without the risk of additional water losses.

The porous Navajo sandstone that forms the bulk of the container for the Lake Powell reservoir at Glen Canyon Dam absorbs an estimated 4% of the river's overall annual discharge as *bank storage*, water that is not retrievable in any practical sense. The higher the lake level, the greater the loss into the sandstone. To date, approximately 10 maf have disappeared into this highly permeable rock. In addition, Lake Powell is an open body of water in an extremely arid desert, where hot, dry winds accelerate evaporative losses. Another 4% of the river's overall discharge is lost annually in this manner from Lake Powell alone.

A third aspect of water loss involves the now-permanent sand bars and banks that have stabilized along the course of the regulated river. Water-demanding plants known as *phreatophytes* are now established and extracting an additional 4% of the flow of the river. Combined, these losses total 1.5 maf, or 12% of the long-term average flow—all attributable to Glen Canyon Dam and Lake Powell. Given the chronic deficits in the overall Colorado River budget, we can now seriously question the negative impact of this single project.

Flood Control in 1983

Intense precipitation and heavy snowpack in the Rockies, attributable to the 1982–1983 El Niño (see Focus Study 10–1), led to record-high discharge rates on the Colorado, testing the control ability of one of the most regulated rivers in the world. Federal reservoir managers were not prepared for the high discharge, having set aside Lake Mead's primary purpose (flood control) in favor of competing water and power interests. What ensued was a human-caused flood. Davis Dam, which regulates releases from Hoover Dam, was within 30 cm (1 ft) of overflow by 22 June 1983, a real problem for a structure made partially of earth fill.

The only time the spillways at Hoover Dam had ever operated was during a test more than 40 years before, when the reservoir capacity was artificially raised for the test; now they were opened to release the floodwaters. In addition, Glen Canyon Dam was over capacity and at risk; it actually was damaged by the volume of discharge tearing through its spillways. The decision to increase releases to save these and other upstream facilities doomed towns and homeowners along the river, especially in subdivisions near Needles, California.

We might wonder what John Wesley Powell would think if he were alive today to witness such errant attempts to control the mighty and variable Colorado. He foretold such a " . . . heritage of conflict and litigation."

into one sloping surface (see Figure 15–19). Fan formation of any sort is inhibited in humid climates because perennial streams constantly carry away much of the alluvium.

An interesting aspect of an alluvial fan is the natural sorting of materials by size. Near the mouth of the canyon, boulders and gravels are deposited, grading slowly to pebbles and finer gravels with distance out from the mouth. Then sands and silts are deposited, with the finest clays and salts carried in suspension and solution all the way to the valley floor. Minerals dissolved in solution accumulate there as precipitates left by evaporation, forming evaporite deposits.

Well-developed alluvial fans also may be a major source of groundwater. Some cities—San Bernardino, California, for example—are built on alluvial fans and extract their municipal water supplies from them. However, because water resources in an alluvial fan are recharged from surface supplies and these fans are in arid regions, groundwater mining and overdraft beyond recharge rates are common. In other parts of the world, such water-bearing alluvial fans are known as *qanat* (Iran), *karex* (Pakistan), and *foggara* (western Sahara).

Desert Landscapes

Contrary to popular belief, deserts are not wastelands, for they abound in specially adapted plants and animals. Moreover, the limited vegetation, intermittent rainfall, intense insolation, and distant vistas produce starkly beautiful landscapes. And, all deserts are not the same: for example, North American deserts have more vegetation cover than do the generally barren Saharan expanses. The shimmering heat waves and related mirage effects in the desert are products of light refraction through layers of air that have developed a temperature gradient near the hot ground. The desert's enchantment is captured in the book *Desert Solitaire:*

> **Around noon the heat waves begin flowing upward from the expanses of sand and bare rock. They shimmer like transparent, filmy veils between my sanctuary in the shade and all the sun-dazzled world beyond. Objects and forms viewed through this tremulous flow appear somewhat displaced or distorted. . . . the great Balanced Rock floats a few inches above its pedestal, supported by a layer of**

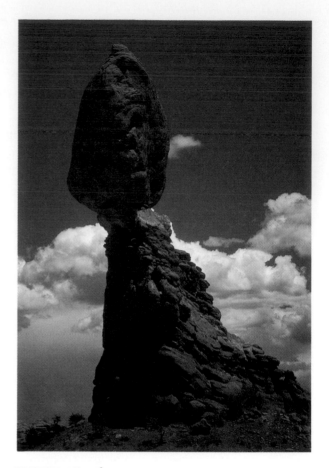

FIGURE 15–16
Balanced Rock. Photo by author.

> **superheated air. The buttes, pinnacles, and fins in the Windows area bend and undulate beyond the middle ground like a painted backdrop stirred by a draft of air.***

The buttes, pinnacles, and mesas characteristic of arid and semiarid landscapes are resistant horizontal rock strata that have eroded differentially. Removal of the less-resistant sandstone strata produces unlikely desert sculptures—arches, pedestals, and delicately balanced rocks (Figure 15–16). Specifically, the upper layers of sandstone along the top of an arch or butte are more resistant to weathering and protect the sandstone rock beneath. The removal of all surrounding rock through differential weathering leaves enormous buttes as residuals on the land-

---

*From Edward Abbey, *Desert Solitaire,* New York: McGraw-Hill Book Co., 1968, p. 154. Copyright © 1968 by Edward Abbey.

scape. If you imagine a line intersecting the tops of the Mitten Buttes shown in the photograph that begins this chapter, you can gain some idea of the quantity of material that has been removed.

In a desert area, weak surface material also may weather to a complex, rugged topography, usually of relatively low and varied relief. Such a landscape is called a **badland**, probably so-named because it offered little economic value and was difficult to traverse in nineteenth-century wagons. The Badlands region of the Dakotas is of this texture, as are portions of Death Valley soils shown in Figure 18–21.

In addition, sand dunes that existed in ancient deserts have lithified, forming sandstone structures that bear the imprint of cross-stratification. When such a dune was accumulating, sand cascaded down its slipface, and distinct bedding planes (layers) were established that remained after the dune lithified (Figure 15–17). Ripple marks, animal tracks, and fossils also are found preserved in these sandstones, which originally were eolian-deposited sand dunes.

Basin and Range. Dry climate, few permanent streams, and interior drainage patterns describe the **Basin and Range Province** of the western United States. A **province** is a large region that shares several homogeneous traits, making appropriate assignment of a regional name. These arid and semiarid basins and ranges lie in the rain shadow of mountains to the west. As the North American plate moved westward, it overrode former oceanic crust and hot spots at such a rapid pace that slabs of subducted material literally were run over. The associated extension and spreading of the crust into basins and ranges created a landscape fractured by many faults.

The present basin-and-range landscape features nearly parallel sequences of *horsts* (upward-faulted blocks or ridges) and *grabens* (downward-faulted blocks or valleys) that cover almost 800,000 km^2 (308,800 mi^2), although the pattern of faults is not that simple, as Figure 15–18 shows.

Basin-and-range relief is abrupt, and rock structures are angular and rugged. As the ranges erode, the transported materials accumulate to great depths in the basins, gradually producing extensive desert plains. The basins average 1220–1525 m (4000–5000 ft) above sea level, with mountain crests rising higher by some 915–1525 m (3000–5000 ft). Death Valley is an extreme, the lowest of these basins, with an elevation of −86 m (−282 ft). However, to the west of the valley, the Panamint Range rises to 3368 m (11,050 ft) at Telescope Peak—over 3 vertical kilometers (2 mi) of desert relief!

Figure 15–19 illustrates a typical **bolson**, which is a slope-and-basin area between the crests of two adjacent ridges in a dry region. The figure also identifies a playa (central salt pan), a bajada (co-

(a)

(b)

FIGURE 15–17
(a) Sandstone cross bedding; (b) sketch demonstrates the bedding pattern in the rock. Photo by author.

(a)

(b)

FIGURE 15–18

(a) Map of the Basin and Range Province in the western United States and northern Mexico; (b) *Landsat* image of the area. (b) Courtesy of National Aeronautics and Space Administration.

alesced alluvial fans), and a **pediment**, the area created as a mountain front retreats through erosional action. A pediment is an area of bedrock that is layered with a thin veneer, or coating, of alluvium. It represents an erosional surface, as opposed to the depositional surface of the bajada. Often the distinction between the two is obscured by the continuous cover of alluvial deposits, but that distinction becomes significant when it is necessary to drill in search of groundwater.

Other mountainous deserts are found in interior Asia, from Iran to Pakistan, and in China and Mongolia. In South America, lying between the ocean and the Andes, is the rugged Atacama Desert. Deserts also occur worldwide as topographic plains, such as the Great Sandy and Simpson deserts of Australia, the Arabian Desert, the Kalahari Desert, and portions of the extensive Takla Makan Desert, which covers some 270,000 km^2 (105,000 mi^2) in the central Tarim Basin of China.

FIGURE 15–19

A bolson in the mountainous desert landscape of the Basin and Range Province. Adapted by permission of Merrill, an imprint of Macmillan Publishing Company, from *Earth Science*, 5th ed., by Edward J. Tarbuck and Frederick K. Lutgens, p. 116. Copyright © 1988 by Merrill Publishing.

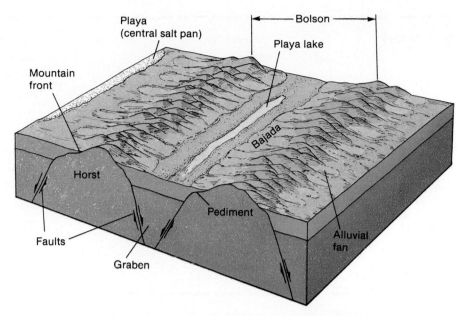

SUMMARY

Winds are produced by the movement of the atmosphere in response to pressure differences. Resulting eolian processes influence the immediate surface of the land. In arid and semiarid climates and along some coastlines where sand is available, wind creates accumulations called dunes, which appear in a variety of forms. Important wind-blown loess deposits occur worldwide and provide a basis for soil development. Wind and water operate together in dry regions. Although water events are infrequent, running water is still the major erosional agent in deserts. Desert landscapes are regions of special plant and animal adaptations and unique rock structures that appear stark and angular, bared of vegetation and other coverings found in the more humid regions of the world. These fragile environments are susceptible to disruption, as evidenced in the Persian Gulf region following the recent war.

KEY TERMS

Alluvial fan	Flash flood
Arroyo	Loess
Badland	Pediment
Bajada	Playa
Basin and Range Province	Province
Blowout depressions	Ripples
Bolson	Sand sea
Deflation	Surface creep
Desert pavement	Ventifacts
Dune	Wadi
Eolian	Wash
Erg desert	Yardangs

REVIEW QUESTIONS

1. Who was Ralph Bagnold? What was his contribution to eolian studies?
2. Explain the term eolian and its application in this chapter. How would you characterize the ability of the wind to move material?
3. Describe the erosional processes associated with moving air.
4. Explain deflation and the evolutionary sequence that produces desert pavement, ventifacts, and yardangs.
5. Differentiate between a dust storm and a sand storm.
6. What is the difference between eolian saltation and fluvial saltation?
7. Explain the concept of fluid threshold.
8. What is the difference between an erg and a reg desert? Which type is a sand sea? Are all deserts covered by sand? Explain.
9. What are the three classes of dune forms? Describe the basic types of dunes within each class. What do you think is the major shaping force for sand dunes?
10. Which form of dune is the mountain giant of the desert? What are the characteristic wind patterns that produce such dunes?
11. How are loess materials generated? What form do they assume when deposited? Name a few examples of significant loess deposits.
12. Characterize desert energy and water balance regimes. What are the significant patterns of these arid landscapes in the world?
13. How would you describe the budget of the Colorado River? What was the basis for agreements regarding distribution of the rivers? Why has thinking about river flows been so optimistic?
14. Describe a desert bolson from crest to crest. Draw a simple sketch with the components of the landscape labeled.
15. Where in the Basin and Range Province do we find the greatest topographic relief?

SUGGESTED READINGS

Abbey, Edward, and the editors of Time-Life Books. *Cactus Country*. American Wilderness Series. New York: Time-Life Books, 1973.

Bagnold, Ralph A. *The Physics of Blown Sand and Desert Dunes*. London: Methuen & Co., 1941, 1954.

Berkman, Richard L., and W. Kip Viscusi. *Damming the West*. New York: Grossman, 1973.

Cooke, R. U., and Andrew Warren. *Geomorphology in Deserts*. Los Angeles: University of California Press, 1973.

Dunbier, Roger. *The Sonoran Desert: Its Geography, Economy, and People*. Tucson: University of Arizona Press, 1968.

Hamblin, W. Kenneth. *The Earth's Dynamic Systems*. New York: Macmillan, 1989.

Hamilton, William J., III. "The Living Sands of the Namib." *National Geographic* 164, no. 3 (September 1983): 364–77.

Hundley, Norris. *Water and the West: The Colorado River Compact and the Politics of Water in the American West*. Berkeley: University of California Press, 1975.

Kahrl, William L. *Water and Power*. Berkeley: University of California Press, 1982.

Krutch, Joseph Wood. *The Voice of the Desert*. Wm. Sloane Associates, 1954.

Lancaster, Nicholas. "On Desert Sand Seas." *Episodes* 11, no. 1 (March 1988): 12–18.

Leopold, A. Starker, and the editors of Life. *The Desert*. Life Nature Library. New York: Time, Inc., 1962.

Mabbutt, J. A. *Desert Landforms*. Cambridge, MA: MIT Press, 1977.

McKee, Edwin D. *A Study of Global Sand Seas*. U.S. Geological Survey Professional Paper 1052. Washington: Government Printing Office, 1979.

Page, Jake, and the editors of Time-Life Books. *Arid Lands*. Planet Earth Series. Alexandria, VA: Time-Life Books, 1984.

Péwé, T. L. *Desert Dust: Origin, Characteristics, and Effects on Man*. Geological Survey Special Paper 186. Washington: Government Printing Office, 1981.

Powell, John Wesley. *The Exploration of the Colorado River and Its Canyons*, formerly titled *Canyons of the Colorado*. 1895. Reprint. New York: Dover Publications, 1961.

Stegner, Wallace. *Beyond the Hundredth Meridian*. New York: Houghton Mifflin, 1954.

United Nations Educational, Scientific, and Cultural Organization (UNESCO). Many publications and studies of interest by the Arid Zone Research Programme Paris.

Wallace, Robert, and the editors of Time-Life Books. *The Grand Canyon*. American Wilderness Series. New York: Time-Life Books, 1972.

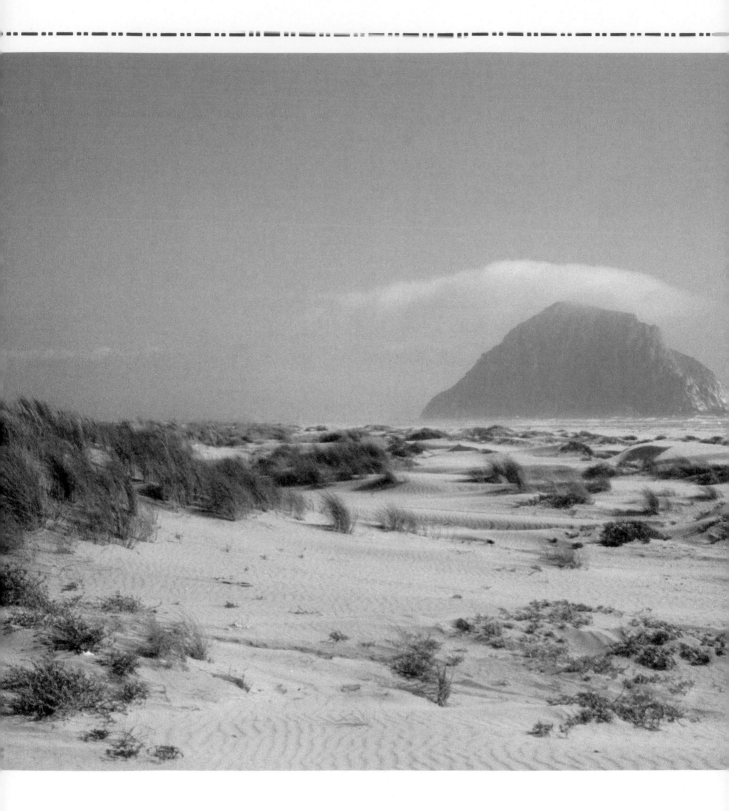

Morro Rock and beach on the Pacific coast. Photo
by author.

16

Coastal Processes and Landforms

The interaction of vast oceanic and atmospheric masses is dramatic along a shoreline. At times, the ocean attacks the coast in a stormy rage of erosive power; at other times, the moist sea breeze, salty mist, and repetitive motion of the water are gentle and calming. Despite such unpredictable variability, the realities of commerce and access to sea routes, fishing, and tourism prompt many people to settle near the ocean. A population survey by the United Nations estimates that about two-thirds of Earth's present population lives in or very near coastal regions. Therefore, an understanding of coastal processes and landforms is critical. And, because the exogenic processes along coastlines produce relatively rapid change there – much faster than the rate of change inland—these processes are essential to consider in planning and development.

You will find coastal processes organized in this chapter into a system of specific inputs (components and driving forces), actions (movements and processes), and outputs (results and consequences). We conclude with a look at the considerable human impact on coastal environments.

COASTAL SYSTEM COMPONENTS

Coastal regions are unique and dynamic environments. Most of Earth's coastlines are relatively new and are the setting for continuous change. The land, ocean, and atmosphere interact to produce tides, waves, erosional features, and depositional features along the continental margins.

The Coastal Environment

The coastal environment is called the **littoral zone**. (Littoral comes from the Latin word for shore.) The littoral zone spans both land and water. Landward, it extends to the highest water line that occurs on shore during a storm. Seaward, it extends to the point at which storm waves can no longer move sediments on the seafloor (usually at depths of approximately 60 m or 200 ft). The specific contact line between the sea and the land is the *shoreline,* and adjacent land is considered the *coast.* Figure 16–1 illustrates the littoral zone and includes specific components that are discussed later in the chapter.

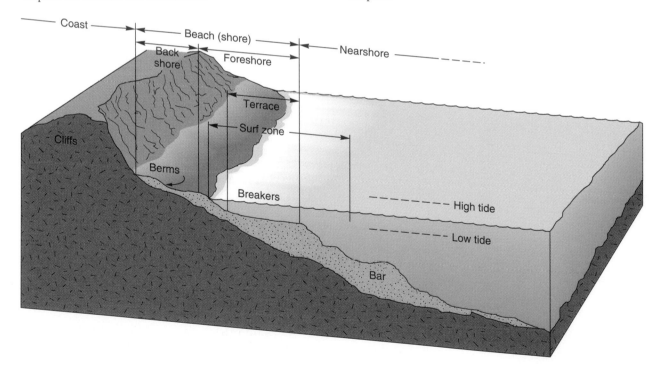

FIGURE 16–1
The littoral zone includes the coast, beach, and nearshore environments.

Because the level of the ocean varies, the littoral zone is redefined naturally from time to time. A rise in sea level is a coastal transgression that causes *submergence* of land, whereas a drop in sea level is a regression that produces coastline *emergence*. Changes in the littoral zone also are initiated by uplift and subsidence of the coast itself. **Mean sea level** is a calculated value based on average tidal levels recorded hourly at a given site over a period of at least 19 years, which is one full lunar tidal cycle.

Mean sea level varies spatially from place to place because of ocean currents and waves, tidal variations, air temperature and pressure differences, and ocean temperature variations. The mean sea level of the U.S. Gulf Coast is about 25 cm (10 in.) higher than that of the east coast of Florida, which has the lowest mean sea level in North America. Sea level rises as one moves northward up the east coast, measuring about 38 cm (15 in.) higher in Maine than in Florida. Along the U.S. west coast, mean sea level in San Diego is about 58 cm (23 in.) higher than it is in Florida, rising to about 86 cm (34 in.) higher in Oregon than in Florida. Overall, the Pacific coast of North America has a mean sea level that averages about 66 cm (26 in.) higher than the average sea level along the Atlantic coast.

At present, the overall mean sea level for the United States is calculated at approximately 40 locations along the coastal margins of the continent, augmented by data from a radar altimeter in orbit (on board the *SEASAT* satellite). Knowing these values precisely assists in calibrating the present sea level rise.

Inputs to the Coastal System

Inputs to the coastal environment include many of the elements already discussed in this text. Solar energy remains the driving force of both the atmosphere and the hydrosphere. Prevailing winds and active weather patterns are produced by conversion of insolation to kinetic and mechanical energy; and in turn, ocean currents and waves are generated principally by atmospheric winds. Climatic regimes are influential in coastal geomorphic processes, and the nature of coastal rock is important in determining rates of erosion and sediment production. All of these inputs occur within the ever-present influence of gravity's pull, not only from Earth but also from the Moon and the Sun. Gravity provides the potential energy of position for materials in motion and produces the tides.

COASTAL SYSTEM ACTIONS

The complex fluctuation of the tides, the movement of winds, waves, and ocean currents, and the occasional impact of storms all are powerful actions within the coastal system. These forces shape landforms from gentle beaches to steep cliffs, and sustain delicate ecosystems.

Tides

Earth's orientation to the Sun and the Moon and the reasons for the seasons are discussed in Chapters 2 and 4. These astronomical relationships produce the pattern of **tides**, the complex daily oscillations in sea level that are experienced to varying degrees around the world. Tides also are influenced by the size, depth, and topography of ocean basins, by latitude, and by shoreline configuration.

Tidal action is a distinctive energy agent for geomorphic change. Significant changes that affect sediment erosion and transportation result from the migration of the shorelines landward and seaward as tides flood (rise) and ebb (fall). Figure 16–2 illustrates the relationship among the Moon, the Sun, and Earth and the generation of variable tidal bulges on opposite sides of the planet.

Tides are produced by the gravitational pull exerted on Earth by both the Sun and the Moon. Although the Sun's influence is only about half that of the Moon (46%) because of the Sun's greater distance from Earth, it still is a significant force. Figure 16–2(a) shows the Moon and the Sun in *conjunction* (lined up with Earth), a position in which their gravitational forces add together. The gravitational effect on the side of Earth facing the Moon and Sun is greater than that experienced by the far side, because gravitational influences decrease with distance. Earth's solid and fluid surfaces actually experience some stretching as a result of this difference in gravitational pull. The water, and even the crust to a slight degree, are pulled outward on the side of Earth directly toward the Moon and Sun.

FIGURE 16–2
Sun, Moon, and Earth
relationships combined to
produce spring tides (a), (b)
and neap tides (c), (d).

The corresponding bulge on Earth's opposite side is primarily the result of the far-side water remaining in position (being left behind) because of its inertia, as the whole planet and near-side water are drawn toward the Moon and Sun. In effect, from this inertial point of view, Earth is being pulled away from the water on the far side, which is left behind due to the slightly weaker gravitational pull. This produces the two opposing tidal bulges on opposite sides of Earth.

(a)

(b)

FIGURE 16–3
Tidal range is high in some bays and estuaries, such as in the Bay of Fundy. Photos by Imagery.

Tides appear to move in and out along the shoreline, but they do not actually do so. Instead, the surface rotates into and out of the relatively "fixed" tidal bulges as Earth changes its position in relation to the Moon and Sun. Every 24 hours and 50 minutes, any given point on Earth rotates through two bulges as a direct result of this rotational positioning. Thus, every day, most coastal locations experience two high (rising) tides known as **flood tides,** and two low (falling) tides known as **ebb tides.** The difference between consecutive high and low tides is considered the *tidal range.*

The combined gravitational effect is strongest in the conjunction alignment and results in the greatest tidal range between high and low tides, known as **spring tides.** (Relative to tides, spring means to "spring forth"; it has no relation to the season of the year.) Figure 16–2(b) shows the other alignment that gives rise to spring tides, when the Moon and Sun are at *opposition.* In this arrangement, the Moon and Sun cause separate tidal bulges, affecting the water nearest each of them. In addition, each bulge is augmented by the left-behind water resulting from the pull of the body on the opposite side.

When the Moon and the Sun are neither in conjunction nor in opposition, but are more-or-less in the positions shown in (c) and (d), their gravitational influences are offset and counteract each other somewhat, producing a lesser tidal range known as **neap tide.**

Tides also are influenced by other factors, so there is a great variety of tidal ranges. For example, some locations may experience almost no difference between high and low tides, whereas the highest tides occur when open water is forced into partially enclosed gulfs or bays. The Bay of Fundy in Nova Scotia records a difference of 16 m (52.5 ft) (Figure 16–3). Tidal prediction is especially important to ships because the entrance to many ports is limited by shallow water and thus high tide is required for passage. At other times, tall-masted ships need a low tide to clear overhead bridges. Tides also exist in large lakes, but the tidal range is small, making tides difficult to distinguish from changes caused by wind. Lake Superior, for instance, has a tidal variation of only about 5 cm (2 in.).

A tidal current entering a bay or river is visible as a **tidal bore** and can cause turbulence as it moves within a narrowing, constricting estuary or bay, or against an opposing tidal current, or against a river's discharge. Such a current entering the mouth of the Amazon River can raise the water level in the river channel a meter or more. Exceptional tidal bores occur in the Chian Tang Kiang River in China, measuring 2.5–3.5 m (8.2–11.5 ft) higher than normal river level.

Tidal Power. The fact that sea level changes daily with the tides suggests an opportunity: could these predictable flows be harnessed to produce electric-

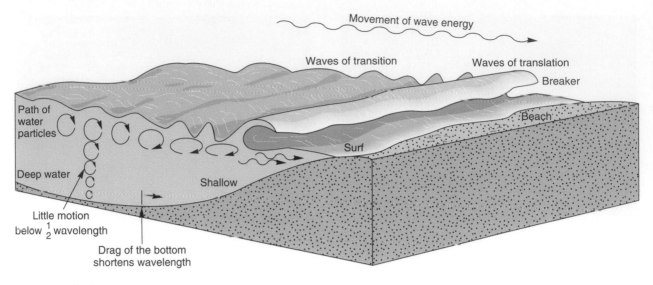

FIGURE 16–4

Wave structure and the orbiting tracks of water particles which change from deep waves of transition to shallow waves of translation.

ity? The answer is yes, given the right conditions. The bay or estuary under consideration must have a narrow entrance suitable for the construction of a dam with gates and locks, and it must experience a tidal range of flood and ebb tides large enough to turn turbines, at least a 5 m (16 ft) range.

About 30 locations in the world are suited for tidal power generation, although at present only two of them are actually producing electricity—an experimental 1-megawatt station in the Soviet Union at Kislaya–Guba Bay, on the White Sea, since 1969, and a facility in the Rance River estuary on the Brittany coast of France since 1967. The tides in the Rance estuary fluctuate up to 13 m (43 ft), and power production has been almost continuous there since 1967, providing an electrical generating capacity of a moderate 240 megawatts (about 20% of the electrical capacity generated at Hoover Dam). According to studies completed by the Canadian government, the present cost of tidal power at ideal sites is economically competitive with that of fossil fuels, although certain environmental concerns must be addressed.

Waves

Undulations of ocean water called **waves** travel in *wave trains,* or groups of waves. In simplest terms, a moving boat creates a wake, which consists of waves;

in more complex terms storms around the world generate large groups of wave trains. A stormy area at sea is called a *generating region* for these large waves, which radiate outward from their formation center. As a result, the ocean is crisscrossed with intricate patterns of waves traveling in all directions. The waves seen along a coast may be the product of a storm center thousands of kilometers away.

Regular patterns of smooth, rounded waves are called **swells**—the undulations of the open ocean. When a storm center subsides or waves leave the wind field of the generating region, wave energy continues to run in these swells, which can range from small ripples to very large flat-crested waves. One moonlit night in 1933, the U.S. Navy tanker *Ramapo,* running downwind in the direction of wave travel for stability, was en route to San Diego and reported a wave higher than their mainmast at about 34 m (112 ft)!

A deep-water wave leaving a generating region tends to extend its wavelength many meters. Water within such a wave is not really migrating but is transferring energy through the water in simple cyclic undulations, which form *waves of transition* (Figure 16–4). Individual water particles move forward only slightly, forming a vertically circular pattern. The diameter of the paths formed by the orbiting water particles decreases with depth. As a

FIGURE 16–5

The process of coastal straightening brought about by wave refraction, converging and diverging in response to headlands.

deep-ocean wave approaches the shoreline and enters shallower water (10–20 m or 30–65 ft), the drag of the bottom shortens the wavelength and slows the speed of the circular motion, flattening the orbital paths to elliptical shapes. As the peak of each wave rises, a point is reached when its height exceeds its vertical stability and the wave falls into a characteristic **breaker**, crashing onto the beach.

In a breaker the orbital motion of transition gives way to form *waves of translation,* in which energy and water actually move forward toward shore as water cascades down from the wave crest. In conjunction with the energy of arriving waves, the slope of the shore determines wave style: plunging breakers indicate a steep bottom profile, whereas spilling breakers indicate a gentle, shallow bottom profile.

As various wave trains move along in the open sea, they sometimes align so that the wave crests and troughs from one wave train are in phase with those of another. When this in-phase wave interference occurs, the heights of the waves are increased. The resulting "killer waves" or "sleeper waves," as they are known along some coastlines, can sweep in unannounced and overtake unsuspecting beachcombers. When wave trains are out of phase with each other, wave heights are dampened. When you observe the breakers along a beach, the changing beat of the surf is actually produced by the patterns of amplified and diminished wave interference generated in far-distant areas of the ocean.

Wave Refraction. Generally, wave action is a process which results in coastal straightening. As waves approach an irregular coast, they bend and focus around **headlands**, protruding landforms generally composed of more resistant rocks (Figure 16–5). Thus, headlands represent a specific point of wave attack along a coastline. Waves tend to disperse their energy in *coves* and *bays* on either side of the headlands. This **wave refraction** (wave bending) along a coastline redistributes wave energy so that different sections of the coastline are subjected to variations in erosion potential. The submarine topography also acts to refract the wave pattern around headlands, which usually extend under the sea, and to dissipate the wave pattern in coves and submerged coastal valleys.

As waves of translation approach the coast, they usually arrive at an angle rather than parallel to it (Figure 16–6). As the current enters shallow water, the waves are refracted and generate a current parallel to the coast, zigzagging back and forth in the prevalent direction of the incoming waves. This **littoral current**, or longshore current, depends on the effective wind and wave direction. Such a current is generated only in the surf zone and works in combination with wave action to transport large amounts of sand, gravels, and shells along the shore. Particles on the beach also are moved along as **littoral drift**, or beach drift, shifting back and forth between water and land in the effective wind and

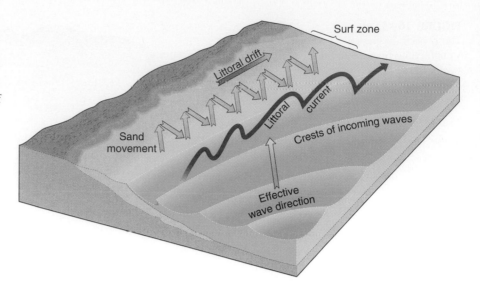

FIGURE 16–6
Longshore currents are produced as waves approach the surf zone and shallower water. Littoral and beach drift result as substantial volumes of material are moved along the shore.

wave direction. These dislodged materials are available for transport and eventual deposition in coves and inlets and can represent a significant volume.

Tsunami, or Seismic Sea Wave. A particular type of wave that greatly influences coastlines is the **tsunami**. Statistically, this seismic sea wave occurs an average of every 3.5 years in the Pacific basin. Specifically for Hawaii, the U.S. Army Corps of Engineers has reported 41 damaging occurrences during the past 142 years. Tsunamis, often incorrectly referred to as tidal waves, are formed by sudden and sharp motions in the seafloor, such as those caused by earthquakes, submarine landslides, or eruptions of undersea volcanoes.

Usually a solitary wave of great wavelength is stimulated, or sometimes a small group of two or three long waves. These waves generally exceed 100 km (60 mi) in wavelength but only a meter or so in height. Because of their great wavelength, tsunamis are affected by the topography of the deep-ocean floor and are refracted by rises and ridges. They travel at great speeds in deep-ocean water—velocities of 600 to 800 kmph (375–500 mph) are not uncommon—but often pass unnoticed because their great length makes the slow rise and fall of water hard to observe.

However, as a tsunami approaches a coast, the shallow water forces the wavelength to shorten. As a result, the wave height may increase up to 15 m (50 ft) or more. The impact of such a wave has the potential for coastal devastation, resulting in property damage and death. In the Alaskan earthquake of 1964, a tsunami was generated by undersea landslides and radiated across the Pacific Ocean. It reached Crescent City, California, in 4 hours 3 minutes; Hilo, Hawaii, in 5 hours 24 minutes, Hokkaido, Japan, in 6 hours 44 minutes; and La Punta, Peru, in 15 hours 34 minutes. Curious onlookers died in both California and Oregon as a result of the distant Alaskan earthquake. Because tsunamis travel such great distances so quickly and are undetectable in the open ocean, a warning system is in operation for nations surrounding the Pacific. Warnings always should be heeded, despite the many false alarms, for the causes lie beneath the ocean and are difficult to monitor in any consistent manner.

Sea Level Changes

Over the long term, sea level fluctuations expose a great range of coastal landforms to tidal and wave processes. Such changes are initiated either by tectonic forces that alter coastlines and basins, or by glacio-eustatic processes that determine the quantity of water stored as ice. Because of complex interactions, it is difficult to isolate individual factors. However, as average global temperatures cycle through cold or warm climatic spells, the quantity of ice locked up in Antarctica, Greenland, mountain glaciers, and the great Pleistocene ice sheets can increase or decrease accordingly.

If Antarctica and Greenland were ever completely ice-free (completely melted), sea level would rise some 65 m (215 ft) or more. Indeed, sea level has risen throughout the most recent epoch, the Holocene (representing the last 10,000 years), and generally ever since the Pleistocene Ice Age, which began about 1.65 million years ago. At the peak of the last Pleistocene glaciation about 18,000 years B.P. (before the present), sea level was about 130 m (430 ft) lower than it is today. Just 100 years ago it was about 10 cm (4 in.) lower.

Chapter 10 presents the climatic implications of a greenhouse-effect warming pattern on high-latitude and polar environments, which are critical areas because they contain enormous deposits of ice. With warming, sea level will rise not only because of melted ice, but also because of the thermal expansion of warmer water (water is densest at 4°C or 39°F, and expands at higher temperatures). Given present warming trends and the predicted climatic change, sea level could rise as much as 6 m (20 ft) during the next century. However, it is generally believed that the rise will be less than 2 m (6.5 ft). Nonetheless, even this smaller amount represents a potentially devastating change for many coastal locations, be-cause a rise of only 0.3 m (1 ft) would cause shorelines worldwide to move inland an average of 30 m (100 ft)!

COASTAL SYSTEM OUTPUTS

Coastlines are the most active portions of continents, with energy and material being continuously delivered to a narrow environment; as a result, these areas are constantly changing and unstable. The action of tides, currents, waves, and sea level changes produces a variety of erosional and depositional landforms.

Erosional Coastal Processes and Landforms

The active margins of the Pacific along the North and South American continents are characteristic coast-lines affected by erosional landform processes. Erosional coastlines tend to be rugged, of high relief, and tectonically active, as expected from their association with the leading edge of a drifting lithospheric plate. Figure 16–7 presents common fea-

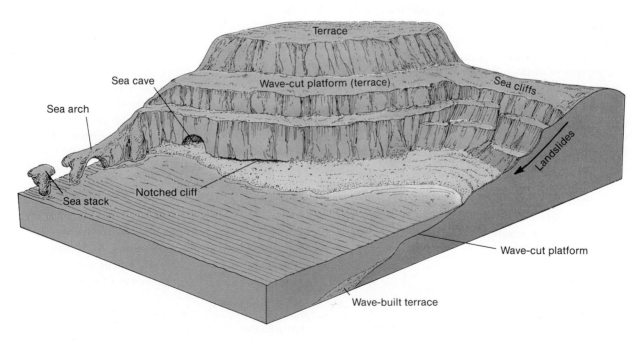

FIGURE 16–7
Characteristic coastal erosional landforms.

FIGURE 16–8
Marine terraces formed as wave-cut platforms along the Pacific coast. Photo by author.

tures found along an erosional coast. *Sea cliffs* are formed along a coastline by the undercutting action of the sea. As indentations are produced at water level, such a cliff becomes notched, leading to subsequent collapse and retreat of the cliff. Other erosional forms evolve along cliff-dominated coastlines, including *sea caves, sea arches,* and *sea stacks.* As erosion continues, arches may collapse, leaving isolated stacks out in the water. The coasts of southern England and Oregon are prime examples of such erosional landscapes.

Wave action can cut a horizontal bench in the tidal zone, extending from a sea cliff out into the sea. Such a structure is called a **wave-cut platform**, or *wave-cut terrace*. If the relationship between the land and sea level has changed over time, multiple platforms or terraces may arise like stairsteps back from the coast. These marine terraces are remarkable indicators of an emerging coastline, with some upper terraces more than 365 m (1200 ft) above sea level. A tectonically active region, such as the California coast, has many examples of multiple wave-cut platforms (Figure 16–8).

Depositional Coastal Processes and Landforms

Depositional coasts generally are located near on-shore plains of gentle relief, where sediments are available from many sources. Such is the case with the Atlantic and Gulf coastal plains of the United States, which lie along the relatively passive, trailing edge of the North American lithospheric plate. These depositional coasts also are influenced by erosional processes and inundation, particularly during storm activity.

Characteristic depositional landforms are illustrated in Figure 16–9. A **spit** consists of sand deposited in a long ridge extending out from a coast; it partially crosses and blocks the mouth of a bay. Classic examples include Sandy Hook, New Jersey (south of New York City), and Cape Cod, Massachusetts. The spit becomes a **baymouth bar** if it completely cuts off the bay from the ocean and forms an inland **lagoon**. Tidal flats and salt marshes are characteristic low-relief features wherever tidal influence is greater than wave action. Spits and bars are made up of materials that have been eroded and transported by littoral drift; for much sand to accumulate, offshore currents must be weak.

A **tombolo** occurs when sand deposits connect the shoreline with an offshore island or sea stack. A tombolo forms when sediments finer than sand accumulate seaward of wave-cut platforms, forming a wave-built terrace that extends out below the water.

Beaches. Of all the features associated with a depositional coastline, beaches probably are the most familiar. Beaches occur in a variety of types and a range of permanence, especially along coastlines

FIGURE 16–9
Characteristic coastal
depositional landforms.

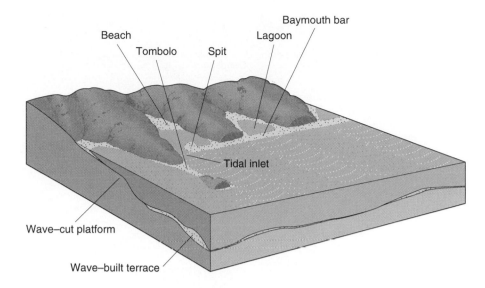

dominated by wave action. Technically, a **beach** is
that place along a coast where sediment is in motion.
Material from the land temporarily resides there
while it is in active transit along the shore. You may
have experienced a beach at some time, along a sea
coast, a lake shore, or even a stream. Perhaps you
have even built our own "landforms" in the sand,
only to see them washed away by tides and waves
(Figure 16–10).

The beach zone ranges, on average, from 5 m (16
ft) above high tide to 10 m (33 ft) below low tide,
although specific definition varies greatly along
individual shorelines. Beaches are dominated by
quartz (SiO_2) because it is the most abundant
mineral on Earth, resists weathering, and therefore
remains after other minerals are removed. In volcanic areas, beaches are derived from wave-processed
lava. These black-sand beaches are found in Hawaii
and Iceland, for example. Other beaches, such as
those in southern France and western Italy, are
composed of pebbles and cobbles—a type of "shin-
gle beach." Finally, some shores have no beaches at
all; scrambling across boulders and rocks may be the
only way to move along the coast. The coast of Maine
is a classic example, composed of resistant granite
rock that is scenically rugged with few beaches.

A beach acts to stabilize a shoreline by absorbing
wave energy, as is evident by the amount of material
that is in almost constant motion (see "sand movement" in Figure 16–6). Some beaches are continuous and stable. Others cycle seasonally; they accumulate during the summer only to be removed by
winter storm waves, and again replaced the following
summer. Protected areas along a coastline tend to
accumulate sand, which can lead to large coastal sand
dunes. Prevailing winds often drag such coastal
dunes inland, sometimes burying trees and highways.

FIGURE 16–10
A familiar beach scene—a continuously moving
inventory of sand. Photo by author.

FIGURE 16–11
Human intervention through various constructions to control littoral drift and longshore currents along a coast.

Locations of the various beach elements—the backshore, berm, and foreshore (which includes the surf zone)—are shown in Figure 16–1. Beach profiles fall into two broad categories, according to prevailing conditions. During quiet, low-energy periods, both backshore and foreshore areas experience net accumulations of sand and are well defined. However, a storm brings increased wave action and high-energy conditions to the beach, usually reducing its profile to a sloping foreshore only. Sediments from the backshore area do not disappear but may be moved by wave action to the nearshore (offshore) area, only to be returned by a later low-energy surf. Therefore, beaches are in constant flux, especially in this era of rising sea levels.

Changes in sand transport can thwart human activities; beaches are lost; harbors closed; coastal highways and beach houses can be inundated with sand. Consequently, various strategies are employed

to interrupt longshore currents and littoral drift, either to halt sand accumulation or to force it. Figure 16–11 illustrates several of the common approaches: *jetties* to block material from harbor entrances; *groins* to slow drift action along the coast, and a **breakwater** to create a zone of still water near the coastline.

However, interrupting the coastal drift that is the natural replenishment for beaches may lead to unwanted changes in sand distribution downcurrent. In addition, enormous energy and materials must be committed to counteract the relentless energy that nature invests along the coast. Years of human effort and expense can be erased by a single storm, such as when offshore islands were completely eliminated by Hurricane Camille along the Gulf Coast in 1969. In Florida, the city of Miami and surrounding Dade County have spent almost $70 million rebuilding their beaches, and needless to say, the effort is continuous. As of 1989, much of the beach-replenishment work done in recent years already had been washed away by winter storms.

A further disadvantage is that unforeseen environmental impacts may accompany the addition of new and foreign sand types to a beach. If the new sands do not match the natural varieties, untold disruption of coastal marine life is possible.

Barrier Chains. The image in Figure 16–12 illustrates the many features of barrier chains. These chains are long, narrow, depositional features, generally of sand, that form offshore roughly parallel to the coast. Common forms are **barrier beaches**, or the broader, more extensive landform, **barrier islands**. Tidal variation in the area usually is moderate to low, with adequate sand supplies coming from nearby coastal plains. On the landward side of a barrier formation are tidal flats, marshes, swamps, lagoons, coastal dunes, and beaches. Barrier beaches appear to adjust to sea level and may naturally shift position from time to time in response to wave action and longshore currents. Breaks in a barrier form inlets, connecting a bay with the ocean.

Barrier beaches and islands are quite common worldwide, lying offshore of nearly 10% of Earth's coastlines. Examples include Africa, India (eastern), Sri Lanka, Australia, the north slope of Alaska, and the shores of the Baltic and Mediterranean Seas. The most extensive chain of barrier islands is along the Atlantic and Gulf Coast states, extending some 5000 km (3100 mi) from Long Island to Texas and Mexico. One particular portion of this chain, shown in Figure 16–12, is the Outer Banks off North Carolina's coast, including Cape Hatteras, across Pamlico Sound from the mainland. The area presently is designated as one of three national seashore reserves supervised by the National Park Service. Figure 16–12 displays key depositional forms: sand spit, island, beach, lagoon, sound, and inlet.

Various ideas have been proposed to explain the formation of barrier beaches. Some may begin as offshore sand bars and then gradually migrate toward shore as the sea level rises. Because many barrier beaches evidence this landward migration today, they are an unwise choice for a homesite or commercial building. Nonetheless, they are a common choice, even though they take the brunt of storm energy and actually act as protection for the

mainland. The hazard represented by the settlement of barrier islands was made graphically clear when Hurricane Hugo assaulted South Carolina.

In 1989, Hurricane Hugo completed its destructive pass over Puerto Rico, the U.S. and British Virgin Islands, the northeastern Caribbean, and then turned northwestward. Hugo attacked the Grand Strand barrier islands off the northern half of South Carolina's coastline (the counties of Horry, Georgetown, and Berkeley), moving hardest over Charleston and the South Strand portion of the islands. On September 20, the storm made landfall as the worst hurricane to strike there in 35 years.

Beachfront houses, barrier beach developments, and millions of tons of sand were swept away; up to 95% of the single-family homes in Garden City alone were destroyed. In addition, the southern portion of Pawleys Island was torn away, and one of every four homes was destroyed. Some homes that had escaped the last major storm, Hurricane Hazel in 1954, were lost to Hugo's peak winds of 209 kmph (130 mph). In comparable dollars, Hugo caused nearly $4 billion in damage, compared to Hazel's $1 billion. The increased damage partially resulted from expanded construction during the intervening years between the two storms. Due to increased development and real estate appreciation, each future storm can be expected to cause ever-increasing capital losses.

Emergent and Submergent Coastlines

Coastlines can be classified in several ways. Most significant is the relationship between rising or falling sea level and whether the continental margins are uplifting or subsiding. Although not entirely satisfactory, in summary we can simplify our discussion to two general types of coasts—emergent and submergent. Please note that coastal landforms may exhibit traits of both emergence and submergence.

Emergent Coastlines. Emergent coastlines are indicated by sea cliffs and wave-cut platforms, such as terraces and benches, that appear raised above water level (Figure 16–7). Such coastlines develop for various reasons. They are common during glaciations, when more of Earth's water is tied up as ice, thus lowering worldwide sea level. Emergent coasts also may be of tectonic origin, especially common around the Pacific Rim; former depositional features

FIGURE 16–12

Landsat image of barrier island chain along the North
Carolina coast. Image courtesy of National Aeronautics
and Space Administration.

thus may appear high above sea level. Other emergent coastlines may be isostatically responding to the melting and retreat of great expanses of ice by uplifting submerged coastal valleys. Areas of northern Canada and Scandinavia are still emerging in this way, at a centimeter (0.4 inch) or so a year, a rather rapid rate, geologically speaking.

Submergent Coastlines. Submergent coasts are drowned by the sea, either because of sea level increases or land subsidence. Characteristically, submergent coastlines form *embayments*, or the inundation of lowlands near the coast by the sea, generally near the mouth of a river. Evidence relates such coastlines to interglacial periods and times of higher temperatures, with corresponding lower volumes of ice and rising sea levels. A Spanish term, *ria* (for "river") is applied to this type of coast, which often is penetrated by a river entering the sea and therefore forming an *estuary*, where seawater and freshwater mix. Familiar estuaries include the flooded lower reaches of the Hudson River around New York City, and Delaware Bay and Chesapeake Bay to the south, which also contain tributary estuaries (Figure 16–13).

If sea level continues to rise over the next century, many estuaries will expand inland, and new ones could form, submerging even more coastline. Despite isostatic rising, many glaciated coasts still feature drowned, glacially carved valleys called *fjords*. Their great depth, to hundreds of meters, precludes any development of depositional features. Fjords exist in South Island of New Zealand, southern Chile, British Columbia, and Scandinavia.

Organic Processes: Coral Formations

Not all coastlines form due to purely physical processes. Some form as the result of organic processes, such as coral growth. A **coral** is a simple marine animal with a cylindrical, saclike body; it is related to other marine invertebrates, such as anemones and jellyfish. Corals secrete calcium carbonate ($CaCO_3$) from the lower half of their bodies, forming a hard external skeleton. Corals function in a *symbiotic* (mutually helpful) relationship with simple red brown-to-green algae—that is, they live in close association with the algae, and are codependent for survival. Corals cannot photosynthesize, but do provide some of their own nourishment, whereas algae perform photosynthesis and convert solar energy into the system, providing the coral with 50–90% of its nutrition and assisting the coral with calcification processes. In return, corals provide the algae with nutrients.

Although both solitary and colonial corals exist, it is the colonial forms that produce enormous structures, varying from treelike and branching forms to round and flat shapes. Through many generations, live corals near the ocean's surface build on the foundation of older corals below, which in turn may rest upon a volcanic seamount or some other submarine feature built up from the ocean floor.

Because corals thrive in warm tropical oceans, the difference in ocean temperature between the western coasts and eastern coasts of continents is critical to their distribution. Western coastal waters tend to be cooler, thereby discouraging coral activity, whereas eastern coastal currents are warmer and thus enhance coral growth. Live colonial corals range from about 30° N to 30° S and occupy a very specific ecological zone: 10–55 m depth (30–180 ft), 27–40‰ salinity, and 18–29°C water temperature (64–85°F). Corals require clear, sediment-free water and consequently do not locate near the mouths of sediment-charged freshwater tributaries (Figure 16–14).

Coral Reefs. An organically derived sedimentary formation of coral rock is called a *reef* and can assume one of several distinctive, and possibly sequential, shapes, principally a fringing reef, a barrier reef, or an **atoll** (Figure 16–15). *Fringing reefs,* sometimes called table reefs, are extensive platforms of coral rock that are attached to land. The largest of these is the Bahama platform in the western Atlantic, covering some 96,000 km² (37,000 mi²). *Barrier reefs* align with the shoreline, separated from the shore by partially enclosed lagoons; they range from 300 to 1000 m wide (1000–3300 ft). The Great Barrier Reef along the shore of Queensland State in Australia exceeds 2025 km length (1260 mi.), 16–145 km width (10–90 mi), and includes at least 200 coral-formed islands and keys. *Atolls* frequently are circular, or ring-shaped, with a central lagoon; they are often built on an undersea volcanic island.

FIGURE 16–13
Landsat image of Chesapeake Bay, showing a *ria* coast of
submergence. Image courtesy of National Aeronautics
and Space Administration.

FIGURE 16–14
Worldwide distribution of living coral formations. Shaded areas also include most of the prolific reef growth and atoll formation. After J. L. Davies, 1973, *Geographical Variation In Coastal Development,* Essex, England: Longman House. Adapted by permission.

Scientists at the University of Puerto Rico and NOAA are tracking an unprecedented bleaching and dying-off of corals worldwide; for instance, the Caribbean, Australia, Japan, Indonesia, Kenya, Florida, Texas, and Hawaii are experiencing this phenomenon. Normally colorful corals have turned stark white as nutrient supplying algae are expelled by the host coral. The bleaching is due to a loss of colorful algae from within and upon the coral itself. Exactly why the coral ejects its living partner is unknown. The 1 to 2C° (1.8–3.6F°) warming of sea-surface temperatures, as stimulated by greenhouse warming of the atmosphere, is thought to be one possible cause, although the condition remains under intense study. Further bleaching through the decade will effect most of Earth's coral-dominated reefs and may be an indicator of serious and enduring environmental trauma.

Salt Marshes and Mangrove Swamps

Narrow, vegetated strips occupy many coastal areas and shorelines along estuaries worldwide. Their great productivity stems from trapped organic matter and sediments that permit a coastal marsh to greatly outproduce a wheat field in raw vegetation per acre. These **wetlands** are productive ecosystems, quite fragile ecologically, and threatened by human tendencies to settle along coasts. In terms of wetland distribution, **salt marshes** tend to form north of the 30th parallel, whereas **mangrove swamps** form equatorward of that point. This is dictated by the occurrence of freezing conditions, which control the survival of mangrove seedlings. Roughly the same latitudinal limits apply in the Southern Hemisphere.

Salt marshes usually form in estuaries and behind barrier beaches and sand spits. An accumulation of

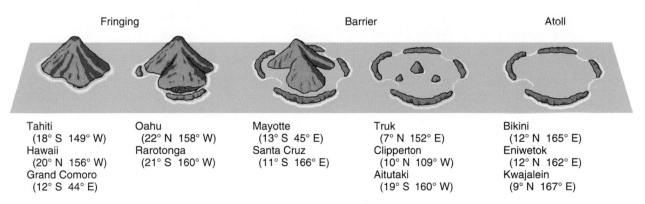

Fringing | Barrier | Atoll

Tahiti
(18° S 149° W)
Hawaii
(20° N 156° W)
Grand Comoro
(12° S 44° E)

Oahu
(22° N 158° W)
Rarotonga
(21° S 160° W)

Mayotte
(13° S 45° E)
Santa Cruz
(11° S 166° E)

Truk
(7° N 152° E)
Clipperton
(10° N 109° W)
Aitutaki
(19° S 160° W)

Bikini
(12° N 165° E)
Eniwetok
(12° N 162° E)
Kwajalein
(9° N 167° E)

FIGURE 16–15
Common coral formations: fringing reefs, barrier reefs, and atolls. After D. R. Stoddart, *The Geographical Magazine* LXIII (1971): 610. Adapted by permission.

FIGURE 16–16
Salt marshes, a productive ecosystem commonly occurring poleward of 30° latitude in both hemispheres. Photo by author.

mud produces a site for the growth of *halophytic* (salt-tolerant) plants. Plant growth then traps additional alluvial sediments and adds to the salt marsh area. Because salt marshes are in the intertidal zone (between the furthest reaches of high and low tides), sinuous, branching channels are produced as tidal waters flood into and ebb from the marsh (Figure 16–16).

Sediment accumulation on tropical coastlines provides the site for growth of mangrove trees, shrubs, and other small trees. The adventurous prop roots of the mangrove are constantly finding new anchorages and are visible above the water line but reach below the water surface, providing a habitat for a multitude of specialized lifeforms. Mangrove swamps often secure and fix enough material to form islands (Figure 16–17).

HUMAN IMPACT ON COASTAL ENVIRONMENTS

The development of modern societies is closely linked to estuaries, wetlands, barrier beaches, and coastlines. Estuaries are important sites of human settlement because they provide natural harbors, a food source, and convenient sewage and waste disposal. Society depends upon daily tidal flushing of estuaries to dilute the pollution created by waste disposal. Thus, the estuarine and coastal environment is vulnerable to abuse and destruction if development is not carefully planned. Nationally, about 50% of estuarine zones now have been altered or destroyed, and by the middle of this decade most barrier islands will have been developed and occupied. A useful comparison is the fact that only 28 of

280 barrier islands in the United States were developed to any degree before World War II.

Barrier islands and coastal beaches do migrate over time. Houses that were more than a mile from the sea in the Hamptons along the southeastern shore of Long Island, New York, are now within only 30 m (100 ft) of it. The time remaining for these homes can be quickly shortened by a single hurricane or by a sequence of storms with the intensity of a series that occurred in 1962. Nonetheless, despite our understanding of beach and barrier island migration, and the effects of storms, and despite warnings from government agencies, coastal planning and zoning goes ahead. Society behaves as though beaches and barrier islands are stable, fixed features, or as though they can be engineered to be permanent. Experience has shown that severe ero-

sion generally cannot be prevented, as we have seen with the cliffs of southern California, the shores of the Great Lakes, the New Jersey shore, the Gulf coast, and the devastation of coastal South Carolina by Hurricane Hugo. One specific case, the New Jersey shore area, is examined in Focus Study 16–1.

Marco Island, Florida: An Example of Impact

An example of intense barrier island development is the Marco Island area on the coast in southwestern Florida, about 24 km (14.9 mi) south of the city of Naples. Figure 16–18(a) shows Marco Island as it was prior to 1952: approximately 5300 acres of subtropical mangrove habitat and barrier beach terrain, formed from river sediments and old shell

FIGURE 16–17
Mangroves, which tend to grow equatorward of 30° latitude, retain sediments and can form anchors for island formations. Photo by author.

(a)

MARCO
ISLAND
1952

RED MANGROVES
BLACK MANGROVES
MIXED MANGROVES
CONVERTED/URBAN
UPLAND
SAND/BEACH
RIVERINE MANGROVES
WATER
MARSH
ROADS

(b)

MARCO
ISLAND
1984

RED MANGROVES
BLACK MANGROVES
MIXED MANGROVES
CONVERTED/URBAN
UPLAND
SAND/BEACH
RIVERINE MANGROVES
WATER
ROADS

(c)

FIGURE 16–18
(a) 1952 and (b) 1984 digitized inventories of Marco
Island, Florida, showing dramatic development; (c) 1984
color-infrared photograph of the island. From Samuel
Patterson, *Mangrove Community Boundary
Interpretation and Detection of Areal Changes on Marco
Island, Florida,* Biological Report 86(10), for National
Wetlands Research Center, U.S. Fish and Wildlife Service,
August 1986, p. 23, 46, 49.

An Environmental Approach to Shoreline Planning

Coastlines are places of specific constraints as well as opportunities. Unfortunately, poor understanding of this resource and a lack of environmental analysis often go hand in hand. The New Jersey shore provides an excellent example of how proper understanding of a coastal environment could have avoided problems encountered during major storms and from human development and settlement. Much of what is presented here is characteristic of similar coastal areas elsewhere. Figure 1 illustrates the New Jersey shore from ocean to back bay and describes the tolerance to human occupation and use of each specific zone.

Sand beaches are the primary natural defense against the ocean in such an environment; they act as bulwarks against the pounding of a stormy sea. Ocean and beach environments also are useful for recreation and other purposes. But they are susceptible to pollution and require environmental protection to control ocean dumping of dangerous materials near the shore and inappropriate construction on the beach itself. The shoreline is not tolerant of construction because of its shifting, changing nature during storms, daily tidal fluctuations, and the potential effects of a rising sea level. In recent years, high water levels in the Great Lakes and along the southern California coast certainly attest to the vulnerability of shorelines to erosion and inundation. Wave erosion attacks cliff formations and undermines, bit

Ocean	Beach	Primary dune	Trough	Secondary dune	Backdune	Bayshore	Bay
Tolerant	Tolerant	Intolerant	Relatively tolerant	Intolerant	Tolerant	Intolerant	Tolerant
Intensive recreation	Intensive recreation	No passage, breaching, or building	Limited recreation	No passage, breaching, or building	Most suitable for development	No filling	Intensive recreation
Subject to pollution controls	No building		Limited structures				
	Intolerant of construction						

FIGURE 1
Coastal environment—a planning perspective from ocean to bay. After *Design with Nature* by Ian McHarg. Copyright © 1969 by Ian L. McHarg. Adapted by permission of Doubleday, a division of Bantam Doubleday Dell Publishing Group, Inc.

FIGURE 2
Coastal erosion and a failed house foundation. Photo by author.

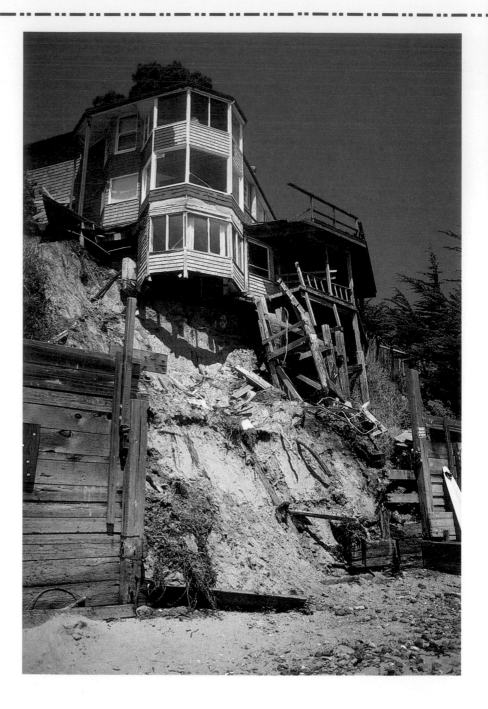

by bit, the foundations of houses and structures that were built too close to the edge (Figure 2).

The front, or primary, dune along a coast also is intolerant of heavy use, even the passage of people trekking to the beach; it is fragile, easily disturbed, and vulnerable to erosion. Fragile plants work to hold the sand in place, eventually resulting in a gradual increase in dune height. Primary dunes are like the dikes in the Netherlands: they are the primary defense against the sea, and they should not be the sites of development or heavy traffic. Carefully controlled access points to the beach should be enforced, for even foot traffic can cause destruction.

The trough behind the primary dune is relatively tolerant, with limited recreation and building potential. The plants that fix themselves to the surface send roots down to fresh groundwater reserves. Thus, if construction should inhibit the surface recharge of that supply, the natural protective ground cover could fail and destabilize the environment; or subsequent saltwater intrusion might contaminate well water. Clearly, groundwater resources and the location of recharge aquifers must be considered in planning.

Behind the trough is the secondary dune, a second line of defense against the sea. It, too, is tolerant of some uses yet is vulnerable to destruction through improper use. The backdune along the New Jersey shore is a more suitable location for development than any of the zones between it and the sea. Further inland are the bayshore and the bay, where ideally there should be no dredging and filling and only limited dumping of treated wastes and toxics, because these activities damage the environment. In reality, the opposite of these ideals prevails.

Prior to an intensive study completed in 1962, no analysis had been completed outside academic circles. Thus, what is common knowledge to botanists, biologists, ecologists, and geographers in the classroom and laboratory still has not filtered through to the general planning and political processes. As a result, on the New Jersey shore and along much of the Atlantic and Gulf coasts, improper development of the fragile coastal zone led to extensive destruction during the storms of March 1962 and the numerous hurricanes of this century. In 1962, damage was massive from the Carolinas to Massachusetts.

When the storms of 1962 hit, Robert Dolan was researching the area. He forecast a strong probability that the Cape May portion of the New Jersey shore could be 90 m (295 ft) further inland by the year 2010, less than two decades from now. There is a 15% chance that the shore could be twice as far inland (180 m or 600 ft) by that time. Similar estimates persist along the entire east coast, making it clear that society must develop in a manner that synthesizes ecology and economics if these coastal environments are to be sustained. Ian McHarg, ecologist and landscape architect, summarizes the situation well:

> May it be that these simple ecological lessons will become known and incorporated into ordinance [law] so that people can continue to enjoy the special delights of life by the sea.*

---

*From *Design with Nature* by Ian McHarg, p. 17. Copyright © 1969 by Ian L. McHarg. Published by Doubleday, a division of Bantam Doubleday Dell Publishing Group, Inc.

and reef fragments. Then development began in 1962, despite tropical storms that occasionally assault the area, as Hurricane Donna had done in 1960. Development included artificial landfill for housing sites and general urbanization of the entire island. Scientists regarded the island and estuarine habitat as highly productive and unique in its mixture of mangrove species, endangered bird species, and a productive fishery. Consequently, numerous challenges and court actions attempted to halt development. But construction activities continued, and today Marco Island is completely developed—Figure 16–18(b).

In 1984 a GIS approach (Geographic Information System is described in Chapter 2) was used to study Marco Island, incorporating existing maps, remote-sensing techniques, and a knowledge of the ecology of an estuarine mangrove community.* Using the three images in Figure 16–18, you can compare the 1952 and 1984 digitized inventory maps and a 1984 color-infrared aerial photograph. With these tools, it is possible to visually and quantitatively compare this island over time to see the impact of urban development. Note that the only remaining natural community is restricted to a very limited portion along the extreme perimeter.

---

*Samuel G. Patterson, *Mangrove Community Boundary Interpretation and Detection of Areal Changes in Marco Island, Florida: Application of Digital Image Processing and Remote Sensing Techniques.* Biological Report 86(10) for the U.S. Fish and Wildlife Service, August 1986.

SUMMARY

Coastal processes are a system of physical environment inputs, tidal and wave actions, and erosional, depositional, and organic outputs, with the fluid agents of air and water delivering energy to perform work. The coastal environment exists where the tide-driven, wave-driven sea confronts the land. Coastal margins have been subjected to numerous changes in sea level throughout geologic time, with each change altering wave and tidal processes. Terraces, embayments, lagoons, cliffs, and beaches are the result.

The coastline also is the location of organic formations built by corals, salt marshes, and mangroves, all of which help to produce reefs and islands. Presently, sea level is rising and sea-surface temperatures are increasing. The varied interaction of ocean and continent produces a fragile balance that can be both hazardous to humans and can be damaged by human occupation. The coastal zone, like the river floodplain, requires careful assessment of natural factors for sound use. Ideally, ecological planning can enhance long-term economic goals and our enjoyment of Earth's wonderful coastal environments.

KEY TERMS

Atoll	Mangrove swamps
Barrier beaches	Mean sea level
Barrier islands	Neap tide
Baymouth bar	Salt marshes
Beach	Spit
Breaker	Spring tides
Breakwater	Swells
Coral	Tidal bore
Ebb tides	Tides
Flood tides	Tombolo
Headlands	Tsunami
Lagoon	Wave-cut platform
Littoral current	Wave refraction
Littoral drift	Waves
Littoral zone	Wetlands

REVIEW QUESTIONS

1. What are the key terms used to describe the coastal environment?
2. Define mean sea level. How is this value determined? Is it constant or variable around the world? Explain.
3. What interacting forces generate the pattern of tides?
4. What characteristic tides are expected during a new Moon or a full Moon? During the first-quarter and third-quarter phases of the Moon? What is meant by a flood tide? An ebb tide?
5. Is tidal power being used anywhere to generate electricity? Explain briefly how such a plant would utilize the tides to produce electricity.
6. What is a wave? How are waves generated, and how do they travel across the ocean? Discuss the process of wave formation and transmission.
7. Describe the refraction process that occurs when waves reach an irregular coastline.
8. Define the components of littoral drift.
9. Explain how seismic sea waves attain such tremendous velocities.
10. What is meant by an erosional coast? What are the expected features of such a coast?
11. What are some of the depositional features encountered along a coastline?
12. How do people attempt to modify littoral drift? What are the positive and negative impacts of these actions?
13. Describe a beach—its form, composition, function, and evolution.
14. What success has Miami had with beach replenishment? Is it a practical strategy?
15. Based on the information in the text and any other sources at your disposal, do you think barrier islands and beaches should be used for settlements? If so, under what conditions? If not, why not?
16. After the Grand Strand off South Carolina was destroyed by Hurricane Hazel in 1954, settlements were rebuilt, only to be hit by Hurricane Hugo 35 years later, in 1989. Why do these recurring events happen to human populations? Compare the impact of the two storms.

17. Describe the western and eastern coasts of North America as emergent or submergent coastlines.
18. How are corals able to construct reefs and islands?
19. Why are the wetlands poleward of 30° N and S latitude different from those that are equatorward? Describe the differences.
20. Describe the condition of Marco Island at the present time. Do you think a rational model was used to assess the environment prior to development?
21. What type of environmental analysis is needed for rational development and growth in a region like the New Jersey shore?

SUGGESTED READINGS

Bernstein, Joseph. "Tsunamis." *Scientific American* (August 1954): 60–64.

Blockson, Charles L. "Sea Change in the Sea Islands." *National Geographic* (December 1987): 735–63.

Dolan, Robert, Bruce Hayden, and Harry Lins. "Barrier Islands." *American Scientist* 68, no. 1 (January–February 1980): 16–25.

Greely, A. W. "Hurricanes on the Coast of Texas." *National Geographic* (November 1900): 442–45.

McGee, W. J. "The Lessons of Galveston." *National Geographic* (October 1900): 377–83.

McHarg, Ian L. *Design with Nature.* Garden City, NY: Doubleday, 1969.

Ross, David A. *Introduction to Oceanography.* Englewood Cliffs, NJ: Prentice Hall, 1988.

Rowntree, Rowan A. "Coastal Erosion: The Meaning of a Natural Hazard in the Cultural and Ecological Context." *Natural Hazards.* Edited by Gilbert F. White. New York: Oxford University Press, 1974.

Sackett, Russell, and the editors of Time-Life Books. *Edge of the Sea.* Planet Earth Series. Alexandria, VA: Time-Life Books, 1983.

Schwartz, Maurice L. *The Encyclopedia of Beaches and Coastal Environments.* Stroudsburg, PA: Hutchinson Ross, 1982.

Stowe, Keith S. *Essentials of Ocean Science.* New York: John Wiley & Sons, 1987.

Thurman, Harold V. *Introductory Oceanography.* 5th ed. Columbus: Macmillan, 1988.

Whipple, A. B. C., and the editors of Time-Life Books. *Restless Oceans.* Alexandria, VA: Time-Life Books, 1983.

Ziegler, J. M. "Origin of the Sea Islands of the Southeastern United States." *Geographical Review* 49 (1959): 222–37.

Mountains and glacial ice. Photo by author.

17

Glacial Processes and Landforms

A large measure of the freshwater on Earth is frozen, with the bulk of that ice sitting restlessly in just two places—Greenland and Antarctica. The remaining ice covers various mountains and fills alpine valleys. More than 29 million km³ (7 million mi³) of water is tied up as ice, or about 77% of all freshwater. These deposits of ice, laid down over several million years, provide an extensive frozen record of Earth's climatic history and perhaps some clues to its climatic future. This chapter focuses on Earth's extensive ice deposits—their formation, movement, and the ways in which they function to produce various erosional and depositional landforms. Glaciers are transient landforms themselves and leave in their wake a variety of landform features. The fate of glaciers is intricately tied to changes in global temperature and ultimately concerns us all.

RESERVOIRS OF ICE

A **glacier** is a large mass of perennial ice, resting on land or floating shelflike in the sea adjacent to land. Glaciers form by the accumulation and recrystallization of snow. They move under the pressure of their own mass and the pull of gravity. Today, about 11% of Earth's land area is dominated by these slowly flowing streams of ice. During colder episodes in the past, as much as 30% of continental land was covered by glacial ice because below-freezing temperatures prevailed at lower latitudes, allowing snow to accumulate.

A **snowline** is the elevation where snow ends in a snow-covered area, specifically the lowest line where winter snowfall persists throughout the summer. Glaciers form in such areas of permanent snow, both at high latitudes and high elevations, even forming on some high mountains along the equator. In equatorial mountains, the snowline is around 5000 m (16,400 ft); on midlatitude mountains, such as the European Alps, snowlines average 2700 m (8850 ft); and in southern Greenland snowlines are down to 600 m (1970 ft).

Types of Glaciers

Glaciers are as varied as the landscape itself. They fall within two general groups, based on their form, size, and flow characteristics: alpine glaciers and continental glaciers.

Alpine Glaciers. With few exceptions, a glacier in a mountain range is called an **alpine glacier**, or *mountain glacier*. It occurs in several subtypes. One prominent type is a **valley glacier**, an ice mass constricted within the confines of a valley. Such glaciers range in length from only 100 m (325 ft) to over 100 km (62 mi). The snowfield that feeds the glacier with new snow is at a higher elevation. In Figure 17–1, at least a dozen valley glaciers are identifiable in the *Landsat* image of the Alaska Range.

As a valley glacier flows slowly downhill, the mountains, canyons, and river valleys beneath its mass are profoundly altered by its passage. Some of the glacial debris created by its erosive excavation can be seen as dark streaks on the ice as the glacier transports the debris for deposition elsewhere.

An erosional landform that appears as a scooped-out recess at the head of a valley is called a **cirque**. Most alpine glaciers originate in a mountain snowfield that is confined in such a bowl-shaped recess. A glacier that forms in a cirque is called a *cirque glacier*. Several cirque glaciers may feed a valley glacier, as shown on the topographic map in Figure 17–2. The Kuskulana Glacier, which flows off the map to the west, is supplied by numerous cirque glaciers feeding several tributary valley glaciers.

Wherever several valley glaciers extend beyond their confining valleys and coalesce at the base of a mountain range, a *piedmont glacier* is formed and spreads freely over the nearby lowlands. A *tidal glacier,* such as the Columbia Glacier on Prince William Sound in Alaska, ends in the sea and breaks off as floating ice.

Continental Glaciers. A continuous mass of ice is known as a **continental glacier** and in its most extensive form is called an **ice sheet**. Two additional types of continuous ice cover associated with mountain locations are designated as *ice caps* and *ice fields*. Most glacial ice exists in the snow-covered ice sheets that blanket 80% of Greenland (1.8 million km³, or 0.43 million mi³) and 90% of Antarctica (13.9 million km³, or 3.3 million mi³). Antarctica alone has 91% of all the glacial ice on the planet.

These two ice sheets represent such an enormous volume of ice that large portions of each landmass

FIGURE 17–1
Valley glaciers in the Alaska Range of Denali National
Park. *Landsat* image courtesy of National Aeronautics and
Space Administration.

FIGURE 17–2
Numerous cirque glaciers on a topographic map of a portion of Alaska (McCarthy C–7 Quadrangle, 1:63,360 scale, 100 ft contour interval). The blue areas represent active glaciers. Courtesy of U.S. Geological Survey.

(a)

FIGURE 17–3

(a) The Vatnajökull ice cap in southeastern Iceland; (b) the southern Patagonian ice field of Argentina. (a) *Landsat* image courtesy of National Aeronautics and Space Administration; (b) photo by Cosmonauts G. M. Greshko and Yu V. Romanenko, *Salyut 6.*

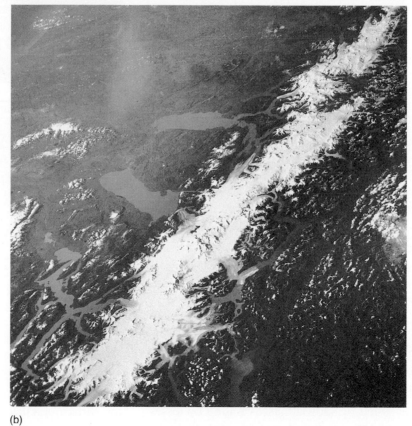

(b)

have been isostatically depressed more than 2000 m (6500 ft) below sea level. Each ice sheet is more than 3000 m (9800 ft) deep, burying all but the highest peaks. An ice sheet is the most extensive form of continental glacier. Both ice caps and ice sheets completely bury the underlying landscape, although an **ice cap** is roughly circular and covers an area of less than 50,000 km² (19,300 mi²). The island country of Iceland features several ice caps—Figure 17–3(a).

The photograph in Figure 17–3(b) shows ridges and peaks visible above the buried terrain of the Andes in the southern Patagonian **ice field**, which extends in a characteristic elongated pattern in a mountainous region. The term *nunatak* refers to peaks that are visible above the glaciers. The Patagonian ice field is one of Earth's largest, stretching 360 km (224 mi) in length between 46° and 51° S latitude and up to 90 km (56 mi) in width. An ice field is not extensive enough to form the characteristic dome of an ice cap.

Continuous ice sheets, caps, and fields are drained by **outlet glaciers** that form around their periphery. An outlet glacier flows out from a continental glacier, taking the form of a valley glacier, sometimes moving down a rock-walled valley to the sea or to lowlands. Outlet glaciers flow radially from the edges of a dome-shaped ice mass. Figure 17–4 shows the Jakobshavns Isbrae outlet glacier flowing from the ice sheet in western Greenland. The glacier's path is about 150 km (93 mi) in length, ending in the sea, where large pieces break off and drift away as **icebergs**. The flooded valley in the *Landsat* image forms a *fjord* and is inundated by the sea.

GLACIAL PROCESSES

A glacier is an example of an open system, with interacting components that result in a mass balance readily visible in the glacier's physical stature. A glacier's budget consists of net gains or losses of glacial ice, which determine whether the glacier expands or retreats.

Glacial Mass Balance

A glacier is fed by snowfall and is wasted by losses from its upper and lower surfaces and along its

FIGURE 17–4
Jakobshavns Isbrae outlet glacier on the west coast of Greenland. The glacier is visible as a distinct ice stream. *Landsat* image courtesy of National Aeronautics and Space Administration.

margins. A snowline called a **firn line** is visible across the surface of a glacier, indicating where the winter snows and ice accumulation survived the summer

FIGURE 17–5

(a) Mass balance of a glacial system showing the relationship between accumulation and ablation and the location of the equilibrium line; (b) cross section of a typical alpine glacier.

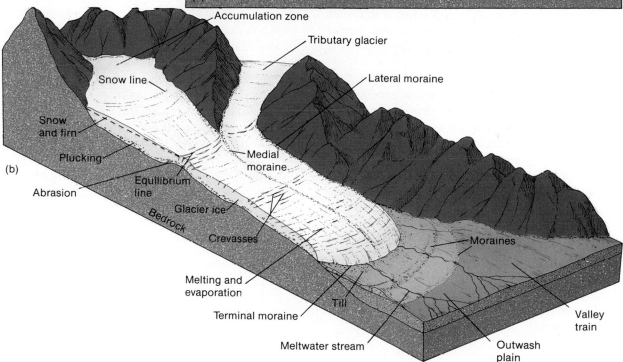

melting season. In a glacier's lower elevation, losses of mass occur because of surface melting, internal and basal melting, sublimation (recall from Chapter 7 that this is the direct evaporation of ice), wind removal by deflation (see Chapter 15), and the calving, or breaking off, of ice blocks. The combined effect of these losses is called **ablation**.

A glacier's area of excessive accumulation is, logically, at colder higher elevations. The zone where accumulation gain ends and loss begins is the **equilibrium line**—Figure 17–5(a). This area of a glacier generally coincides with the firn line, except in subpolar glaciers where frozen surface water occurs below the firn line. Glaciers achieve a positive net balance of mass—grow larger—during colder periods with adequate precipitation. In warmer times, the equilibrium line migrates to a higher elevation and the glacier retreats—grows smaller—

due to its negative net balance. Internally, a glacier can continue to move forward even though its lower terminus is in retreat—Figure 17–5(b).

Formation of Glacial Ice. The essential input to a glacier is precipitation that accumulates in a snowfield, a glacier's formation site. Snowfields usually are at the highest elevation of an ice sheet, ice cap, or head of a valley glacier, usually in a cirque. Highland snow accumulation sometimes is the product of orographic processes as discussed in Chapter 8. Avalanches from surrounding mountain slopes can add to the snowfield. As the snow accumulation deepens in sedimentary-like layers, the increasing thickness results in increased weight and pressure on underlying portions. Rain and summer snowmelt then contribute water, which stimulates further melting, and that meltwater seeps down into the snowfield and refreezes.

Snow that survives the summer and into the following winter begins a slow transformation into glacial ice. Air spaces among ice crystals are pressed out as snow packs to a greater density. The ice crystals recrystallize under pressure and go through a process of regrowth and enlargement. In a transition step to glacial ice, snow becomes **firn**, which has a granular texture. As this process continues, many years pass before denser **glacial ice** is produced. Formation of glacial ice is analogous to formation of metamorphic rock: sediments (snow and firn) are pressured and recrystallized into a dense metamorphic rock (glacial ice). In Antarctica, glacial ice formation may take 1000 years because of the dryness of the climate (minimal snow input), whereas in wet climates that time is reduced to just a few years because of the volume of new snow constantly being added to the system.

Glacial Movement

We generally think of ice as those small, brittle cubes from the freezer, but glacial ice is quite different. In fact, glacial ice behaves in a plastic manner, for it distorts and flows in response to weight and pressure from above and the degree of slope below. Rates of flow range from almost nothing to more than 300 m (980 ft) a year. Ice can move on a steep slope at a rate of a kilometer or two per year. The rate of accumulation of snow in the formation area is critical to the pace of glacial movement.

The movement of a glacier is not simply a process of ice slipping downhill. The greatest movement within a valley glacier occurs below the rigid surface layer, which fractures as the underlying zone moves forward. At the same time, the base creeps and slides along, varying its speed with temperature and the presence of any water beneath the ice. This *basal slip* usually is much less rapid than the internal flow of the glacier (Figure 17–6). In addition, unevenness in the landscape beneath the ice may produce varia-

FIGURE 17–6
Cross section of a glacier, showing the nature of its forward motion.

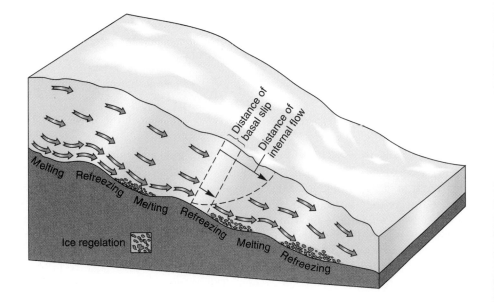

FIGURE 17–7

Muir Glacier, Alaska, fed by tributary glaciers and marked by patterns of transported debris entering Glacier Bay inlet. Photo by Stephen Cunha.

tions in pressure, melting some of the basal ice at one moment, only to have it refreeze a moment later; this process is called *ice regelation*—an important factor in glacial downslope movement. The basal ice layer, which can extend tens of meters above the base of the glacier, has a much greater debris content than occurs in the glacial ice above.

The flowing movement of a compound valley glacier (i.e., one with tributaries) is different from that of a river with tributaries. As the picture of the Muir Glacier in Alaska demonstrates (Figure 17–7), tributary glaciers flow into a main glacier, but merge alongside one another by extending and thinning rather than blending and mixing as do rivers. You can clearly see this because they maintain their own patterns of transported debris (dark streaks).

A flowing glacier can develop vertical cracks known as **crevasses**. These result from friction with valley walls, or tension from extension as the glacier passes over convex slopes, or compression as the glacier passes over concave slopes. Such crevasses are in evidence in Figure 17–5(b). Traversing a glacier, whether an alpine glacier or an ice sheet, is dangerous because a thin veneer of snow sometimes masks the presence of a crevasse. One type of crevasse occasionally associated with a cirque glacier is known as a *bergschrund,* or *headwall crevasse;* it forms when snow and firn compact and pull away from ice that remains frozen to rock at the head of the glacier.

Glacial Surges. Although today many glaciers are in retreat, perhaps due to a global warming trend, some will lurch forward with little or no warning in a **glacial surge**. (This is not quite as abrupt as it sounds; in glacial terms, a surge can be tens of meters per day.) The Jakobshavn Glacier in Greenland, for example, is known to move between 7 and 12 km (4.3 and 7.5 mi) a year. Such surges appear more common today than previously thought. In the spring of 1986 Hubbard Glacier and its tributary Valerie Glacier surged across the mouth of Russell Fjord in Alaska, cutting it off from contact with Yakutat Bay. This area in southeast Alaska is fed by annual snowfall in the Saint Elias mountain range that averages more than 850 cm (335 in.) a year. Consequently, the surge event had been predicted for some time, but the rapidity of the surge, largely

FIGURE 17–8
The Hubbard Glacier surged across Russell Fjord, effectively damming the fjord and temporarily creating Russell Lake. Photo by Stephen Cunha.

triggered by rapid flows from Valerie, was surprising. The fjord was dubbed Russell Lake for the time it remained isolated (Figure 17–8). The glacier's movement exceeded 34 m (112 ft) per day during the peak surge, an enormous increase over its normal rate of 15 cm per day (6 in.).

The exact cause of such a glacial surge is still being studied. Some surge events result from a buildup of water pressure in the basal layers of the glacier. Sometimes that pressure is enough to actually float the glacier slightly during the surge. As a surge begins, icequakes are detectable, and ice faults are visible along the margins that separate the glacier from the surrounding stationary terrain.

Glacial Erosion. Glacial erosion is similar to a large excavation project, with the glacier hauling debris from one site to another for deposition. As rock structure fails along joint planes, the passing glacier mechanically *plucks* the material and carries it away. There is evidence that rock pieces actually freeze to the basal layers of the glacier and, once embedded, allow the glacier to scour and sandpaper the landscape as it moves, a process called *abrasion*. This abrasion and gouging produces a smooth surface on exposed rock, which shines with glacial polish when the glacier retreats. Larger rocks in the glacier act much like chisels, working the underlying surface to produce glacial striations parallel to the flow direction (Figure 17–9).

GLACIAL LANDFORMS

The processes of glacial erosion and deposition produce unique landform features. You might expect all glaciers to create the same landforms, but this is not so. Alpine and continental glaciers each generate their own characteristic landscapes.

Erosional Landforms Created by Alpine Glaciation

W. M. Davis characterized the stages of a valley glacier in a set of drawings published in 1906 and

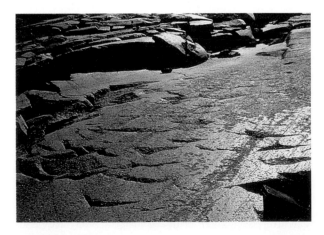

FIGURE 17–9
Glacial polish and striations, examples of glacial erosion. Photo courtesy of U.S. Geological Survey.

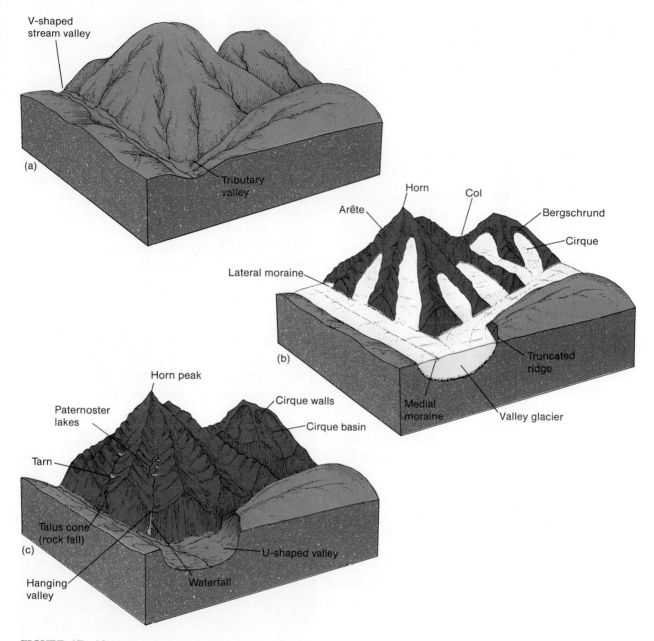

FIGURE 17–10

(a) Before, (b) during, and (c) after a valley glacier occupies a formerly stream-cut valley. After W. M. Davis.

redrawn here in Figure 17–10. Illustration (a) shows a typical river valley with characteristic **V**-shape and stream-cut tributary valleys that exist before glaciation. Illustration (b) shows that same landscape

during a later period of active glaciation. Glacial erosion and transport are actively removing much of the regolith (weathered bedrock) and the soils that covered the preexisting valley landscape. Illustration

(c) shows the same landscape at a later time when climates have warmed and ice has retreated. The glaciated valleys now are U-shaped, greatly changed from their previous stream-cut form. You can see the oversteepened sides and the straightened course of the valley. The physical weathering associated with a freeze-thaw cycle has loosened much rock along the steep cliffs, falling to form talus cones along the valley sides during the postglacial period. Reexamine the cover photo for this text—can you identify this glacially-produced shape?

The cirques where the valley glacier originated are clearly visible in the figure. As the cirque walls have worn away, an *arête* has formed, a sharp ridge that divides two cirque basins. Arêtes ("knife-edge" in French) become the sawtooth and serrated ridges in glaciated mountains. When two eroding cirques reduce an arête to form a pass or saddlelike depression, the term *col* is used. A pyramidal peak called a *horn* results when several cirque glaciers gouge an individual mountain summit from all sides. The most famous example is the Matterhorn in the Swiss Alps, but many others occur worldwide.

A small mountain lake, especially one that collects in a cirque basin behind risers of rock material, is called a *tarn*. Small, circular, stair-stepped lakes in a series are called *paternoster lakes* because they look like a string of rosary (religious) beads forming in individual rock basins aligned down the course of a glaciated valley. Such lakes may have been formed by the differing resistance of rock to glacial processes or by the damming effect of glacial deposits.

Figure 17–10(c) also shows tributary valleys left stranded high above the glaciated valley floor, following the removal of previous slopes and stream courses by the glacier. These *hanging valleys* are the sites of spectacular waterfalls as streams plunge down the steep cliffs (Figure 17–11).

When a glacial trough enters the sea, the glacier can erode even the landscape below sea level. As the glacier retreats, the trough floods and forms a **fjord** in which the sea extends inland, filling the lower reaches of the steep-sided valley. The fjord may be flooded further by a rising sea level or by changes in the elevation of the coastal region. All along the glaciated coast of Alaska, glaciers now are in retreat, thus opening many new fjords that previously were blocked by ice. Coastlines with notable fjords include Norway, Greenland, Chile, the South Island of New Zealand, Alaska, and British Columbia (Figure 17–12).

Depositional Landforms Created by Alpine Glaciation

You have just seen how glaciers excavate tremendous amounts of material and create fascinating landforms in the process. Glaciers also produce a whole other set of landforms when they melt and deposit their eroded and transported debris cargo, which is carried toward the *terminus* (end) of the glacier. Where the glacier melts, this debris accumulates to mark the former margins of the glacier—the end and sides. *Glacial drift* is the general term for all glacial deposits. Direct deposits appear unstratified and unsorted and are called *till*.

Glacial till moving downstream in a glacier can form a marginal unsorted deposit known as a **moraine**—Figure 17–13. A *lateral moraine* forms along each side of a glacier. If two glaciers with lateral moraines join, their point of contact becomes a *medial moraine*. Eroded debris that is dropped at the glacier's farthest extent is called a *terminal*

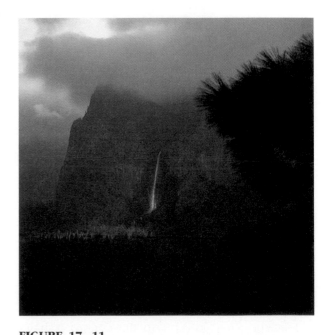

FIGURE 17–11
A hanging valley and waterfall, part of the scenic beauty of Yosemite National Park with its glaciated stream valleys. Photo by author.

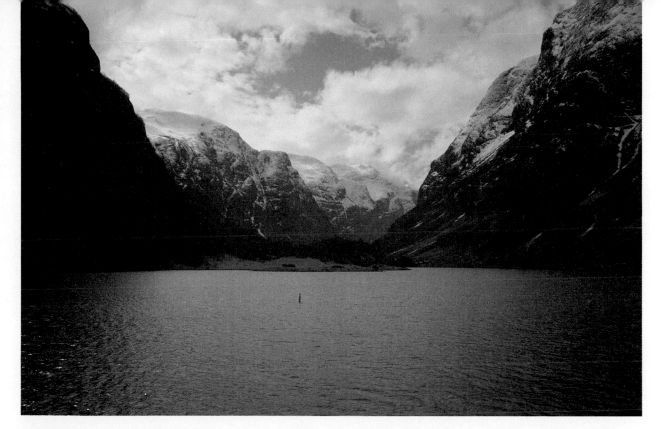

FIGURE 17–12
A typical fjord on the coast of Norway, near Bergen. Photo by Gael Summer-Hebdon.

moraine. However, there also may be *end moraines,* formed wherever a glacier pauses after reaching a new equilibrium. If a glacier is in retreat, individual deposits are called *recessional moraines*. And finally, a deposition of till generally spread across a surface is called a *ground moraine*. Lakes may form behind terminal and end moraines following a glacier's retreat, with the moraine acting as an earthen dam.

In contrast, sorted and stratified glacial drift, characteristic of stream-deposited material, is called *outwash* and forms an *outwash plain* of glacio-fluvial deposits across the landscape. The outwash plain downvalley from Peyto Glacier in British Columbia continues into Peyto Lake as a *valley train deposit* below the *terminal moraine*—Figure 17–14(a). Distributary stream channels appear braided across its surface. The picture also shows the milky meltwater associated with glaciers, charged with finely ground rock ("rock flour") that becomes milky when mixed with water. Meltwater is produced by glaciers at all times, not just when they are retreating.

FIGURE 17–13
Morainal deposits in Jasper Park, Canada. Photo by Stephen Cunha.

(a)

(b)

FIGURE 17–14
(a) Peyto Glacier in Alberta, Canada, valley train, and glacial meltwater; (b) glacial erratics in Maine. (a) Photo by author; (b) photo by Chuck Breitsprecher.

Retreating glaciers also leave behind large rocks (sometimes house-sized) and smaller boulders that are different in composition and origin from the ground on which they have been deposited. These **erratics** were an early clue to researchers that blankets of ice had once covered the land—Figure 17–14(b). Louis Agassiz (1807–1873), a professor of natural history and the author of two books on glaciers, was an important early proponent of a continental glaciation hypothesis as the explanation for many of Europe's erosional and depositional features. Today his grave is marked by one of the clues—appropriately, a polished and striated erratic boulder.

Erosional and Depositional Features of Continental Glaciation

The extent of continental glaciation in North America and Europe is portrayed in Figure 17–19. When these huge sheets of ice advanced and retreated, they produced many of the erosional and depositional features characteristic of alpine glaciation. However, because continental glaciers formed under different circumstances, the intricately carved alpine features, lateral moraines, and medial moraines all are lacking in continental glaciation. Table 17–1 compares the erosional and depositional features of the two types of glaciers.

With the retreat of the glaciers, many relatively flat plains of unsorted coarse till were formed behind terminal moraines. Low, rolling relief and deranged drainage patterns are characteristic of these **till plains**. Beyond the morainal deposits, glacio-fluvial outwash plains of sorted materials feature stream channels that are meltwater-fed, braided, and overloaded with debris. Figure 17–15 illustrates some of the most common erosional and depositional features associated with the passage of a continental glacier.

A sinuously curving, narrow ridge of coarse sand and gravel is called an **esker**. It forms along the channel of a meltwater stream that flows beneath a glacier, or in an ice tunnel, or between ice walls

TABLE 17–1
Valley- and continental-glaciation features compared.

Features	Alpine (Valley) Glacier	Continental Glaciation
Striations, polish, etc.	Common	Common
Cirques	Common	Absent
Horns, arêtes, cols	Common	Absent
U-shaped valleys, truncated spurs, hanging valleys	Common	Rare
Fjords	Common	Absent
Till	Common	Common
Terminal moraines	Common	Common
Recessional moraines	Common	Common
Ground moraines	Common	Common
Lateral moraines	Common	Absent
Medial moraines	Common, easily destroyed	Absent
Drumlins	Rare or absent	Locally common
Erratics	Common	Common
Stratified drift	Common	Common
Kettles	Common	Common
Eskers, crevasse fillings	Rare	Common
Kames	Common	Common
Kame terraces	Common	Present in hilly country

SOURCE: Adapted from L. Don Leet, Sheldon Judson, and Marvin Kauffman. *Physical Geology*, 5th ed., © 1978, p. 317. Reprinted by permission of Prentice-Hall, Inc., Englewood Cliffs, NJ.

FIGURE 17–15
Common depositional landforms produced by glaciers.

beneath the glacier. As a glacier retreats, the steep-sided esker is left behind in a pattern roughly parallel to the path of the glacier. The ridge may not be continuous and at times may even appear branched along the course set by the subglacial watercourse.

Another feature of the outwash plain in Figure 17–15 is a **kame**, a small hill, knob, or mound of poorly sorted sand and gravel that is deposited directly by water, or by ice in crevasses, or from ice-caused indentations in the surface. Kames can also be found in deltaic forms and in terraces along the valley walls.

An isolated large mass of ice that persists after a glacier retreats will leave a pit behind as it melts. The steep-sided basin or pit that remains is called a *kettle*. To understand how a kettle forms, imagine an isolated block of ice more than a kilometer in diameter left on an outwash plain or valley floor. Perhaps 20 to 30 years would be required for it to melt. In the interim, material would continue to

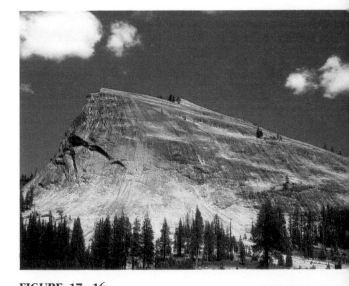

FIGURE 17–16
Roche moutonnée, as exemplified by Lambert Dome in the Tuolumne Meadows area of Yosemite National Park. Photo by author.

FIGURE 17–17

Topographic map of Williamson, New York, featuring numerous drumlins (7.5-minute series quadrangle map of 1:24,000 scale, 10 ft contour interval). Courtesy of U.S. Geological Survey.

accumulate around the melting ice block, finally leaving behind a steep-sided indentation or hole vacated by the melted ice. Such a feature then frequently fills with water. Thoreau's Walden Pond, mentioned in the quotation in Chapter 7, is such a kettle.

Two other landforms created by glacial action are streamlined hills, one erosional (called a roche moutonnée) and the other depositional (called a drumlin). A **roche moutonnée** is an asymmetrical hill of exposed bedrock. Its gently sloping upstream side (stoss side) has been polished smooth by glacial action, whereas its downstream side (lee side) is abrupt and steep where rock was plucked by the glacier (Figure 17–16).

A **drumlin** is deposited till that has been streamlined in the direction of continental ice movement, blunt end upstream and tapered end downstream.

Multiple drumlins (called *swarms*) occur in fields in New York and Wisconsin, among other areas. Sometimes their shape is that of an elongated teaspoon bowl, lying face down. They attain lengths of 100 to 5000 m (330 ft to 3.1 mi) and heights up to 200 m (650 ft). Figure 17–17 shows a portion of a topographic map for the area south of Williamson, New York, which experienced continental glaciation. Study the map; can you identify the numerous drumlins? In what direction do you think the continental glaciers moved across this region?

THE PLEISTOCENE ICE AGE EPOCH

The most recent episode of cold climatic conditions began about 1.65 million years ago, launching the Pleistocene epoch. At the height of the Pleistocene, ice sheets and glaciers covered 30% of Earth's land area, amounting to more than 45 million km² (17.4 million mi²). The Pleistocene is thought to have been one of the more prolonged cold periods in Earth's

history. At least 18 expansions of ice occurred over Europe and North America, each obliterating and confusing the evidence from the one before. The term **ice age** is applied to any such extended period of cold, even though an ice age is not a single cold spell. Instead, it is a period of generally cold climate, called a *glacial,* interrupted by brief warm spells, known as *interglacials.* Each glacial and interglacial is given a name that is usually based on the location where evidence of the episode is prominent, for example "Wisconsinan glacial."

Traditionally, four major glacials and three interglacials were acknowledged for the Pleistocene epoch. (The glacials were named the Nebraskan, Kansan, Illinoian, and Wisconsinan.) In Europe, similar episodes coincided with those in North America, but were given different names. Now, modern techniques have opened the way for a new chronology and understanding. Currently, glaciologists acknowledge the Illinoian glacial and Wisconsinan glacial, and the Sangamon interglacial between them. These span the past 300,000 years (Figure 17–18). The Illinoian is believed to have had two glacials (designated stages 6 and 8 in the figure), as did part of the Wisconsinan (stages 2 and 4), which is dated at 10,000 to 35,000 years ago. The stages on the chart are numbered back to stage 23 at approximately 900,000 years ago.

The continental ice sheets over Canada, the United States, Europe, and Asia are illustrated on the polar map projection in Figure 17–19. In North America, the Ohio and Missouri river systems mark the southern terminus of continuous ice. The edge of an ice sheet is not even; instead it expands and retreats in ice lobes. Such lobes were positioned where the Great Lakes are today and contributed to the formation of these lakes when the ice retreated from their gouged basins.

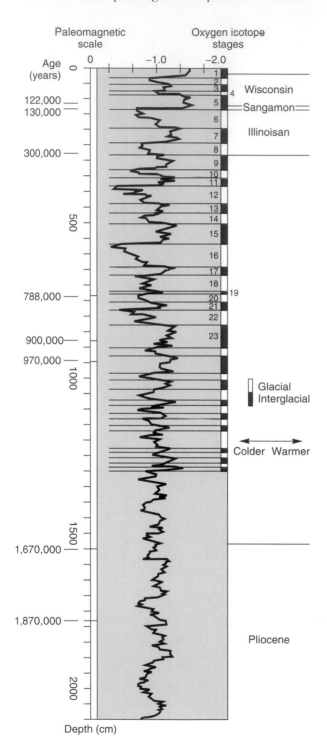

FIGURE 17–18 (left)
Pleistocene temperatures, determined by oxygen isotope fluctuations in planktonic foraminifera from deep-sea cores. Twenty-three stages cover 900,000 years, with names assigned for the past 300,000 years. After N. J. Shackelton and N. D. Opdyke, *Oxygen-Isotope and Paleomagnetic Stratigraphy of Pacific Core V28–239, Late Pliocene to Latest Pleistocene,* Geological Society of America Memoir 145. Copyright © 1976 by the Geological Society of America. Adapted by permission.

FIGURE 17–19

Extent of Pleistocene glaciation
in the Northern Hemisphere
18,000 years ago viewed from a
polar perspective. From Andrew
McIntyre, *CLIMAP Project,*
Lamont-Dougherty Geological
Observatory. Copyright © 1981
by the Geological Society of
America. Reprinted by
permission.

Scientists worldwide participated in the ten-year
long CLIMAP project (*C*limate: *L*ong-Range *I*nvestiga-
tion, *M*apping, *a*nd *P*rediction). Using oxygen isotope ratios from sea-floor core samples, they constructed sea-surface temperature maps for a time
18,000 years ago, when the last major extent of
Pleistocene glacial ice prevailed. Sea levels at that
time were approximately 100 m (330 ft) lower than
today because so much of Earth's water was frozen
and tied up in the glaciers. In fact, sea ice extended
southward into the North Atlantic and Pacific, and
northward in the Southern Hemisphere about 50%
farther than it does today. Sea-surface temperatures
during February were 1.4C° (2.5F°) lower than
today's global average, and in August were 1.7C°
(3.1F°) lower. Figure 17–20, prepared by the CLIMAP
group, depicts estimated conditions 18,000 years
ago. It presents average ocean temperatures and

shows the extent of continental glaciers and sea ice.
The extended coastlines of 18,000 years ago also are
illustrated and reflect the reduced sea level at that
time.

As these glaciers retreated, they exposed a drastically altered landscape: the rocky soils of New
England, the polished and scarred surfaces of Canada's Maritime Provinces, the sharp crests of the
Sawtooth Range and Tetons of Idaho and Wyoming,
the scenery of the Canadian Rockies and the Sierra
Nevada, the Great Lakes of the United States and
Canada, the fjords of New Zealand, Norway, and
Chile, the Matterhorn of Switzerland, and much
more. Study of these glaciated landscapes is important, so that we can better understand paleoclimatology (past climates) and discover the mechanisms that
produce ice ages and climatic change. Focus Study
17–1 elaborates on this.

FIGURE 17–20
Ocean temperatures and the location of glaciers 18,000 years ago. From Andrew McIntyre, *CLIMAP Project,* Lamont-Dougherty Geological Observatory. Copyright © 1981 by the Geological Society of America. Reprinted by permission.

Pluvial Lakes

The term **pluvial** refers to past periods of greater moisture availability or moisture efficiency than we experience today, principally during the Pleistocene epoch. During these periods, lake levels increased in arid and seasonally arid regions. The drier periods between these significantly moist climatic times are called *interpluvial*. Interpluvials are marked with **lacustrine deposits** of lake sediments that form terraces along former shorelines, or with deltas from tributary streams, or with wave-cut cliffs and beaches.

Earlier researchers were tempted to correlate pluvial and glacial ages, given their coincidence during the Pleistocene. However, few sites actually demonstrate such a simple relationship. For example, in the western United States, the estimated ice melt amounts to only a small portion of the water volume in pluvial lakes of the region, and these lakes tend to predate glacial times and correlate instead with wetter periods. Evidence of pluvial lakes exists in North and South America, Africa, Asia, and Australia, although much remains to be discovered about these lakes.

The largest inland sea in the world is the Caspian Sea in southern USSR, with a present surface level about 30 m (100 ft) below mean world sea level. The

FOCUS STUDY 17–1

Deciphering Past Climates: Paleoclimatology

The Vikings were favored by a medieval warming episode as they sailed out into the less-frozen North Atlantic to settle Iceland and Greenland between A.D. 900 and 1200, and inadvertently ventured onto the North American continent as well. However, between the years 1250 and 1740, a "little ice age" took place, freezing many key mountain passes of Europe under deep ice, and lowering snowlines about 200 m (650 ft). Evidence that Earth's climate has fluctuated in and out of warm and cold ages now is being traced back through ice cores, layered deposits of silts and clays, changing paleomagnetism, the geologic sequences locked in the rocks, and radioactive dating methods. And one interesting fact is emerging from these studies: humans (*Homo erectus* and *Homo sapiens* of the last 1.9 million years) have never experienced Earth's *normal* climate, that is, the climate most characteristic of Earth's 4.6 billion-year span.

Apparently, Earth's climates were quite moderate until the last 1.2 billion years, when cyclic temperature patterns of 200 to 300 million years became more pronounced. The most recent cold episode was the Pleistocene epoch, which began in earnest 1.65 million years ago, and through which we may still be progressing. The Holocene epoch began 10,000 years ago and may either represent an end to the Pleistocene, or merely an interglacial (Figure 1). Several episodes of less severe cold have occurred during the Holocene.

The mechanisms that bring on an ice age are the subject of much research and debate. Because past occurrences of low temperatures appear to have followed a pattern, researchers have looked for causes that also are cyclic in nature. They have identified a complicated mix of interacting variables that appear to influence long-term climatic trends.

Galactic and Earth-Sun Relationships

As our Solar System revolves around the distant center of the Milky Way, it crosses the plane of the galaxy approximately every 32 million years. At that time Earth's plane of the ecliptic aligns parallel to the galaxy's plane, and we pass through periods of increased interstellar dust and gas, which may have some climatic effect.

Other possible astronomical factors were proposed by Milutin Milankovitch (1879–1954), a Yugoslavian astronomer who compiled extensive formulations about Earth-Sun orbital relationships. Milankovitch

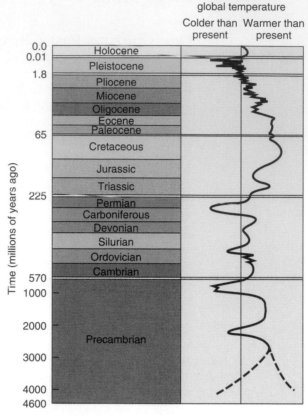

Estimated mean
global temperature

Colder than | Warmer than
present | present

FIGURE 1

A generalization of Earth's surface temperatures over
geologic time. After L. A. Frakes, *Climates Throughout
Geologic Time,* Amsterdam: Elsevier, 1979. Adapted by
permission.

wondered whether the development of an ice age was related to seasonal
astronomical factors—Earth's revolution, rotation, and tilt—extended over
a longer time span. Earth's elliptical orbit about the Sun varies by more than
17.7 million km (11 million mi) during a 100,000-year cycle, stretching out
to an extreme ellipse—Figure 2(a). In addition, Earth's axis "wobbles"
through a 26,000-year cycle, in a movement much like that of a spinning top
winding down. Earth's slowly wobbling axis inscribes an inverted cone, a
movement called *precession*—Figure 2(b).

Finally, Earth's present axial tilt of 23.5° varies from 22° to 24° during a
40,000-year period—Figure 2(c). Milankovitch calculated, without the aid of

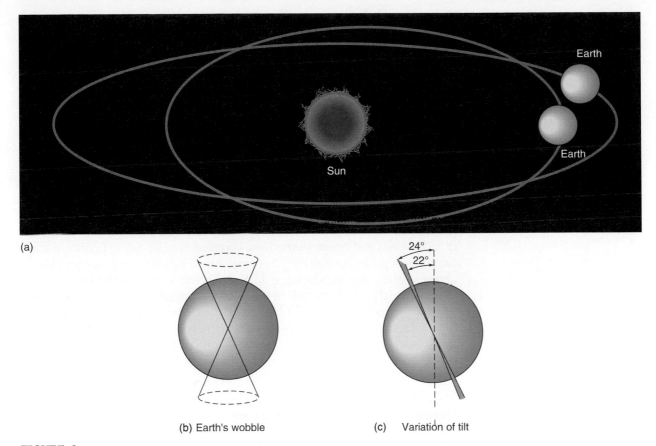

(a)

(b) Earth's wobble

(c) Variation of tilt

FIGURE 2
Astronomical factors that possibly affect broad climatic cycles. (a) Earth's elliptical orbit varies widely during a 100,000-year cycle, stretching out to an extreme ellipse. (b) Earth's 26,000-year wobble. (c) Variation in Earth's axial tilt every 40,000 years.

today's computers, that the interaction of these Earth-Sun relationships creates a 96,000-year climatic cycle in keeping with Earth's orbital eccentricities. His model assumes that changes in astronomical relationships affect the amounts of insolation received.

Milankovitch died in 1954, his ideas still not accepted by a skeptical scientific community. Now, in the era of modern computers, remote sensing satellites, and worldwide efforts to decipher past climates, Milankovitch's work is widely accepted as an important explanation of broad climatic cycles.

Solar Variability

If the Sun significantly varies its output over the years, as some other stars do, that change would seem a convenient and plausible cause of ice-age timing. However, the lack of evidence that the Sun behaves in a significantly variable manner over long cycles argues against this hypothesis. Nonetheless, inquiry about the Sun's variability actively continues. We do know that fluctuations in solar wind have some effect on Earth's weather. Analyzing the effects of the solar maximum of 1990 and early 1991 may further our understanding in this area.

Geophysical Factors

Major glaciations also can be associated with the migration of landmasses to higher latitudes. Chapters 11 and 12 explain that the shape and orientation of landmasses and ocean basins have changed greatly during Earth's history. Continental plates have drifted from equatorial locations to polar regions and vice versa, thus producing cold and warm episodes on the land. Gondwana (the southern half of Pangaea) experienced extensive glaciation that left its mark on the rocks of parts of Africa, South America, India, Antarctica, and Australia. Landforms in the Sahara, for example, bear the markings of even earlier glacial activity, partly explained by the fact that portions of Africa were centered near the South Pole during the Ordovician period (450 million years ago).

Geographical-Geological Factors

These factors include the terrestrial sources of atmospheric variables, increases in relief associated with mountain building, changes in sea level, and alterations of oceanic circulation patterns. One such terrestrial source, a volcanic eruption, might produce lower summer temperatures for a year or two, leading to a buildup of permanent snow cover at high latitudes. Those high-albedo surfaces would then reflect more insolation away from Earth, to further enhance cooling. The climatic effects of the volcanic eruptions in Japan (Mount Unzen) and the Philippines (Mount Pinatubo) in the summer of 1991 are being closely studied for any possible climatic effects.

Furthermore, episodes of mountain building over the past billion years have forced mountain summits above the snowline, where snow has collected beyond the summer melt. Mountain chains can affect downwind weather patterns and jet stream circulation, which in turn guides weather systems. In addition, oceanic circulation patterns have changed; for example, the Isthmus of Panama formed about 3 million years ago and effectively separated the circulation of the Atlantic and Pacific oceans. Such changes in

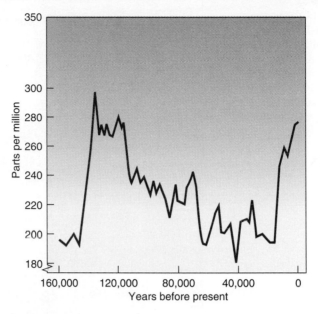

FIGURE 3

Previous CO_2 levels revealed in air bubbles trapped in Antarctic ice removed from the Vostok ice core. Adapted by permission from *Nature* 329, p. 408. Copyright © 1988 Macmillan Magazines Ltd.

circulation and ocean basin configuration affect air mass formation and temperature patterns.

Also linked to a terrestrial source is the fluctuation of atmospheric greenhouse gases, triggering higher or lower temperatures. An ice core taken at Vostok, the Soviet research station near the geographic center of Antarctica, contained trapped air samples from approximately 160,000 years ago and showed carbon dioxide levels varying from more than 290 ppm to a low of nearly 180 ppm (Figure 3). Higher levels of CO_2 are generally thought to correlate with each interglacial. During the 1990s, CO_2 will reach 360 ppm, or higher than at any time in the past 160,000 years, principally due to anthropogenic sources.

Thus, Earth's climate is unfolding as a multicyclic system controlled by an interacting set of cooling and warming processes, all founded on galactic and Earth-Sun relationships. Within this complex of change, scientists are attempting to decipher past climatic rhythms and present and future climatic trends.

FIGURE 17–21

Pluvial lakes of the western
United States. Lake Lahontan
and Lake Bonneville are the
largest. After R. F. Flint, *Glacial
and Pleistocene Geology*.
Copyright © 1957 by John Wiley
& Sons. Adapted by permission
of John Wiley & Sons, Inc.

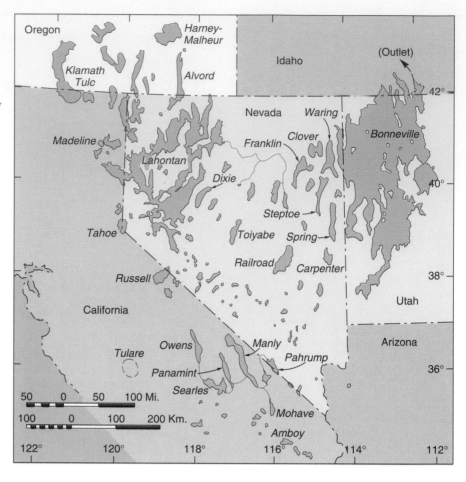

Caspian Sea has visible pluvial shorelines about 80 m
(265 ft) above today's lake level. In North America,
two of the largest pluvial lakes were in the Basin and
Range Province. They attained their greatest extent
15,000 to 25,000 years ago. Since then, climatic shifts
have been too small to be formally identified as
another pluvial cycle. Figure 17–21 portrays these
two lakes, the largest pluvials in the United States—
Lake Bonneville and Lake Lahontan—and other
pluvial lakes at their highest levels. Pluvial lakes also
occurred in Mexico, with one even larger than Lake
Bonneville.

The Great Salt Lake, near Salt Lake City, Utah, and
the Bonneville Salt Flats in western Utah are significant remnants of Lake Bonneville, which covered
more than 50,000 km² (19,500 mi²) in the greatest of
its three expansions and reached depths of 300 m

(985 ft), spilling over into the Snake River drainage
area to the north.

Although not specifically qualifying as a pluvial
stage, the Great Salt Lake experienced a dynamic
trend change during the 1980s. The lake level began
rising in 1981, following periods of increased rainfall
and snowpack input from the adjoining Wasatch
Mountains. Definite shifts in storm tracks which
favored increased precipitation were a feature of the
1982–1986 period. Evaporation is the lake's only
output in the closed Salt Lake basin system. By the
mid-1980s, the lake achieved its highest level in
historic times, flooding an interstate highway, trans-
continental railroad tracks, resorts, and thousands of
acres of grazing land. Fortunately, the two summers
following the highest lake level had high evaporation
rates and precipitation dropped to 50% below

average. Predictions indicate that lake levels will again rise to even greater record depths in the early 1990s. These ancient pluvial lakes associated with past glacial times remain today as climatic reminders of environmental change.

ARCTIC AND ANTARCTIC REGIONS

Climatologists use environmental criteria to define the Arctic and the Antarctic regions (Figure 17–22). For the Arctic area, the 10°C (50°F) isotherm for July is used; it coincides with the visible treeline, which is the boundary between the northern forests and tundra climates. The Arctic Ocean is a sea covered by floating sea ice (frozen seawater), as well as by glacier ice (frozen freshwater) in icebergs that formed on the surrounding land. This ice pack thins in the summer months and sometimes breaks up.

The criterion used for Antarctica is the *antarctic convergence*, a narrow zone that extends around the continent as a boundary between colder antarctic water and warmer subantarctic water. This boundary follows roughly the 10°C (50°F) February isotherm and is located near 60° S latitude. The Antarctic region that is covered with sea ice represents an area greater than North America, Greenland, and Western Europe combined.

The Antarctic Ice Sheet

In contrast to the Arctic, Antarctica is much colder overall and is underlain by a continent-sized landmass instead of a sea. In simplest terms, Antarctica can be thought of as a single enormous glacier, although it contains distinct regions such as the East Antarctic and West Antarctic ice sheets, which respond differently to slight climatic variations. (That portion of the ice sheet located in West Antarctica and discussed in Chapter 10 can be identified on the illustration.) The edges of the ice sheet that enter bays along the coast form extensive **ice shelves**, with sharp ice cliffs rising up to 30 m (100 ft) above the sea. Large tabular islands of ice are formed when sections of the shelves break off and move out to sea. These ice islands can be very large; one in the late 1980s exceeded the size of Rhode Island.

Individual glaciers operate around the periphery of the ice sheet in Antarctica. The Byrd Glacier moves

(a)

(b)

FIGURE 17–22
The Arctic and Antarctic regions. From Central Intelligence Agency *Atlas of Polar Regions.*

through the Transantarctic Mountains, draining an area of more than 1,000,000 km² (386,000 mi²) as it flows 180 km (112 mi) and drops 5000 m (3.1 mi) in elevation. This glacier exceeds 22 km in width (13.7 mi) at its narrowest point and, where its thickness is 1000 m (3280 ft), it maintains a flow rate of approximately 840 m (2755 ft) per year, as fast as some surging glaciers. Numerous outlet glaciers operate around the periphery of Antarctica.

SUMMARY

More than 77% of the freshwater on Earth is frozen. Ice presently covers about 11% of Earth's surface. Alpine glaciers occupy high-elevation and high-latitude sites in mountains. Extensive valley glaciers have shaped mountains worldwide. Larger masses of ice cover Greenland and Antarctica as ice sheets. A glacier is an open system with inputs and outputs that can be analyzed through observation of the expansion and wasting of the glacier itself. The passage of ice over a landscape excavates and creates profound changes, producing many erosional and depositional landforms.

Modern active glaciers and the present physical landscape as molded by past glaciers provide evidence for the study of past climates. During the Pleistocene epoch, extensive continental ice covered portions of the United States, Canada, Europe, and the Soviet Union. Glaciers worldwide are being watched more closely than ever because of the direct link they have to changing trends in world climates and the potential increase in mean sea level that can be brought on by the melting of even small amounts of stored ice. Antarctica remains as one of Earth's last great frontiers, in scientific terms.

REVIEW QUESTIONS

1. Where is most of the freshwater on Earth today? Describe.
2. What is a glacier? What is implied about existing climate patterns in a glacial region?
3. Differentiate between an alpine glacier and a continental glacier.
4. Name the three types of continental glaciers. What is the basis for dividing continental glaciers into types? What type is Antarctica?
5. How is an iceberg generated?
6. What is meant by glacial mass balance? What are the basic inputs and outputs underlying that balance?
7. Trace the evolution of glacial ice from fresh fallen snow.
8. What is meant by a glacial surge? What do scientists think produces surging episodes?
9. How does a glacier accomplish erosion?
10. Describe the evolution of a V-shaped stream valley to a U-shaped glaciated valley. What kinds of features are visible after a glacier has retreated?
11. Differentiate between two forms of glacial drift—till and outwash.
12. What is a morainal deposit? What specific forms of moraines are created by alpine and continental glaciers?
13. What are some common depositional features encountered in a till plain?
14. Compare a roche moutonnée and a drumlin.
15. What is paleoclimatology? Describe Earth's past climatic patterns. Are we experiencing a normal climate pattern in this era or have scientists noticed any significant trends?
16. Define an ice age. When was the last one on Earth? How is the modern age classified? Explain "glacial" and "interglacial" in your answer.
17. Give an overview of what science has learned about the various causes of ice ages by listing and explaining at least four possible climate change factors.
18. What did the CLIMAP project determine and plot on the map shown in Figure 17–20?

19. What criteria define the Arctic and Antarctic regions? Is there any coincidence in these criteria and the distribution of Northern Hemisphere forests on the continents?

SUGGESTED READINGS

"The Antarctic." *Scientific American* 207, no. 3 (September 1962): Entire issue.

Bailey, Ronald H., and the editors of Time-Life Books. *Glacier*. Planet Earth Series. Alexandria, VA: Time-Life Books, 1982.

Budyko, M. I. *The Earth's Climate: Past and Future*. International Geophysics Series, vol. 29. New York: Academic Press, 1982.

Chorlton, Windsor, and the editors of Time-Life Books. *Ice Ages*. Planet Earth Series. Alexandria, VA: Time-Life Books, 1983.

Central Intelligence Agency. *Atlas of Polar Regions*. Washington: CIA—National Foreign Assessment Center, 1978.

CLIMAP Project Members. "Seasonal Reconstructions of the Earth's Surface at the Last Glacial Maximum." CLIMAP Project, compiled by Andrew McIntyre. Geological Society of America Map and Chart Series MC–36. Palisades, NY: 1981.

———. "The Surface of the Ice-Age Earth." *Science* 191, No. 4232 (19 March 1976): 1131–37.

Embleton, C., and C. A. M. King. *Glacial Geomorphology*. London: Edward Arnold, 1975.

Estes, John E., ed. *Manual of Remote Sensing*. Vol. 2. *Interpretation and Applications*. Falls Church, VA: American Society of Photogrammetry, 1983, 1762–1910.

Flint, R. F. *Glacial and Quaternary Geology*. New York: John Wiley & Sons, 1971.

Griffiths, John F., and Dennis M. Driscoll. *Survey of Climatology*. Columbus: Macmillan, 1982.

Kamb, Barclay, C. F. Raymond, W. D. Harrison, Herman Englehardt, K. A. Echelmeyer, N. Humphrey, M. M. Brugman, and T. Pfeffer. "Glacier Surge Mechanism: 1982–1983 Surge of Variegated Glacier, Alaska." *Science* 227, No. 4686 (1 February 1985): 469–79.

Lamb, H. H. *Climate: Present, Past and Future*. Vol. 2. *Climate History and the Future*. London: Methuen, 1977.

Ley, Wiley, and the editors of Time-Life Books. *The Poles*. Life Nature Library. New York: Time, 1962.

Oliver, John E., and Rhodes W. Fairbridge, eds. *Encyclopedia of Earth Sciences*. Vol. 11. *The Encyclopedia of Climatology*. New York: Van Nostrand Reinhold, 1987.

Radok, Uwe. "The Antarctic Ice." *Scientific American* 253, no. 2 (August 1985): 98–105.

Short, Nicholas M., and Robert W. Blair, Jr., eds. *Geomorphology from Space—A Global Overview of Regional Landforms*. Washington: National Aeronautics and Space Administration, 1986.

Stanley, Steven M. *Earth and Life Through Time*. 2d ed. New York: W. H. Freeman, 1989.

Swithinbank, Charles. *Antarctica*. U.S. Geological Survey Professional Paper 1386–B. Washington: Government Printing Office, 1988.

Earth is the home of the only known biosphere in the Solar System—a unique, complex, and interactive system of abiotic and biotic components working together to sustain a tremendous diversity of life. Energy enters the system through conversion of solar energy by photosynthesis in the leaves of plants. Soil is the essential interface among the lithosphere, plants, and the rest of Earth's physical systems. Thus, soil helps sustain life. Life is organized into a feeding (trophic) hierarchy from producers to consumers, ending with decomposers. Taken together, the soils, plants, animals, and all abiotic components produce functioning aquatic and terrestrial ecosystems, known as biomes. Specific questions facing society today relate to preservation of the diversity of life in the biosphere and the survival of the biosphere itself—important applied topics considered in Part 4.

PART FOUR

Soils, Ecosystems, and Biomes

Photo by author.

Soils of Kansas. Photo by author.

18

The Geography of Soils

Earth's landscape is generally covered with **soil**. Soil is a dynamic natural body comprised of fine materials in which plants grow, and which is composed of both mineral and organic matter. If you have ever planted a garden, tended a house plant, or been concerned about famine and soil loss, this chapter will interest you. A knowledge of soil is at the heart of agriculture and food production.

Particular climatic regimes produce specific soil-forming processes which contribute to soil development. Physical and chemical weathering in the upper lithosphere provides the raw mineral ingredients for soil formation. These are called *parent materials,* and their composition, texture, and chemical nature are important. Soils subsequently acquire physical properties that can distinguish them from their source materials as they develop through time, and in response to local relief and topography. Air, water, nutrients, and solid materials all cycle through soil and provide a critical sustaining structure for plants, animals, and human life.

Specific soil conditions determine *soil fertility,* which is the ability of soil to support plant productivity. Soil can do this because it contains organic substances and clay minerals that absorb water and certain elemental ions needed by plants. Much effort and many dollars are expended to create fertile soil conditions, yet we live in an era when the future of Earth's most fertile soils is threatened. Soil erosion is increasing worldwide. Some 35% of farmlands are losing soil faster than it can form—a loss exceeding 22.75 billion metric tons per year (25 billion tons). Increases in production resulting from artificial fertilizers and new crop designs partially mask this effect, but such compensations for soil loss are nearing an end. Soil depletion and loss are at record levels from Iowa to China, Peru to Ethiopia, the Middle East to the Americas. The impact on society could be significant.

Soil science is interdisciplinary, involving principles of physics, chemistry, biology, mineralogy, hydrology, taxonomy, climatology, and cartography. Physical geographers are interested in spatial patterns formed by soil types and the physical factors that interact to produce them. As an integrative science, physical geography is well suited for this task.

People most often are interested in the characteristics and fertility of soils where they live. In many locales, an agricultural extension service can provide this information and perform a detailed soil analysis. Soil survey analyses and local soil maps also are available for most counties in the United States and the Canadian provinces. (Contact the U.S. Department of Agriculture, Agriculture Canada's Canada Soil Survey, or the appropriate department at a local college or university.) An important reality is that soil science must deal with a substance that varies from kilometer to kilometer, and even centimeter to centimeter.

SOIL CHARACTERISTICS

Simple theoretical models are the basis for soil classifications. The following model represents general assumptions about soil structure, soil properties, and soil formation processes.

Soil Profiles

Soils should not be studied at the surface level only, for this can produce a misleading picture. Instead, a soil profile selected for study should extend from the surface to the lowest extent of plant roots, or to the point where regolith or bedrock is encountered. Such a profile, known as a **pedon**, is imagined as a hexagonal column encompassing from 1 m^2 to 10 m^2 in surface area (Figure 18–1). At the sides of the pedon, the various layers of the soil profile are visible in cross section.

A pedon is the basic sampling unit in soil surveys. Many pedons together in one area comprise a **polypedon**, which has distinctive characteristics differentiating it from surrounding polypedons. These polypedons are the essential soil individuals, constituting an identifiable series of soils in an area. A polypedon has a minimum dimension of about 1 m^2 and no specified maximum size. It is the soil unit used in preparing local soil maps.

Soil Horizons

Each layer exposed in a pedon is a **soil horizon**. A horizon is roughly parallel to the pedon's surface

FIGURE 18–1
Schematic illustration of a
pedon (sampling unit) as
derived from a polypedon
(mapping unit).

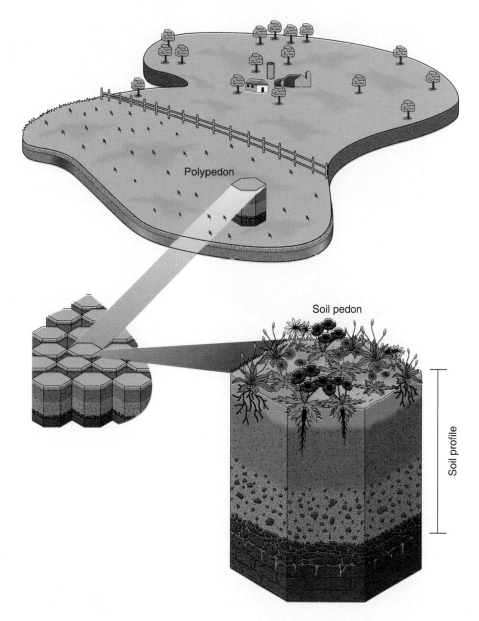

Polypedon

Soil pedon

Soil profile

and has characteristics distinctly different from horizons directly above or below. The boundary between horizons usually is visible in the field, using the properties of color, texture, structure, consistence, porosity, the presence or absence of certain minerals, moisture, and chemical processes. Soil horizons are the building blocks of soil classification. However, because they can vary in an almost endless number of ways, a simple model is useful. Figure 18–2 presents an ideal pedon and soil profile, with letters assigned to each soil horizon and numbers added for further delineation.

At the top of the soil profile is the O horizon, composed of organic material derived from plant and animal litter that was deposited on the surface and transformed into humus. **Humus** is a mixture of decomposed organic materials in the soil and is usually dark in color. Consumer and decomposer microorganisms are busy at work on this organic debris, performing a portion of the humification

FIGURE 18–2

Typical O, A, B, C, and R soil horizons within a developed soil pedon.

(humus-making) process. The O horizon is 20–30% or more organic matter.

At the bottom of the soil profile is the R horizon, representing either unconsolidated material or consolidated bedrock of granite, sandstone, limestone, or other rock. When bedrock weathers to form regolith, it may or may not contribute to overlying soil horizons.

The A, B, and C horizons mark differing mineral strata between O and R; these middle layers are composed of sand, silt, clay, and other weathered by-products. (In Chapter 11, Table 11–3 presents a description of grain sizes of these weathered soil particles.)

In the A horizon, the presence of humus and clay particles is particularly important, for they provide essential chemical links between soil nutrients and plants. An A1 horizon usually is darker and richer in organic content than lower horizons; the A2 is a bit lighter or paler and is made up of coarse sand and silts. Clays and oxides of aluminum and iron are leached (removed) from the A2 horizon and are translocated to lower horizons by internal washing in the soil. This process of rinsing through upper horizons and removing finer particles and minerals is termed **eluviation**. The greater the precipitation in an area, the higher the rate of eluviation in the A2 horizon. A transition horizon, A3, may or may not be present.

In contrast to the A horizons, B horizons demonstrate an accumulation of clays, aluminum, iron, and possibly humus. These horizons are dominated by **illuviation**, a depositional process. B horizons may exhibit reddish or yellowish hues because of the illuviated presence of mineral and organic oxides. Some materials occurring in this soil zone also may have formed in place from weathering processes rather than translocation. In the humid tropics, these layers often develop deeply. The combination of the A horizon with its eluviation removals and the B horizon with its illuviation accumulations is designated the **solum**, considered the true soil of the pedon. The A and B horizons are most representative of active soil processes.

The C horizon is weathered bedrock or weathered parent material, excluding the bedrock itself. This zone is identified as *regolith* (although this term sometimes is used to include the solum as well). The C horizon is not much affected by soil operations in the solum and lies outside the biological influences experienced in the shallower horizons. Plant roots and soil microorganisms are rare in the C horizon; it lacks clay concentrations and generally is made up of carbonates, gypsum, or soluble salts, or of iron and silica, which form cemented soil structures. In dry climates, calcium carbonates commonly form the cementing material of these hardened layers.

Soil scientists use an array of letters to further define special conditions within each soil horizon. Of some two dozen conditions, Table 18–1 samples the suffixes that soil scientists add to horizon designations (for example, Ap or Bt). Soil horizons are the basis for soil classification. Let's look at soil properties before we return for more on actual soil horizons.

Soil Properties

Let's examine briefly some essential soil properties. It is helpful if you can observe an actual soil body to help identify some of these soil properties. A good place to see a soil profile is at a current construction site or excavation, perhaps on your campus.

TABLE 18–1

Soil horizon suffixes.

Symbol	Description
b	A buried soil horizon; soils that resulted from some previous set of soil-forming processes and that are now buried beneath another solum
ca	Accumulation of carbonates of alkaline elements, commonly calcium; usually lime ($CaCO_3$) or dolomite ($MgCO_3$) in concentrations greater than that of the parent material; may appear in A, B, or C horizons
f	Frozen soil; applied to layers in the soil that are permanently frozen, as in permafrost
g	Strong gleying; condition usually found in stagnant, poorly aerated water; usually dull gray or blue-gray because of the removal of oxygen from iron compounds; may appear mottled or spotted
h	Accumulations of humus; distinct coating of sand and silt particles with illuviated humus or organic material in the B horizon; appears as a dark coating
ir	Accumulation of iron; distinct coating or cementing of silts and sands with illuviated iron; appears reddish in color
m	Strong cementation; applied to hardened horizons that are more than 90% continuous, such as caliche or desert hardpan; not applicable to the firmness associated with fragipans
p	Plowing or other disturbance; used with the A horizon to designate disturbance by cultivation or pasturage; Ap used even when plowing may have gone into what was once B horizon
sa	Accumulation of soluble salts; similar to ca, in that salts are present in an amount greater than that of parent materials
t	Illuviated clay; accumulations of silicate clays as masses, pore fillings, or coatings on sand and silt within B horizons; transported there from horizons above
x	Fragipan formation; firmness or brittleness found in A2, B, and C horizons

SOURCE: *Soil Taxonomy*, Agricultural Handbook No. 436, U.S. Department of Agriculture, 1975.

Soil Color. Color is important, for it sometimes reflects composition and chemical makeup. If you look at exposed soil in a garden or along a highway roadcut, color may be the most obvious trait. Among the many possible hues are the red and yellow soils of the southeastern United States, which are high in ferric (iron) oxides; the black-prairie, richly organic soils of the Soviet Ukraine and portions of U.S. grain-growing regions; or the white-to-colorless soils attributable to the presence of aluminum oxides and silicates. Color can be deceptive too, for soils of high humus content are often dark, yet clays of warm-temperate and tropical regions with less than 3% organic content may be some of the blackest soils.

To eliminate as much subjectivity as possible and to improve consistency, soil scientists describe a soil's color by comparing it to a *Munsell Color Chart* (from the Munsell Color Company). These charts are in a loose-leaf binder, with 175 different colors arranged by *hue* (the dominant spectral color, like red), *value* (degree of darkness or lightness), and *chroma* (purity and strength of the color saturation, which increases with decreasing grayness). Color is identified by a name and a Munsell notation, and checked at various depths within a pedon.

Soil Texture. Soil texture, perhaps its most permanent attribute, refers to the size and organization of particles in the soil. Individual mineral particles are called *soil separates;* those smaller than 2 mm in diameter (0.08 in.), such as very coarse sand, are considered part of the soil, whereas larger particles are identified as pebbles, gravels, or cobbles.

Figure 18–3 shows a diagram of soil textures with sand, silt, and clay concentrations distributed along each side of a triangle. Each corner of the triangle represents a soil comprised solely of the particle size noted (rarely are true soils composed of a single separate). The figure includes the common designation **loam**, which is a mixture of sand, silt, and clay in almost equal shares. A sandy loam with clay content below 30% usually is considered ideal by farmers because of its water-holding characteristics and cultivation ease.

For example, one soil type found in Indiana is called the Miami silt loam. Samples from it are

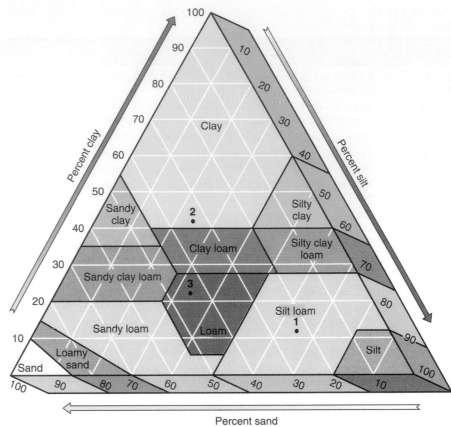

FIGURE 18–3

Soil texture diagram. The points marked 1, 2, and 3 designate samples taken at three different horizons in the Miami silt loam of Indiana. After USDA Soil Conservation Service, *Soil Taxonomy*, Agricultural Handbook No. 436, Washington: Government Printing Office, 1975.

plotted on the soil texture triangle (Figure 18–3). A sample taken near the surface in the A horizon is recorded at point 1; in the B horizon at point 2; and in the C horizon at point 3. Textural analyses of these samples are summarized in Table 18–2; note that silts dominate the surface, clays the B horizon, and sands the C horizon.

The USDA *Soil Survey Manual* presents guidelines for estimating soil texture by feel, a relatively accurate method when used by an experienced individual. However, laboratory methods using graduated sieves and separation by sedimentation in

water allow more precise measurements. Soil texture is important in determining water retention and water transmission traits. Figures 9–5 and 9–6 illustrate the water-holding capacity and availability for various soil textures (field capacity minus wilting point).

Soil Structure. Soil structure refers to the aggregation or arrangement of primary soil particles. Structure is important because it can partially modify or overcome aspects of soil texture. The term *ped* describes an individual unit of soil aggregates; it is a

TABLE 18–2

Textural analysis of Miami silt loam.

Sample Points	% Sand	% Silt	% Clay
1 = A horizon	21.5	63.4	15.0
2 = B horizon	31.1	25.0	43.4
3 = C horizon	42.4	34.0	23.5

SOURCE: Joseph E. Van Riper, *Man's Physical World,* p. 570. Copyright © 1971 by McGraw-Hill. Adapted by permission.

natural lump or cluster, with clay and humus holding the particles together (Figure 18–4). Peds separate from each other along zones of weakness, creating voids that are important for moisture storage and drainage. Spheroidal peds (peds that are rounded) have more pore space and greater permeability. They are therefore more productive for plant growth than are coarse, blocky, prismatic, or platy peds, despite comparable fertility. Terms used to describe soil structure include fine, medium, or coarse, with structural grades of adhesion within aggregates ranging from weak, to moderate, to strong.

Soil Consistence. A corollary to texture and structure is *soil consistence,* which is the cohesion in soil and its resistance to mechanical stress and

Crumb or granular

Platy

Blocky

Prismatic or columnar

FIGURE 18–4
Types of soil structure.

manipulation under varying moisture conditions. Wet soils are variably sticky when held between the thumb and forefinger, ranging from a little adherence to either finger, to sticking to both fingers, to stretching when the fingers are moved apart. *Plasticity,* the quality of being molded, is roughly measured by rolling a piece of soil between your fingers and thumb to see whether it rolls into a thin strand.

Moist soil implies that it is filled to about half of field capacity, and its consistence grades from loose (noncoherent), to *friable* (easily pulverized), to firm (not crushable between thumb and forefinger). Finally, a dry soil is typically brittle and rigid, with consistence ranging from loose, to soft, to hard, to extremely hard.

The cementation that occurs in various soil horizons is a function of consistence and usually is described as continuous or discontinuous. Soils are variously noted as weakly or strongly cemented or *indurated* (hardened). Calcium carbonate, silica, and oxides or salts of iron and aluminum can all serve as cementing agents.

Soil Porosity. Soil porosity, permeability, and soil moisture storage are discussed in Chapter 9. Pores, or voids, in the soil body control the flow of water, its intake and drainage, and air ventilation. In addition to pore size, other factors are important: *continuity* (whether the pores are interconnected), *orientation* (whether tubular pore spaces are vertical, horizontal, or random), *shape* (whether spherical, irregular, or tubular), and *pore location* (whether within or between soil peds). Porosity is increased and improved by the biotic actions of plant roots, animal activity, and human intervention through soil management strategies. When you work in the garden or with house plants, much of your soil preparation work probably will involve improving soil porosity.

Soil Moisture Regimes. Recall that plants operate most efficiently when the soil body is at field capacity, and that field capacity is determined by knowing the particular soil type. The effective rooting depth of a plant determines the amount of soil moisture to which the plant's roots are exposed. As soil moisture is drawn down below field capacity, soil moisture utilization rates become increasingly inefficient until a wilting point is reached, beyond

TABLE 18–3

Principal soil moisture regimes.

Name	Description
Aquic (L. *aqua*, "water")	The groundwater table lies at or near the surface, so the soil is almost constantly wet, as in bogs, marshes, and swamps, a reducing environment with virtually no dissolved oxygen present. Commonly, groundwater levels fluctuate seasonally; a small borehole will produce standing, stagnant water.
Aridic (torric) (L. *aridis*, "dry," and L. *torridus*, "hot and dry")	Soils in this regime are dry more than half the time, with soil temperatures at a depth of 50 cm (19.7 in.) above 5°C (41°F). In some or all areas, soils are never moist for as long as 90 consecutive days. This regime is mainly in arid climates, although where surface structure inhibits infiltration and recharge or where soils are thin and therefore dry, this regime occurs in semiarid regions as well.
Udic (L. *udus*, "humid")	Soils have little or no moisture deficiency throughout the year, specifically during the growing season. The water balance exhibits soil moisture surpluses that flush through the soils during one season of the year. If the water balance exhibits a moisture surplus in all months of the year, the regime is called *perudic*, with adequate soil moisture always available to plants.
Ustic (L. *ustus*, "burnt," implying dryness)	This regime is intermediate between aridic and udic regimes; it includes the semiarid and tropical wet-dry climates. Moisture is available but is limited, with a prolonged deficit period following the period of soil moisture utilization during the early portion of the growing season. Temperature is important in determining the moisture efficiency of available water in this borderline moisture regime.
Xeric (L. *xeros*, "dry")	This regime applies to those few areas that experience a Mediterranean climate, dry and warm summer, rainy and cool winter. There are at least 45 consecutive dry days during the four months following the summer solstice in every 6 of 10 years. Winter rains effectively leach the soils.

SOURCE: *Soil Taxonomy*, Agricultural Handbook No. 436, U.S. Department of Agriculture, 1975.

which plants cannot pump the water bonded to each soil particle.

Soil moisture regimes and their associated climate types shape the biotic and abiotic properties of the soil more than any other factor. Based on Thornthwaite's water-balance principles (Chapter 9), the U.S. Soil Conservation Service recognizes five soil moisture regimes (Table 18–3).

Soil Chemistry

Water present in soil pores is the *soil solution*. It is the medium in the soil for chemical reactions. This solution is critical to plants as their source of nutrients and is the foundation of soil fertility. The atmospheric component of soil is mostly nitrogen, oxygen, and carbon dioxide. Nitrogen concentrations in the soil are about the same as in the atmosphere. Oxygen is lower and carbon dioxide is higher because of ongoing respiration processes. Both carbon dioxide and various organic constituents combine with water to produce carbonic and organic acids, respectively. These acids are then active in soil processes, as are dissolved alkalies and salts.

To understand how the soil solution behaves, let's go through a quick chemistry review. An *ion* is an atom, or group of atoms, that carries an electrical charge. An ion has either a positive charge or a negative charge. For example, when NaCl (sodium chloride) is dissolved in solution, it separates into two ions: Na^+, a *cation* (positively charged ion), and Cl^-, an *anion* (negatively charged ion). Some ions in soil carry single charges, whereas others carry double or even triple charges (e.g., SO_4^{--}, Al^{+++}).

Soil colloids are important for retention of ions in soil. These tiny clay and organic particles carry a negative electrical charge and consequently are attracted to any positively charged ions in the soil (Figure 18–5). The positive ions, many metallic, are critical to plant growth. If it were not for the soil colloids, the positive ions would be leached away by the soil solution and thus would be unavailable to plant roots.

Clay colloids and organic colloids have different levels of chemical activity. Individual clay colloids are

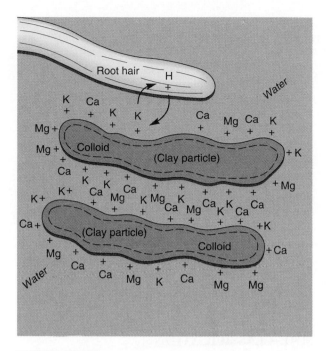

FIGURE 18–5
Typical soil colloids retaining mineral ions by adsorption.

thin and platelike, with parallel surfaces that are negatively charged. They are more chemically active than silt and sand particles but less active than organic colloids. Colloids can exchange cations between their surfaces and the soil solution, a measured ability called **cation-exchange capacity (CEC)**. A high CEC means that the soil colloids can store or exchange more cations from the soil solution, an indication of good soil fertility (unless there is a complicating factor, such as the soil being too acid). Cations attach to the surfaces of the

colloids by **adsorption**, that is, the metallic cations are adsorbed by the soil colloids.

Soil Acidity and Alkalinity. A soil solution may contain a significant quantity of hydrogen ions (H^+), or cations that indirectly stimulate acid formation. These ions will displace critical base cations (such as calcium, magnesium, potassium, and sodium). The result is a soil rich in hydrogen ions: an *acid soil*. On the other hand, a soil high in base cations is a *basic* or *alkaline* soil. Such acidity and alkalinity is expressed on the pH scale (Figure 18–6).

Pure water is nearly *neutral,* with a pH of 7.0. Readings below 7.0 represent increasing acidity. Readings above 7.0 represent increasing alkalinity. Acidity usually is regarded as strong at 5.0 or lower on the pH scale, whereas 10.0 or above is considered strongly alkaline. Today, the major contribution of acidity to soil is through acid deposition by precipitation. Acid rain actually has been measured below pH 3.0—an incredibly low value for natural precipitation. Because most crops are sensitive to specific pH levels, acid soils below pH 6.0 require the addition of bases in the form of minerals that are rich in base cations, usually lime (calcium carbonate). Increased acidity in the soil solution accelerates the chemical weathering of mineral nutrients and increases their depletion rates.

Soil Formation and Management

There are three soil-forming factors: dynamic (climatic and biologic), passive (parent material, topography and relief, and time), and the human component, for it is woven throughout. These three factors work together as a system to form soils. The role of

FIGURE 18–6
Soil pH scale, measuring acidity (lower pH) and alkalinity (higher pH).

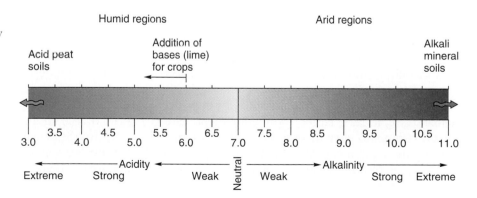

parent material in providing weathered minerals to form soils was discussed earlier in this chapter. The role played by climate is considered in the soil order discussions that follow.

Worldwide, soil types show a close correlation to climate types. The moisture, evaporation, and temperature regimes associated with varying climates determine the chemical reactions, organic activity, and eluviation rates of soils. Not only is the present climate important, but many soils exhibit the imprint of past climates, sometimes over thousands of years. Most notable is the effect of glaciations, which produced the loess soil materials that have been transported thousands of miles to their present locations.

The organic content of soil is determined in part by the vegetation growing in that soil, as well as by animal and bacterial activity. The chemical makeup of the vegetation contributes to the acidity or alkalinity of the soil solution. For example, broadleaf trees tend to increase alkalinity, whereas needleleaf trees tend to produce higher acidity. Thus, when civilization moves into new areas and alters the natural vegetation by logging or plowing, the affected soils are likewise altered and in many cases changed forever.

Landforms also affect soil formation, mainly through slope and orientation. Slopes that are too steep do not have full soil development, but slopes that are slight may inhibit soil drainage. As for orientation, in the Northern Hemisphere a southern slope exposure is warmer (slope faces the southern Sun), which affects water balance relationships. North-facing slopes are colder and the slower snow-melt and lower evaporation rates result in more moisture for plants.

Human intervention has a significant impact on slope utilization. Millennia ago, farmers in every culture on Earth learned to plow slopes "on the contour"—to plow around a slope at the same elevation, not vertically up and down the slope. The reasons behind contour plowing are to reduce sheet erosion (which removes topsoil) and simultaneously to improve water retention in the soil. However, the demands of today's larger farm equipment have forced farmers to alter their practice of contouring because the equipment cannot make the tight turns such proper plowing requires. Other effective soil-holding practices, such as wind breaks, shelter belts, and terracing, also are being eliminated to accommodate the larger equipment.

All of these identified factors require time to operate. A few centimeters thickness of prime farmland soil may require *500 years* for maturation. Yet these same soils are being lost at a few centimeters *per year* to sheetwash and gullying. Over the same period, exposed soils may be completely leached of needed cations, thereby losing their fertility. The U.S. General Accounting Office recently estimated that from 3 to 5 million acres of prime farmland are lost each year through mismanagement or conversion to nonagricultural uses. It is estimated that about half of U.S. cropland is experiencing excessive rates of soil erosion (Figure 18–7). Human influence is ever-present, and our spatial impact on this precious resource must be recognized, with appropriate protective measures taken.

SOIL CLASSIFICATION

Classification of soils is complicated by the variety of interactions that create thousands of distinct soils—about 13,000 soil types in the United States alone. Not surprisingly, a number of different classification systems are in use worldwide. The United States, Canada, the United Kingdom, Germany, Australia, the Soviet Union, and the United Nations Food and Agricultural Organization (FAO) all have their own soil classification systems. In particular, the National Soil Survey Committee of Canada developed a system appropriately suited to its great expanses of boreal forest, tundra, and cooler climatic regimes.

A Brief History of Soil Classification

In 1883 the Russian soil scientist V. V. Dukuchaev published a monograph that organized soils into rough groups based on observable properties, most of which resulted from climatic and biological soil-forming processes. The Russians contributed greatly to modern soil classification because they were first to consider soil an independent natural body with a definite individual soil genesis that was recognizable in an orderly global pattern. By contrast, American and European scientists were still considering soil as a geologic product, or simply a mixture of chemical compounds. The Russians were first to demonstrate broad interrelationships among

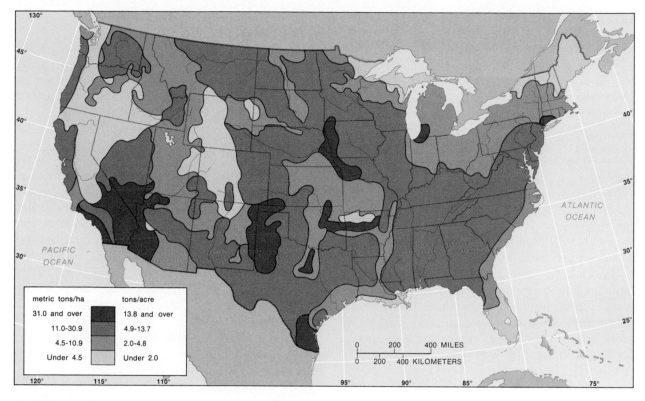

FIGURE 18–7
Soil erosion on cropland in the United States (gully erosion not included). Erosion
rates that exceed 11 metric tons/hectare (4.9 tons/acre) reduce productivity. Adapted
from U.S. Department of Agriculture, Soil Conservation Service.

the physical environment, vegetation, and soils, at a
time when a scientific revolution was beginning to
sweep their country in various fields.

The first formal American system is credited to Dr.
Curtis F. Marbut, who derived his approach from
Dukuchaev. Published in the 1935 *Atlas of American
Agriculture* and the *Yearbook of Agriculture—1938*,
Marbut's classification of great soil groups recognized the importance of soils as products of dynamic
natural processes. However, as the database of soil
information grew, the inadequacies of this system
became apparent. In 1951 the U.S. Soil Conservation
Service began researching a new soil classification
system. That process went through seven exhaustive
reviews, culminating in what is called the "seventh
approximation" in 1960 and 1964. The current

version was published in 1975: *Soil Taxonomy—A
Basic System of Soil Classification for Making and
Interpreting Soil Surveys*, generally called simply **Soil
Taxonomy**. Much of the information in this chapter
is derived from that keystone publication.

The Soil Taxonomy system is based on observable
soil properties actually seen in the field. Thus, it is
open to addition, change, and modification as the
sampling database grows. The classification system
divides soils into six categories, creating a hierarchical sorting system (Table 18–4). Each *soil series* (the
smallest, most detailed category) ideally includes
only one polypedon, but may include portions
continuous with adjoining polypedons in the field.
An important aspect of the system is that it recognizes
the interaction of humans and soils and the changes

TABLE 18–4

Soil taxonomy.

Category	Number Included
Soil orders	10
Suborders	47
Great groups	230
Subgroups	1200
Families	6000
Soil series	13,000

that humans have introduced, both purposely and inadvertently.

Prior to the Soil Taxonomy system, **pedogenic regimes** were used, featuring specific soil-forming processes keyed to climatic regions. Such climate-based regimes are convenient for relating climate and soil processes, and so we discuss them within the related soil orders. However, this system recognizes that there is great uncertainty in basing an entire classification on such inferences.

Diagnostic Soil Horizons

To identify specific soil types within the Soil Taxonomy classification, the U.S. Soil Conservation Service describes "diagnostic horizons" in a pedon. A diagnostic horizon often reflects a physical property (color, texture, structure, consistence, porosity, moisture), or a dominant soil process.

In the solum, two diagnostic horizons may be identified: the epipedon and the subsurface. The **epipedon** (literally, over the soil) is the diagnostic horizon that forms at the surface and may extend through the A horizon, even including all or part of an illuviated B horizon. The epipedon is visibly darkened by organic matter and sometimes is leached of minerals. There are six recognized epipedons. (Alluvial deposits, eolian deposits, or areas disturbed by cultivation are not included in the epipedon because soil-forming processes have not had sufficient time to erase these transient characteristics.)

The second type of diagnostic horizon is the **subsurface diagnostic horizon**. It originates below the surface at varying depths and may include part of the A and/or B horizons. Many subsurface diagnostic horizons have been identified.

Table 18–5 summarizes both types of diagnostic horizons and their general characteristics, as described in the Soil Taxonomy. The presence or absence of either diagnostic horizon usually distinguishes a soil for classification.

The Ten Soil Orders of Soil Taxonomy

At the heart of the Soil Taxonomy are ten general soil orders, which are listed in Table 18–6 along with other helpful information. Their worldwide distribution is shown in Figure 18–8. You will find this table and map useful in the following descriptions. Because the taxonomy evaluates each soil order on its own characteristics, there is no priority to the classification. You will find the 10 orders loosely arranged by latitude, beginning along the equator.

Oxisols. The remarkable moisture and temperature intensity and uniform daylength of equatorial latitudes greatly affect soils. These generally old landscapes, exposed for many thousands to hundreds of thousands of years, are deeply developed and have greatly altered minerals (except for certain newer volcanic soils in Indonesia). The soil structure at these latitudes usually lacks distinct horizons where these soils are well drained. These soils are called **Oxisols** (tropical soils) and feature five suborders. Oxisols are found in South America, Africa, India, Southeast Asia, and Hawaii. Related vegetation (discussed in Chapter 20) is the luxuriant and diverse tropical and equatorial rain forest.

The large amount of precipitation leaches soluble minerals and soil constituents (such as silica) from the A horizons at equatorial latitudes. Typical soils are reddish and yellowish from the iron and aluminum oxides left behind, with a weathered clay-like texture, sometimes in a granular soil structure that is easily broken apart. The high degree of eluviation removes base cations and colloidal material to illuviated horizons. Thus, Oxisols are low in CEC (cation-exchange capacity) and fertility, except in regions augmented by alluvial or volcanic materials. Figure 18–9 illustrates **laterization**, the leaching process described here, which operates in well-drained soils in warm, humid climates.

TABLE 18–5
Major features of diagnostic horizons.

	Epipedons	Derivation	Characteristics
Surface Horizons	Mollic (A)	*mollis,* "soft"	Thick, dark-colored, humus-rich; base cation dominant; strong structure
	Anthropic (A)	*anthropos,* "having to do with humans"	Mollic-like, modified by long and continued human use and addition of water to soil
	Umbric (A)	*umbra,* "shade," hence "dark"	Identical to mollic, except low base cation content; dark coloration not related to humus
	Histic (O)	*histos,* "tissue"	Thin layer of peat or muck in wet places; saturated at least 30 consecutive days a year
	Plaggen (A)	*plaggen,* "sod"	Human-made >50 cm thick, produced by long, continuing manuring (W. Europe)
	Ochric (A)	*ochros,* "pale"	Light color, low humus content, hard and massive when dry, too marginal to fit other epipedons

	Name	Derivation	Characteristics
Subsurface Horizons	Argillic	*argilla,* "white clay"	Illuviated clay accumulation
	Agric	*ager,* "field"	Illuvial layer formed below cultivation
	Natric	*natrium,* "sodium"	Argillic that is high in sodium
	Spodic	*spodos,* "wood ashes"	Accumulation of organic matter, Fe, and Al
	Placic	*plax,* "flatstone"	Thin, dark-reddish, iron-cemented pan
	Cambic	*cambiare,* "to exchange"	Altered or changed by physical movement or chemistry
	Oxic	*oxide,* "oxide"	Highly weathered mix of Fe, Al oxides; >30 cm thick
	Duripan	*durus,* "hard"	Silica-cemented hard pan; not softened by water
	Fragipan	*fragilis,* "brittle"	Weakly cemented brittle pan; loam texture
	Albic	*albus,* "white"	Light, pale layer lacking clay, Fe, and Al
	Calcic	Calcium	Accumulation of $CaCO_3$ or $CaCO_3 \cdot MgCO_3$
	Gypsic	Gypsum	Accumulation of gypsum (hydrous calcium sulfate)
	Salic	Salts	Accumulation of >2% soluble salts
	Plinthite	*plinthos,* "brick"	Iron-rich, humus-poor; becomes ironstone under repeated wet/dry cycles

SOURCE: *Soil Taxonomy,* Agricultural Handbook No. 436, U.S. Department of Agriculture, 1975.

The subsurface diagnostic horizon in such a soil is highly weathered, containing iron and aluminum oxides, at least 30 cm (12 in.) thick and within 2 m (6.5 ft.) of the surface (Figure 18–10). If these horizons are exposed and subjected to repeated wetting and drying sequences, an *ironstone hardpan* soil structure called a **plinthite** results (from the Greek *plinthos,* meaning "brick"). This form of soil, also called a *laterite,* can be quarried in blocks and used as a building material (Figure 18–11).

Simple agricultural activities can be conducted in these soils with care. Earlier slash-and-burn shifting cultivation practices were adapted to these soil conditions and formed a unique style of crop rotation. The scenario went like this: people in the tropics cut down (slashed) and burned the rain forest in small tracts, cultivated the land with stick and hoe. After several years the soil lost fertility, and the people moved on to the next tract to repeat the process. After many years of movement from tract to

TABLE 18-6

Soil taxonomy soil orders.

Order	Pronunciation	Derivation of Term	Marbut Equivalent (Canadian)	General Location and Climate	World Land Area (%)	U.S. Land Area (%)	Description
Oxisols	*ox*	Fr. *oxide*, "oxide" Gr. *oxide*, "acid or sharp"	Latosals, lateritic soils	Tropical soils, hot humid areas	9.2	0.02	Maximum weathering and eluviation, oxic horizon, continuous plinthite layer
Aridisols	*arid*	L. *aridos*, "dry"	Reddish desert, gray desert, sierozems	Desert soils, hot dry areas	19.2	11.5	Limited alteration of parent material, low climate activity, ochric epipedon, light color, subsurface illuviation of carbonates
Mollisols	*mollify*	L. *mollis*, "soft"	Chestnut, chernozem (Chernozemic)	Grassland soils; subhumid, semiarid lands	9.0	24.6	Noticeably dark with organic material, base saturation high, mollic epipedon; friable surface with well-structured horizons
Alfisols	*alf*alfa	Invented syllable	Gray-brown podzolic, degraded chernozem (Luvisol)	Moderately weathered forest soils, humid temperate forests	14.7	13.4	B horizon high in clays, moderate to high degree of base saturation, orgillic or natric horizons, no pronounced color change with depth
Ultisols	*ulti*mate	L. *ultimus*, "last"	Red-yellow podzolic, reddish-yellow lateritic	Highly weathered forest soils, subtropical forests	8.5	12.9	Similar to Alfisols, B horizon high in clays, generally low amount of base saturation, strong weathering in orgillic horizons, redder than Alfisols, warmer soil temperature regime

Order	Mnemonic	Derivation	Equivalent	Occurrence			Characteristics
Spodosols	*odd*	Gr. *spodos* or L. *spodus*, "wood ash"	Podzols, brown podzolic (Podzol)	Northern conifer forest soils, cool humid forests	5.4	5.1	Illuvial B2 horizon of Fe/Al clays, humus; without structure, partially cemented; spodic horizon; highly leached, strongly acid; coarse texture of low bases
Entisols	*recent*	Invented syllable	Azonal soils, tundra	Recent soils, profile undeveloped, all climates	12.5	7.9	Limited development, inherited properties from parent material; ochric epipedon; few specific properties; young soils, lacking horizons
Inceptisols	*inception*	L. *inceptum*, "beginning"	Ando, subarctic brown forest, lithosols, some humic gleys (Brunisol, Cryosol with permafrost, Gleysol wet)	Weakly developed soils, humid regions	15.8	18.2	Intermediate development; embryonic soils, cambic horizon, but few diagnostic features; further weathering possible
Vertisols	*invert*	L. *verto*, "to turn"	Grumusols (1949), tropical black clays	Expandable clay soils; subtropics, tropics; sufficient dry period	2.1	1.0	Forms large cracks on drying, self-mixing action, contains >30% in swelling clays, ochric epipedon
Histosols	*histology*	Gr. *histos*, "tissue"	Peat, muck, bog (Organic)	Organic soils, wet places	0.8	0.5	Peat or bog, >20% organic matter, muck with clay >40 cm thick, surface organic layers, no diagnostic horizons

SOURCE: *Soil Taxonomy*, Agricultural Handbook No. 436, U.S. Department of Agriculture, 1975.

FIGURE 18-8

Worldwide distribution of the 10 soil orders. Adapted from U.S. Soil Conservation Service and by permission of Merrill, an imprint of Macmillan Publishing Company, from *Geography and Development,* James S. Fisher, 3d ed., pp. 58–59. Copyright © 1989 by Merrill Publishing.

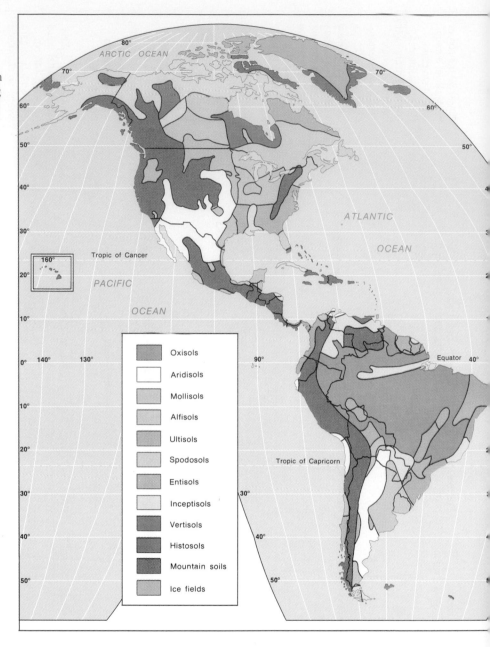

tract, the group returned to the first patch to begin the cycle again. This practice protected the limited fertility of the soils somewhat, allowing periods of recovery to follow active production.

However, this orderly native pattern of land rotation was halted by the invasion of foreign plantation interests, development by local govern-

ments, vastly increased population pressures, and conversion of vast new tracts to pasturage. Permanent tracts of cleared land put tremendous pressure on the remaining tracts of forest and brought disastrous consequences. When Oxisols are disturbed in such a regime of high temperature and moisture, soil loss can exceed a thousand tons per

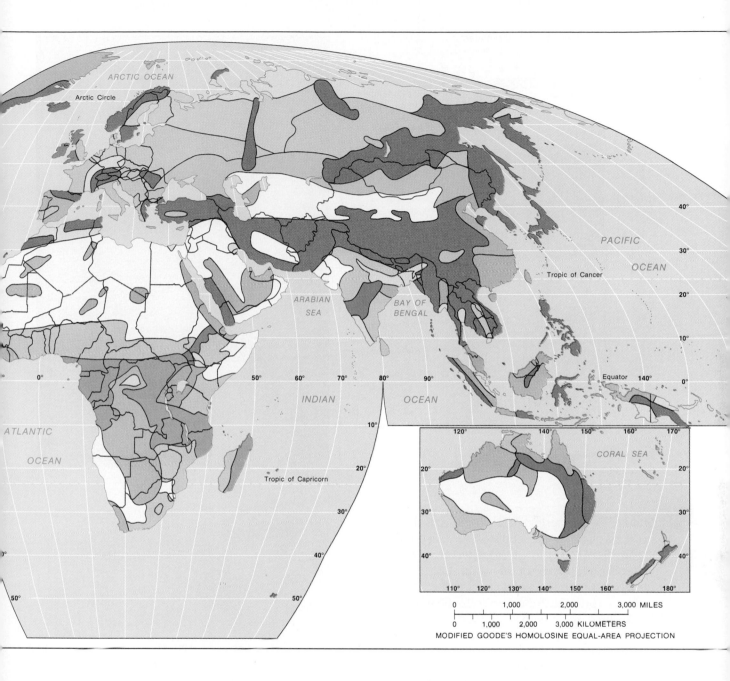

square kilometer per year, not to mention the greatly increased rate of extinction of plant and animal species that accompanies such destruction. The regions dominated by the Oxisols and rain forests are rightfully the focus of much worldwide environmental attention at this time.

Aridisols. The largest single soil order occurs in the world's deserts. **Aridisols** (desert soils) occupy approximately 19% of Earth's land surface and some 12% of the land in the United States. Figure 18–8 shows the distribution of the Aridisols in the southwestern United States and Mexico, Africa and the

FIGURE 18–9
Laterization process in Oxisols.

(a)

(b)

FIGURE 18–10
(a) Deeply weathered Oxisol profile in central Puerto Rico and (b) general map of worldwide distribution. Photo from Marbut Collection, Soil Science Society of America, Inc.

Middle East, interior Asia and the Gobi Desert, portions of Argentina, southwest Africa, and Australia. A pale, light soil color is characteristic of the *ochric epipedon,* and a large variety of subsurface diagnostic horizons can be identified (Figure 18–12).

Not surprisingly for desert soils, the water balance in Aridisol regions has marked periods of soil moisture deficit and generally inadequate soil moisture for plant growth. High annual values of potential evapotranspiration and low amounts of precipitation produce very shallow development of soil horizons. Usually there is no period greater than three months when the soils have adequate moisture. Aridisols also lack organic matter of any consequence. Although the soils are not leached often, because of the low

FIGURE 18–11
Plinthite quarrying activity in
India. Photo by Henry D. Foth.

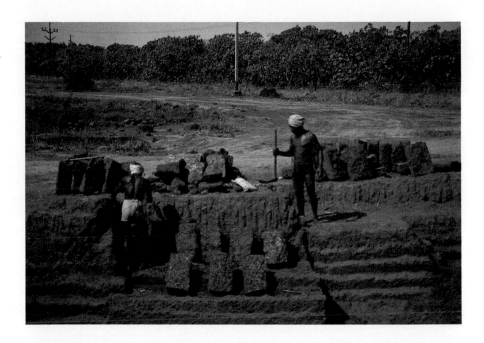

precipitation, they are leached easily when exposed to excessive water for they lack a significant colloidal structure.

A soil process that occurs in Aridisols and nearby soil orders is salinization. **Salinization** results from excessive POTET rates in the deserts and semiarid regions of the world. Salts dissolved in soil water are brought to surface horizons and deposited there as surface water evaporates. These deposits appear as subsurface *salic horizons,* which will damage and kill plants when they occur near the root zone. The salt accumulation associated with a desert playa is an example of extreme salinization (see Chapter 15).

Obviously, salinization complicates farming in Aridisols. The introduction of irrigation water may either waterlog poorly drained soils or lead to salinization. Nonetheless, vegetation does grow where soils are better drained and lower in salt content. In the Nile and Indus valleys, for example, Aridisols are intensively farmed with a careful balance of these environmental factors, although thousands of acres of once-productive land not so carefully treated now sit idle and salt-encrusted.

If large capital investments are made in water, drainage, and fertilizers, Aridisols possess much agricultural potential. However, certain agricultural practices may present problems of a different sort. In the Imperial, Coachella, and San Joaquin valleys of California and in the Wellton-Mohawk district of Arizona, soils were actually dug up, underlain with drainage tiles, and then replaced (Figure 18–13). The installed tiles and pipes drain away excess irrigation water, which has been purposely applied to leach salts away from plant roots in the soil. The problem then is to find a place to dump the salt- and chemical-laden drainage water.

Such practices have fueled political and environmental struggles all over the West. From the 1950s through the early 1970s, Colorado River water was made so salty by drainage water that it killed many irrigated crops across the border in Mexico. In California, the Kesterson Wildlife Refuge, protected by international agreements, was reduced to a toxic waste dump in the early 1980s. Focus Study 18–1 elaborates on the Kesterson tragedy. Ironically, in the Wellton-Mohawk district, if the government had

Selenium Concentrations in Western Soils

I rrigated agriculture has increased greatly since 1800, when only 8 million hectares (19.8 million acres) were irrigated worldwide. Today approximately 200 million hectares (494 million acres) are irrigated, and that figure may double by the turn of the century. Two related problems common in irrigated lands are salinization and waterlogging, especially in arid lands that are poorly drained. Representing about 15% of Earth's agricultural land, irrigated lands account for nearly 35% of the food and fiber harvest. However, in many areas production has decreased and even ended because of the salt buildup in the soils. A few examples include areas along the Tigris and Euphrates Rivers, the Indus River valley, sections of South America and Africa, and the western United States.

About 95% of the irrigated acreage in the United States is west of the 98th meridian. This region is increasingly troubled with salinization and waterlogging problems. But at least nine sites in the West, particularly California's western San Joaquin valley, are experiencing related contamination of a more serious nature—increasing selenium concentrations. The soils in these areas were derived from former marine sediments that formed shales in the adjoining Coast Ranges. As parent materials weathered, selenium-rich alluvium washed into the semiarid valley, forming the soils that needed only irrigation water to become productive. After 1960, such large-scale irrigation efforts intensified, resulting in subtle initial increases in selenium concentrations in soil and water. In trace amounts, selenium is a dietary requirement for animals and humans, but in higher amounts it is toxic to both. Toxic effects were reported during the 1980s in some domestic animals grazing on grasses grown in selenium-rich soils in the Great Plains.

Drainage of agricultural wastewater poses a particular problem in semiarid and arid lands where river discharge is inadequate to dilute and remove field runoff. One solution to prevent salt accumulations and waterlogging is to place field drains beneath the soil at depths of 3 or 4 meters (10–13 feet). Such drains collect gravitational water from fields that have been purposely overwatered to keep salts away from the effective rooting depth of the crops.

But drainage water must go somewhere, and for the San Joaquin Valley this problem triggered a 15-year controversy. Potential drain outlets to the ocean, San Francisco Bay, and an area in the central portion of the valley north of the irrigated lands all failed to pass environmental impact assessments under Environmental Policy Act requirements. Nonetheless, by the late 1970s, about 80 miles of the San Luis drain were finished, even though no formal plan nor adequate funding had been completed for its outlet! In the absence of such a plan, large-scale irrigation continued, supplying the field drains with selenium-laden runoff that made its way to the Kesterson National Wildlife Refuge in the northern portion of the San Joaquin Valley east of San Francisco, where the drain terminated.

It only took three years before the wildlife refuge was officially declared a contaminated toxic waste site. Aquatic life-forms (e.g., marsh plants, plankton, and insects) had taken in the selenium, which thus made its way into the diets of higher life-forms in the refuge. According to U.S. Fish and Wildlife Service scientists, the toxicity moved through the food chain and genetically damaged and killed wildlife. For example, birth defects and death were widely reported in all varieties of birds that nested at Kesterson; approximately 90% of the exposed birds perished or were injured. Because this wildlife refuge was a major migration flyway and stopover point for birds from throughout the hemisphere, this destruction of the refuge also violated several multinational protection treaties.

Such damage to wildlife presents a real warning to human populations—remember where we are in the food chain. The field drains were sealed and removed in 1986, following court action that forced the federal government to uphold existing laws. But irrigation water immediately began backing up in the fields, producing both waterlogging and selenium contamination. This contamination of groundwater is of course an ongoing concern. Since 1985 more than 0.6 million hectares (1.5 million acres) of irrigated Aridisols have gone out of production in California, marking the end of several decades of irrigated farming in climatically marginal lands. Severe cutbacks in irrigated acreage no doubt will continue, underscoring the need to preserve remaining prime farmlands elsewhere.

Frustrated agricultural interests have asked the federal government to finish the drain, either to San Francisco Bay or to the ocean. However, neither option appears capable of passing an environmental impact analysis. Another strategy was to allow irrigated lands to start pumping again into the former wildlife refuge because it was now a declared toxic dump site. Government agencies are wrestling with these options. At the very least we must acknowledge that soil processes are complex. Certainly, much remains to be learned to avoid environmental tragedies such as the one at Kesterson—remember there are nine such threatened sites in the West, and Kesterson was first to fail.

(a)

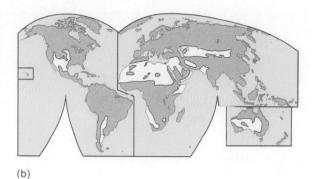

(b)

FIGURE 18–12
(a) Aridisol profile from central Arizona and (b) general map of worldwide distribution. Photo from Marbut Collection, Soil Science Society of America, Inc.

FIGURE 18–13
Agriculture in arid lands. Providing drainage to accommodate excessive water applications in such areas is often necessary to prevent salinization of the rooting zone. Photo by author.

taken out of production the irrigated land that receives subsidized federal water, it actually would have been cheaper than desalinization of the drainage water that eventually contaminates the Colorado River.

Mollisols. Mollisols (grassland soils) are some of Earth's most significant agricultural soils. There are seven recognized suborders, not all of which bear the same degree of fertility. The dominant diagnostic horizon is called the *mollic epipedon,* which is a dark, organic surface layer some 25 cm (10 in.) thick. As the Latin name implies, Mollisols are soft, even when dry, with granular or crumbly peds, loosely arranged when dry. These humus-rich organic soils are high in base cations (calcium, magnesium, and potassium) and have a high CEC (Figure 18–14). In terms of water balance, we can think of mollic soils as representing an intermediate stage between humid and arid soil moisture budgets.

(a)

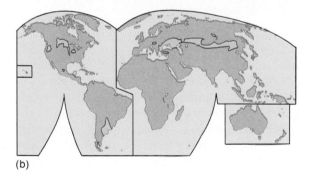

(b)

FIGURE 18–14

(a) Mollisol profile from central Iowa and (b) general map of worldwide distribution. Photo from Marbut Collection, Soil Science Society of America, Inc.

The steppes and prairies of the world belong to this soil group: the North American Great Plains, the Pampas of Argentina, and the region from Manchuria in China through to Europe. Agriculture ranges from large-scale commercial grain farming to grazing along the drier portions of the soil order. With fertilization or soil-building cultural practices, high crop yields are common. The soils of the fertile triangle of the Soviet Union, which stretches from the Caspian Sea westward in a widening triangle toward Europe, are known as the *chernozem* in other soil classification systems (Figure 18–15). The remainder of the Soviet landscape presents varied soils of lower fertility and productivity.

In North America, the Great Plains straddle the 98th meridian, which is coincident with the 51 cm (20 in.) isohyet of annual precipitation—wetter to the east and drier to the west. The Mollisols also mark the historic division between the short and tall grass prairies. The relationship among Mollisols, Aridisols (to the west), and Alfisols (to the east) is shown in Figure 18–16. The illustration also presents some of the important graduated changes that denote these different soil regions, including the level of pH concentration and the depth of available lime.

A soil process characteristic of portions of the Mollisols, and adjoining marginal areas having Aridisols, is calcification. **Calcification** involves the illuviated accumulation of calcium carbonate or magnesium carbonate in the B and C horizons, forming a subsurface diagnostic calcic horizon, which is thickest along the boundary between dry and humid climates (Figure 18–17). When cemented or hardened, these deposits are called **caliche**, or *kunkur;* they occur in widespread soil formations in central and western Australia, the Kalahari region of interior Southern Africa, and the High Plains of the west-central United States, among other places. (Soils dominated by the processes of calcification and salinization were formerly known as pedocals.)

Alfisols. Alfisols (moderately weathered forest soils) are the most wide-ranging of the 10 soil orders, extending in five suborders from near the equator to high latitudes. Representative Alfisol areas include Boromo and Burkina Faso (interior western Africa); Fort Nelson, British Columbia; the states near the Great Lakes, and the valleys of central California.

FIGURE 18–15

The fertile triangle of the Soviet Union. After P. W. English and J. A. Miller, *World Regional Geography: A Question of Place,* p. 183. Copyright © 1989 by John Wiley & Sons. Adapted by permission of John Wiley & Sons, Inc.

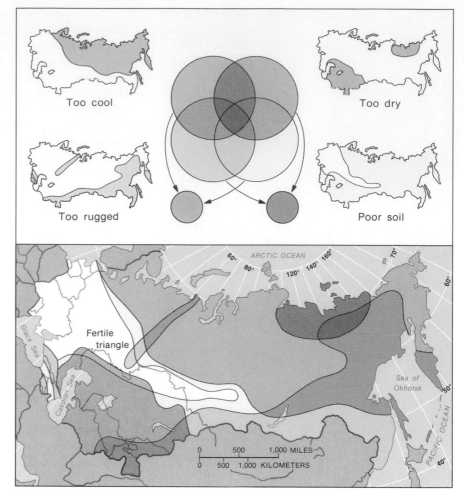

Most Alfisols have a grayish brown-to-reddish ochric epipedon and are considered moist versions of the Mollisol soil group. Moderate eluviation is present, as well as a subsurface *argillic horizon* of illuviated clays because of the increased precipitation pattern (Figure 18–18).

Alfisols have moderate-to-high reserves of base cations and are fertile. However, productivity depends on specific patterns of moisture and temperature. Alfisols usually are supplemented by a moderate application of lime and fertilizers in areas of active agriculture. Some of the best farmland in the United States stretches from Illinois, Wisconsin, and Minnesota through Indiana, Michigan, and Ohio to Pennsylvania and New York. This land produces grains, hay, and dairy products. The soil here is an

Alfisol subgroup called *Udalfs,* which are characteristic of humid continental/hot summer climates.

The *Xeralfs,* another Alfisol subgroup, are associated with the moist winter/dry summer pattern of the Mediterranean climate. These naturally productive soils are farmed intensively for subtropical fruits, nuts, and special crops that can grow only in a few locales worldwide (e.g., grapes, citrus, artichokes, almonds, figs).

Ultisols. Farther south in the United States are the **Ultisols** (highly weathered forest soils) and their five designated suborders. An Alfisol might degenerate into an Ultisol, given time and exposure to increased weathering under moist conditions. These soils tend to be reddish because of residual iron and aluminum oxides in the A horizon (Figure 18–19).

The increased precipitation in Ultisol regions means greater mineral alteration, more eluvial leaching, and therefore a lower level of base cations, leading to infertility. Fertility is further reduced by certain cultural practices and the effects of growing soil-damaging crops such as cotton and tobacco, which deplete nitrogen and expose soil to erosion. However, these soils respond well if subjected to good management—for example, crop rotation that protects fixed nitrogen and cultivation practices that prevent sheetwash and soil erosion. Much needs to

be done relative to sustainable management of these soils.

Spodosols. The Spodosols (northern coniferous forest soils) and their four related suborders occur generally to the north and east of the Alfisols. They are in cold and forested moist regimes (Dfb climates) in northern North America and Eurasia, Denmark, the Netherlands, and southern England. Because there are no comparable climates in the Southern Hemisphere, this soil type is not identified there. Spodosols form from sandy parent materials,

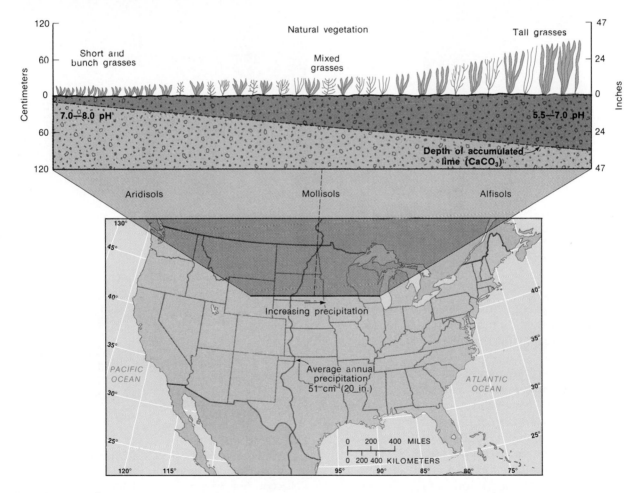

FIGURE 18–16
Aridisols, Mollisols, and Alfisols—a soil continuum in the north-central United States. Adapted with permission of Macmillan Publishing Company from *Nature and Properties of Soils,* by Nyle C. Brady, p. 444. Copyright © 1984 by Macmillan Publishing Company.

POTET equal to or
greater than PRECIP

Dark
color,
high in
bases

O Dense sod
A cover of
interlaced
grasses
and roots

Calcic
horizon;
possible
formation
of caliche

B Accumulation
of excess
calcium
carbonate

C

FIGURE 18–17
Calcification process in Aridisol/Mollisol soils in climatic
regimes with precipitation equal to or less than potential
evapotranspiration.

(a)

(b)

FIGURE 18–18
(a) Alfisol profile from central California and (b) general
map of worldwide distribution. Photo from Marbut
Collection, Soil Science Society of America, Inc.

shaded under evergreen forests of spruce, fir, and
pine. Spodosols with more moderate properties
form under mixed or deciduous forests—Figure
18–20(a, b).

Spodosols lack humus and clay in the A horizons.
An eluviated *albic horizon,* sandy and leached of
clays and irons, lies in the A horizon instead and
overlies a *spodic horizon* of illuviated organic matter
and iron and aluminum oxides. The surface horizon
receives organic litter from base-poor, acid-rich
trees, which contribute to acid accumulations in the
soil. The low pH (acid) soil solution effectively
removes clays, iron, and aluminum, which are passed
to the upper diagnostic horizon. An ashen-gray color
is common in these subarctic forest soils and is
characteristic of a formation process called **podzolization**—Figure 18–20(c).

(a)

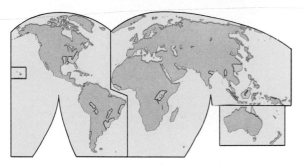

(b)

FIGURE 18–19
(a) Ultisol profile from the upper coastal plain of central North Carolina and (b) general map of worldwide distribution. Photo from Marbut Collection, Soil Science Society of America, Inc.

The low base-cation content of Spodosols requires the addition of nitrogen, phosphate, and potash, and perhaps crop rotation as well, if agriculture is to be attempted. A soil amendment such as limestone can significantly increase crop production in these acidic soils. For example, the yields of several crops grown in specific Spodosols in New York State were increased by as much as 33% with the application of 1.8 metric tons (2 tons) of limestone per 6-year rotation. Table 18–7 summarizes the results of this experiment.

Entisols. The Entisols (recent, undeveloped soils) lack vertical development of horizons. The five suborders of this soil group are based on differences in parent materials and climatic conditions, although the presence of Entisols is not climate dependent, for they occur in many climates worldwide. Entisols are true soils that have not had sufficient time to generate the usual horizons.

Entisols generally are poor agricultural soils, although those formed from river silt deposits are quite fertile. The processes that have inhibited complete development also have prevented adequate fertility—too much or too little water, poor structure, and insufficient accumulation of weathered nutrients. Active slopes, alluvium-filled floodplains, poorly drained tundra, tidal mud flats, dune sands and erg (sandy) deserts, and plains of glacial outwash all are characteristic of these soils (Figure 18–21).

Inceptisols. Inceptisols (weakly developed soils) and their six related suborders are inherently infertile. They are weakly developed young soils, although they are more developed than the Entisols. Inceptisols include a wide variety of different soils, all holding in common a lack of maturity with evidence of weathering just beginning. Inceptisols are associated with moist soil regimes and are regarded as eluvial because they demonstrate a loss of soil constituents throughout their profile but retain some weatherable minerals. This soil group has no distinct illuvial horizons and therefore shows no evidence of the translocation of clays, organics, or other minerals.

Inceptisols include the soils of most of the arctic tundra; volcanic lava soils on Pacific islands and in Washington, Oregon, and Idaho; glacially derived till

(a)

(b)

Cool and moist
climate

Humus
rich

Albic
horizon

Sandy,
bleached

Spodic
horizon

low pH soil
solution

O — Acidic organic
litter

A — Eluviation of
bases, oxides,
clays

B — Illuviation of
oxides and
clays

C — Loss of base
to water table

Water table

(c)

FIGURE 18–20
(a) Spodosol profile from northern New York and (b) general map of worldwide distribution; (c) podzolization process, typical in cool and moist climatic regimes. (a) from Marbut Collection, Soil Science Society of America, Inc.

TABLE 18–7
Yield increases for selected crops.

Crop	Increase per Acre
Corn (silage)	1.0 tons
Oats	5.8 bushels
Wheat	7.8 bushels
Hay	1.8 tons

SOURCE: N. C. Brady, "The Northeast," *1957 Yearbook of Agriculture: Soil,* p. 604, U.S. Department of Agriculture.

and outwash materials from New York down through the Appalachians; and alluvium on the Mekong and Ganges floodplains.

Vertisols. Vertisols (expandable clay soils) are heavy clay soils that contain more than 30% swelling clays (clays that swell significantly when they absorb water), such as *montmorillonite*. They are located in regions experiencing highly variable soil moisture balances through the seasons. These soils occur in subhumid-to-semiarid moisture conditions and under moderate-to-high temperature patterns. Vertisols frequently form under savanna and grassland vegetation in tropical and subtropical climates and are sometimes associated with a distinct dry season following a wet season. Individual Vertisol units are limited in extent, although their overall occurrence is widespread.

Vertisol clays are black when wet (but not because of organic content) and range to brown and dark gray. These deep clays swell when moistened and shrink when dried. In the process, vertical cracks form, which widen and deepen as the soil dries, producing cracks 2–3 cm wide (0.8–1.2 in.) and up to 40 cm deep (16 in.). Loose material falls into these cracks, only to disappear when the soil again expands and the cracks close. After many such cycles, soil contents tend to invert (Figure 18–22).

Despite the fact that clay soils are plastic and heavy when wet, with little available soil moisture for plants, Vertisols are high in bases and nutrients and thus are some of the better farming soils wherever they occur. For example, they occur in a narrow zone along the coastal plain of Texas and in a section along the Deccan region of India. Vertisols often are planted with grain sorghums, corn, and cotton.

Histosols. Histosols (organic soils), including four suborders, are formed from accumulations of thick organic matter. In the midlatitudes, when conditions are right, beds of former lakes may turn into Histosols, with water gradually replaced by organic material to form a bog and layers of peat. (Lake succession and bog/marsh formation are discussed in Chapter 19.) Histosols also form in small, poorly drained depressions. Here, Histosols have produced

FIGURE 18–21
Zabriski Point, Death Valley, a characteristic Entisol. Photo by author.

(a)

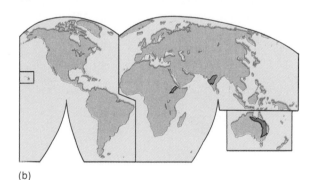

(b)

FIGURE 18–22

(a) Vertisol profile in the Lajas Valley of Puerto Rico and
(b) general map of worldwide distribution. Photo from
Marbut Collection, Soil Science Society of America, Inc.

conditions ideal for significant deposits of *sphagnum peat* to form. This material can be cut, baled, and sold as a soil amendment. In other locales, dried peat has served for centuries as a low-grade fuel. The area southwest of Hudson Bay in Canada is typical of Histosol formation.

SUMMARY

Soil is the portion of the land surface in which plants can grow. Soils range from a few millimeters to many meters in depth; in some locales they may be completely absent. The basic soil individual is the polypedon, made up of numerous sampling units called pedons. Soils are structured in vertical horizons that permit analysis and classification. The key properties for analysis are color, texture, structure, consistence, porosity, moisture regimes, and chemistry. Environmental factors that affect soil formation include parent materials, climate, vegetation, topography, and time. Human influence has also proven to have great impact.

The past century has been marked by repeated efforts to develop soil classifications. Presently, the Soil Taxonomy classification system is used in the United States, built around an analysis of various diagnostic horizons and 10 soil orders. The fertility of soils is critical to society. Essential soils for agriculture and their fertility are threatened by mismanagement, misuse, and conversion to other uses. Much soil loss can be stopped through the application of known technologies, improved agricultural practices, and sustainable government policies.

KEY TERMS

Adsorption	Oxisols
Alfisols	Pedogenic regimes
Aridisols	Pedon
Calcification	Plinthite
Caliche	Podzolization
Cation-exchange capacity (CEC)	Polypedon
Eluviation	Salinization
Entisols	Soil
Epipedon	Soil colloids
Histosols	Soil horizon
Humus	Soil Taxonomy
Illuviation	Solum
Inceptisols	Spodosols
Laterization	Subsurface diagnostic horizon
Loam	Ultisols
Mollisols	Vertisols

REVIEW QUESTIONS

1. Consider this statement: Soils provide the foundation for animal and plant life and therefore are critical to Earth's ecosystems. Why is this true?
2. Explain some of the details supporting concern for the loss of our most fertile soils.
3. Define polypedon and pedon, the basic units of soil.
4. Characterize the principal aspects of each soil horizon. Where does the main accumulation of organic material occur? The formation of humus? What is the eluviated layer? The illuviated layer? The solum, or true soil?
5. How is soil color analyzed and compared?
6. Define a soil separate. What are the various sizes of particles in soil? What is loam? Why is loam regarded so highly by agriculturalists?
7. What is a possible method for determining soil consistence?
8. Summarize the five soil moisture regimes common in mature soils.
9. What are soil colloids? How do they relate to cations and anions in the soil? Explain the cation-exchange capacity.
10. Briefly describe the contribution of the following factors and their effect on soil formation: parent material, climate, vegetation, landforms, time, and humans.
11. Summarize the brief history of soil classification described in this chapter. What led soil scientists to develop the new Soil Taxonomy classification system?
12. What is the basis of the Soil Taxonomy system? How many orders, suborders, great groups, subgroups, families, and soil series are there?
13. Define an epipedon and a subsurface diagnostic horizon. Give a simple example of each.

14. Locate each soil order on the map as you give a general description of it.
15. How was slash-and-burn cultivation, as practiced in the past, a form of crop and soil rotation and a conserving of soil properties?
16. Describe the salinization process in arid and semiarid soils. What associated soil horizons develop?
17. Which of the soil orders are associated with Earth's most productive agricultural areas?
18. Describe the podzolization process associated with northern coniferous forest soils. What characteristics are associated with the surface horizons? What strategies might enhance these soils?
19. Which soil order is associated with the formation of a bog? Explain its use as a low-grade fuel.
20. Why has a selenium contamination problem arisen in western soils? Explain the impact of cultural practices, and tell why you think this is or is not a serious problem.

SUGGESTED READINGS

Botkin, Daniel B., and Edward A. Keller. *Environmental Studies—Earth as a Living Planet*. Columbus: Macmillan, 1987.

Brady, Nyle C. *The Nature and Properties of Soils*. 10th ed. New York: Macmillan, 1990.

Bunting, Brian T. *The Geography of Soil*. Chicago: Aldine, 1966.

Fairbridge, Rhodes W., and Charles W. Finkl, Jr. *The Encyclopedia of Soil Science*. Part 1, Physics, Chemistry, Biology, Fertility, and Technology. *Encyclopedia of Earth Sciences,* vol. 12. Stroudsburg, PA: Dowdon, Hutchinson and Ross, 1979.

Finkl, Charles W., Jr., ed. "Soil Classification." *Benchmark Papers in Soil Science,* vol. 1. Stroudsburg, PA: Hutchinson Ross, 1982.

Foth, Henry D. *Fundamentals of Soil Science*. 8th ed. New York: John Wiley & Sons, 1990.

Gersmehl, Philip J. "Soil Taxonomy and Mapping." *Annals of the Association of American Geographers* 67 (September 1977): 419–28.

Gibbons, Boyd. "Do We Treat Our Soil like Dirt?" *National Geographic* (September 1984): 350–89.

Harris, Tom. "Selenium: The Poisoning of America." Trouble in Paradise Series. *Sacramento Bee* (December 6–10, 1987).

Hausenbuiller, R. L. *Soil Science—Principles and Practices*. Dubuque, IA: W. C. Brown, 1985.

Singer, Michael J., and Donald N. Munns. *Soils—An Introduction*. New York: Macmillan, 1987.

Soil Survey Staff. *Soil Taxonomy—A Basic System of Soil Classification for Making and Interpreting Soil Surveys*. Agricultural Handbook No. 436. Washington: Government Printing Office, 1975.

Soil Survey Staff and the Agronomy Department of Cornell University. *Keys to Soil Taxonomy*. Soil Management Support Services Technical Monograph 6. Ithaca, NY: Cornell University, 1987.

Steila, Donald. *The Geography of Soils*. Englewood Cliffs, NJ: Prentice Hall, 1976.

——————— and Thomas E. Pond. *The Geography of Soils—Formation, Distribution, and Management*. Savage, MD: Rowman & Littlefield, 1989.

The intricate web of life. Photo by author.

19
Ecosystem Essentials

ECOSYSTEM COMPONENTS AND CYCLES

Communities
Plants: The Essential Biotic Component
Abiotic Ecosystem Components
Biotic Ecosystem Operations

STABILITY AND SUCCESSION

Ecosystem Stability
Ecological Succession

SUMMARY

Key Terms
Review Questions
Suggested Readings

The interaction of the atmosphere, hydrosphere, and lithosphere produces conditions within which the biosphere exists. This sphere of life and organic activity extends from the floor of the ocean to a height of about 8 km (5 mi) in the atmosphere. The biosphere is composed of myriad ecosystems from simple to complex, each operating within general spatial boundaries. An **ecosystem** is a self-regulating association of living plants and animals and their nonliving physical environment. Earth itself is an ecosystem within the natural boundary of the atmosphere. Natural ecosystems are open systems in terms of energy and matter, with almost all ecosystem boundaries functioning as transition zones rather than sharp demarcations. Various smaller ecosystems—for example, forests, seas, mountain tops, deserts, beaches, islands, lakes, ponds—make up the larger whole.

Ecology is the study of the relationships between organisms and their environment and among the various ecosystems in the biosphere. **Biogeography**, essentially a spatial ecology, is the study of the distribution of plants and animals and the diverse spatial patterns they create across Earth. The degree to which modern society understands ecosystems will help determine our success as a species and the long-term survival of a habitable Earth. Gilbert White, a well-known geographer, researcher, and activist, and Mostafa Tolba, director of the United Nations Environment Programme, joined in this assessment:

> The time is ripe to step up and expand current efforts to understand the great interlocking systems of air, water, and minerals nourishing the Earth. . . . Moreover, without vigorous action toward that goal, nations will be seriously handicapped in trying to cope with proven and suspected threats to ecosystems and to human health and welfare resulting from alterations in the cycles of carbon, nitrogen, phosphorus, sulfur, and related materials. . . . Society depends upon this life-support system of planet Earth.*

ECOSYSTEM COMPONENTS AND CYCLES

An ecosystem is a complex of many variables, all functioning independently yet in concert, with complicated flows of energy and matter.

> Life devours itself: everything that eats is itself eaten; everything that can be eaten is eaten; every chemical that is made by life can be broken down by life; all the sunlight that can be used is used. . . . The web of life has so many threads that a few can be broken without making it all unravel, and if this were not so, life could not have survived the normal accidents of weather and time, but still the snapping of each thread makes the whole web shudder, and weakens it. . . . You can never do just one thing: the effects of what you do in the world will always spread out like ripples in a pond.**

An ecosystem is composed of both biotic and abiotic components. Nearly all depend upon an input of solar energy; the few limited ecosystems that exist in dark caves or on the ocean floor depend upon chemical reactions. Ecosystems are divided into subsystems, with the biotic portion composed of producers, consumers, and decomposers. Gaseous and sedimentary nutrient cycles comprise the abiotic flows in an ecosystem. Figure 19–1 illustrates the essential elements of an ecosystem.

Communities

A convenient biotic subdivision within an ecosystem is a **community**, which is formed by interacting populations of living animals and plants in an area. An ecosystem is the interaction of a community with the abiotic physical components of its environment. Many communities are included in an ecosystem. For example, in a forest ecosystem a specific community may exist on the forest floor, whereas another functions in the canopy of leaves high above. Similarly, within a lake ecosystem, the plants and animals that flourish in the bottom sediments form one community, whereas those near the surface of

---

*Gilbert F. White and Mostafa K. Tolba, "Global Life Support Systems," *United Nations Environment Programme Information*, no. 47, Nairobi: United Nations, 1979, p. 1.

**Friends of the Earth and Amory Lovins, The United Nations Stockholm Conference: *Only One Earth*. London: Earth Island Limited, 1972, p. 20.

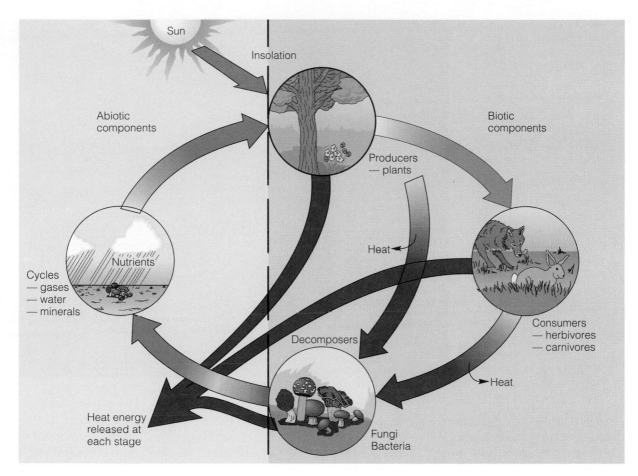

FIGURE 19–1
Abiotic and biotic components that comprise ecosystems.

the lake form another. A community is identified in several ways—by the physical appearance of the community, the number of species and the abundance of each, and the trophic (feeding) structure of the community.

Within a community, two concepts are important: habitat and niche. **Habitat** is the specific physical location of an organism, the place in which it resides or is biologically suited to live. In terms of physical and natural factors, most species have specific habitat parameters (with definite limits) and a specific regimen of sustaining nutrients. **Niche** refers to the function, or occupation, of a life-form within a given community; it is the way an organism obtains and

sustains its living. An individual species must satisfy several aspects in its niche; among these are a habitat niche, a trophic (food) niche, and a reproductive niche. A principle of *competitive exclusion* states that no two species can occupy the same niche successfully in a stable community; in other words, each species operates to reduce competition.

Some species have *symbiotic relationships,* an arrangement that mutually benefits and sustains each organism. For example, *lichen* (pronounced "liken") is made up of algae and fungus. The algae is the producer and food source, and the fungus provides structure and support. Their mutually beneficial relationship allows the two to occupy a niche in

FIGURE 19–2
Lichen, an example of a symbiotic relationship. Photo by Bobbé Christopherson.

which neither could survive alone. Lichen developed from an earlier parasitic relationship in which the fungi broke into algae cells directly. Today the two organisms have evolved into a supportive harmony (Figure 19–2). However, unsuccessful parasitic relationships eventually may kill the host plant, thus destroying the parasite's own niche and habitat. An example is mistletoe (*Phoradendron*) which lives on and may kill various kinds of trees. Some scientists are questioning whether or not our human society and Earth constitute a global form of a symbiotic relationship.

Plants: The Essential Biotic Component

Plants are the critical biotic link between life and solar energy. Ultimately the fate of the biosphere rests on the success of plants and their ability to capture sunlight.

Beginnings. The first plants were simple single- or multiple-celled structures known as *cyanobacteria* (ural blue-green algae). They began the harvest of

sunlight and release of free oxygen about 3.3 billion years ago. Long, slow development ensued as these early aquatic plants migrated from the subsurface to the water surface, increasingly protected by the evolving functional layers of the atmosphere.

The last billion years of Earth's history have seen an explosion of biotic diversity. The move to rocky coastlines and then inland required complicated and integrated multicellular organisms possessing evolved structures for survival on land. Specialized tissues developed for food production, body support, and anchorage. Land plants and animals became common about 430 million years ago, according to fossilized remains. The *equisetum* plant is one example of an ancient plant of primitive structure that still survives today (Figure 19–3).

Vascular plants have internal fluid and material flowing through conductive tissues. At present there are almost 250,000 species of them. This great diversity and complexity compounds the difficulty of classification for convenient study.

Leaf Activity. Photochemical reactions take place within plant leaves; flows of carbon dioxide, water, light, and oxygen enter and exit the surface of each leaf.

The upper and lower surfaces of a leaf are covered with a layer of epidermal cells that protect the inner workings of the leaf and reduce water loss. The largest concentration of light-sensitive cells rests below the upper layers of the leaf. These are called chloroplast bodies, and within each resides a green, light-sensitive pigment called **chlorophyll**. Within this pigment, light stimulates photochemistry.

Gases flow into and out of a leaf through small pores called **stomata** (singular: stoma), which usually are more numerous on the lower side of the leaf. A stoma is surrounded by small guard cells that open and close the pore, depending on the plant's needs at the moment. Veins in the leaf bring in water and nutrient supplies and carry off the sugars produced by photosynthesis. The veins in each leaf connect to

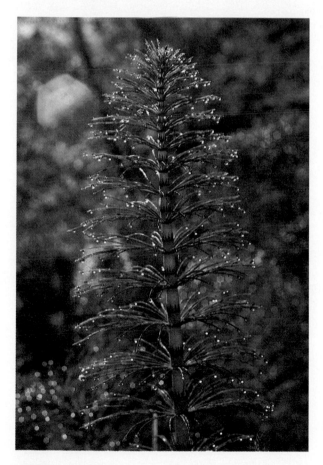

FIGURE 19–3
Equisetum, a plant of ancient origins. Photo by author.

the stems and branches of the plant and thus to the main circulation system. Water that moves through a plant exits the leaves and evaporates from leaf surfaces, thereby assisting heat regulation within the plant. As water evaporates from the leaves, a pressure deficit is created that allows water to move up through the plant all the way from the roots— another example of the presence of atmospheric pressure. We can only imagine the complex operation of a 100 m (328 ft) tree!

Photosynthesis and Respiration. Photosynthesis unites carbon dioxide and oxygen (derived from water in the plant) under the influence of certain wavelengths of visible light, subsequently releasing oxygen and producing energy-rich organic material. (The name is descriptive: *photo-* refers to sunlight, and *-synthesis* describes the bringing together of materials to form compounds and produce reactions within plant leaves.) Only about one-quarter of the light energy arriving at the surface of a leaf is useful to the light-sensitive chlorophyll. Chlorophyll absorbs only the orange-red and violet-blue wavelengths for photochemical operations, and reflects predominantly green hues (and some yellow). (That is why trees and other vegetation look green.) Understandably, competition for light is a dominant factor in the formation of plant communities. This competition is expressed in their height, orientation, and structure.

Photosynthesis essentially follows this equation:

$$CO_2 + H_2O \xrightarrow{\text{(Light)}} (CH_2O) + O_2$$

| (Carbon dioxide) | (Water) | (Light) | (Carbohydrate) | (Oxygen) |

From the equation, you can see that photosynthesis removes carbon (in the form of CO_2) from Earth's atmosphere. The quantity is prodigious: approximately 91 billion metric tons per year (100 billion tons). Carbohydrates, the organic result of the photosynthetic process, are combinations of carbon, hydrogen, and oxygen. They can form simple sugars, such as glucose ($C_6H_{12}O_6$). Glucose, in turn, is used by plants to build starches, which are more complex carbohydrates and the principal food storage substance in plants. *Primary productivity* refers to the rate at which energy is stored in such organic substances.

Plants derive energy for their other operations through **respiration**, which is essentially a reverse of the photosynthetic process:

$$(CH_2O) + O_2 \longrightarrow CO_2 + H_2O + energy$$

(Carbo-   (Oxygen)   (Carbon   (Water)   (Heat)
hydrate)   dioxide)

In respiration, plants oxidize stored energy, releasing carbon dioxide, water, and energy as heat.

The overall growth of a plant depends on a surplus of carbohydrates beyond what is lost through plant respiration. Figure 19–4 presents a simple schematic of this process, which produces plant growth.

The *compensation point* is the break-even point between the production and consumption of organic material. Each leaf operates above the compensation point, with unproductive leaves being eliminated by the plant; something each of us has no doubt

FIGURE 19–4
The relationship of photosynthesis and respiration.

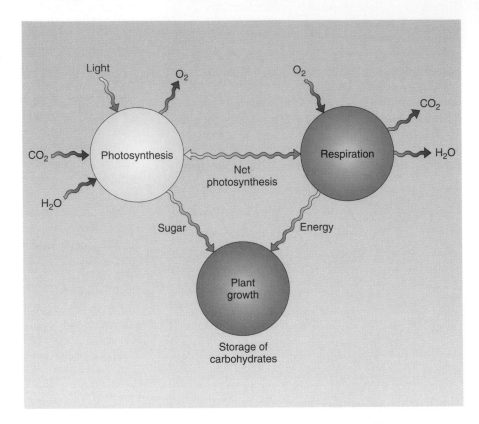

FIGURE 19–5
Energy receipt, utilization, and disposition by green plants.

FIGURE 19–6
Worldwide net primary productivity in grams of carbon per square meter per year
(approximate values). After D. E. Reichle, *Analysis of Temperate Forest Ecosystems,*
Heidelberg, West Germany: Springer-Verlag, 1970. Adapted by permission.

experienced with a house plant that received inadequate water or lighting. The difference between photosynthetic production and respiration loss is called *net photosynthesis*. The amount varies, depending on controlling environmental factors such as light, water, temperature, soil fertility, landforms, and the plant's site, elevation, and competition from other plants and animals.

Plant productivity increases as light availability increases—up to a point. When the light level is too high, *light saturation* occurs and most plants actually reduce their output in response. Individual species differences also are involved. Some plants adapt better to shade, whereas others flourish in full sunlight. Crops such as rice, wheat, and sugar cane do well with high light intensity. Figure 19–5 portrays the general disposition of energy received by green plants, how it is utilized, and the final disposition of net primary production.

Net Primary Productivity. The net photosynthesis for an entire community is its **net primary productivity**. This is the amount of useful chemical energy (biomass) that the community "fixes" in the ecosystem. **Biomass** is the net dry weight of organic material; it is biomass that feeds the food chain.

Primary productivity is mapped in terms of *fixed carbon per square meter per year* (Figure 19–6). On land, primary production tends to be highest in the tropics at sea level and decreases toward higher latitudes. But precipitation also affects productivity, as evidenced on the map by the correlations of abundant precipitation/high productivity and reduced precipitation/low productivity, the latter observable in subtropical deserts. Even though deserts are high in solar radiation, other controlling factors are important, namely water availability and soil conditions. In the oceans, productivity is limited by differing nutrient levels. Regions with nutrient-rich upwelling currents generally are the most productive. The map shows that the tropical ocean and areas of subtropical high pressure are quite low in productivity.

In temperate and high latitudes, the rate at which carbon dioxide is fixed by vegetation varies seasonally, increasing in spring and summer as plants flourish and decreasing in late fall and winter. Rates in the tropics are high throughout the year, and turnover in the photosynthesis-respiration cycle is faster, exceeding by many times the rates experienced in a desert environment or in the far northern limits of the tundra. A lush acre of sugar cane in the tropics might fix 18 metric tons (20 tons) of carbon dioxide in a year, whereas desert plants in an equivalent area might achieve only 1% of this amount.

Table 19–1 lists various ecosystems, their net primary productivity per year, and an estimate of net total biomass (primary production) worldwide. Net primary productivity is estimated at 170 billion tons of dry organic matter per year. Please compare the various ecosystems, especially cultivated land, with most of the natural communities. Net productivity is generally regarded as the most important aspect of

TABLE 19–1

Net primary production and plant biomass on Earth.

Ecosystem	Area $(10^6 km^2)$*	Net Primary Productivity per Unit Area $(g/m^2/yr)$**		World Net Primary Production $(10^9 t/yr)$***
		Normal Range	Mean	
Tropical rain forest	17.0	1000–3500	2200	37.4
Tropical seasonal forest	7.5	1000–2500	1600	12.0
Temperate evergreen forest	5.0	600–2500	1300	6.5
Temperate deciduous forest	7.0	600–2500	1200	8.4
Boreal forest	12.0	400–2000	800	9.6
Woodland and shrubland	8.5	250–1200	700	6.0
Savanna	15.0	200–2000	900	13.5
Temperate grassland	9.0	200–1500	600	5.4
Tundra and alpine region	8.0	10–400	140	1.1
Desert and semidesert scrub	18.0	10–250	90	1.6
Extreme desert, rock, sand, ice	24.0	0–10	3	0.07
Cultivated land	14.0	100–3500	650	9.1
Swamp and marsh	2.0	800–3500	2000	4.0
Lake and stream	2.0	100–1500	250	0.5
Total continental	149.0		773	115.17
Open ocean	332.0	2–400	125	41.5
Upwelling zones	0.4	400–1000	500	0.2
Continental shelf	26.6	200–600	360	9.6
Algal beds and reefs	0.6	500–4000	2500	1.6
Estuaries	1.4	200–3500	1500	2.1
Total marine	361.0		152	55.0
Grand total	510.0		333	170.17

*1 km^2 = 0.39 mi^2
**1 g per m^2 = 8.9 lbs. per acre
***1 metric ton = 1.1023 tons

SOURCE: Robert H. Whittaker, *Communities and Ecosystems,* Heidelberg: Springer-Verlag, 1975, p. 224. Reprinted by permission.

any type of community, and the distribution of productivity over Earth's surface is an important subject of biogeography.

Abiotic Ecosystem Components

Critical in each ecosystem is the flow of energy and the cycling of matter in the form of nutrients and water. Nonliving abiotic components set the stage for ecosystem operations.

Light, Temperature, Water, and Climate. The pattern of solar energy receipt is crucial in both terrestrial and aquatic ecosystems. Solar energy enters an ecosystem by way of photosynthesis, with heat dissipated from the system at many points. Of the total energy intercepted at Earth's surface and available for work, only about 1.0% is actually fixed by photosynthesis as chemical energy (energy stored in the formation of organic compounds and the release of oxygen).

The duration of Sun exposure is the **photoperiod**. Along the equator, days are almost always 12 hours in length; however, with distance from the equator, seasonal effects become pronounced, as discussed in Chapter 4. Plants have adapted their flowering and seed germination to seasonal changes in insolation. Some plant seeds germinate only when daylength reaches a certain number of hours. A plant that responds in the opposite manner is the poinsettia (*Euphorbia pulcherrima*), which requires at least two months of 14-hour nights to start flowering.

Other components also are important to ecosystem processes. Air and soil temperatures determine the rates at which chemical reactions proceed. Average air temperatures (see Chapter 5) are considered in addition to the seasonal variation, duration, and pattern of minimum and maximum values. Operations of the hydrologic cycle and water availability depend on rates of precipitation/evaporation and their seasonal distribution (see Chapters 7 to 9). Water quality is important—its mineral content, salinity, and levels of pollution and toxicity. Also, daily weather patterns over time create regional climates (see Chapter 10), which in turn affect the pattern of vegetation and ultimately influence soil development. All of these factors work together to establish the parameters for ecosystems that may develop in a given location.

Figure 19–7 illustrates the general relationship among temperature, precipitation, and vegetation. In general terms, can you identify the characteristic vegetation type and related temperature and moisture relationship that fits the area of your town, or school?

Alexander von Humboldt (1769–1859), an explorer, geographer, and scientist, deduced that plants and animals recur in related groupings wherever similar conditions occur in the abiotic environment. He described the similarities and dissimilarities of vegetation and the associated uneven distribution of various organisms. After several years of study in the Andes Mountains of Peru, he described a distinct relationship between altitude and plant communities, his life zone concept. As he climbed the mountains, he noticed that the experience was similar to that of traveling away from the equator toward higher latitudes (Figure 19–8).

This zonation of plants with altitude is noticeable on any trip from lower valleys to higher elevations. Each **life zone** possesses its own temperature, precipitation, and insolation relationships and therefore its own biotic communities. The Grand Canyon in Arizona provides a good example. The inner gorge at the bottom of the canyon (<610 m or <2000 ft in elevation) has life-forms characteristic of the lower Sonoran Desert of northern Mexico. However, the north rim of the canyon (>2134 m or >7000 ft in elevation) is dominated by ecosystems similar to those of southern Canadian forests. On the summits of the nearby San Francisco Mountains (3600 m or 12,000 ft in elevation), the vegetation is similar to the arctic tundra of northern Canada.

Beyond these general conditions, each ecosystem further produces its own *microclimate,* specific to individual sites. For example, in forests the insolation reaching the ground is reduced. A pine forest cuts light by 20 to 40%, whereas a birch-beech forest reduces it by as much as 50 to 75%. Forests also are about 5% more humid than nonforested landscapes, have warmer winters and cooler summers, and experience reduced winds.

Slope and exposure are important, too, for they translate into differences in temperature and moisture efficiency, especially in middle and higher latitudes. With all other factors equal, slopes facing away from the Sun's rays tend to be more moist and vegetated. Such highly localized *microecosystems* are

ECOSYSTEM ESSENTIALS • **555**

FIGURE 19–7
Abiotic climate control of ecosystem types: the generalized relationship among
rainfall, temperature, and vegetation. Adapted from Daniel D. Chiras, *Environmental
Science,* 2d ed. Copyright © 1988, Benjamin/Cummings Publishing Co., Menlo Park,
CA. Reprinted by permission.

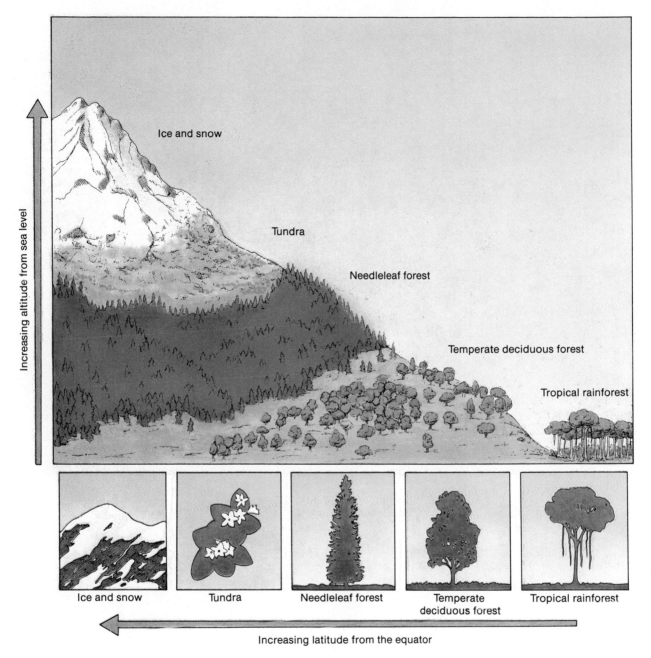

FIGURE 19–8
Progression of plant community life zones with increasing altitude or latitude.

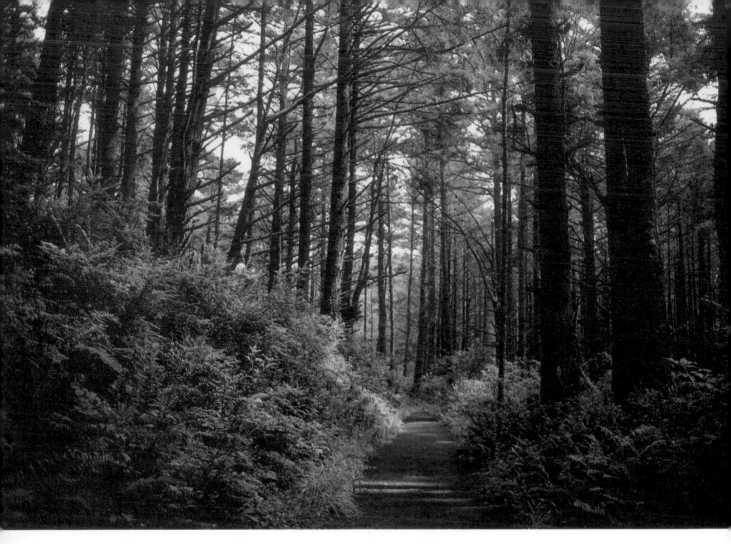

FIGURE 19–9
Fern and forest trail, indicative of a microclimate. Photo by author.

evident along a mountain trail, where changes in exposure and moisture can be easily seen (Figure 19–9).

Gaseous and Sedimentary Cycles. The most abundant natural elements in living matter are hydrogen (H), oxygen (O), and carbon (C). Together, these elements comprise more than 99% of Earth's biomass; in fact, all life contains hydrogen and carbon. In addition, nitrogen (N), calcium (Ca), potassium (K), magnesium (Mg), sulfur (S), and phosphorus (P) are important nutrients, elements necessary for the growth and development of a living organism.

Several important cycles of chemical elements operate. Oxygen, carbon, and nitrogen form gaseous cycles involving atmospheric phases. Other elements form sedimentary cycles principally involved in mineral and solid phases (such as phosphorus, calcium, and sulfur). Some elements combine the two cycles. The various processes are called **biogeochemical cycles**, indicating the occurrence of chemical reactions in both life and Earth systems. The chemical elements themselves recycle over and over again in life processes.

We consider the carbon and oxygen cycles together because they are so closely intertwined through photosynthesis and respiration (Figure 19–

10). The atmosphere is the principal reserve of available oxygen (the larger reserves of oxygen in Earth's crust are tied up in combination with other elements). Unoxidized reserves of photosynthetic fossil fuels and sediments also contain oxygen.

The greatest pool of carbon is in the ocean—about 39,000 billion tons, or about 93% of the total. The ocean absorbs carbon dioxide through the functioning of small phytoplankton organisms. The atmosphere, which serves as the integrating link in the cycle, contains only about 700 billion tons at any one moment, far less even than fossil fuels and oil shales (12,000 billion tons) or dead and living organic matter (2275 billion tons). Carbon dioxide in the atmosphere is produced by the respiration of plants and animals, volcanic activity, and fossil fuels combustion by industry and transportation. In organic matter, carbon is stored in carbohydrate molecules; in fossil fuels it appears in hydrocarbon molecules. Carbon also is stored in certain carbonate minerals, such as limestone.

The carbon dioxide dumped into the atmosphere by human activity constitutes a vast geochemical experiment, using the real-time atmosphere as a laboratory. Since 1970, we have added to the atmospheric pool an amount of carbon equivalent to more than 25% of the total amount added since 1880. Annual emissions by society have now reached 6 billion tons and represent our greatest single atmospheric waste product. Furthermore, about 50% of the carbon dioxide emitted since the beginning of the industrial revolution and not absorbed by oceans and organisms is still in the atmosphere—an important topic of discussion in Chapters 3, 5, and 10.

Nitrogen is the major constituent of the atmosphere, 78.084% of each breath we take. It also is important in the makeup of organic molecules, especially proteins, and therefore is essential to living processes. However, this vast atmospheric reservoir is inaccessible directly to most organisms. The key link in the nitrogen cycle is provided by nitrogen-fixing bacteria, which live principally in the

FIGURE 19–10
A simplified diagram of the carbon and oxygen cycles.

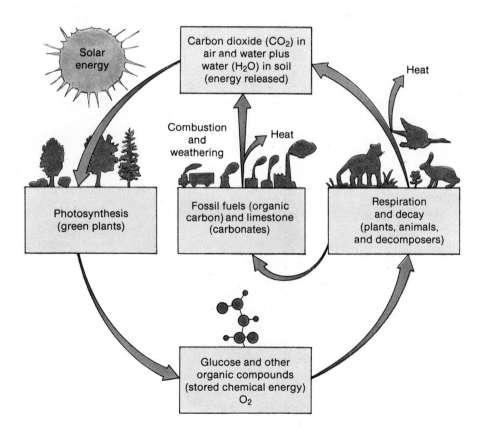

soil and are associated with the roots of certain plants—for example, legumes such as clover, alfalfa, soybeans, peas, beans, and peanuts. The bacteria reside in nodule colonies on the legume roots and fix the nitrogen from the air in the form of nitrates (NO_3) and ammonia (NH_3). Plants that have used these fixed forms of nitrogen and have produced organic matter then provide a source of nitrogen to anyone or anything feeding on the plants. Finally, the nitrogen in the organic wastes of these consuming organisms is freed by denitrifying bacteria, which recycle it back to the atmosphere.

Today, artificial nitrogen fixation, principally for the production of inorganic fertilizers, far exceeds natural denitrification. The surplus of usable nitrogen accumulates in Earth's ecosystems. Some is present as excess nutrients in our waterways. In addition, nitrogen dioxide is a component in acid deposition and photochemical smog production (see Chapter 3).

Limiting Factors. These abiotic components set the limits for organisms that live in various ecosystems. The term **limiting factor** identifies the one physical or chemical component that most inhibits biotic operations, through its lack or excess. A few examples include the low temperatures at high elevations, the lack of water in a desert, the excess water in a bog, the amount of iron in ocean surface environments, the phosphorus content of soils in the eastern United States, or the general lack of active chlorophyll above 6100 m (20,000 ft). In most ecosystems, precipitation is the limiting factor, although variation in temperatures and soil characteristics certainly affect vegetation patterns.

Each organism possesses a range of tolerance for each variable in its environment. Figure 19–11 illustrates the range for two types of trees and two species of birds. The coast redwood (*Sequoia sempervirens*) is limited to a narrow section of the Coast Ranges in California, covering barely 9500 km² (3667 mi²) and concentrated in areas that receive summer advection fog. The red maple (*Acer rubrum*), on the other hand, covers more than 4,000,000 km² (1,544,000 mi²) under varying conditions of moisture and temperature, thus demonstrating a broader tolerance to environmental variations.

The mallard duck (*Anas platyrhynchos*) and the snail kite (*Rostrhámus sociabilis*) also demonstrate a

(a)

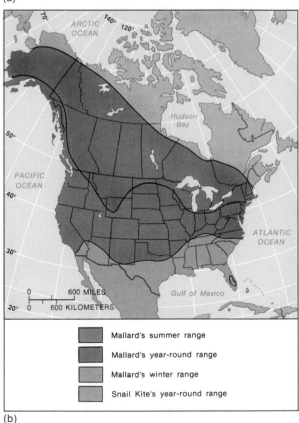

(b)

FIGURE 19–11
(a) Distribution of coast redwood and red maple; (b) distribution of mallard ducks and snail kites.

variation in tolerance and range. The mallard feeds from widely diverse sources, is easily domesticated, and is found throughout most of North America. Snail kites are South American birds that feed only on one specific type of snail; this single food source, then, is their limiting factor. Note their small habitat area near Lake Okeechobee in Florida.

Biotic Ecosystem Operations

The abiotic components of energy, atmosphere, water, weather, climate, and minerals support the biotic components of each ecosystem and their constituent soils, vegetation, and life-forms.

Producers, Consumers, and Decomposers. Organisms that are capable of using carbon dioxide as their sole source of carbon are called **autotrophs**, or *producers;* they fix carbon through photosynthesis. Organisms that depend on autotrophs for their source of carbon are called **heterotrophs**, or *consumers*. Autotrophs are the essential producers in an ecosystem—capturing light and heat energy and converting it to chemical energy, fixing carbon, forming new plant tissue and biomass, and freeing oxygen, all as a part of photosynthesis. From these producers, which manufacture their own food, energy flows through the system along a circuit called the **food chain**, reaching consumers and eventually decomposers. Ecosystems are generally structured as a **food web**, a complex network of interconnected food chains, in which consumers participate in several different food chains. Organisms that share the same basic foods are said to be at the same *trophic* (feeding, nutrition) *level.*

As an example, look at the food web based on krill, a shrimplike crustacean about 7 cm (2.8 in.) long that is a major food source for an interrelated group of organisms, including whales, fish, seabirds, seals, and squid in the Antarctic region (Figure 19–12). All of these organisms participate in numerous other food chains as well, some consuming and some being consumed.

Phytoplankton begin this chain by harvesting solar energy in photosynthesis. The krill, along with other organisms, feed on the phytoplankton. Krill in turn are fed upon by the next trophic level. Because they are a protein-rich, plentiful food, krill are increas-

ingly sought by factory ships, such as those from Japan and the Soviet Union; the annual harvest currently surpasses half a million tons. The impact on the food web of further harvest increases is uncertain, for little is known about krill—their reproduction rate, life span, availability, and any effects of increases in ultraviolet radiation resulting from the seasonal hole in the ozone layer above Antarctica. All of these interrelationships should be clarified before further exploitation continues.

Primary consumers feed on producers, which are always plants, so the primary consumer is called an **herbivore**, or plant eater. A **carnivore** is a *secondary consumer* and primarily eats meat for sustenance. A *tertiary consumer* eats secondary consumers and is referred to as the "top carnivore" in the food chain. A consumer that feeds on both producers (plants) and consumers (meat) is called an **omnivore**—a role occupied by humans, among others.

Any assessment of world food resources depends on the level of consumer being targeted. For example, wheat harvested from an acre of land produces about 810 kg (1800 pounds) of grain. However, if herbivores eat that grain, only 82 kg (180 pounds) of biomass is produced (Figure 19–13). In terms of energy, only about 10% of the calories in plant matter survive from the primary to the secondary trophic level. If, in turn, humans consume the meat derived, there is a further loss of biomass and added inefficiency. And more energy is lost to the environment at each progressive step in the food chain. An omnivorous diet is quite expensive in terms of biomass.

Food chain concepts are becoming politicized as world food issues grow more critical. Today, approximately half of the cultivated acreage in the United States is planted for animal consumption, which includes more than 80% of the annual corn and nonexported soybean harvest. In addition, some lands cleared of rain forests in Central and South America have been converted to pasture to produce beef for export to some restaurants, stores, and fast-food outlets in the United States. Thus, life-style decisions and dietary patterns in the United States and Canada are perpetuating certain food chains, not to mention the destruction of valuable resources, both here and overseas.

The **decomposers** are the final links in the chain; they are the microorganisms—bacteria, fungi, in-

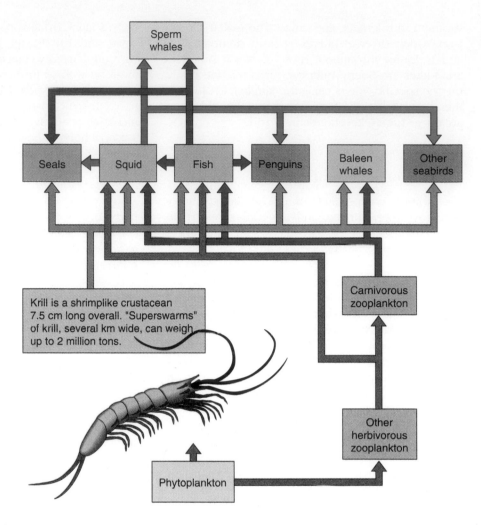

FIGURE 19–12

The food web of the krill in Antarctic waters. After *State of the Ark* by Lee Durrell. Copyright © 1986 by Gaia Books Ltd. Adapted by permission of Doubleday, a division of Bantam Doubleday Dell Publishing Group, Inc.

Krill is a shrimplike crustacean 7.5 cm long overall. "Superswarms" of krill, several km wide, can weigh up to 2 million tons.

Potential human biomass

82 kg beef (180 lbs)

810 kg wheat (1800 lbs.)

810 kg wheat (1800 lbs.)

Consumption of grain-fed beef

Direct consumption of grain

FIGURE 19–13

Biomass pyramids illustrating the difference between direct and indirect consumption of grain.

sects, worms, and others—that digest and recycle the organic debris and waste in the environment. Waste products, dead plants and animals, and other organic remains are their principal food source. And then the chain begins again.

Clearly, some food chains are exceptionally simple, such as eating grains directly. Others are more complex than the Antarctic krill food chain. And familiarity is no indication of simplicity: the home gardener's tomatoes may be eaten by a tomato hornworm, which is then plucked off by a passing robin, which is later eaten by a hawk—and so it goes, in seemingly endless cycles.

Ecological Relationships. A study of food chains and webs is a study of who eats what, and where they do the eating. Figure 19–14 demonstrates the summertime distribution of various populations in specific grassland and temperate forest ecosystems. Such a stepped population pyramid is characteristic of summer conditions in such ecosystems. You can see the decreasing number of organisms at each successive trophic level. The base of the temperate forest pyramid is narrow, however, because most of the producers are large, highly productive trees and shrubs, which are outnumbered by the consumers they can support in the chain.

A basic approach to ecological studies is an analysis of a community's metabolism, the way in which it uses energy and produces food for continued operation. In 1957 H. T. Odum completed one of the best-known studies of this type for Silver Springs, Florida. Figure 19–15, an adaptation of his work, shows that the energy source for the community is insolation. The amount of light absorbed by the plants for photosynthesis is 24% of the total amount arriving at the study site. Of the absorbed amount, only 5% is transformed into the community's gross photosynthetic production—only 1.2% of the total insolation input. The net plant production (gross production minus respiration) is 42.4% of the gross production, or 8833 kilocalories per square meter per year. That amount then moves through the food chain of herbivores and carnivores, with the top carnivores receiving only 0.24% of the net plant production of the system.

You can see that at each trophic level some of the biomass flows to the decomposers. The largest portion is exported from the system in the form of community respiration. In addition, a relatively small portion leaves the system as organic particles by downstream export. The biomass pyramid shown in Figure 19–15(b) portrays the same community from a standing crop perspective, which is an important structural way of examining the existing biomass at each trophic level.

Concentration in Food Chains. When a typical ecosystem of producers and consumers has certain chemical pesticides applied to the producer population, the food web functions to concentrate some of these chemicals. Many chemicals are degraded or diluted in air and water and thus are rendered relatively harmless. However, other chemicals are long-lived, stable, and soluble in the fatty tissues of consumers. These become concentrated at each higher trophic level. Figure 19–16 illustrates this *biological amplification*.

DDT is one such chemical; although banned in some countries, it still is widely used in others. In addition, some other organic and synthetic chemicals, radioactive debris, and exotics such as lead and mercury become concentrated in the food chain, with only small amounts being lost to the environment through respiration and excretion. Thus, a food chain can efficiently poison the organism at the top.

FIGURE 19–14
Ecological pyramids for 0.1 hectares (0.25 acres) of grassland and forest in the summer. Numbers of consumers are shown. Primary, secondary, and tertiary consumers are designated C_1, C_2, and C_3 respectively; producers are labeled P. Adaptation of Figure 3–15a from *Fundamentals of Ecology* by Eugene P. Odum, copyright © 1971 by Saunders College Publishing, a division of Holt, Rinehart and Winston, Inc. Adapted by permission of the publisher.

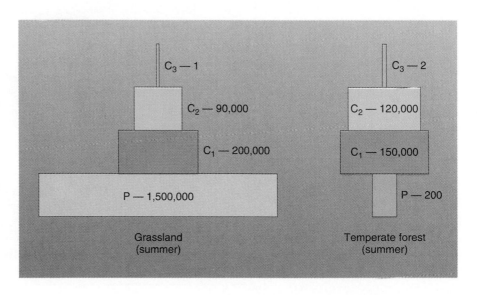

Grassland (summer): C_3 — 1; C_2 — 90,000; C_1 — 200,000; P — 1,500,000

Temperate forest (summer): C_3 — 2; C_2 — 120,000; C_1 — 150,000; P — 200

FIGURE 19–15

(a) Analysis of community metabolism and (b) standing crop distribution (gm/m²) for Silver Springs, Florida. Adaptation of Figures 6–1a and 6–1b from *Fundamentals of Ecology* by Eugene P. Odum, copyright © 1971 by Saunders College Publishing, a division of Holt, Rinehart and Winston, Inc. Adapted by permission of the publisher.

(a)

(b)

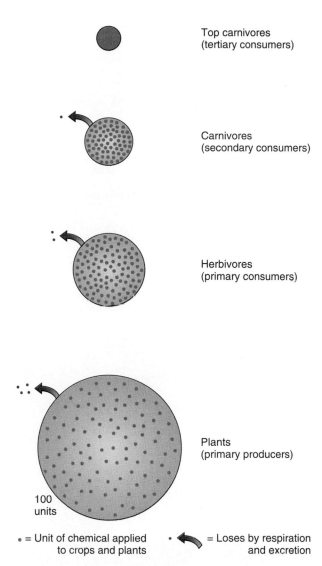

Top carnivores
(tertiary consumers)

Carnivores
(secondary consumers)

Herbivores
(primary consumers)

Plants
(primary producers)

100
units

• = Unit of chemical applied   = Loses by respiration
    to crops and plants   and excretion

FIGURE 19–16
Chemical concentration in food chains. Chemical
residues are passed along a simple food chain to the top
carnivores.

Many species are threatened in this manner, and, of
course, humans are the world's top carnivore.

STABILITY AND SUCCESSION

A rough equilibrium is achieved in an ecosystem by
the constant interplay of increasing and decreasing
factors, both biotic and abiotic, that affect growth

(Figure 19–17). Far from static, Earth's ecosystems
are dynamic and ever-changing. Over time, communities of plants and animals have adapted to such
variation, evolved, and in turn shaped their environments.

Ecosystem Stability

In a given ecosystem, a community moves toward
maximum biomass and relative stability. However,
inertial stability, the tendency for birth and death
rates to balance and the composition of species to
remain stable, does not necessarily foster *resilience,*
or the ability to recover from change. A redwood
forest, a pine forest at a high elevation, and a tropical
rain forest near the equator are all stable communities, yet cleared tracts recover slowly (if ever) and
therefore have poor resilience (Figure 19–18). In
contrast, a midlatitude grassland is low in stability;
yet when burned, its resilience is high, and the
community recovers rapidly.

Another aspect related to stability is *diversity.* The
more diverse the species population (in number of
species and quantity of members in each species),
the more risks are spread over the entire community,
because several food sources exist at each trophic
level. In other words, greater diversity in an ecosystem results in greater stability. An artificially produced monoculture community, such as a field of
wheat, is singularly vulnerable to failure or attack. In
some regions, a simple return to multiple crops
brings more stability to the ecosystem.

Climate Change. Certainly the distribution of plant
species is affected by changes in climate. Plant
communities have survived wide climate swings in
the past. Consider the beginning of the Tertiary
period, 75 million years ago. Warm, humid conditions and tropical forests dominated northward to
southern Canada, pines were in the Arctic, and
deserts were few. Then, between 15 and 50 million
years ago, deserts began developing in the southwestern United States, and mountain-building processes were creating higher elevations, with leeward
slopes and rain-shadow aridity affecting plants.

The key question is: as temperature patterns
change, how fast can plants either adapt to the new
conditions, or migrate to remain within their specific
habitats? As an example, the present transition area

FIGURE 19–17
Ecosystem and population
balance in a community.

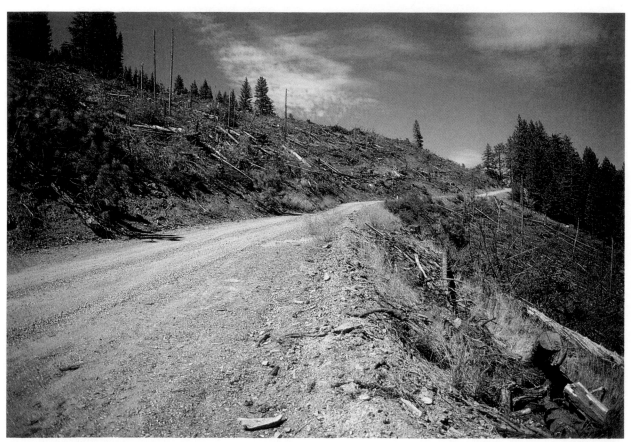

FIGURE 19–18
Logged pine forest—an example of timber harvesting disrupting a stable community. Photo by author.

between prairie grasslands and northern forests is in central Minnesota, but with increasing temperatures this transition is expected to move northward 400–600 km (250–375 mi) over the next 40 years, an adjustment of more than 100 km (60 mi) per decade. Agricultural lands producing wheat, corn, soybeans, and other commodities also will shift. Society will have to adapt to new cropping patterns.

A study completed by scientist Margaret Davis on North American forests suggests that trees will have to respond quickly if temperatures increase. Shifts in the location of certain species could reach 100 to 400 kilometers (60–250 mi) during the next 100 years. Some species, such as the sugar maple, may disappear in the United States, except for Maine, and migrate northward into eastern Ontario and Quebec. Davis prepared a map of the possible impact of rising temperatures on the distribution of beech trees (Figure 19–19).

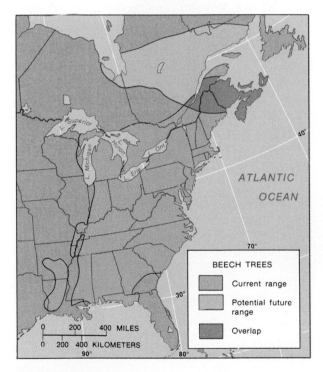

FIGURE 19–19
Present and predicted future distribution of beech trees in North America, given estimated climate changes. After *Science* 243 (10 February 1989): 735. "How Fast Can Trees Migrate?" by Leslie Roberts, 1989. Copyright 1989 by the AAAS.

Ecological Succession

Ecological succession occurs when different communities of plants and animals (usually more complex) replace older communities (usually simpler). Each temporary community of species modifies the physical environment in a manner suitable for the establishment of a later set of species. Changes apparently move toward a more stable and mature condition, which is optimum for a specific environment. This end product in an area is called the *ecological climax,* with plants and animals forming a *climax community*—a stable, self-sustaining, and symbiotically functioning community with balanced birth, growth, and death.

A generalized ecological succession provides a convenient model with which to characterize community development. However, given the complexity of natural ecosystems, real succession involves much more than a series of predictable stages ending with a specific "monoclimax" community. Instead, there may be several final stages, or a "polyclimax" condition. Climax communities are properly thought of as being in dynamic equilibrium, and at times may even be out of phase with the immediate physical environment.

Succession often requires an initiating disturbance, such as strong winds or a storm, or a practice such as prolonged overgrazing. Early communities may actually inhibit the growth of other species, but when existing organisms are disturbed or removed, new communities can emerge. At such times of transition, the interrelationships among species produce elements of chance, and species having an adaptive edge will succeed in the competitive struggle for light, water, nutrients, space, time, reproduction, and survival. The *principle of individuality* holds that each species succeeds in its own way, according to its own requirements relating to both the physical environment and the complex interactions with other species. Thus, the succession of plant and animal communities is an involved process with many interactive variables.

Terrestrial Succession. An area of bare rock and soil without any vestige of a former community can be a site for primary succession. The initial community is called a **pioneer community**. It may be found in sites such as a new surface created by mass movements of land, in areas exposed by a retreating glacier, on devastated lands such as those north of Mount Saint Helens in Washington State, or on cooled lava flows, or on lands disturbed by human activities (e.g., strip mining and land development). Mosses and lichens often begin primary succession on bare rock. However, succession from a previously functioning community is more common. An area whose natural community has been destroyed or disturbed, but still has the underlying soil intact, may experience *secondary succession*.

In terrestrial ecosystems, regrowth begins with pioneer species and further soil development (Figure 19–20). As succession progresses, a different set of plants and animals with different niche require-

ments may adapt. Further niche expansion follows as the community matures. For example, when a previously farmed field is permitted to lie fallow (unused), it initially fills with annual weeds, then perennial weeds and grasses, followed by shrubs and perhaps trees. Figure 19–21 illustrates such a sequence for the southeastern United States. But despite the convenient categorization shown in the figure, think of succession as continuous, with succeeding communities overlapping in time and space.

One challenge in a study of plant and animal communities is to devise criteria that allow researchers to distinguish between a successional community and a climax community. The relative stability of the

FIGURE 19–20
Succession and recovery of pioneer species in the devastated area north of Mount Saint Helens, three years after the 1980 eruption. Photo by author.

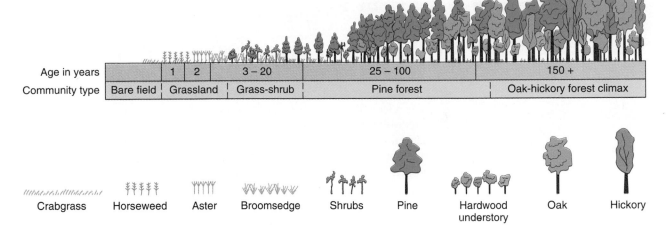

| Age in years | | 1 | 2 | 3 – 20 | 25 – 100 | 150 + |
| Community type | Bare field | Grassland | | Grass-shrub | Pine forest | Oak-hickory forest climax |

| Crabgrass | Horseweed | Aster | Broomsedge | Shrubs | Pine | Hardwood understory | Oak | Hickory |

FIGURE 19–21

Typical secondary succession of principal plants in the southeastern United States. Adaptation of Figure 9–4 from *Fundamentals of Ecology* by Eugene P. Odum, copyright © 1971 by Saunders College Publishing, a division of Holt, Rinehart and Winston, Inc., adapted by permission of the publisher.

community appears to be the key factor. Unfortunately, however, that stability must be assessed over a period of time that greatly exceeds human life spans and may be interrupted from time to time by various phenomena. Focus Study 19–1 addresses fire, which is one such natural interruption.

Aquatic Succession. The biotic aspects of lakes and ponds exhibit another form of ecological succession. A lake experiences successional stages as it fills with nutrients and sediment and as aquatic plants take root and grow, capturing more sediment and adding organic debris to the system (Figure 19–22). This gradual enrichment through various stages in water bodies is known as **eutrophication**. For example, in moist climates a floating mat of vegetation grows outward from the shore as the lake fills, to form a bog. Cattails and other marsh plants become established, and partially decomposed organic material accumulates in the basin, with additional vegetation bordering the remaining lake surface. A meadow may form as the peat bog solidifies; willow trees follow, and perhaps cottonwood trees; and eventually the lake may evolve into a locally appropriate forest community.

The progressive stages in lake succession are named *oligotrophic* (low nutrients), *mesotrophic* (medium nutrients), and *eutrophic* (high nutrients). Each stage is marked by an increase in primary productivity and resultant decreases in water transparency so that photosynthesis becomes concentrated near the surface. Energy flow shifts from production to respiration in the eutrophic stage, with oxygen demand exceeding oxygen availability.

The nutrient levels vary in different areas of a lake: oligotrophic conditions occur in deep water, and eutrophic conditions along the shore, in shallow bays, or in locations where sewage or other nutrient inputs occur. Even large water bodies may have eutrophic areas along the shore. As society dumps sewage and pollution in waterways, the nutrient load is enhanced beyond the cleansing ability of natural biological processes, thus producing *cultural eutrophication* and hastening the succession in aquatic systems.

SUMMARY

Earth's biosphere is unique in the Solar System; its ecosystems are the heart of life. Plants and animals

Fire Ecology and Succession

It is estimated that one-fourth of Earth's land area experiences fire each year. Over the past 50 years, *fire ecology* has been the subject of much scientific research and experimentation. Today, fire is recognized as a natural component of most ecosystems and not the enemy of nature it once was popularly considered to be. In fact, in many forests, undergrowth is purposely burned in controlled "cool fires" to remove fuel that could enable a catastrophic and destructive "hot fire." Ironically, when fire suppression and prevention strategies are rigidly followed, they can lead to abundant undergrowth accumulation, which allows the potential total destruction of a forest by a major fire.

Fire ecology imitates nature by recognizing fire as a dynamic ingredient in community succession. The Forest Service first recognized the principle of fire ecology in the early 1940s and formally implemented the practice in 1972. Their challenge became one of regulating and controlling fires to secure reproduction and to prevent accumulation of forest litter and brush. Controlled ground fires are now widely regarded as wise forest management practice and are used across the country (Figure 1).

FIGURE 1
Fire ecology practices of the U.S. Forest Service; controlled burning has been used for several decades. Photo by author.

The geographer Carl Sauer (1889–1975) wrote in 1955 about the use of fire throughout the history of civilization. He stated that people often used fire deliberately to increase the food supply on their land:

> Mature woody growth provides less food for humans and ground animals than do fire-disturbed sites, with protein-rich young growth and stimulated seed production . . . game yields are usually greatest where the vegetation is kept in an immediate state of ecological succession.*

One of Sauer's major contentions was that the origin and preservation of grasslands, which occur in various climates and over a wide range of soil types, are mainly the result of human-set fires. The origin of grasslands is still controversial and is the subject of continued discussion.

The history behind modern society's demand for fire prevention to protect property goes back to European forestry of the 1800s. Fire prevention became an article of faith for forest managers in North America. However, in studies of the longleaf pine forests that stretch in a wide band from the eastern coastal plain around to Texas, fire was discovered to be an integral part of their regrowth following lumbering activity. In fact, in some species of pine, such as the knobcone pine, seed dispersal does not occur without the assistance of a forest fire. Heat from the fire opens the cones, releasing seeds so they can fall to the ground for germination.

Nonetheless, after some 72,000 forest fires in the western United States in 1988, especially those that charred portions of the highly visible Yellowstone National Park, an outcry was heard from forestry and recreational interests. The demand was for the Forest Service and ecologists to admit they were wrong and to abandon fire ecology. Critics called fire ecology practices the government's "let burn" policy. In truth, about 20% of the acreage in Yellowstone Park actually burned (180,000 out of 900,000 hectares, or 440,000 of 2,200,000 acres). Only about half of that area experienced the worst fire damage (Figure 2). This expanse was far less than originally reported in the media.

In its final report on the fire, a government interagency task force concluded that, "an attempt to exclude fire from these lands leads to major unnatural changes in vegetation . . . as well as creating fuel accumulation that can lead to uncontrollable, sometimes very damaging wildfire." Thus, participating federal land managers and others reaffirmed their stand that fire ecology is a fundamentally sound concept.

---

*Carl O. Sauer, "Agency of Man on Earth," in *Man's Role in Changing the Face of the Earth,* ed. William L. Thomas. Chicago: University of Chicago Press, 1956, p. 54.

FIGURE 2
(a) Yellowstone National Park
fire of 1988; (b) map of burned
vegetation in Park. Courtesy of
the National Park Service.

(a)

(b)

FIGURE 19–22
Lake bog succession.

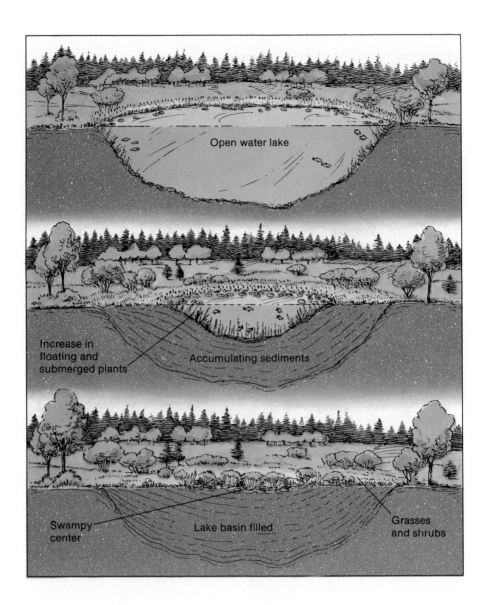

Open water lake

Increase in
floating and
submerged plants

Accumulating sediments

Swampy
center

Lake basin filled

Grasses
and shrubs

have developed complex community relationships centered in specific habitats and niches. Plants perform photosynthesis, providing energy and oxygen to drive biological processes. Net primary productivity may be the most important aspect of any community. Light, temperature, water, and the cycling of gases and nutrients constitute the supporting abiotic components of ecosystems. Biomass and population pyramids characterize the flow of energy and the numbers of producers, consumers, and decomposers operating in an ecosystem. Ecosystems appear to move toward stability and balance, with individual terrestrial and aquatic ecosystems capable of progressing through a succession of stages. Human activity intercedes at almost every level of natural ecosystem operations.

KEY TERMS

Autotrophs	Habitat
Biogeochemical cycles	Herbivore
Biogeography	Heterotrophs
Biomass	Life zone
Carnivore	Limiting factor
Chlorophyll	Net primary productivity
Community	Niche
Decomposers	Omnivore
Ecological succession	Photoperiod
Ecology	Photosynthesis
Ecosystem	Pioneer community
Eutrophication	Respiration
Food chain	Stomata
Food web	Vascular plants

- --

REVIEW QUESTIONS

1. What is the relationship between the biosphere and an ecosystem? Define ecosystem and give some examples.
2. What does biogeography include? Describe its relationship to ecology.
3. Briefly summarize what an ecosystem is and what its operation implies about the complexity of life.
4. Define a community within an ecosystem.
5. What do the concepts of habitat and niche refer to? Relate them to some specific plant and animal communities.
6. Describe symbiotic and parasitic relationships in nature. Draw an analogy between those relationships and human societies on our planet. Explain.
7. What are the principal abiotic components in terrestrial ecosystems?
8. Describe what Humboldt found that led him to propose the life-zone concept. What are life zones? Explain the interaction among altitude, latitude, and the types of communities that develop.
9. What are biogeochemical cycles? Describe several of the essential cycles.
10. What role is played in an ecosystem by autotrophs and heterotrophs?
11. What is a limiting factor? How does it function to control the spatial distribution of plant and animal species? Explain the concept of tolerance in this connection.
12. Define a vascular plant. How many species are there on Earth?
13. How do plants function to link the Sun's energy to living organisms? What is formed within the light-activated cells of plants?
14. What is net photosynthesis? How does the concept relate to photosynthesis, respiration, or net primary productivity?
15. What are the broad spatial patterns worldwide of net primary productivity?
16. Describe the relationship among producers, consumers, and decomposers in an ecosystem. What is the trophic nature of an ecosystem? What is the place of humans in a trophic system?
17. What are biomass and population pyramids? Describe how these models help explain the nature of food chains and communities of plants and animals.

18. Follow the flow of energy and biomass through the Silver Springs, Florida, ecosystem. Describe the pathways in Figure 19–15(a).

19. What is meant by ecosystem stability?

20. How does ecological succession proceed? What are the relationships that exist between existing communities and new, invading communities? Explain the principle of competition.

21. Summarize the process of succession in a body of water. What is meant by cultural eutrophication?

22. Discuss the concept of fire ecology in the context of the Yellowstone National Park fires of 1988. What were the findings of the government task force?

SUGGESTED READINGS

Berner, Robert A., and Antonio C. Lasaga. "Modeling the Geochemical Carbon Cycle." *Scientific American* 260 (March 1989): 74–84.

Bolin, Bert. "The Carbon Cycle." *Scientific American* 223 (September 1970): 124–32.

Bormann, F. Herbert, and Gene E. Likens. "The Nutrient Cycles of an Ecosystem." *Scientific American* (October 1970): 92–101.

Christensen, Norman L. *Ecological Consequences of the 1988 Fires in the Greater Yellowstone Area—Final Report*. Denver: Greater Yellowstone Coordinating Committee, 1988.

Clapham, W. B., Jr. *Natural Ecosystems*. 2d ed. New York: Macmillan, 1983.

Delwiche, C. C. "The Nitrogen Cycle." *Scientific American* 223 (September 1970): 136–46.

Ehrlich, Paul R. and Anne H., and John P. Holdren. *Ecoscience—Population, Resources, Environment*. San Francisco: W. H. Freeman, 1977.

Farb, Peter, and the editors of Life. *Ecology*. Life Nature Library. New York: Time, 1963.

Findley, Rowe. "Will We Save Our Own [Forests]?" *National Geographic* 78, no. 3 (September 1990): 106–36.

Jeffery, David. "Yellowstone—The Great Fires of 1988." *National Geographic* 175, no. 2 (February 1989): 255–73.

Jensen, William A., and Frank B. Salisbury. *Botany: An Ecological Approach*. Belmont, CA: Wadsworth, 1972.

Miller, G. Tyler, Jr. *Living in the Environment*. 6th ed. Belmont, CA: Wadsworth, 1990.

Mills, Susan M., and the Greater Yellowstone Coordinating Committee, eds. *The Greater Yellowstone Postfire Assessment*. Yellowstone National Park: National Park Service, 1989.

Northington, David K., and J. R. Goodin. *The Botanical World*. St. Louis: Times Mirror/Mosby, 1984.

Odum, Eugene P. *Ecology and Our Endangered Life-Support Systems*. Sunderland, MD: Sinauer Associates, 1989.

————. *Fundamentals of Ecology*. 3d ed. Philadelphia: W. B. Saunders, 1971.

Odum, Howard T. *Systems Ecology—An Introduction*. New York: John Wiley & Sons, 1983.

Wessells, Norman K., and Janet L. Hopson. *Biology*. New York: Random House, 1988.

Whittaker, Robert H. *Communities and Ecosystems*. 2d ed. New York: Macmillan, 1975.

Williams, Ted. "Incineration of Yellowstone." *Audubon* (January 1989): 38–85.

Coastal meadow in spring. Photo by author.

20
Terrestrial Biomes

Comparison of two quotations, written little more than a century apart in 1874 and 1989, suggests the danger of ignoring the interrelationships among plants and their environment:

> We have now felled forest enough everywhere, in many districts far too much. Let us now restore this one element of material life to its normal proportions, and devise means of maintaining the permanence of its relations to the fields, the meadows, and the pastures, to the rain and the dews of heaven, to the springs and rivulets with which it waters the Earth.*

> The grim state of much of the world's forests and the consequences of their destruction came to light in numerous, unsettling ways in 1988. Hills stripped of trees in the Himalayan watershed contributed to the worst flooding on record in Bangladesh. . . . Twenty million acres of Brazilian rain forest were burned in 1987 alone. . . . In industrial countries, forests continued to deteriorate under the stresses of air pollution and acid rain. . . . Only vigorous efforts to protect the remaining forests and to plant billions more trees each year can reverse these worrisome trends.**

Both these selections offer a warning; unfortunately, the first one has been ignored. In 1874, Marsh perceived the need for understanding and managing the environment at a time when Earth was viewed as offering few limits to human activity. In contrast, scientists today measure ecosystems carefully and build elaborate computer models simulating the evolving human-environment experiment. Humans are Earth's major biotic agent, influencing all ecosystems. What will the quotation about Earth's plant communities be like in another century? In this chapter we explore Earth's major terrestrial ecosystems, their appearance and structure, location, and the present status of related plants, animals, and environment.

---

*George Perkins Marsh, *The Earth as Modified by Human Action,* New York: Scribner's, 1874, p. 385–86.

**Lester R. Brown, *State of the World—1989,* New York: W. W. Norton, 1989, p. 180. Reprinted by permission.

BIOGEOGRAPHICAL REALMS

Biogeographical realms of plants and animals are geographic regions where groups of species evolved. From these centers, species migrate worldwide according to their niche requirements, reproductive success, competition, and climatic and topographic barriers. Recognition that such distinct regions of flora and fauna exist was an early beginning of biogeography as a discipline.

The map in Figure 20–1 illustrates the botanical (plant) and zoological (animal) regions forming these biogeographical realms. Each realm contains many distinct ecosystems that distinguish it from other realms. The Australian realm is perhaps the most unique, giving rise to 450 species of *Eucalyptus* among its plants and 125 species of marsupials among its animals. Australia's uniqueness is the result of its early isolation from the other continents. During critical evolutionary times, Australia drifted away from Pangaea (see Chapter 11) and was never reconnected by a land bridge, even when sea level lowered during repeated glacial ages. New Zealand's detached isolation from Australia also undoubtedly explains why there are no native marsupials in New Zealand.

Alfred Wallace (1823–1913), the first scholar of zoogeography, used the stark contrast in animal species that occurs among islands in present-day Indonesia to delimit a boundary between the Oriental and Australian realms, thinking that deep water in the straits had prevented species crossover. However, his boundary line actually is the site of a wide transition zone, where one region grades into the other. Wallace himself recognized the existence of such transition zones. Boundaries between natural systems are, more accurately, "zones of shared traits." Thus, despite distinctions among individual realms, it is best to think of their borders as transition zones of mixed identity and composition, rather than as rigidly defined boundaries.

Aquatic Ecosystems

The major ecosystems that comprise biogeographical realms are conveniently divided for study into aquatic and terrestrial ecosystems. Oceans, estuaries, and freshwater bodies comprise the **aquatic ecosys-**

FIGURE 20–1

Biogeographical realms of plants and animals. Adaptation of Figure 14–1 from *Fundamentals of Ecology* by Eugene P. Odum, copyright © 1971 by Saunders College Publishing, a division of Holt, Rinehart and Winston, Inc., adapted by permission of the publisher.

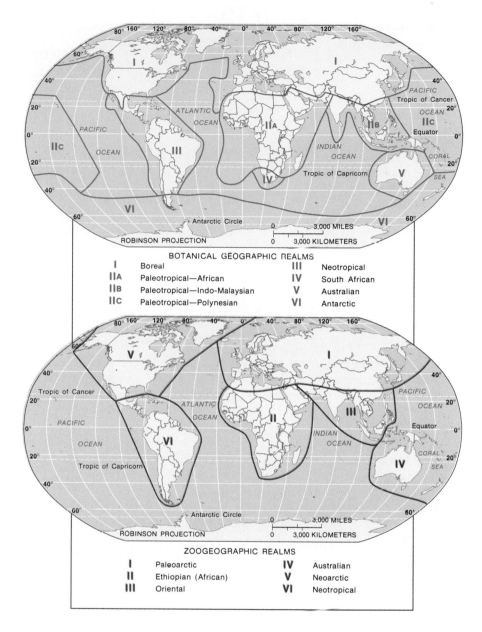

BOTANICAL GEOGRAPHIC REALMS

I	Boreal	III	Neotropical
IIA	Paleotropical—African	IV	South African
IIB	Paleotropical—Indo-Malaysian	V	Australian
IIC	Paleotropical—Polynesian	VI	Antarctic

ZOOGEOGRAPHIC REALMS

I	Paleoarctic	IV	Australian
II	Ethiopian (African)	V	Neoarctic
III	Oriental	VI	Neotropical

tems. The *photic layer* is the active zone that receives light for *phytoplankton,* which arc the one-celled, chlorophyll-based primary producers suspended in the water. Primary consumers include innumerable herbivores, some as small as 0.2 mm (0.008 in.). These form the basis of a productive and complex food chain.

Kenneth Sherman and Lewis Alexander, scientists with the U.S. Marine Fisheries Service Laboratory and the University of Rhode Island, respectively, conceived the designation of a **large marine ecosystem (LME)** to foster better management and improve the sustainability of both biotic and abiotic aquatic resources. An LME is an oceanic region with unique organisms, floor topography, currents, areas of nutrient-rich upwelling circulations, or areas of significant predation, including human (Figure 20–2). Such areas also include nearly enclosed bodies of

FIGURE 20–2

Large marine ecosystems (LMEs): a. Oyashio Current; b. Kuroshio Current; c. Yellow Sea; d. Gulf of Thailand; e. Great Barrier Reef; f. Tasman Sea; g. Insular Pacific; h. East Bering Sea; i. Gulf of Alaska; j. California Current; k. Humboldt Current; l. Antarctic; m. Gulf of Mexico; n. Southeast Continental Shelf; o. Northeast Continental Shelf; p. East Greenland Sea; q. Barents Sea; r. Baltic Sea; s. North Sea; t. Iberian Coastal; u. Gulf of Guinea; v. Benguela Current. After Kenneth Sherman, *World Resources 1988–89,* New York: World Resources Institute, 1988, p. 147. Adapted by permission.

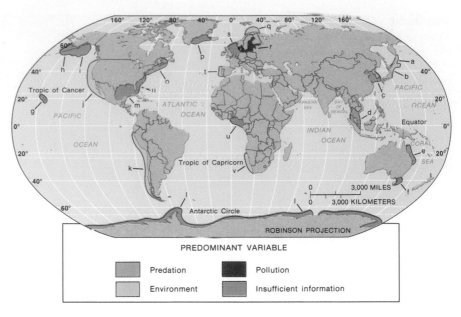

water, such as the Baltic and Mediterranean seas, which are so threatened by pollution.

Thirty LMEs with more than 200,000 km² (77,200 mi²) in each have been identified. The goal of such ecosystem designations is to encourage resource planners and managers to consider complete ecosystems, not just a targeted species. Poor understanding of aquatic ecosystems has led to the failure of some species, such as herring in the Georges Bank area of the Atlantic. Also, LMEs are designated to insure long-term data collection for management and further research.

Terrestrial Ecosystems

Plants are the most visible part of the biotic landscape, a key aspect of Earth's terrestrial ecosystems. In their growth, form, and distribution, plants reflect Earth's key systems: its energy patterns; atmospheric composition; temperature and winds; air masses; water quantity, quality, and seasonal timing; soils; regional climates; geomorphic processes; and ecosystem dynamics. Here is a review of ways that plants are grouped for geographic analysis: A *community* is formed by interacting populations of plants and animals in an area. An *ecosystem* involves the interplay between a community of plants and animals

and its abiotic physical environment. Each plant occupies an area in which it is biologically suited to live—its *habitat;* and within that habitat it performs a basic operational function—its *niche.*

Large vegetation units, the floristic component of a terrestrial ecosystem characterized by a dominant plant community, are called plant **formation classes.** Each formation includes numerous plant communities, and each community includes innumerable plant habitats. Within those habitats, Earth's diversity is expressed in 250,000 plant species. Despite this intricate complexity, we can generalize the size, shape, and structure of these wide-ranging species into six general formations: forest, woodland, grassland, shrubland, desert, and tundra.

Other more specific systems are used for structural classifications, such as *life-form* designations based on the outward physical properties of individual plants or the general form of a vegetation cover. These physical life-forms include trees (larger woody main trunk, perennial, usually above 3 m or 10 ft); lianas (woody climbers and vines); shrubs (smaller woody plants; branching stems at ground); herbs (small plants without woody stems above ground); bryophytes (mosses, liverworts); epiphytes (plants growing above the ground, on other plants); and thallophytes, which lack true leaves, stems, or roots (bacteria, fungi, algae, lichens) (Figure 20–3).

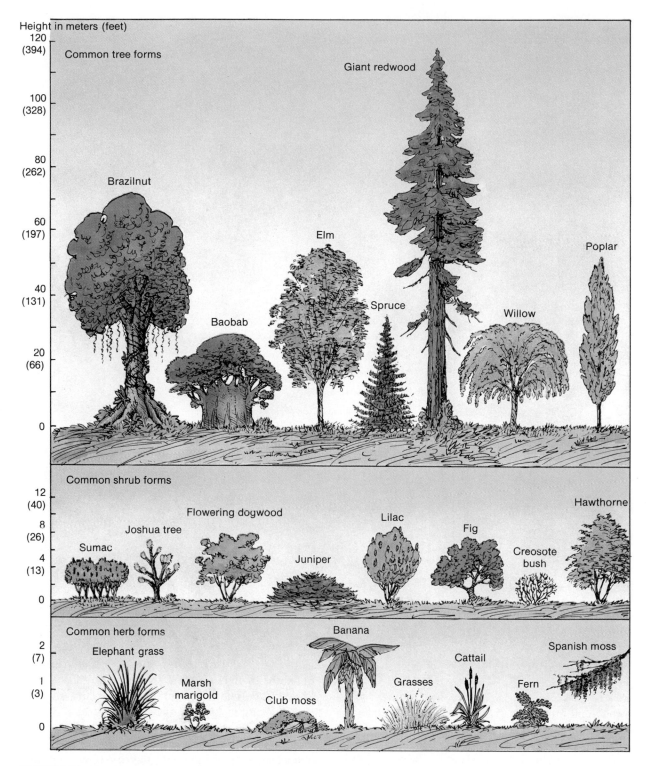

FIGURE 20–3

Examples of plant life-forms. After William M. Marsh and J. Dozier, *Landscape: An Introduction to Physical Geography,* copyright © 1981, p. 273. Adapted by permission of John Wiley & Sons, Inc.

FIGURE 20–4
The 10 major global terrestrial biomes.

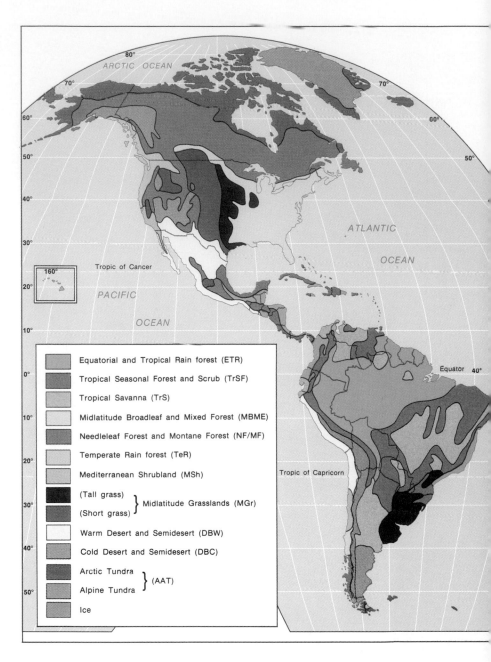

Equatorial and Tropical Rain forest (ETR)

Tropical Seasonal Forest and Scrub (TrSF)

Tropical Savanna (TrS)

Midlatitude Broadleaf and Mixed Forest (MBME)

Needleleaf Forest and Montane Forest (NF/MF)

Temperate Rain forest (TeR)

Mediterranean Shrubland (MSh)

(Tall grass)
                    } Midlatitude Grasslands (MGr)
(Short grass)

Warm Desert and Semidesert (DBW)

Cold Desert and Semidesert (DBC)

Arctic Tundra
                } (AAT)
Alpine Tundra

Ice

EARTH'S MAJOR BIOMES

A large, stable **terrestrial ecosystem** is known as a biome. A **biome** is characterized by specific plant and animal communities and their interrelationship with the physical environment. Each biome is usually named for its dominant vegetation. Because plant distributions are responsive to environmental conditions and reflect variation in climatic and other abiotic factors, the biome-related world climate map in Figure 10–5 is a helpful reference for this chapter.

Few truly natural communities of plants and animals remain; most biomes have been greatly altered by human intervention. Thus, the "natural vegetation" identified on many biome maps reflects ideal climax vegetation potential, given the physical factors prevalent in a region. Even though human practices have greatly altered these ideal forms, it is valuable to

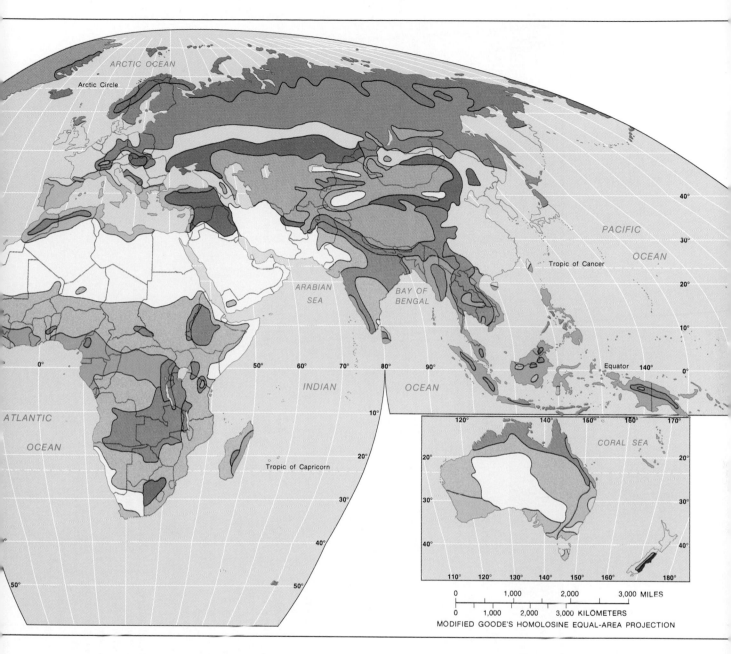

MODIFIED GOODE'S HOMOLOSINE EQUAL-AREA PROJECTION

study the natural biomes to better understand the natural environment and to assess the extent of human-caused alteration. In addition, knowing the ideal in a region guides us to a closer approximation of natural vegetation in the plants we introduce. In the United States and Canada, we humans tend to perpetuate a type of transition community, somewhere between a grassland and a forest. We plant trees and lawns and then must invest energy, water, and capital to sustain such artificial modifications of nature.

Table 20–1 describes each biome and summarizes other pertinent information from throughout this text—a synthesis of many chapters. Earth's biomes provide an appropriate point of synthesis of the environment and biosphere. The pages that follow expand on the table and describe each biome in greater detail. The global distribution of Earth's major terrestrial biomes is portrayed in Figure 20–4.

TABLE 20–1

Major terrestrial biomes and their characteristics.

Biomes and Ecosystems (map symbol)	Vegetation Characteristics	Soil Class	Köppen Climate Designation	Annual Precipitation Range	Temperature Patterns	Water Balance
Equatorial and Tropical Rain Forest (ETR): Evergreen broadleaf forest Selva	Leaf canopy thick and continuous; broadleaf evergreen trees; vines (lianas), epiphytes, tree ferns, palms	Oxisols Ultisols (on well-drained uplands)	Af Am (limited dry season)	180–400cm (>6 cm/mo)	Always warm (21°–30°C; avg. 25°C)	Surpluses all year
Tropical Seasonal Forest and Scrub (TrSF): Tropical monsoon forest Tropical deciduous forest Scrub woodland and thorn forest	Transitional between rain forest and grasslands; broadleaf, some deciduous trees; open parkland to dense undergrowth; acacias and other thorn trees in open growth	Oxisols Ultisols Vertisols (in India) Some Alfisols	Am Aw Borders BS	130–200 cm (<40 rainy days during 4 driest months)	Variable, always warm (>18°C)	Seasonal surpluses and deficits
Tropical Savanna (TrS): Tropical grassland Thorn tree scrub Thorn woodland	Transitional between seasonal forests, rain forests, and semiarid tropical steppes and desert; trees with flattened crowns, clumped grasses, and bush thickets; fire association important	Alfisols (dry: Ustalfs) Ultisols Oxisols	Aw BS	90–150 cm, seasonal	No cold weather limitations	Tends toward deficits, therefore fire and drought susceptible
Midlatitude Broadleaf and Mixed Forest (MBME): Temperate broadleaf Midlatitude deciduous Temperate needleleaf	Mixed broadleaf and needleleaf trees; deciduous broadleaf, losing leaves in winter; southern and eastern evergreen pines demonstrating fire association	Ultisols Some Alfisols	Cfa Cwa Dfa	75–150 cm	Temperate, with cold season	Seasonal pattern with summer maximum PRECIP and POTET; no irrigation needed

Ecosystem	Vegetation	Soils	Köppen Classification	Annual Precipitation	Temperature	Water Balance
Northern Needleleaf Forest and Montane Forest (NF/MF): Taiga, Boreal forest, Other Montane forests and highlands	Needleleaf conifers, mostly evergreen pine, spruce, fir; Russian larch, a deciduous needleleaf	Spodosols Histosols Inceptisols (permafrost) Alfisols (Boralfs: cold)	Subarctic Dfb Dfc Dfd	35–100 cm	Short summer, cold winter	Low POTET, moderate PRECIP, moist soils, some waterlogged and frozen in winter; no deficits
Temperate Rain Forest (TeR): West coast forest Coast redwoods (U.S.)	Narrow margin of lush evergreen and deciduous trees on windward slopes; redwoods, tallest trees on Earth	Spodosols Inceptisols (mountainous environs)	Cfb Cfc	150–500 cm	Mild summer and mild winter for latitude	Large surpluses and runoff
Mediterranean Shrubland (MSh): Sclerophyllous shrubs Australian Eucalyptus forest	Short shrubs, drought adapted, trending to grassy woodlands; chaparral	Alfisols (Xeralfs) Mollisols (Xerolls)	Csa Csb	25–65 cm	Hot, dry summers, cool winters	Summer deficits, winter surpluses
Midlatitude Grasslands (MGr): Temperate grassland Sclerophyllous shrub	Tall grass prairies and short grass steppes, highly modified by human activity; major areas of commercial grain farming; plains, pampas, and veld	Mollisols Aridisols	Cfa Dfa	25–75 cm	Temperate continental regimes	Soil moisture utilization and recharge balanced; irrigation and dry farming in drier areas
Warm Desert and Semidesert (DBW): Subtropical desert and scrubland	Bare ground graduating into xerophytic plants, including succulents, cacti, and dry shrubs	Aridisols Entiso's (sand dunes)	BWh BWk	<2 cm	Average annual temperature around 18°C, highest temperatures on Earth	Chronic deficits, irregular precipitation events, PRECIP < ½ POTET
Cold Desert and Semidesert (DBC): Midlatitude desert, scrubland, and steppe	Cold desert vegetation includes short grass, and dry shrubs	Aridisols Entisols	BSh BSk	2–25 cm	Average annual temperature around 18°C	PRECIP > ½ POTET
Arctic and Alpine Tundra (AAT)	Treeless; dwarf shrubs, stunted sedges, mosses, lichens, and short grasses; alpine, grass meadows	Inceptisols Histosols Entisols (permafrost)	ET Dwd	15–80 cm	Warmest month <10°C, only 2 or 3 months above freezing	Not applicable most of the year, poor drainage in summer
Ice			EF			

Equatorial and Tropical Rain Forest

Earth is girdled with a lush biome—the **equatorial and tropical rain forest**. In a climate of consistent daylength (12 hours), high insolation, and average annual temperatures around 25°C (77°F), plant and animal populations have responded with the most diverse body of life on the planet. The Amazon region is the largest tract of equatorial and tropical rain forest, also called the *selva*. Rain forests also cover equatorial regions of Africa, Indonesia, the margins of Madagascar and Southeast Asia, the Pacific coast of Ecuador and Colombia, and the east coast of Central America, with small discontinuous patches elsewhere. The cloud forests of western Venezuela are such tracts of rain forest at high elevation, perpetuated by high humidity and cloud cover. Undisturbed tracts of rain forest are rare.

In his 1960 book, *The Forest and the Sea,* Marston Bates compared rain forests and oceans and found remarkable similarities in these very dissimilar environments. Biomass in a rain forest is concentrated high up in the canopy, that dense mass of overhead leaves, just as life in the sea is concentrated in the photic layer near the ocean's surface. The sea and the forest both have a vertical distribution of life, dependent on a competitive struggle for sunlight. The treetops of the rain forest are generally analogous to the upper (pelagic) region of the sea, where photosynthesis fixes energy for a dependent food chain. The floor of the rain forest and the floor of the ocean also are roughly parallel in that both are dark or dimly lit, relatively barren, and a place of fewer life-forms. This general analogy by Bates gives an idea of the structure of life in the rain forest.

Rain forests feature ecological niches distributed by height rather than horizontally because of the competition for light. The canopy has a rich variety of plants and animals. Lianas (vines) branch from tree to tree, binding them together, with cords that can reach 20 cm (8 in.) in diameter. Epiphytes flourish there, too: such plants as small orchids, bromeliads, and ferns that live completely above ground, supported physically but not nutritionally by the structures of other plants.

The rain forest canopy usually is considered to have three levels: a high level that averages 50 to 60 m (165–200 ft) above ground, a middle level from 20 to 40 m (65–130 ft), and a lower level from 5 to 15 m (15–50 ft). Of the three, the middle canopy is the most continuous, with its broad leaves blocking much of the light and creating a darkened forest floor (Figure 20–5).

The upper level is not continuous but features tall trees whose high crowns rise above the middle canopy. The lower level is composed of small seedlings, ferns, bamboo, and the like, leaving the ground level in deep shade and fairly open. A look at aerial photographs of a rain forest, aided by the false Hollywood-movie imagery of the jungle, makes it difficult to imagine the shadowy environment of the actual rain forest floor, which receives only about 1% of the sunlight arriving at the canopy. The constant moisture, rotting fruit and moldy odors, strings of

FIGURE 20–5
The three levels of a rain forest canopy.

60 m/200 ft ⎤
            ⎬ High-level canopy
50 m/165 ft ⎦

40 m/130 ft ⎤
            ⎬ Middle-level canopy (heaviest of the three)
20 m/65 ft ⎦

15 m/50 ft ⎤
            ⎬ Lower-level canopy
5 m/15 ft   ⎦

FIGURE 20–6
The forest floor of an equatorial rain forest along the Amazon River. Photo by Stephen Cunha.

thin roots and vines dropping down from above, and echoing sounds of life in the trees, together create a unique environment (Figure 20–6).

The smooth, slender trunks of rain forest trees are covered with thin bark and buttressed by large wall-like flanks that grow out from the trees to brace their trunks. These buttresses form angular open enclosures, a ready habitat for various animals. There are usually no branches for at least the lower two-thirds of the tree trunks. The wood of many rain forest trees is extremely hard, heavy, and dense—in fact, some species will not even float. (Exceptions are balsa and a few others, which are very light.) Varieties of trees include mahogany, ebony, and rosewood. Logging is difficult because individual species are widely scattered; a species may occur only once or twice per square kilometer.

Rain forests represent approximately one-half of Earth's remaining forests, occupying about 7% of the total land area worldwide. This biome is stable in its natural state, resulting from the long-term residence of these continental plates near equatorial latitudes and their escape from glaciation. The soils, principally Oxisols, are essentially infertile, yet they support rich vegetation. The trees have adapted to these soil conditions with root systems able to capture nutrients from litter decay at the soil surface. High precipitation and temperatures work to produce deeply weathered and leached soils characteristic of the laterization process, with a clayey texture sometimes breaking up into a granular structure, and

lacking in nutrients and colloidal material. With use of fertilizers, pesticides, and machinery these soils can be productive.

The animal and insect life of the rain forest is widespread, ranging from small decomposers (bacteria) working the surface to many animals living almost completely in the upper stories of the trees. These tree dwellers are referred to as *arboreal,* from the Latin for tree, and include sloths, monkeys, lemurs, parrots, and snakes. Beautiful birds of many colors, tree frogs, lizards, bats, and a rich insect life that includes over 500 species of butterflies are found in rain forests. The present human assault on Earth's rain forests has put this diverse fauna and the varied flora at risk. It also jeopardizes an important recycling system for atmospheric carbon dioxide.

Deforestation of the Tropics. When Space Shuttle astronauts or Soviet cosmonauts in orbit look down on the rain forests at night, they see thousands of human-set fires burning through the forests. During the day the lower atmosphere in these regions appears choked with the smoke from these fires (Figure 20–7). Such fires are used to clear land for agriculture, which is intended to feed the domestic population as well as to produce cash exports of beef, rubber, coffee, and other commodities. Edible fruits are not abundant in an undisturbed rain forest, but cultivated clearings produce bananas (plantains), mangos and jack fruit, guava, and starch-rich roots such as manioc and yams.

FIGURE 20–7
The Shuttle *Discovery* captured
these photographs 1 October
1988: Thousands of fires
obscure the Amazon rain forest
(right); a widespread smoke
plume rising from a single fire
greater in area than the 1988
Yellowstone fire (left). Courtesy
of National Aeronautics and
Space Administration.

However, poor soil fertility means that the lands are quickly exhausted under intensive use and are then generally abandoned in favor of newly burned and cleared lands, unless the lands are artificially maintained. Unfortunately, the dominant trees require from 100 to 250 years to establish themselves after major disturbances. Following clearing, the massive growth of low bushes intertwined with vines and ferns means that a new forest may be slow to follow.

Burning is more common than logging because of the scattered distribution of specific types of trees mentioned earlier. Every year approximately 6.1 million hectares (15.25 million acres) are thus destroyed, and more than 4 million hectares (10 million acres) are selectively logged. This total is a loss of 0.6% of tropical rain forest each year, at which rate they will be removed completely in fewer than 180 years, if the destruction continues unabated. The United Nations Food and Agricultural Organization (FAO) estimates that about one-third of these forests are now gone: more than 50% of those in Africa, over

40% of those in Asia, and 40% of those in Central and South America. Brazil, Colombia, Mexico, and Indonesia lead the list of lesser-developed countries that are removing their forests at record levels, although these statistics are disputed by the countries in question.

To prevent this continuing catastrophe, the Tropical Forestry Action Plan was initiated in 1985 by the Food and Agriculture Organization (FAO), the UN Development Programme, World Bank, and the World Resources Institute. That plan instituted a more accurate worldwide survey of the rate of deforestation, using orbiting satellites for remote measurement. Workshops also have been conducted to stimulate action. But we have barely begun to reverse the trend. To accelerate data gathering in the rain forest, a Rapid Assessment Program (RAP) Team is operating under the auspices of Conservation International, as of 1991. Focus Study 20–1 looks more closely at efforts to curb increasing rates of species extinction, much of which is directly attributable to the loss of rain forests.

FOCUS STUDY 20–1

Biosphere Reserves

As more is learned about Earth's ecosystems and their related communities of plants and animals, more is known of their value to civilization. They are a major source of new foods, new chemicals, new medicines and drugs, specialty woods, and of course they are indicators of a healthy, functioning biosphere. Various international efforts are underway to study and preserve specific segments of the biosphere: Natural World Heritage Sites, Wetlands of International Importance, the Nature Conservancy, and Biosphere Reserves are just a few. The motivation to set aside natural sanctuaries is directly related to concern over the increase in the rate of species extinctions. We are facing a loss of genetic diversity that may be unparalleled in Earth's history.

Table 1 summarizes the known and estimated species on Earth. Scientists have classified only 1.4 million species of an estimated 10 to 30 million overall; the latter figure represents an increase in what scientists once thought to be the diversity of life on Earth. And those yet-to-be-discovered species represent a potential future resource for society. However, in 1982, the *Global 2000 Report,* prepared by the Council of Environmental Quality under a commission originally established by President Carter, stated that between 15–20% of all plant and animal species on Earth would be extinct by the year 2000. That percentage equals 500,000 to 2 million species, many of which have not yet even been identified. This rate of species extinction averages one every 30 minutes! The effects of pollution, loss of wild habitats, excessive grazing, poaching, and collecting are at the root of such devastation. Approximately 60% of the extinctions are attributable to the clearing and loss of rain forests alone.

TABLE 1
Known and estimated species on Earth.

Form of Life	Known Species	Estimated Total Species
Insects and other invertebrates	989,761	30 million (insect species), extrapolated from surveys in forest canopy in Panama; most believed unique to tropical forests
Vascular plants	248,400	At least 10–15% of all plants believed to be undiscovered
Fungi and algae	73,900	Not available
Microorganisms	36,600	Not available
Fishes	19,056	21,000 based on assumption that 10% of fish remain undiscovered (the Amazon and Orinoco Rivers alone may account for 2000 additional species)
Birds	9040	Probably 2% more (98% of total known)
Reptiles and amphibians	8962	Probably less than 5% more (95% of total known)
Mammals	4000	
Miscellaneous chordates	1273	Not available
TOTAL	1,390,992	10 million, considered a conservative count; with insect estimates perhaps over 30 million

SOURCE: Reprinted from *State of the World 1988,* Worldwatch Institute Report on Progress Toward a Sustainable Society, Project Director: Lester R. Brown, by permission of W. W. Norton & Company, Inc. Copyright © 1988 by Worldwatch Institute.

Both black rhinos and white rhinos in Africa exemplify species in jeopardy. Rhinos once grazed over much of the savanna grasslands and woodlands. Today, they survive only in protected districts and are threatened even there (Figure 1). There are approximately two dozen white rhinos in existence now in sub-Saharan ranges; the number of black rhinos is approximately 8000, but that is less than half the number in 1980. These large land mammals are nearing extinction and will be survived only by a dwindling zoo population. The limited genetic pool that remains complicates further reproduction.

To slow this type of extinction, formal natural reserves are a possible strategy. Setting up such a *biosphere reserve* involves principles of *island biogeography*. Island communities are special places for study because of their spatial isolation and the relatively small number of species present. They resemble natural experiments because the impact of individual factors, such as civilization, can be more easily assessed on islands than over larger continental areas.

For example, when the first European settlers landed in the Hawaiian Islands in the late 1700s, 43 species of birds were counted. Today, 15 of those species are extinct, and 19 more are threatened or endangered, with only one-fifth of the original species relatively healthy. According to the equilibrium theory of island biogeography, developed by biogeographers R. H. MacArthur and E. O. Wilson, the number of species should increase with the size of the island, decrease with increasing distance from the nearest continent, and remain about the same over time, even though composition may vary. Clearly, human intervention throws natural balances out of equilibrium.

Studies of islands also can assist in the study of mainland ecosystems, for in many ways a park or biosphere reserve, surrounded by modified areas and artificial boundaries, is like an island. Indeed, a biosphere reserve is conceived as an ecological island in the midst of change. The intent is to establish a core in which genetic material is protected from outside disturbances, surrounded by a buffer zone that is, in turn, surrounded by a transition zone and experimental research areas.

It is predicted that new, undisturbed reserves will not be possible after A.D. 2010, because pristine areas will be gone. Coordinated by the Man and the Biosphere (MAB) Programme of UNESCO, nearly 300 such biosphere reserves are now operated voluntarily in almost 70 countries. Some of the best examples have been in operation since the late 1970s and range from the struggling Everglades National Park in Florida, to the Changbai Reserve in China, to the Tai forest on the Ivory Coast. Added to these efforts, the work of the Nature Conservancy in acquiring land for preservation is a valuable part of the reserve process.

The ultimate goal, about half achieved, is to establish at least one reserve in each of the 194 distinctive biogeographical communities presently identified. In addition, the United Nations list of national parks and protected areas, compiled by the International Union for Conservation of Nature and Natural Resources, has more than 3500 protected areas totaling 4.25 million km^2 (1.64 million mi^2). Even though these are not all set aside to the degree of biosphere reserves, they do demonstrate progress toward preservation of our planet's plant and animal heritage.

FIGURE 1

(a) Failing population and receding distribution of both the black and white rhinos of Africa. (b) Black rhino and young in Tanzania. (a) Illustration after *State of the Ark* by Lee Durrell. Copyright © 1986 by Gaia Books Ltd. Adapted by permission of Doubleday, a division of Bantam Doubleday Dell Publishing Group, Inc. (b) Photo by Gael Summer-Hebdon.

(a)

(b)

Tropical Seasonal Forest and Scrub

A varied biome on the margins of the rain forest is the **tropical seasonal forest and scrub**, which occupies regions of lower and more erratic rainfall. The shifting intertropical convergence zone (ITCZ) brings precipitation with the seasonally shifting high Sun and dryness with the low Sun, producing a seasonal pattern that features areas of moisture deficits, some leaf loss, and dry season flowering. The term *semideciduous* is appropriate. Areas of this biome have fewer than 40 rainy days during their four consecutive driest months, yet heavy monsoon downpours characterize their summers (see Chapter 6, especially Figure 6–20). The Köppen climates Am and Aw apply to these transitional communities between rain forests and tropical grasslands.

Portraying such a varied biome is difficult. In many areas the natural biome is disturbed by human clearing, so that the savanna grassland adjoins the rain forest directly. The biome does include a gradation from wetter to drier areas: monsoonal forests, open woodlands and scrub woodland, thorn forests, and finally drought-resistant scrub species. The monsoonal forests average 15 m (50 ft) high with no continuous canopy of leaves, graduating into open orchardlike parkland with grassy openings or into areas choked by dense undergrowth. In more open tracts, a common tree is the acacia, with its flat-topped appearance and usually thorny stems. Most of the trees in this biome have the look of the fabric portion of an upside-down umbrella, as do trees in the tropical savanna (Figure 20–8).

Local names are given to these communities: the *caatinga* of northeast Brazil, *chaco* area of Paraguay and northern Argentina, the *brigalow scrub* of Australia, and the *dornveld* of southern Africa. The map in Figure 20–4 shows areas of this biome in Africa, extending from eastern Angola through Zambia to Tanzania; in southeast Asia and portions of India, from interior Myanmar (Burma) through northeast Thailand; and in parts of Indonesia. The trees throughout most of this biome are not good for lumber but may be valuable for fine cabinetry; the

FIGURE 20–8
Open thorn forest and savanna. Photo by Gael Summer-Hebdon.

teak wood is especially valuable. In addition, some of the plants with dry season adaptations produce usable waxes and gums, such as carnauba and palm-hard waxes. Animal life varies from the koalas and cockatoos of Australia to elephants, rodents, and ground-dwelling birds (such as the cassowary) elsewhere.

Tropical Savanna

Large expanses of grassland interrupted by trees and shrubs aptly describe the **tropical savanna**, a transitional biome between the rain forests and tropical seasonal forests, and between semiarid tropical steppes and deserts. The savanna biome also includes treeless tracts of grasslands, and in very dry savannas, grasses grow discontinuously in clumps, with bare ground between them. The trees of the savanna woodlands are characteristically flat topped, although one relatively rare tree associated with the tropical savanna, the baobab tree, is not.

Savannas covered more than 40% of Earth's land surface before human intervention but were especially modified by human-caused fire. Elephant grasses averaging 5 m (16 ft) high and forests are assumed to have once penetrated much farther into the dry regions, for they are known to survive there when protected. Savanna grasslands are much richer in humus content than the wetter tropics and are better drained, therefore providing a better base for agriculture and grazing. Sorghums, wheat, and groundnuts (peanuts) are common commodities.

Tropical savannas receive their precipitation during less than six months of the year, when they are influenced by the ITCZ. The rest of the year they are under the drier influence of shifting subtropical high-pressure cells. Savanna shrubs and trees are frequently *xerophytic,* or drought resistant, with various adaptations to protect them from the dryness: small, thick leaves; rough bark; or waxy surfaces. Africa has the largest region of this biome, including the famous Serengeti Plains and the ever-expanding Sahel. Sections of Australia, India, and South America also are part of the savanna biome. Some of the localized names for these lands include the *Llanos* in Venezuela, stretching along the coast and inland east of Lake Maricaibo and the Andes; the *Campo Cerrado*

of Brazil and Guiana; and the *Pantanal* of southwest Brazil.

Particularly in Africa, savannas are the home of large land mammals that graze on savanna grasses or feed upon the grazers themselves: lions, cheetahs, zebras, giraffes, buffalo, gazelles, wildebeests, antelopes, rhinoceroses, and elephants. However, in our lifetime we may see the reduction of these animal herds to zoo stock only, because of poaching and habitat losses. The loss of both the black and white rhino is discussed in Focus Study 20–1. Establishment of large tracts of savanna as biosphere reserves is critical for the preservation of natural aspects of this biome and its associated fauna (Figure 20–9).

Midlatitude Broadleaf and Mixed Forest

Moist continental climates support a mixed broadleaf and needleleaf forest in areas of warm-to-hot summers and cool-to-cold winters. This **midlatitude broadleaf and mixed forest** includes several distinct communities. Relatively lush evergreen broadleaf forests are found in narrow margins along the Gulf of Mexico, southern Japan, eastern and northeastern China, southeastern Australia, New Zealand, and central Chile. Northward of this broadleaf margin in the Northern Hemisphere are mixed deciduous and evergreen needleleaf stands, specifically associated with sandy soils and burning. When areas are given fire protection, broadleaf trees quickly take over.

Pines of several varieties—longleaf, shortleaf, pitch, and loblolly—predominate in the southeast and Atlantic coastal plains of the United States. Farther north into New England and westward in a narrow belt to the Great Lakes, white and red pines and eastern hemlock are the principal evergreens, mixed with deciduous varieties of oak, beech, hickory, maple, elm, chestnut, and many others. These mixed stands contain valuable timber for logging, but their distribution has been greatly altered by human activity. Native stands of white pine in Michigan and Minnesota were removed before 1910; only later reforestation sustains their presence today.

This biome is quite consistent in appearance from continent to continent and at one time represented the principal vegetation of the Cfa, Cfb, Cwa, and

FIGURE 20–9
Savanna landscape of the Serengeti Plains, with wildebeest, zebras, and thorn forest.
Photo by Stephen Cunha.

Cwb climatic regions of North America, Europe, and Asia. However, the level of deforestation in historic times is well illustrated by the two maps in Figure 20–10. The first shows the forest cover estimated for Europe in A.D. 900. The second map is the distribution of the remaining forest, A.D. 1900. The comparison dramatically illustrates human modification of natural vegetation. This deforestation was principally for agricultural purposes, as well as construction materials and fuel.

A wide assortment of mammals, birds, reptiles, and amphibians is distributed throughout this biome. Representative animals (some migratory) include fox, squirrel, deer, opossum, bear, rodents, and a great variety of birds. To the north of this biome, poorer soils and colder climates favor stands of coniferous trees and a gradual transition to the northern needleleaf forests.

Northern Needleleaf Forest and Montane Forest

Stretching from the east coast of Canada and the Maritimes westward to Alaska and continuing from Siberia across the entire extent of the Soviet Union to the European Plain is the **northern needleleaf forest**, also called the **taiga** (a Russian word) or **boreal forest**. The Southern Hemisphere, lacking D climates except in mountainous locales, has no biome designated as such. However, **montane forests** of needleleaf trees exist worldwide at high elevation (Figure 20–11).

Boreal forests of pine, spruce, fir, and larch occupy most of the subarctic climates on Earth. Although similar in formation, individual species vary between North America and Eurasia. The larch (*Larix*) is interesting because it loses its needles in the winter months, perhaps as a defense against the extreme cold of its native Siberia (see the Verkhoyansk climograph in Figures 10–17 and 5–8); larches also occur in North America. The Sierra Nevada, Rocky Mountains, Alps, and Himalayas have similar forest communities occurring at lower latitudes. Douglas and white fir are associated with the western mountains in the United States and Canada. Economically, these forests are important for lumbering, with saw timber occurring in the southern margins of the biome and pulpwood throughout the middle and northern portions.

Certain regions of the northern needleleaf biome experience the permafrost and gelifluction (solifluction) discussed in Chapter 13. When coupled with rocky and poorly developed soils, these conditions generally limit the existence of trees to only those with shallow rooting systems. The summer thaw of surface layers results in muskeg (moss-covered) bogs and Histolic soils of poor drainage and stability.

Representative fauna in this biome include wolf, moose, bear, lynx, wolverine, martin, small rodents, and migratory birds during the brief summer season. About 50 species of insects particularly adapted to the presence of coniferous trees inhabit the biome.

Temperate Rain Forest

The **temperate rain forest** biome is recognized by its lush forests at middle and high latitudes, occurring only along narrow margins of the Pacific Northwest in North America, with some similar types in southern China, small portions of southern Japan, Nev

(a)

(b)

FIGURE 20–10

A comparison of Europe's forests in (a) A.D. 900 and (b) A.D. 1900. From H. C. Darby, "The Clearing of the Woodland in Europe," in *Man's Role in Changing the Face of the Earth,* edited by William L. Thomas, Jr. Copyright © 1956 by the University of Chicago Press. All rights reserved. Reprinted by permission.

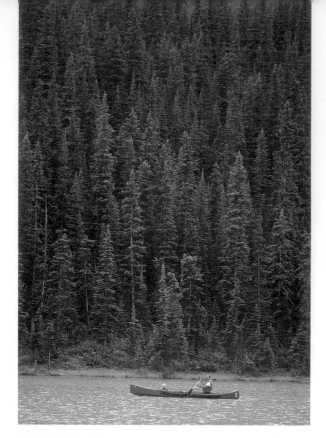

FIGURE 20–11
Boreal forest of Canada. Photo by author.

Zealand, and a few areas of Chile. This biome contrasts with the diversity of the equatorial and tropical rain forest in that only a few species make up the bulk of the trees. The rain forest of the Olympic Peninsula in Washington State is a mixture of broadleaf and needleleaf trees, huge ferns, and thick undergrowth. Precipitation approaching 400 cm (160 in.) per year on the western slopes, moderate air temperatures, summer fog, and an overall maritime influence produce this extreme moisture example.

The tallest trees in the world are found in this biome—the coastal redwoods (*Sequoia sempervirens*). Their distribution is shown on the map in Figure 19–11(a). These trees can exceed 1500 years of age and typically range in height from 60 to 90 m (200 to 300 ft), with some exceeding 100 m (330 ft). Virgin stands of other representative trees—Douglas fir, spruce, cedar, and hemlock—have been reduced to a few remaining valleys in Oregon and Washington (Figure 20–12). Animals include bear, badger, deer, wild pig, wolf, bobcat, and fox as predators.

The trees are home to numerous birds and mammals.

Mediterranean Shrubland

The **Mediterranean shrubland**, also referred to as a temperate shrubland, occupies those regions poleward of the shifting subtropical high-pressure cells. As those cells shift poleward with the high Sun, they cut off available storm systems and moisture. Their stable high-pressure presence produces the characteristic dry summer climate—Köppen's Csa—and establishes conditions susceptible to numerous fires. Plant ecologists think that this biome is well adapted to frequent fires, for many of its characteristically deep-rooted plants have the ability to resprout from their roots after a fire.

The dominant shrub formations that occupy these regions are short, stunted, and tough in their ability to withstand hot-summer drought. The vegetation is called *sclerophyllous* (from *sclero* for "hard" and *phyllos* for "leaf"); it averages a meter or two in height and has deep, well-developed roots, leathery leaves, and uneven low branches.

Typically, the vegetation varies between woody shrubs covering more than 50% of the ground and grassy woodlands with 25–60% coverage. In California, the Spanish word *chaparro* for "scrubby evergreen" gives us the word **chaparral** for this vegetation type (Figure 20–13). This scrubland includes species such as manzanita, toyon, red bud, ceanothus, mountain mahogany, blue and live oaks, and the dreaded poison oak. A counterpart to chaparral in the Mediterranean region, called *maquis,* includes live and cork oak trees, as well as pine and olive trees. In Chile, such a region is called *mattoral;* in southwest Australia, *mallee scrub*. Of course, in Australia the bulk of the eucalyptus species is sclerophyllous in form and structure in whichever climate it occurs.

As described in Chapter 10, Mediterranean climates are important in commercial agriculture for specific varieties of subtropical fruits, vegetables, and nuts, with many food types produced only in this biome (e.g., artichokes, olives, almonds). Larger animals, such as different types of deer, are grazers and browsers, with coyote, wolf, and bobcat as predators. Many rodents, other small animals, and a variety of birds also proliferate.

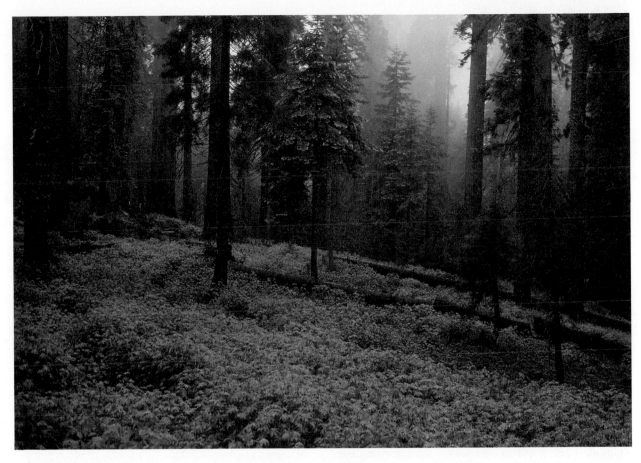

FIGURE 20–12
Temperate rain forest in the northwestern United States. Photo by author.

FIGURE 20–13
Chaparral vegetation associated with the Mediterranean dry-summer climate. Photo by author.

Midlatitude Grasslands

Of all the natural biomes, the **midlatitude grasslands** are most modified by human activity. Here are the world's "breadbaskets"—regions of grain and livestock production (Figure 20–14). In these regions, the only trees were deciduous broadleafs along streams and other limited sites. These regions are called grasslands because of the original predominance of grasslike plants:

> In this study of vegetation, attention has been devoted to grass because grass is the dominant feature of the Plains and is at the same time an index to their history. Grass is the visible feature which distinguishes the Plains from the desert. Grass grows, has its natural habitat, in the transition area between timber and desert . . . . The history of the Plains is the history of the grasslands.*

In North America, tall grass prairies once rose to heights of 2 m (6.5 ft) and extended westward to about the 98th meridian, with shortgrass prairies farther west. The 98th meridian is roughly the location of the 50 cm (20 in.) isohyet, with wetter conditions to the east and drier to the west. The deep sod of those grasslands posed problems for the first settlers, as did the climate. The self-scouring steel plow, introduced in 1825 by John Deere, allowed the interlaced grass sod to be broken apart, freeing the

---

*From *The Great Plains* by Walter Prescott Webb, p. 32. © Copyright 1959 by Walter Prescott Webb, published by Ginn and Company. Used by permission of Silver, Burdett & Ginn, Inc.

soils for agriculture. Other eastern inventions were critical to opening this region and solving its unique spatial problems: barbed wire (the fencing material for a treeless prairie), well-drilling techniques developed by Pennsylvania oil drillers but used for water wells, windmills for pumping, and railroads to conquer the distances.

Few patches of the original prairies (tall grasslands) or steppes (short grasslands) remain within this biome. For prairies alone, the reduction of natural vegetation went from 100 million hectares (250 million acres) down to a few areas of several hundred hectares each. A "prairie national park" was considered as a preservationist effort along the Kansas-Oklahoma border, but plans were put on hold. The map in Figure 20–4 shows the natural location of these former prairie and steppe grasslands.

Outside North America, the Pampas of Argentina and Uruguay and the Soviet Union's Ukraine are characteristic midlatitude grassland biomes. Figure 18–15 portrays the interacting variables of the physical environment that produce the grasslands and Mollisols of the Soviet fertile triangle.

In each region of the world where these grasslands have occurred, human development of them was critical to territorial expansion. This is the home of large grazing animals including deer, antelope, pronghorn, and bison, although the almost complete annihilation of the latter is part of American history. Grasshoppers and other insects feed on the grasses and crops as well, and gophers, prairie dogs, grouse,

FIGURE 20–14
Midlatitude grassland under cultivation. Photo by author.

and prairie chickens are on the land; predators include coyote and badger, and birds of prey such as hawks.

Desert Biomes

Earth's **desert biomes** cover more than one-third of its land area. In Chapter 15 we examined desert landscapes and in Chapter 10, desert climates. On a planet with such a rich biosphere, the deserts stand out as unique regions where fascinating adaptations for survival have developed. Much as a group of humans in the desert might behave with short supplies, plant communities also compete for water and site advantage. Some desert plants, called *ephemerals,* wait years for a rainfall event, at which time their seeds quickly germinate, develop, flower, and produce new seeds, which then rest again until the next rainfall event. The seeds of some xerophytic species open only when fractured by the tumbling, churning action of flash floods cascading down a desert arroyo, and of course such an event produces the moisture that a germinating seed needs.

Desert plants employ other strategies, such as long, deep tap roots (e.g., the mesquite), succulence (i.e., thick, fleshy, water-holding tissue such as that of cacti), spreading root systems to maximize water availability, waxy coatings and fine hairs on leaves to retard water loss, leafless conditions during dry periods (e.g., palo verde and ocotillo), reflective surfaces to reduce leaf temperatures, and tissue that tastes bad to discourage herbivores.

The creosote bush (*Lorrea divaricata*) sends out a wide pattern of roots and contaminates the surrounding soil with toxins that prevent the germination of other creosote seeds, possible competitors for water. When a creosote bush dies and is removed, surrounding plants or germinating seeds work to occupy the abandoned site, but must rely on infrequent rains to remove the toxins.

The faunas of both warm and cold deserts are limited by the extreme conditions and include few resident large animals, an exception being the desert bighorn sheep. Some representative desert animals are the ring-tail cat, kangaroo rat, lizards, scorpions, and snakes. Most of these animals are quite secretive and become active only at night, when temperatures are lower. In addition, various birds have adapted to desert plants and other available food sources—for example, roadrunners, thrashers, ravens, wrens, hawks, grouse, and nighthawks.

Desertification. We are witnessing an unwanted expansion of the desert biome. This is due principally to poor agricultural practices (overgrazing and inappropriate agricultural activities), improper soil-moisture management, erosion and salinization, deforestation, and the ongoing climatic change. A process known as **desertification** is now a worldwide phenomenon along the margins of semiarid and arid lands.

As a result, the southward expansion of the Sahara through portions of the Sahel region has left many African peoples on land that no longer experiences the rainfall patterns of just two decades earlier. Regions at risk of desertification stretch from Asia and central Australia to portions of North and South America. The United Nations estimates that desertified lands have covered some 800 million hectares (2 billion acres) since 1930; many millions of additional hectares are added each year. The process is increasing at an alarming rate. Of course, an immediate need is to improve the database for a more accurate accounting of the problem.

Figure 20–15 is based on a map prepared during the 1977 U.N. Conference on Desertification held in Nairobi, Kenya, and their "Plan of Action to Combat Desertification." Desertification areas are ranked: moderate hazard with an average 10–25% drop in agricultural productivity; high hazard with a 25–50% drop; and very high, representing more than a 50% decrease. Because human activities and economies appear to be the major cause, especially unwise grazing practices, actions to slow the desertification process are readily available with leadership.

Earth's deserts are subdivided into desert and semidesert associations, to distinguish those with expanses of bare ground from those covered by xerophytic plants of various types. The two broad associations are further separated into warm deserts, principally tropical and subtropical, and cold deserts, principally midlatitude.

Warm Desert and Semidesert. Earth's warm desert and semidesert biomes are associated with the presence of dry air and low precipitation from subtropical high-pressure cells. These areas are very dry, as evidenced by the Atacama Desert of northern

FIGURE 20–15

Worldwide desertification estimates by the United Nations. Data from U.N. Food and Agriculture Organization (FAO), World Meteorological Organization (WMO), United Nations Educational, Scientific, and Cultural Organization (UNESCO), Nairobi, Kenya, 1977; adapted by permission of Merrill, an imprint of Macmillan Publishing Company, from *The Cultural Landscape,* by James M. Rubenstein, p. 468–69. Copyright © 1989 by Merrill Publishing.

Chile, where only a minute amount of rain has ever been recorded—a 30-year annual average of only 0.05 cm (0.02 in.)! Like the Atacama, the true deserts of the Köppen BW classification are under the influence of the descending, drying, and stable air of high pressure from 8 to 12 months of the year.

Most warm deserts receive less than 2 cm (1 in.) of precipitation each year, with vegetation almost totally lacking, whereas the warm semideserts receive less than 25 cm (10 in.) of precipitation each year and have various xerophytic shrubs, succulents, and thorn tree forms. A few of the subtropical deserts—such as those in Chile, Western Sahara, and Namibia—are right on the sea coast and are influenced by cool offshore ocean currents. As a result, these true deserts experience summer fog that mists the plant and animal populations with needed moisture.

The equatorward margin of the subtropical high-pressure cell is a region of transition to savanna, thorn tree, scrub woodland, and tropical seasonal forest. Poleward of the warm deserts, the subtropical cells shift to produce the Mediterranean dry summer regime along west coasts and may grade into cool deserts elsewhere.

Cold Desert and Semidesert. The cold desert and semidesert biomes tend to occur at higher latitudes where seasonal shifting of the subtropical high is of some influence less than six months of the year. Specifically, interior locations are dry due to their distance from moisture sources or location in rain shadow areas on the lee side of mountain ranges, such as the Sierra Nevada of the western United States, Himalayas, and Andes. The combination of interior location and rain shadow positioning produces the cold deserts of the Great Basin of western North America.

Winter snows, although light, are possible in the cold deserts. Summers are hot, with highs from 30° to 40°C (86°–104°F) (Figure 20–16). Nighttime lows, even in the summer, can cool 10°–20C° (18°–36F°) from the daytime high. The dryness, lack of cloud cover, and sparse vegetation lead to high radiative heat loss and the cool evenings. Many areas of these cold deserts that are covered by sagebrush and scrub vegetation were actually dry shortgrass regions in the past, before extensive grazing forever altered their appearance. The look of the deserts in the upper Great Basin is the result of more than a century of such cultural practices.

Arctic and Alpine Tundra

The **arctic tundra** is found in the extreme northern area of North America and the Soviet Union, bordering on the Arctic Ocean and generally north of the 10°C (50°F) isotherm for the warmest month. Most of the region was covered by ice during all of the Pleistocene glaciations, with only a few portions of Alaska and Siberia escaping glaciation. There is no comparable tundra biome on the continent of Antarctica, except on an extreme portion of Palmer Land across Drake Passage from the tip of South America. Winters in this biome, an ET climate classification, are long and cold; cool summers are brief.

FIGURE 20–16
Cold desert scene in the interior Great Basin of the western United States. Photo by author.

Winter is governed by intensely cold continental polar air masses and stable high-pressure conditions. A growing season of sorts lasts only 60 to 80 days, and even then frosts can occur at any time. Vegetation is fragile in this flat, treeless world; soils are poorly developed periglacial surfaces, which are continually underlain by permafrost. In the summer months only the surface horizons thaw, thus producing a mucky surface of poor drainage. Roots can penetrate only to the depth of thawed ground, usually about a meter. The surface is shaped by freeze-thaw cycles that create the frozen ground phenomena and gelifluction (solifluction) processes discussed in Chapter 13.

Tundra vegetation is characterized by low, ground-level herbaceous plants and some woody plants. Representative are sedges, mosses, arctic meadow grass, snow lichen, and dwarf willow. Animals of the tundra include, among others, musk ox, caribou, wolf, small rodents (important food for the larger carnivores), fox, weasel, rabbit, reindeer, snowy owls, ptarmigans, polar bears, and of course mosquitoes (Figure 20–17).

Alpine tundra is similar in many ways to arctic tundra, although alpine tundra can occur at lower latitudes because it is associated with high elevation. This biome usually is described as above the timberline, that elevation above which trees cannot grow. Alpine timberlines increase in elevation with decreasing latitude in both hemispheres. Alpine tundra communities occur in the Andes near the equator, the White Mountains of California, the Rockies, the Alps, and Mount Kilimanjaro of equatorial Africa, as well as mountains through the Middle East to Asia. Alpine tundra can experience permafrost in the rocky soils and periglacial soil flow. Alpine meadows feature grasses and stunted shrubs, such as willows and heaths. Because alpine locations are frequently windy sites, many plants appear sculpted by the wind.

Because the tundra biome is fragile, any disturbance such as tire tracks and mineral exploitation leave marks that persist for hundreds of years. As development continues, the region will face even greater challenges from oil spills, contamination, and disruption. The controversy over Alaska's Arctic National Wildlife Refuge (ANWR) is growing daily as political and corporate pressures mount to begin oil exploration there. The resulting disruption would be devastating and the cost of development would price the oil at well over $30 per barrel, far above market prices. Consequently, economic activities should be weighed against the nature of the biome itself, its limitations and its uniqueness.

SUMMARY

Earth is the only planet in the Solar System with a biosphere. The interplay among supporting physical factors within Earth's ecosystem determines the distribution of plant and animal communities. Plants assume adaptive associations and forms as each plant competes for individual success in its habitat and niche. Biomes are Earth's major terrestrial ecosystems and the 10 major biomes are named for their dominant plant community, and specifically plant life-form and formation class. Ideally a biome represents a climax community of natural vegetation. In reality, few undisturbed biomes exist in the world. Many of Earth's plant and animal communities are experiencing an accelerated rate of change that could produce dramatic alterations within our lifetime. Efforts are underway worldwide to set aside and protect remaining representative sites.

FIGURE 20–17
Tundra in Denali Park, Mount McKinley. Photo by Stephen Cunha.

KEY TERMS

Aquatic ecosystems	Mediterranean shrubland
Arctic tundra	Midlatitude broadleaf and mixed forest
Biogeographical realms	Midlatitude grasslands
Biome	Montane forests
Boreal forest	Northern needleleaf forest
Chaparral	Taiga
Desert biomes	Temperate rain forest
Desertification	Terrestrial ecosystem
Equatorial and tropical rain forest	Tropical savanna
Formation class	Tropical seasonal forest and scrub
Large marine ecosystem (LME)	

REVIEW QUESTIONS

1. Reread the two opening quotations in this chapter. What clues do you have to the path ahead? Is our future direction controllable? Explain.
2. What is a biogeographical realm? How is the world subdivided according to plant and animal types?
3. What is a large marine ecosystem? What is the purpose of such a classification?
4. Distinguish between plant associations and formation classes as a basis for spatial classification.
5. Define biome. What is the basis of the designation?
6. Use the integrative chart in Table 20–1 and the world map in Figure 20–4 to correlate the description of each biome with its specific spatial distribution.
7. Describe the equatorial and tropical rain forests. Why is the rain forest floor somewhat clear? Why are logging activities for specific species so difficult there?
8. What are the issues surrounding deforestation of the rain forest? What is the impact of these losses on the rest of the biosphere?
9. What do caatinga, chaco, brigalow, and dornveld refer to? Explain.
10. Describe the role of fire or fire ecology in the tropical savanna biome and the midlatitude broadleaf and mixed forest biome.
11. Why does the northern needleleaf forest biome not exist in the Southern Hemisphere? Where is this biome located in the Northern Hemisphere, and what is its relationship to climate type?
12. In which biome do we find Earth's tallest trees? Which biome is dominated by small, stunted plants, lichens, and mosses?
13. What type of vegetation predominates in the Mediterranean dry summer climates? Describe the adaptation necessary for these plants to survive.
14. What is the significance of the 98th meridian in terms of North American grasslands? What types of eastern inventions were necessary for humans to cope with the grasslands?
15. Describe some of the unique adaptations found in a desert biome.
16. What is desertification? Explain its impact.
17. What physical weathering processes are specifically related to the tundra biome? What types of plants and animals are found there?
18. What is the relationship between island biogeography and biosphere reserves? Describe a biosphere reserve.

SUGGESTED READINGS

Attenborough, David, Philip Whitfield, Peter D. Moore, and Barry Cox. *The Atlas of the Living World*. Boston: Houghton Mifflin, 1989.

Barbour, Michael G., and William D. Billings, eds. *North American Terrestrial Vegetation*. Cambridge: Cambridge University Press, 1988.

Brown, James H., and Arthur C. Gibson. *Biogeography*. St. Louis: C. V. Mosby Company, 1983.

Colinvaux, Paul A. "The Past and Future Amazon." *Scientific American* 260, no. 5 (May 1989): 102–8.

Dansereau, Pierre. *Biogeography: An Ecological Perspective*. New York: Ronald Press, 1957.

Durcell, Lee, and the International Union for Conservation of Nature and Natural Resources staff. *State of the Ark*. Garden City, NY: Doubleday, 1986.

Editors of Time-Life Books. *Grasslands and Tundra*. Alexandria, VA: Time-Life Books, 1985.

Ellis, William S. "Rondônia: Brazil's Imperiled Rain Forest." *National Geographic* 174, no. 6 (December 1988): 772–79.

Mabbutt, J.A. "Desertification of the World's Rangelands." *Desertification Control Bulletin* 12 (1984): 1–11.

————. "A New Global Assessment of the Status and Trends of Desertification." *Environmental Conservation* 11, no. 2 (1984): 106.

Northington, David K., and J. R. Gordon. *The Botanical World*. St. Louis: Times Mirror/Mosby, 1984.

Page, Jake, and the editors of Time-Life Books. *Forest*. Alexandria, VA: Time-Life Books, 1983.

Perry, Donald R. "The Canopy of the Tropical Rain Forest." *Scientific American* 251, no. 5 (November 1984): 138–47.

Postal, Sandra, and Lori Heise. *Reforesting the Earth*. Worldwatch Paper 83. Washington: Worldwatch Institute, 1988.

Raven, Peter H., and Ray F. Evert. *Biology of Plants*. New York: Worth, 1976.

Sherman, Kenneth, Lewis M. Alexander, and Barry D. Golds, Eds. *Large Marine Ecosystems—Patterns, Processes, and Yields*. Washington: American Association for the Advancement of Science, No. 90–30S, 1990.

Taylor, J. A., ed. *Biogeography—Recent Advances and Future Directions*. Totowa, NJ: Barnes & Noble, 1984.

Tucker, Compton J., John R. G. Townshend, and Thomas E. Goff. "African Land-Cover Classification Using Satellite Data." *Science* 227 (25 January 1985): 369–76.

Watts, David. *Principles of Biogeography*. New York: McGraw-Hill, 1971.

White, Peter T. "Tropical Rain Forests: Nature's Dwindling Treasures." *National Geographic* 163, no. 1 (January 1983): 2–47.

Paddy rice farming in Dali, China. Photo by
Stephen Cunha.

21
The Human Denominator

During my space flight, I came to appreciate my profound connection to the home planet and the process of life evolving in our special corner of the Universe, and I grasped that I was part of a vast and mysterious dance whose outcome will be determined largely by human values and actions.*

Earth is observed and studied from many profound vantage points, as this astronaut experienced on the 1969 *Apollo IX* mission. Our vantage point here is physical geography; we have examined Earth through its energy, atmosphere, water, weather, climate, endogenic and exogenic systems, soils, ecosystems, and biomes—all of which leads to an examination of the planet's most abundant large animal, *Homo sapiens*.

Because human influence is so pervasive, we can consider the totality of our impact the human denominator. Just as the denominator in a fraction tells how many parts a whole is divided into, so the growing human population and the increasing demand for resources suggest how much the whole Earth system must adjust. Yet, Earth's resource base remains relatively fixed.

In his 1980 book and PBS television series, astronomer Carl Sagan asked:

> What account would we give of our stewardship of the planet Earth? We have heard the rationales offered by the nuclear superpowers. We know who speaks for the nations. But who speaks for the human species? Who speaks for Earth?**

We might attempt an answer: perhaps those who have studied Earth and know the operations of the global ecosystem should speak for Earth. Or perhaps those who have the greatest stake in the continued functioning of that ecosystem—the children—should address the importance of a workable future. Moreover, we might question whether such human issues belong in an introductory physical geography class. Some might say that these questions of technology, politics, and future thinking belong outside of academia, that our job here is merely to learn how Earth's processes work. But biologist/ecologist Marston Bates addressed this line of thought in 1960:

Then we came to humans and their place in this system of life. We could have left humans out, playing the ecological game of "let's pretend humans don't exist." But this seems as unfair as the corresponding game of the economists, "let's pretend nature doesn't exist." The economy of nature and the ecology of humans are inseparable and attempts to separate them are more than misleading, they are dangerous. Human destiny is tied to nature's destiny and the arrogance of the engineering mind does not change this. Humans may be a very peculiar animal, but they are still a part of the system of nature.***

Interestingly, most of the environmental problems being studied by scientists have known solutions or strategies for positive resolution, if only this knowledge could be applied in a constructive way. Unfortunately, many of these solutions are not being acted upon by society. We do not seem to assess the true long-term costs of present economic thinking. A role for physical geography is to explain Earth systems through spatial analysis and through a synthesis of information from many disciplines—hopefully to assist an informed citizenry toward more considered judgments.

AN OILY DUCK

At first glance the chain of events that exposes wildlife to oil contamination seems to stem from a technological problem. An oil tanker splits open at sea and releases its petroleum cargo, which is moved by ocean currents toward shore, where it coats coastal waters, beaches, and animals (Figure 21–1). In response, concerned citizens mobilize and try to save as much of the spoiled environment as possible, but the real problem goes far beyond the physical facts of the spill.

On 24 March 1989, in Prince William Sound off the southern coast of Alaska in clear weather and calm seas, a single-hulled supertanker operated by Exxon Corporation, an international energy corporation, struck a reef that was outside the normal shipping lane and spilled 41.64 million liters (11

---

*Rusty Schweickart, "Our Backs Against the Bomb, Our Eyes on the Stars," *Discovery* (July 1987): 62. Reprinted by permission.

**Carl Sagan, *Cosmos,* New York: Random House, 1980, p. 329. Reprinted by permission.

***Marston Bates, *The Forest and the Sea,* New York: Random House, 1960, p. 247. Reprinted by permission.

FIGURE 21–1
Oil-contaminated cormorant at Prince William Sound. Photo by Gary Braasch.

million gallons) of oil. It took only 12 hours for the *Exxon Valdez* to spill its contents, yet a reasonable cleanup will take years and billions of dollars.

Because contingency emergency plans were not in place, and promised equipment was unavailable, response by the oil industry took 10 to 12 hours to activate, about the same time that it took the ship to empty. Eventually, almost 2415 km (1500 mi) of coastline were ruined; for perspective, had this spill occurred farther south, enough spilled to blacken every beach and bay along the Pacific Coast from southern Oregon to the Mexican border (Figure 21–2).

The Larger Picture

The immediate effect on wildlife was contamination and death, but the issues involved are bigger than these damaged ecosystems. Let's ask some fundamental questions: Why was the oil tanker there in the first place? Why is petroleum being imported into the continental United States from Alaska and abroad in such quantity? Is the demand for petroleum products based on real need, or does it reflect the business strategies of transnational oil corporations? The U.S. demand for oil is higher per capita than the demand of any other country; is this demand inflated by waste and inefficiency?

A great many factors influence our demand for oil. Well over half of our imported oil goes for transportation. Despite this enormous and expensive consumption, the 1980s saw a major rollback of auto efficiency standards, a reduction in gasoline prices, large reductions in funding for rapid transit development, and the continuing slow demise of America's passenger rail system. Thus, waste, low prices, and a lack of alternatives all combined to spur demand for petroleum. In addition, land-use patterns continue to foster diffuse sprawl, thereby placing stress upon existing transportation systems.

The demand for fossil fuels in the 1980s also was affected by the slowing of domestic conservation programs, the elimination of research for energy alternatives such as solar and wind, and even delays of a law requiring small appliances to be more energy efficient. Conservation plans again were politically blocked in the Department of Energy in 1990 and early 1991. All of these manipulations increased fossil fuel demand (5.4 billion barrels a year for oil alone) at a time (1989) when the U.S. Geological Survey estimated that the United States had only 16 years of actual and expected domestic petroleum reserves remaining, or 32 years if 50% of U.S. consumption is imported. Increased imports make the United States more vulnerable to foreign conflicts and war.

FIGURE 21-2
Track of the Alaskan oil spill.

FIGURE 21-3
Location of visible oil slicks worldwide. Data from Organization for Economic
Cooperation and Development, *The State of the Environment,* Paris: Organization for
Economic Cooperation and Development, 1985.

Thus, the immediate problem of cleaning oil off a duck in Prince William Sound becomes a national and international concern with far-reaching significance. And while we search for answers, oil slicks continue their contamination. In just one year following the *Exxon Valdez* disaster, nearly 76 million liters (20 million gallons) of oil were spilled in 10,000 accidents worldwide (Figure 21–3). In addition, an even greater spill than that of the *Exxon Valdez* occurred off the coast of Morocco (in northwest Africa) on 19 December 1989, following an explosion aboard the tanker *Khark 5*. Into those warm tropical waters that vessel spilled 140.6 million liters (37 million gallons) of oil. And, the Persian Gulf is now flowing with as much as 10 times this amount of oil as a result of the recent war.

GAIA HYPOTHESIS

Just as the abiotic spheres affect the biosphere, so do living processes affect abiotic functions. All of these interactive effects in concert influence Earth's overall ecosystem. In essence, the planetary ecosystem sets the physical limits for life, which in turn evolves and helps to shape the planet. Thus, Earth can be viewed as one vast, self-regulating organism. This controversial concept, called the *Gaia hypothesis,* was proposed in 1979 by James Lovelock, a British astronomer and inventor, and defined and elaborated by American biologist Lynn Margulis.

Gaia is the ultimate synergistic relationship, in which the whole greatly exceeds the sum of the individual interacting components. The hypothesis contends that life processes control and shape inorganic physical and chemical processes, with the biosphere so interactive that a very small mass can affect a very large mass. Thus, Lovelock and Margulis think that the material environment and the evolution of species are tightly joined; as the species evolve through natural selection, they in turn affect their environment. The present composition of the atmosphere is given as proof of this coevolution of living and nonliving systems.

From the perspective of physical geography, Gaia permits a view of all of Earth and the spatial interrelationships among systems. One disturbing aspect of this unity, however, is that any biotic threats to the operation of an ecosystem tend to move toward extinction themselves, thus preserving the system overall. Earth systems operation and feedback tend to eliminate offensive members. The degree to which humans represent a planetary threat, then, becomes a topic of great concern, for Earth (Gaia) will prevail, regardless of the outcome of the human experiment.

> The maladies of Gaia do not last long in terms of her life span. Anything that makes the world uncomfortable to live in tends to induce the evolution of those species that can achieve a new and more comfortable environment. It follows that, if the world is made unfit by what we do, there is the probability of a change in regime to one that will be better for life but not necessarily better for us.[*]

The debate is vigorous regarding the true applicability of this hypothesis to nature, although it remains philosophically intriguing in its portrayal of the relationship between humans and Earth.

BIOSPHERE II

Another approach to understanding the spatial interplay of human and Earth systems is taking place in Arizona. Forty miles north of Tucson, a living model of Earth's tropical ecosystems is now operational— climates, communities, and biomes. Called Biosphere II, it is partially illustrated in Figure 21–4. This $100 million venture was conceived by the Institute of Ecotechnics in London and implemented by Space Biospheres Ventures. In September 1991, eight people entered it and began a two-year stay in the Earth habitat model with 3800 species of plants and animals, living and working with desert, savanna, freshwater and saltwater marsh, ocean, tropical, and agricultural ecosystems contained in a glass terrarium 165 m (540 ft) long. Several successful test runs were completed in a smaller facility in 1989 and early 1990. According to Carl Hodges, director of the University of Arizona's Environmental Research Lab,

---

[*]James Lovelock, *The Ages of Gaia—A Biography of Our Living Earth,* New York: W. W. Norton, 1988, p. 178. Used with permission.

FIGURE 21–4
Biosphere II illustration. After Space Biospheres Ventures, copyright © 1985.

> **If Biosphere II causes people to look inward and understand the Earth and to look outward and understand the potential of the universe, it will be the most significant scientific project of all time. . . . Biosphere II represents a serious attempt to model interactive ecosystems.***

Whether or not this prediction proves true, Biosphere II represents a serious attempt to model the interactive ecosystems of Biosphere I—Earth itself.

THE NEED FOR INTERNATIONAL COOPERATION

We already have seen dramatic examples of international environmental problem solving: the Limited Test Ban Treaty of 1962, which bans atmospheric testing of nuclear weapons, and the 1987 and 1990 treaties proposing bans on ozone-destroying chlorofluorocarbons are examples. The International Geosphere-Biosphere Program (IGBP) represents another integrative effort; its goal is to improve understanding of the entire natural system and discern how the many and varied subsystems interact. All major concerns need cooperative attention. For example, the escalating greenhouse effect that

---

*Carl Hodges, "Biosphere II," *Discover* 8, no. 5 (May 1987): 28–43.

weaves through this entire text and influences so many global systems certainly deserves international research and mutual action. (See Appendix B for the numerous groups, organizations, and international action programs related to geography and environmental concerns.)

An important corollary to international efforts is the linkage of academic disciplines. A positive step in that direction is an *Earth systems science* approach, as illustrated in this text. Exciting progress toward an integrated understanding of Earth's formation and the operation of its physical systems is happening right now, driven by insights drawn from our remote-sensing capabilities. Never before has society been able to monitor Earth's physical geography so thoroughly.

Modern warfare has had a definite impact on the environment: scars from World Wars I and II are still visible on the landscapes of Europe and Asia, and more than 33 million bomb craters and massive defoliation provide environmental evidence of the more recent struggle in Vietnam. In fact, during the Vietnam War more than 2.3 times the equivalent tonnage of all munitions used in World War II were detonated. In terms of more modern weapons, just one of the planned MX missiles carries 3 megatons of explosive power, the equivalent of all the firepower of all the combatants in World War II! The 1991

Persian Gulf War subjected the desert biome to 88,500 tons of bombs. We must consider the environmental consequences of such weapons to air, water, land, and biotic systems, especially in light of the conventional war devastation in the Middle East, where the ecosystems were damaged extensively—with unknown long-term consequences.

In 1975, the National Academy of Sciences published its study on the impact of modern nuclear war on stratospheric ozone. This was the beginning of public awareness that modern technological warfare is not an environmentally sound activity. This NAS study states that a nuclear war could lead to the reduction of 40 to 70% of the stratospheric ozone, a catastrophic change for food chains, plants, animals, and humans.

In addition, from 1982 to the present, scientific publications have described the *nuclear winter hypothesis*. This hypothesis indicates that a "nuclear winter" might follow the detonation of only modest numbers of nuclear weapons. As you have learned, the atmosphere is very dynamic. The soot and ash from the many urban fires following a nuclear exchange would enter the atmosphere and raise Earth's albedo, absorbing insolation in the stratosphere and upper troposphere, reradiating it to space, and thus cooling Earth's surface to below freezing even during midlatitude summer months. As the nuclear winter hypothesis developed, others suggested a milder "nuclear autumn" as a consequence, although still significant.

Refinement of the nuclear winter hypothesis continues, but the warnings are clear. One of the original nuclear winter study teams recently affirmed that "severe environmental anomalies—possibly leading to more human casualties globally than the direct effects of nuclear war—would be not just a remote possibility, but a likely outcome."* These potential consequences demand that the methods of conflict resolution chosen by nations should preferably be those more acceptable within the limiting factors imposed by Earth's ecosystems. Modern technological conflicts can no longer be thought of only in a regional context. The real need for international cooperation should consider our sym-

biotic relationships with each other and with Earth's resilient, yet fragile, life-support systems.

WHO SPEAKS FOR EARTH?

Public awareness and education is a growing positive force on Earth. Geography awareness is improving each year. The National Geographic Society is even conducting an annual National Geography Bee to promote geography education and global awareness. The global concern about environmental impacts prompted *Time* magazine in 1989 to deviate from its 60-year tradition of naming a prominent citizen as its person of the year, and instead *Time* named Earth the "Planet of the Year." The magazine featured 33 pages of Earth's physical and human geography, covering the endangered status of many ecosystems and cultures (Figure 21–5). Importantly, *Time* also offered positive policy strategies for consideration. We seem on the brink of a new age in Earth awareness.

Carl Sagan answered his question "who speaks for Earth" with this perspective:

> We have begun to contemplate our origins: starstuff pondering the stars; organized assemblages of ten billion billion billion atoms considering the evolution of atoms; tracing the long journey by which, here at least, consciousness arose. Our loyalties are to the species and the planet. We speak for Earth. Our obligation to survive is owed not just to ourselves but also to that Cosmos, ancient and vast, from which we spring.**

May we all perceive our spatial importance within Earth's ecosystems and do our part to maintain a life-supporting Earth far into the future.

**Carl Sagan, *Cosmos*, New York: Random House, 1980, p. 345. Reprinted by permission.

---

*R. P. Turco, O. B. Toon, T. P. Ackerman, J. B. Pollack, and C. Sagan, "Climate and Smoke: An Appraisal of Nuclear Winter," *Science* (12 January 1990): 174.

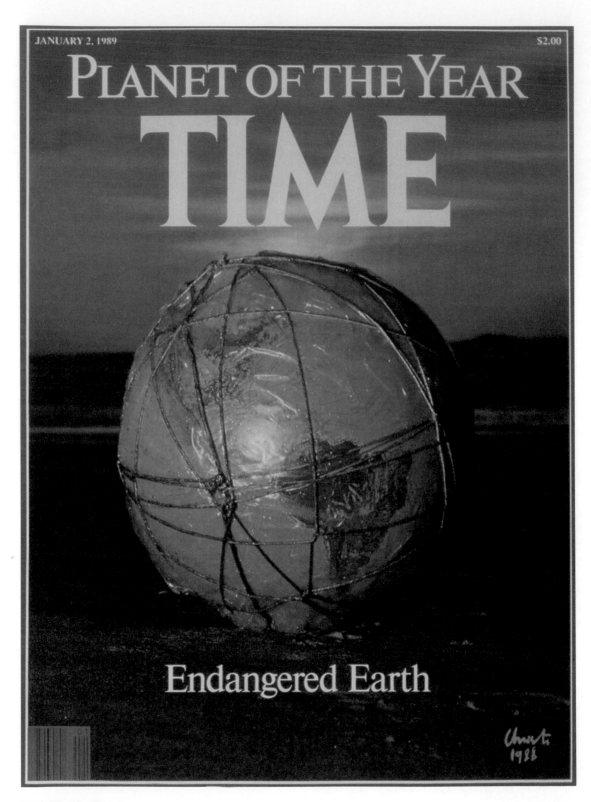

FIGURE 21–5
Time magazine cover for 2 January 1989, naming Earth "Planet of the Year."
Copyright © 1988 The Time Inc. Magazine Company. Reprinted by permission.

REVIEW QUESTIONS

1. What part do you think technology, politics, and future thinking should play in an academic physical geography course?
2. According to the discussion in the chapter, what factors led to the crash of the *Exxon Valdez?* Describe the complexity of that event.
3. What is meant by the Gaia hypothesis? Describe any concepts from this text that pertain to this hypothesis.
4. Relate the content of the various chapters in this text to the integrative Earth systems science concept. How does the Biosphere II project relate to this Earth systems science model?
5. Explain the potential spatial impact of nuclear warfare on the environment. What is the nuclear winter hypothesis, and what are its potential implications for the environment?
6. This chapter states that we already know many of the solutions to the problems we face. Why do you think we are not implementing those solutions?
7. Who speaks for Earth?

SUGGESTED READINGS

Association of American Geographers. "Statement of the Association of American Geographers on Nuclear War." Adopted by AAG, 6 May 1986, Minneapolis.

Barney, Gerald O. *The Global 2000 Report to the President*. New York: Penguin Books, 1982.

Blackburn, Anne M. *Pieces of the Global Puzzle—International Approaches to Environmental Concerns*. Golden, CO: Fulcrum, 1986.

Brown, Lester R. *State of the World—1989*. Worldwatch Institute Series. New York: W. W. Norton, 1989.

Canby, Thomas Y. "The Persian Gulf After the Storm." *National Geographic* 180, no. 2 (August 1991): 2–35.

The Conservation Foundation. *State of the Environment: A View Toward the Nineties*. Washington: Conservation Foundation and the Charles Stewart Mott Foundation, 1987.

Ehrlich, Paul R., Carl Sagan, Donald Kennedy, and Walter Orr Roberts. *The Cold and the Dark—The World After Nuclear War*. New York: W. W. Norton, 1984.

Hodgson, Bryan. "Alaska's Big Spill—Can the Wilderness Heal?" *National Geographic* 177, no. 1 (January 1990): 5–43.

Kates, Robert W., and Ian Burton, eds. *Geography, Resources, and Environment*. Vol. 1. *The Selected Writings of Gilbert F. White*. Vol. 2. *Themes from the Work of Gilbert F. White*. Chicago: University of Chicago Press, 1986.

Lee, Douglas B. "Tragedy in Alaskan Waters." *National Geographic* 176, no. 2 (August 1989): 260–63.

Lovelock, James. *The Ages of Gaia—A Biography of Our Living Earth*. New York: W. W. Norton, 1988.

———. *Gaia—A New Look at Life on Earth*. London: Oxford Press, 1979.

National Academy of Sciences and National Research Council. *Long-Term Worldwide Effects of Multiple Nuclear-Weapons Detonations*. A report for the Department of Transportation's Climatic Impact Assessment Program. Washington: National Academy of Sciences, 1975.

Sagan, Carl. *Cosmos*. New York: Random House, 1980.

Schell, Jonathan. *The Fate of the Earth*. New York: Alfred A. Knopf, 1982.

Turco, R. P., O. B. Toon, T. P. Ackerman, J. B. Pollack, and C. Sagan, "Climate and Smoke: An Appraisal of Nuclear Winter." *Science* 247, no. 4939 (12 January 1990): 166–74.

——————. "The Climatic Effects of Nuclear War." *Scientific American* 251, no. 2 (August 1984): 33–43.

——————. "Nuclear Winter: Global Consequences of Multiple Nuclear Explosions." *Science* 222, no. 4630 (23 December 1983): 1283–92.

U.N. Environment Programme. *Environmental Data Report*. New York: Basil Blackwell, 1987.

Westing, Arthur H., and E. W. Pfeiffer. "The Cratering of Indochina." *Scientific American* 226, no. 5 (May 1972): 21–29.

Appendix A
Climate and Water Balance Data

Tropical Rain Forest (Af)

Belau (Palau), Caroline Islands (Af): pop. 9400, lat.7° 30′ N, long. 134° 30′ E, elev. 32 m (104 ft)													
	Jan	Feb	Mar	Apr	May	Jun	Jul	Aug	Sep	Oct	Nov	Dec	Annual
Temperature°C	26.9	26.7	27.2	27.8	27.5	27.5	26.9	27.5	27.2	27.2	27.2	26.9	27.2
(°F)	(80.5)	(80.0)	(81.0)	(82.0)	(81.5)	(81.5)	(80.5)	(80.5)	(81.0)	(81.0)	(81.0)	(80.5)	(81.0)
PRECIP cm	38.8	23.9	17.3	19.3	39.4	31.5	50.5	35.6	39.9	37.6	30.0	32.3	395.5
(in.)	(15.3)	(9.4)	(6.8)	(7.6)	(15.5)	(12.4)	(19.9)	(14.0)	(15.7)	(14.8)	(11.8)	(12.7)	(155.7)
POTET cm	14.0	12.4	14.5	14.7	15.2	15.0	14.7	14.7	14.2	14.5	13.7	14.0	172.0
(in.)	(5.5)	(4.9)	(5.7)	(5.8)	(6.0)	(5.9)	(5.8)	(5.8)	(5.6)	(5.7)	(5.4)	(5.5)	(67.7)

Salvador (Bahia), Brazil (Af): pop. 1,507,000, lat. 12° 59′ S, long. 38° 31′ W, elev. 9 m (30 ft)													
	Jan	Feb	Mar	Apr	May	Jun	Jul	Aug	Sep	Oct	Nov	Dec	Annual
Temperature°C	26.0	26.3	26.3	25.8	24.8	23.8	23.0	22.9	23.6	24.5	25.1	25.6	24.8
(°F)	(78.8)	(79.3)	(79.3)	(78.4)	(76.6)	(74.8)	(73.4)	(73.2)	(74.5)	(76.1)	(77.2)	(78.1)	(76.6)
PRECIP cm	7.4	7.9	16.3	29.0	29.7	19.6	20.6	11.2	8.4	9.4	14.2	9.9	183.6
(in.)	(2.9)	(3.1)	(6.4)	(11.4)	(11.7)	(7.7)	(8.1)	(4.4)	(3.3)	(3.7)	(5.6)	(3.9)	(72.3)
POTET cm	13.7	12.4	13.5	11.9	10.7	9.1	8.6	8.6	9.4	11.2	11.9	13.2	134.4
(in.)	(5.4)	(4.9)	(5.3)	(4.7)	(4.2)	(3.6)	(3.4)	(3.4)	(3.7)	(4.4)	(4.7)	(5.2)	(52.9)

Tropical Monsoon (Am)

Rangoon, Burma (Am): pop. 2,459,000, lat. 16° 47′ N, long. 96° 10′ E, elev. 23 m (76 ft)													
	Jan	Feb	Mar	Apr	May	Jun	Jul	Aug	Sep	Oct	Nov	Dec	Annual
Temperature°C	24.3	25.2	27.2	29.8	29.5	27.8	27.6	27.1	27.6	28.3	27.8	25.0	27.3
(°F)	(75.7)	(77.4)	(81.0)	(85.6)	(85.1)	(82.0)	(81.7)	(80.8)	(81.7)	(82.9)	(82.0)	(77.0)	(81.1)
PRECIP cm	0.8	0.5	0.5	1.8	25.9	52.3	49.3	57.4	39.9	20.8	3.3	0.3	252.7
(in.)	(0.3)	(0.2)	(0.2)	(0.7)	(10.2)	(20.6)	(19.4)	(22.6)	(15.7)	(8.2)	(1.3)	(0.1)	(99.5)
POTET cm	9.4	9.9	14.5	16.8	17.8	16.0	16.3	15.2	14.7	15.2	13.7	10.2	169.4
(in.)	(3.7)	(3.9)	(5.7)	(6.6)	(7.0)	(6.3)	(6.4)	(6.0)	(5.8)	(6.0)	(5.4)	(4.0)	(66.7)

Mayagüez, Puerto Rico (Am): pop. 101,000, lat. 18° 12′ N, long. 67° 9′ W, elev. 24 m (79 ft)													
	Jan	Feb	Mar	Apr	May	Jun	Jul	Aug	Sep	Oct	Nov	Dec	Annual
Temperature°C	23.7	23.4	23.6	24.4	25.3	25.9	26.0	26.2	26.4	26.2	25.3	24.5	25.1
(°F)	(74.7)	(74.1)	(74.5)	(75.9)	(77.5)	(78.6)	(78.8)	(79.2)	(79.5)	(79.2)	(77.5)	(76.1)	(77.2)
PRECIP cm	4.7	6.1	8.1	14.3	19.0	23.6	29.2	27.9	26.6	24.3	15.7	6.3	205.8
(in.)	(1.9)	(2.4)	(3.2)	(5.6)	(7.5)	(9.3)	(11.5)	(11.0)	(10.5)	(9.6)	(6.2)	(2.5)	(81.0)
POTET cm	8.9	8.0	9.4	10.7	12.9	13.6	14.2	14.3	13.6	13.0	10.8	9.9	139.3
(in.)	(3.5)	(3.1)	(3.7)	(4.2)	(5.1)	(5.4)	(5.6)	(5.6)	(5.4)	(5.1)	(4.3)	(3.9)	(54.8)

Tropical Savanna (Aw)

Calcutta, India (Aw): pop. 3,305,000, lat. 22° 32′ N, long. 88° 20′ E, elev. 6 m (20 ft)													
	Jan	Feb	Mar	Apr	May	Jun	Jul	Aug	Sep	Oct	Nov	Dec	Annual
Temperature°C	20.2	23.0	27.9	30.1	31.1	30.4	29.1	29.1	29.2	27.9	24.0	20.6	26.8
(°F)	(68.4)	(73.4)	(82.2)	(86.2)	(88.0)	(86.7)	(84.4)	(84.4)	(84.6)	(82.2)	(75.2)	(69.1)	(80.2)
PRECIP cm	1.3	2.3	2.8	4.3	12.2	25.9	30.2	30.5	29.0	16.0	3.6	0.3	159.0
(in.)	(0.5)	(0.9)	(1.1)	(1.7)	(4.8)	(10.2)	(11.9)	(12.0)	(11.4)	(6.3)	(1.4)	(0.1)	(62.6)
POTET cm	5.3	7.4	15.2	17.3	19.3	18.8	18.3	17.5	16.3	14.7	8.9	5.6	163.3
(in.)	(2.1)	(2.9)	(6.0)	(6.8)	(7.6)	(7.4)	(7.2)	(6.9)	(6.4)	(5.8)	(3.5)	(2.2)	(64.3)

Managua, Nicaragua (Aw): pop. 682,000, lat. 12° 09′ N, long. 86° 17′ W; elev. 56 m (184 ft)													
	Jan	Feb	Mar	Apr	May	Jun	Jul	Aug	Sep	Oct	Nov	Dec	Annual
Temperature°C	25.7	26.7	27.3	28.2	28.3	27.0	26.7	27.1	27.0	26.2	26.4	26.3	26.8
(°F)	(78.3)	(80.0)	(81.1)	(82.8)	(82.9)	(80.6)	(80.1)	(80.8)	(80.6)	(79.2)	(79.5)	(79.3)	(80.4)
PRECIP cm	0.3	0.3	0.5	0.3	15.0	23.4	12.7	11.2	21.3	30.5	4.1	1.0	120.4
(in.)	(0.1)	(0.1)	(0.2)	(0.1)	(5.9)	(9.2)	(5.0)	(4.4)	(8.4)	(12.0)	(1.6)	(0.4)	(47.4)
POTET cm	11.7	12.2	14.5	15.5	16.3	14.7	15.0	15.0	14.0	12.7	12.2	12.5	166.4
(in.)	(4.6)	(4.8)	(5.7)	(6.1)	(6.4)	(5.8)	(5.9)	(5.9)	(5.5)	(5.0)	(4.8)	(4.9)	(65.5)

Miami, Florida (Aw): pop. 385,000, lat. 25° 46′ N, long. 80° 11′ W, elev. 2 m (7 ft)													
	Jan	Feb	Mar	Apr	May	Jun	Jul	Aug	Sep	Oct	Nov	Dec	Annual
Temperature°C	19.4	20.0	21.1	23.3	25.6	27.2	27.8	27.8	27.2	25.6	22.2	20.0	23.9
(°F)	(67.0)	(68.0)	(70.0)	(74.0)	(78.0)	(81.0)	(82.0)	(82.0)	(81.0)	(78.0)	(72.0)	(68.0)	(75.0)
PRECIP cm	5.1	4.8	5.8	9.9	16.3	18.8	17.3	17.8	24.1	20.8	7.1	4.3	151.9
(in.)	(2.0)	(1.9)	(2.3)	(3.9)	(6.4)	(7.4)	(6.8)	(7.0)	(9.5)	(8.2)	(2.8)	(1.7)	(59.8)
POTET cm	4.7	5.0	7.4	10.3	13.6	16.2	17.2	16.8	14.7	11.6	7.6	5.7	130.8
(in.)	(1.9)	(2.0)	(2.9)	(4.1)	(5.4)	(6.4)	(6.8)	(6.6)	(5.8)	(4.6)	(3.0)	(2.2)	(51.5)

Humid Subtropical (Cfa)

New Orleans, Louisiana (Cfa): pop. 557,000, lat. 29° 57′ N, long. 90° 04′ W, elev. 3 m (9 ft)													
	Jan	Feb	Mar	Apr	May	Jun	Jul	Aug	Sep	Oct	Nov	Dec	Annual
Temperature°C	13.3	14.4	17.2	21.1	24.4	27.8	28.3	28.3	26.7	22.8	16.7	13.9	21.1
(°F)	(56.0)	(58.0)	(63.0)	(70.0)	(76.0)	(82.0)	(83.0)	(83.0)	(80.0)	(73.0)	(62.0)	(57.0)	(70.0)
PRECIP cm	12.2	10.7	16.8	13.7	13.7	14.2	18.0	16.3	14.7	9.4	10.2	11.7	161.3
(in.)	(4.8)	(4.2)	(6.6)	(5.4)	(5.4)	(5.6)	(7.1)	(6.4)	(5.8)	(3.7)	(4.0)	(4.6)	(63.5)
POTET cm	2.2	2.6	4.9	8.4	12.7	16.8	18.0	17.1	13.9	8.8	4.0	2.4	111.8
(in.)	(0.9)	(1.0)	(1.9)	(3.3)	(5.0)	(6.6)	(7.1)	(6.7)	(5.5)	(3.5)	(1.6)	(0.9)	(44.0)

Milan, Italy (Cfa): pop. 1,561,000, lat. 45° 28′ N, long. 9° 12′ S, elev. 116 m (381 ft)													
	Jan	Feb	Mar	Apr	May	Jun	Jul	Aug	Sep	Oct	Nov	Dec	Annual
Temperature°C	1.3	2.9	7.9	12.2	17.0	20.7	23.0	22.0	18.4	12.4	6.6	3.0	12.3
(°F)	(34.3)	(37.2)	(46.2)	(54.0)	(62.6)	(69.3)	(73.4)	(71.6)	(65.1)	(54.3)	(43.9)	(37.4)	(54.1)
PRECIP cm	4.8	5.6	7.6	7.1	7.9	10.9	5.8	8.4	6.4	12.7	10.9	8.9	97.0
(in.)	(1.9)	(2.2)	(3.0)	(2.8)	(3.1)	(4.3)	(2.3)	(3.3)	(2.5)	(5.0)	(4.3)	(3.5)	(38.2)
POTET cm	0.3	0.5	2.8	5.3	9.7	12.7	15.0	13.0	8.9	4.6	1.8	0.8	73.7
(in.)	(0.1)	(0.2)	(1.1)	(2.1)	(3.8)	(5.0)	(5.9)	(5.1)	(3.5)	(1.8)	(0.7)	(0.3)	(29.0)

Marine West Coast (Cfb)

Valdivia, Chile (Cfb): pop. 103,000, lat. 39° 48′ S, long. 73° 14′ W, elev. 13 m (43 ft)													
	Jan	Feb	Mar	Apr	May	Jun	Jul	Aug	Sep	Oct	Nov	Dec	Annual
Temperature°C	16.5	16.2	14.6	11.6	9.7	7.9	7.6	8.0	9.1	11.6	14.0	16.2	11.9
(°F)	(61.7)	(61.2)	(58.3)	(52.9)	(49.5)	(46.2)	(45.7)	(46.4)	(48.4)	(52.9)	(57.2)	(61.2)	(53.4)
PRECIP cm	10.2	5.1	9.7	17.8	43.9	37.1	45.2	31.8	22.6	9.9	8.1	7.1	248.9
(in.)	(4.0)	(2.0)	(3.8)	(7.0)	(17.3)	(14.6)	(17.8)	(12.5)	(8.9)	(3.9)	(3.2)	(2.8)	(98.0)
POTET cm	6.4	6.4	6.9	5.6	5.1	3.8	3.8	3.8	3.8	4.8	5.3	6.1	60.2
(in.)	(2.5)	(2.5)	(2.7)	(2.2)	(2.0)	(1.5)	(1.5)	(1.5)	(1.5)	(1.9)	(2.1)	(2.4)	(23.7)

Edinburgh, Scotland (Cfb): pop. 445,000, lat. 55° 57′ N, long. 3° 13′ W, elev. 134 m (439 ft)													
	Jan	Feb	Mar	Apr	May	Jun	Jul	Aug	Sep	Oct	Nov	Dec	Annual
Temperature°C	3.0	3.0	5.0	7.6	10.1	12.7	14.7	14.3	12.5	9.7	6.5	4.8	8.7
(°F)	(37.4)	(37.4)	(41.0)	(45.7)	(50.2)	(54.9)	(58.5)	(57.7)	(54.5)	(49.5)	(43.7)	(40.6)	(47.7)
PRECIP cm	4.8	3.6	3.3	3.3	4.8	4.6	8.9	9.1	4.8	5.1	6.1	7.4	65.8
(in.)	(1.9)	(1.4)	(1.3)	(1.3)	(1.9)	(1.8)	(3.5)	(3.6)	(1.9)	(2.0)	(2.4)	(2.9)	(25.9)
POTET cm	1.3	1.3	2.8	4.8	7.6	10.2	11.7	10.2	7.4	4.8	2.8	1.8	64.8
(in.)	(0.5)	(0.5)	(1.1)	(1.9)	(3.0)	(4.0)	(4.6)	(4.0)	(2.9)	(1.9)	(1.1)	(0.7)	(25.5)

Mediterranean Dry, Warm Summer (Csa)

Sacramento, California (Csa): pop. 330,000, lat. 38° 35′ N, long. 121° 21′ W, elev. 11 m (36 ft)													
	Jan	Feb	Mar	Apr	May	Jun	Jul	Aug	Sep	Oct	Nov	Dec	Annual
Temperature°C	8.4	11.2	12.9	15.6	19.1	22.3	24.8	24.2	22.7	18.5	12.6	8.6	16.8
(°F)	(47.1)	(52.2)	(55.3)	(60.1)	(66.3)	(72.2)	(76.6)	(75.6)	(72.9)	(65.3)	(54.7)	(47.5)	(62.2)
PRECIP cm	10.7	7.4	5.6	3.6	1.0	0.3	0.3	0.3	0.8	2.3	5.8	7.6	45.5
(in.)	(4.2)	(2.9)	(2.2)	(1.4)	(0.4)	(0.1)	(0.1)	(0.1)	(0.3)	(0.9)	(2.3)	(3.0)	(17.9)
POTET cm	1.5	2.3	4.0	6.0	8.5	11.5	14.0	12.7	9.9	6.6	3.0	1.5	81.5
(in.)	(0.6)	(0.9)	(1.6)	(2.4)	(3.3)	(4.5)	(5.5)	(5.0)	(3.9)	(2.6)	(1.2)	(0.6)	(32.1)

Athens, Greece (Csa): pop. 3,000,000, lat. 37° 58′ N, long. 23° 43′ E, elev. 107 m (351 ft)													
	Jan	Feb	Mar	Apr	May	Jun	Jul	Aug	Sep	Oct	Nov	Dec	Annual
Temperature°C	9.3	9.9	11.3	15.3	20.0	24.6	27.6	27.4	23.5	19.0	14.7	11.0	17.8
(°F)	(48.7)	(49.8)	(52.3)	(59.5)	(68.0)	(76.3)	(81.7)	(81.3)	(74.3)	(66.2)	(58.5)	(51.8)	(64.0)
PRECIP cm	6.1	3.6	3.8	2.3	2.3	1.3	0.5	0.8	1.5	5.1	5.6	7.1	39.9
(in.)	(2.4)	(1.4)	(1.5)	(0.9)	(0.9)	(0.5)	(0.2)	(0.3)	(0.6)	(2.0)	(2.2)	(2.8)	(15.7)
POTET cm	1.5	1.8	2.8	5.1	9.7	14.5	18.3	16.8	11.2	6.9	3.8	2.0	93.0
(in.)	(0.6)	(0.7)	(1.1)	(2.0)	(3.8)	(5.7)	(7.2)	(6.6)	(4.4)	(2.7)	(1.5)	(0.8)	(36.6)

Mediterranean Dry, Cool Summer (Csb)

Portland, Oregon (Csb): pop. 366,000, lat. 45° 31′ N, long. 122° 40′ W, elev. 9 m (30 ft)													
	Jan	Feb	Mar	Apr	May	Jun	Jul	Aug	Sep	Oct	Nov	Dec	Annual
Temperature°C	4.4	6.7	8.9	12.2	15.0	17.8	20.0	20.0	17.8	13.3	8.3	5.6	12.8
(°F)	(40.0)	(44.0)	(48.0)	(54.0)	(59.0)	(64.0)	(68.0)	(68.0)	(64.0)	(56.0)	(47.0)	(42.0)	(55.0)
PRECIP cm	13.7	12.4	10.7	6.1	4.8	4.1	1.0	1.5	4.6	8.9	15.2	18.0	101.3
(in.)	(5.4)	(4.9)	(4.2)	(2.4)	(1.9)	(1.6)	(0.4)	(0.6)	(1.8)	(3.5)	(6.0)	(7.1)	(39.9)
POTET cm	0.9	1.7	3.4	5.4	8.1	10.1	12.3	11.3	8.1	5.4	2.6	1.3	70.6
(in.)	(0.4)	(0.7)	(1.3)	(2.1)	(3.2)	(4.0)	(4.8)	(4.4)	(3.2)	(2.1)	(1.0)	(0.5)	(27.8)

Capetown, Union of South Africa (Csb): pop. 1,108,000, lat. 33° 55′ S, long. 18° 22′ E, elev. 53 m (173 ft)													
	Jan	Feb	Mar	Apr	May	Jun	Jul	Aug	Sep	Oct	Nov	Dec	Annual
Temperature°C	20.3	20.0	18.8	16.1	14.0	12.6	11.6	12.3	13.7	15.0	17.6	19.3	15.9
(°F)	(68.5)	(68.0)	(65.8)	(61.0)	(57.2)	(54.7)	(52.9)	(54.1)	(56.7)	(59.0)	(63.7)	(66.7)	(60.7)
PRECIP cm	1.0	1.5	1.3	5.3	8.9	8.4	8.4	7.4	4.6	3.0	1.8	1.0	52.6
(in.)	(0.4)	(0.6)	(0.5)	(2.1)	(3.5)	(3.3)	(3.3)	(2.9)	(1.8)	(1.2)	(0.7)	(0.4)	(20.7)
POTET cm	7.6	7.1	7.9	6.4	5.6	4.8	4.3	4.3	4.8	5.1	6.1	6.9	70.6
(in.)	(3.0)	(2.8)	(3.1)	(2.5)	(2.2)	(1.9)	(1.7)	(1.7)	(1.9)	(2.0)	(2.4)	(2.7)	(27.8)

Humid Continental Hot Summer (Dfa)

Belgrade, Yugoslavia (Dfa): pop. 1,204,000, lat. 44° 50′ N, long. 20° 30′ E, elev. 139 m (456 ft)

		Jan	Feb	Mar	Apr	May	Jun	Jul	Aug	Sep	Oct	Nov	Dec	Annual
Temperature	°C	−0.2	1.6	6.2	12.2	17.1	20.5	22.6	22.0	18.3	12.5	6.8	2.5	11.8
	(°F)	(31.6)	(34.9)	(43.2)	(54.0)	(62.8)	(68.9)	(72.7)	(71.6)	(64.9)	(54.5)	(44.2)	(36.5)	(53.2)
PRECIP	cm	4.8	4.6	4.6	5.3	7.4	9.7	6.1	5.6	5.1	5.6	6.1	5.6	70.4
	(in.)	(1.9)	(1.8)	(1.8)	(2.1)	(2.9)	(3.8)	(2.4)	(2.2)	(2.0)	(2.2)	(2.4)	(2.2)	(27.7)
POTET	cm	0	0.3	2.0	5.3	9.7	12.7	14.7	13.2	8.9	4.8	1.8	0.5	72.9
	(in.)	(0)	(0.1)	(0.8)	(2.1)	(3.8)	(5.0)	(5.8)	(5.2)	(3.5)	(1.9)	(0.7)	(0.2)	(28.7)

Chicago, Illinois (Dfa): pop. 3,005,000, lat. 41° 47′ N, long. 87° 45′ W, elev. 186 m (610 ft)

		Jan	Feb	Mar	Apr	May	Jun	Jul	Aug	Sep	Oct	Nov	Dec	Annual
Temperature	°C	−3.9	−2.8	2.3	8.6	14.4	20.0	23.2	22.4	18.7	12.4	4.6	−1.4	9.9
	(°F)	(25.0)	(27.0)	(36.1)	(47.5)	(57.9)	(68.0)	(73.8)	(72.3)	(65.7)	(54.3)	(40.3)	(29.5)	(49.8)
PRECIP	cm	4.9	4.8	6.8	7.4	9.0	9.3	8.4	8.0	7.6	6.7	5.9	5.0	83.8
	(in.)	(1.9)	(1.9)	(2.7)	(2.9)	(3.5)	(3.7)	(3.3)	(3.1)	(3.0)	(2.6)	(2.3)	(2.0)	(33.0)
POTET	cm	0	0	0.6	3.7	7.9	12.2	14.6	13.2	9.0	5.1	1.2	0	67.5
	(in.)	(0)	(0)	(0.2)	(1.5)	(3.1)	(4.8)	(5.7)	(5.2)	(3.5)	(2.0)	(0.5)	(0)	(26.6)

Humid Continental Mild Summer (Dfb)

Montreal, Quebec, Canada (Dfb): pop. 2,818,000, lat. 45° 30′ N, long. 73° 35′ W, elev. 57 m (187 ft)

		Jan	Feb	Mar	Apr	May	Jun	Jul	Aug	Sep	Oct	Nov	Dec	Annual
Temperature	°C	−10.0	−9.4	−3.3	5.6	13.3	18.3	21.1	19.4	15.0	8.3	0.6	−6.7	6.0
	(°F)	(14.0)	(15.1)	(26.1)	(42.1)	(55.9)	(64.9)	(70.0)	(66.9)	(59.0)	(46.9)	(33.1)	(19.9)	(42.8)
PRECIP	cm	9.6	7.7	8.8	6.6	8.0	8.7	9.5	8.8	9.3	8.7	9.0	9.1	103.8
	(in.)	(3.8)	(3.0)	(3.5)	(2.6)	(3.1)	(3.4)	(3.7)	(3.5)	(3.7)	(3.4)	(3.5)	(3.6)	(40.8)
POTET	cm	0	0	0	2.7	8.1	11.9	13.9	12.1	7.7	3.7	0.2	0	60.3
	(in.)	(0)	(0)	(0)	(1.1)	(3.2)	(4.7)	(5.5)	(4.8)	(3.0)	(1.5)	(0.1)	(0)	(23.7)

Fargo, North Dakota (Dfb): pop. 61,000, lat. 46° 52′ N, long. 96° 47′ W, elev. 273 m (895 ft)

		Jan	Feb	Mar	Apr	May	Jun	Jul	Aug	Sep	Oct	Nov	Dec	Annual
Temperature	°C	−13.9	−11.7	−3.9	5.5	12.8	18.3	21.7	20.6	15.0	7.8	−2.2	−10.6	5.0
	(°F)	(7.0)	(11.0)	(25.0)	(42.0)	(55.0)	(65.0)	(71.0)	(69.0)	(59.0)	(46.0)	(28.0)	(13.0)	(41.0)
PRECIP	cm	1.5	1.8	2.3	4.8	5.6	7.6	5.8	6.9	4.3	3.3	2.3	1.5	47.5
	(in.)	(0.6)	(0.7)	(0.9)	(1.9)	(2.2)	(3.0)	(2.3)	(2.7)	(1.7)	(1.3)	(0.9)	(0.6)	(18.7)
POTET	cm	0	0	0	3.1	7.8	11.9	14.0	12.4	7.6	3.3	0	0	60.1
	(in.)	(0)	(0)	(0)	(1.2)	(3.1)	(4.7)	(5.5)	(4.9)	(3.0)	(1.3)	(0)	(0)	(23.7)

Stockholm, Sweden (Dfb): pop. 1,387,000, lat. 59° 20′ N, long. 18° 03′ E, elev. 52 m (170 ft)

		Jan	Feb	Mar	Apr	May	Jun	Jul	Aug	Sep	Oct	Nov	Dec	Annual
Temperature	°C	−2.9	−3.1	−0.7	4.4	10.1	14.9	17.8	16.6	12.2	7.1	2.8	0.1	6.6
	(°F)	(26.8)	(26.4)	(30.7)	(39.9)	(50.2)	(58.8)	(64.0)	(61.9)	(54.0)	(44.8)	(37.0)	(32.2)	(43.9)
PRECIP	cm	4.3	3.0	2.5	3.0	3.3	4.6	6.1	7.6	6.1	4.8	5.3	4.8	55.6
	(in.)	(1.7)	(1.2)	(1.0)	(1.2)	(1.3)	(1.8)	(2.4)	(3.0)	(2.4)	(1.9)	(2.1)	(1.9)	(21.9)
POTET	cm	0	0	2.5	3.0	7.9	11.7	14.0	11.7	7.4	3.8	1.3	0.3	59.7
	(in.)	(0)	(0)	(1.0)	(1.2)	(3.1)	(4.6)	(5.5)	(4.6)	(2.9)	(1.5)	(0.5)	(0.1)	(23.5)

Subarctic (Dfc, Dwd, Dwc)

Churchill, Manitoba, Canada (Dfc): pop. 1300, lat. 58° 47′ N, long. 94° 11′ W, elev. 35 m (115 ft)													
	Jan	Feb	Mar	Apr	May	Jun	Jul	Aug	Sep	Oct	Nov	Dec	Annual
Temperature°C	−28.3	−26.7	−21.1	−10.0	−1.1	6.1	12.2	11.1	6.1	−2.8	−14.4	−23.9	−8.9
(°F)	(−19.0)	(−16.0)	(−6.0)	(14.0)	(30.0)	(43.0)	(54.0)	(52.0)	(43.0)	(27.0)	(6.0)	(−11.0)	(16.0)
PRECIP cm	1.3	1.5	2.3	2.3	2.3	4.8	5.6	6.9	5.8	3.6	2.5	1.8	40.6
(in.)	(0.5)	(0.6)	(0.9)	(0.9)	(0.9)	(1.9)	(2.2)	(2.7)	(2.3)	(1.4)	(1.0)	(0.7)	(16.0)
POTET cm	0	0	0	0	0	7.1	11.2	9.6	5.1	0	0	0	33.0
(in.)	(0)	(0)	(0)	(0)	(0)	(2.8)	(4.4)	(3.8)	(2.0)	(0)	(0)	(0)	(13.0)

Tobol'sk, Siberia, Soviet Union (Dwc): pop. 72,000, lat. 58° 09′ N, long. 68° 11′ E, elev. 44 m (144 ft)													
	Jan	Feb	Mar	Apr	May	Jun	Jul	Aug	Sep	Oct	Nov	Dec	Annual
Temperature°C	−18.5	−17.3	−10.8	1.3	9.6	16.0	18.1	15.7	9.4	1.6	−9.4	−16.8	−0.1
(°F)	(−1.3)	(0.9)	(12.6)	(34.3)	(49.3)	(60.8)	(64.6)	(60.3)	(48.9)	(34.9)	(15.1)	(1.8)	(31.9)
PRECIP cm	1.8	1.5	2.0	2.5	4.1	6.1	7.4	6.4	5.3	3.8	3.6	2.5	47.0
(in.)	(0.7)	(0.6)	(0.8)	(1.0)	(1.6)	(2.4)	(2.9)	(2.5)	(2.1)	(1.5)	(1.4)	(1.0)	(18.5)
POTET cm	0	0	0	1.3	8.1	13.5	15.2	11.7	6.1	1.0	0	0	56.9
(in.)	(0)	(0)	(0)	(0.5)	(3.2)	(5.3)	(6.0)	(4.6)	(2.4)	(0.4)	(0)	(0)	(22.4)

Oymyakon, Siberia, Soviet Union (Dwd): pop. 3000, lat. 63° 28′ N, long. 142° 49′ E, elev. 726 m (2382 ft)													
	Jan	Feb	Mar	Apr	May	Jun	Jul	Aug	Sep	Oct	Nov	Dec	Annual
Temperature°C	−47.2	−42.9	−34.2	−15.4	1.4	11.6	14.8	10.9	1.6	−16.2	−35.0	−44.0	−16.3
(°F)	(−53.0)	(−45.2)	(−29.6)	(4.3)	(34.5)	(52.9)	(58.6)	(51.6)	(34.9)	(2.8)	(−31.0)	(−47.2)	(2.7)
PRECIP cm	0.8	0.5	0.5	0.3	1.0	3.3	4.1	3.8	2.0	1.3	1.0	0.8	19.3
(in.)	(0.3)	(0.2)	(0.2)	(0.1)	(0.4)	(1.3)	(1.6)	(1.5)	(0.8)	(0.5)	(0.4)	(0.3)	(7.6)
POTET cm	0	0	0	0	3.0	13.5	15.7	10.9	2.3	0	0	0	45.5
(in.)	(0)	(0)	(0)	(0)	(1.2)	(5.3)	(6.2)	(4.3)	(0.9)	(0)	(0)	(0)	(17.9)

Polar Tundra (ET) and Polar Marine (EM)

Barrow, Alaska (ET): pop. 2200, lat. 71° 18′ N, long. 156° 47′ W, elev. 9 m (31 ft)													
	Jan	Feb	Mar	Apr	May	Jun	Jul	Aug	Sep	Oct	Nov	Dec	Annual
Temperature°C	−26.7	−27.8	−26.1	−17.8	−7.8	0.6	3.9	3.3	−1.1	−8.3	−18.3	−23.9	−12.2
(°F)	(−16.0)	(−18.0)	(−15.0)	(0)	(18.0)	(33.0)	(39.0)	(38.0)	(30.0)	(17.0)	(−1.0)	(−11.0)	(10.0)
PRECIP cm	0.5	0.5	0.3	0.3	0.3	1.0	2.0	2.3	1.5	1.3	0.5	0.5	10.9
(in.)	(0.2)	(0.2)	(0.1)	(0.1)	(0.1)	(0.4)	(0.8)	(0.9)	(0.6)	(0.5)	(0.2)	(0.2)	(4.3)
POTET cm	0	0	0	0	0	4.8	8.4	2.8	0	0	0	0	16.0
(in.)	(0)	(0)	(0)	(0)	(0)	(1.9)	(3.3)	(1.1)	(0)	(0)	(0)	(0)	(6.3)

Macquarie Island, Australia (EM): pop. <100, lat. 54° 30′ S, long. 158° 56′ E, elev. 6 m (20 ft)													
	Jan	Feb	Mar	Apr	May	Jun	Jul	Aug	Sep	Oct	Nov	Dec	Annual
Temperature°C	6.4	6.1	5.6	4.4	4.2	2.8	2.8	2.8	3.1	3.3	4.4	5.6	4.3
(°F)	(43.5)	(43.0)	(42.0)	(40.0)	(39.5)	(37.0)	(37.0)	(37.0)	(37.5)	(38.0)	(40.0)	(42.0)	(39.8)
PRECIP cm	10.2	8.9	10.4	9.7	8.4	7.4	8.1	8.1	9.7	4.2	7.1	9.9	106.2
(in.)	(4.0)	(3.5)	(4.1)	(3.8)	(3.3)	(2.9)	(3.2)	(3.2)	(3.8)	(3.3)	(2.8)	(3.9)	(41.8)
POTET cm	8.4	6.6	5.8	4.1	3.3	2.0	2.3	2.8	3.6	4.6	6.1	7.9	56.6
(in.)	(3.3)	(2.6)	(2.3)	(1.6)	(1.3)	(0.8)	(0.9)	(1.1)	(1.4)	(1.8)	(2.4)	(3.1)	(22.3)

Polar Ice Cap (EF)

Amundsen Base, Antarctica (EF): pop. transitory, lat. 90° 00′ S, long. 0° 00′, elev. 2800 m (9186 ft)													
	Jan	Feb	Mar	Apr	May	Jun	Jul	Aug	Sep	Oct	Nov	Dec	Annual
Temperature°C	−28.2	−38.5	−55.2	−57.7	−56.2	−59.0	−58.5	−60.4	−59.3	−51.4	−38.7	−28.1	−49.3
(°F)	(−18.8)	(−37.3)	(−67.4)	(−71.9)	(−69.2)	(−74.2)	(−73.3)	(−76.7)	(−74.8)	(−60.5)	(−37.7)	(−18.6)	(−56.7)
PRECIP cm	0.1	0.2	0	0	0	0	0	0	0	0	0	0.1	0.4
(in.)	(0)	(0.1)	(0)	(0)	(0)	(0)	(0)	(0)	(0)	(0)	(0)	(0)	(0.2)
POTET cm (in.)	←						near zero						→

McMurdo, Antarctica (EF): pop. transitory, lat. 77° 50′ S, long. 166° 36′ W, elev. 45 m (148 ft)													
	Jan	Feb	Mar	Apr	May	Jun	Jul	Aug	Sep	Oct	Nov	Dec	Annual
Temperature°C	−3.6	−8.7	−18.5	−22.5	−23.1	−24.6	−27.2	−29.0	23.4	−20.4	−9.6	−3.7	−17.9
(°F)	(25.5)	(16.3)	(−1.3)	(−8.5)	(9.6)	(−12.3)	(−17.0)	(−20.2)	(−10.1)	(−4.7)	(14.7)	(25.3)	(−0.2)
PRECIP cm	1.1	0.4	0.6	0.6	1.3	0.5	0.5	1.1	(1.2)	0.8	0.6	0.7	9.4
(in.)	(0.4)	(0.2)	(0.2)	(0.2)	(0.5)	(0.2)	(0.2)	(0.4)	(0.5)	(0.3)	(0.2)	(0.3)	(3.7)
POTET cm (in.)	←						near zero						→

Hot Low-Latitude Desert (BWh)

Alice Springs, Northern Territory, Australia (BWh): pop. 24,000, lat. 23° 42′ S, long. 133° 53′ E, elev. 579 m (1901 ft)													
	Jan	Feb	Mar	Apr	May	Jun	Jul	Aug	Sep	Oct	Nov	Dec	Annual
Temperature°C	28.6	27.8	24.7	19.7	15.3	12.2	11.7	14.4	18.3	22.8	25.8	27.8	20.7
(°F)	(83.5)	(82.0)	(76.5)	(67.5)	(59.5)	(54.0)	(53.0)	(58.0)	(65.0)	(73.0)	(78.5)	(82.0)	(69.4)
PRECIP cm	4.3	3.3	2.8	1.0	1.5	1.3	0.8	0.8	0.8	1.8	3.0	3.8	25.1
(in.)	(1.7)	(1.3)	(1.1)	(0.4)	(0.6)	(0.5)	(0.3)	(0.3)	(0.3)	(0.7)	(1.2)	(1.5)	(9.9)
POTET cm	17.8	14.7	11.9	6.6	3.6	2.0	1.8	3.3	5.6	10.2	14.0	17.0	108.5
(in.)	(7.0)	(5.8)	(4.7)	(2.6)	(1.4)	(0.8)	(0.7)	(1.3)	(2.2)	(4.0)	(5.5)	(6.7)	(42.7)

Yuma, Arizona (BWh): pop. 42,000, lat. 32° 40′ N, long. 114° 36′ W, elev. 61 m (200 ft)													
	Jan	Feb	Mar	Apr	May	Jun	Jul	Aug	Sep	Oct	Nov	Dec	Annual
Temperature°C	12.7	14.8	17.9	21.5	25.1	29.6	33.2	32.8	29.7	23.3	17.1	13.2	22.6
(°F)	(54.9)	(58.6)	(64.2)	(70.7)	(77.2)	(85.3)	(91.8)	(91.0)	(85.5)	(73.9)	(62.8)	(55.8)	(72.7)
PRECIP cm	1.1	1.1	0.8	0.3	0.1	0	0.5	1.4	1.0	0.8	0.6	1.2	8.9
(in.)	(0.4)	(0.4)	(0.3)	(0.1)	(0)	(0)	(0.2)	(0.6)	(0.4)	(0.3)	(0.2)	(0.5)	(3.5)
POTET cm	1.3	2.3	4.6	8.2	13.6	18.9	21.1	20.0	16.4	9.0	3.4	1.5	110.3
(in.)	(0.5)	(0.9)	(1.8)	(3.2)	(5.4)	(7.4)	(8.3)	(7.9)	(6.5)	(3.5)	(1.3)	(0.6)	(43.4)

Cold Midlatitude Desert (BWk)

San Luis Potosí, Mexico (BWk): pop. 2,000,000, lat. 22° 09′ N, long. 100° 58′ W, elev. 1877 m (6157 ft)													
	Jan	Feb	Mar	Apr	May	Jun	Jul	Aug	Sep	Oct	Nov	Dec	Annual
Temperature°C	12.9	15.0	17.0	20.4	21.5	20.9	19.5	19.6	18.4	17.2	14.8	13.9	17.6
(°F)	(55.2)	(59.0)	(62.6)	(68.7)	(70.7)	(69.6)	(67.1)	(67.3)	(65.1)	(63.0)	(58.6)	(57.0)	(63.7)
PRECIP cm	1.0	0.6	1.0	0.8	3.3	7.4	5.7	4.3	7.5	2.0	1.0	1.2	35.8
(in.)	(0.4)	(0.2)	(0.4)	(0.3)	(1.3)	(2.9)	(2.2)	(1.7)	(3.0)	(0.8)	(0.4)	(0.5)	(14.1)
POTET cm	3.6	4.5	6.4	8.8	10.4	9.8	9.0	8.9	7.0	6.2	4.4	3.9	82.9
(in.)	(1.4)	(1.8)	(2.5)	(3.5)	(4.1)	(3.9)	(3.5)	(3.5)	(2.8)	(2.4)	(1.7)	(1.5)	(32.6)

Reno, Nevada (BWk): pop. 115,000, lat. 39° 30′ N, long. 119° 47′ W, elev. 1342 m (4404 ft)													
	Jan	Feb	Mar	Apr	May	Jun	Jul	Aug	Sep	Oct	Nov	Dec	Annual
Temperature°C	−0.6	2.2	5.0	8.9	12.8	16.7	21.1	19.4	15.6	10.6	4.4	0.6	10.0
(°F)	(31.0)	(36.0)	(41.0)	(48.0)	(55.0)	(62.0)	(70.0)	(67.0)	(60.0)	(51.0)	(40.0)	(33.0)	(50.0)
PRECIP cm	2.5	2.5	1.8	1.3	1.3	1.0	0.5	0.5	0.5	1.5	1.5	2.3	17.8
(in.)	(1.0)	(1.0)	(0.7)	(0.5)	(0.5)	(0.4)	(0.2)	(0.2)	(0.2)	(0.6)	(0.6)	(0.9)	(7.0)
POTET cm	0	0.7	2.2	4.0	7.1	10.1	13.0	11.7	7.8	4.3	1.7	0.2	62.8
(in.)	(0)	(0.3)	(0.9)	(1.6)	(2.8)	(4.0)	(5.1)	(4.6)	(3.1)	(1.7)	(0.7)	(0.1)	(24.7)

Hot Low-Latitude Steppe (BSh)

Monterrey, Mexico (BSh): pop. 1,702,000, lat. 25° 40′ N, long. 100° 19′ W, elev. 538 m (1765 ft)													
	Jan	Feb	Mar	Apr	May	Jun	Jul	Aug	Sep	Oct	Nov	Dec	Annual
Temperature°C	16.4	17.1	20.3	23.7	26.1	28.1	28.4	28.3	25.9	22.4	17.7	15.6	22.6
(°F)	(61.5)	(63.7)	(68.5)	(74.7)	(79.0)	(82.6)	(83.1)	(82.9)	(78.6)	(72.3)	(63.9)	(60.1)	(72.7)
PRECIP cm	0.8	2.3	1.3	1.8	1.8	5.1	4.0	7.6	11.9	9.4	2.3	0.8	49.0
(in.)	(0.3)	(0.9)	(0.5)	(0.7)	(0.7)	(2.0)	(1.6)	(3.0)	(4.7)	(3.7)	(0.9)	(0.3)	(19.3)
POTET cm	3.6	4.1	6.9	10.7	14.7	17.0	17.8	16.8	12.7	8.6	4.3	3.3	120.4
(in.)	(1.4)	(1.6)	(2.7)	(4.2)	(5.8)	(6.7)	(7.0)	(6.6)	(5.0)	(3.4)	(1.7)	(1.3)	(47.4)

Walgett, New South Wales, Australia (BSh): pop. 2000, lat. 30° 01′ S, long. 148° 07′ E, elev. 133 m (436 ft)													
	Jan	Feb	Mar	Apr	May	Jun	Jul	Aug	Sep	Oct	Nov	Dec	Annual
Temperature°C	28.1	27.2	24.4	19.7	15.0	11.7	10.8	12.5	19.2	20.6	24.4	27.2	20.0
(°F)	(82.5)	(81.0)	(76.0)	(67.5)	(59.0)	(53.0)	(51.5)	(54.5)	(66.5)	(69.0)	(76.0)	(81.0)	(68.0)
PRECIP cm	5.3	4.8	4.1	3.0	3.8	4.1	3.3	2.8	2.5	3.0	3.8	4.3	45.0
(in.)	(2.1)	(1.9)	(1.6)	(1.2)	(1.5)	(1.6)	(1.3)	(1.1)	(1.0)	(1.2)	(1.5)	(1.7)	(17.7)
POTET cm	17.8	14.5	11.9	6.6	3.6	1.8	1.8	2.5	6.3	8.4	12.7	17.3	104.6
(in.)	(7.0)	(5.7)	(4.7)	(2.6)	(1.4)	(0.7)	(0.7)	(1.0)	(2.5)	(3.3)	(5.0)	(6.8)	(41.2)

N'Djamena (Fort Lamy), Chad (BSh): pop. 512,000, lat. 12° 08′ N, long. 15° 02′ E, elev. 300 m (984 ft)													
	Jan	Feb	Mar	Apr	May	Jun	Jul	Aug	Sep	Oct	Nov	Dec	Annual
Temperature°C	23.5	25.9	30.1	32.7	32.3	30.5	27.5	26.2	27.1	28.6	27.1	24.1	27.9
(°F)	(74.3)	(78.6)	(86.2)	(90.9)	(90.1)	(86.9)	(81.5)	(79.2)	(80.8)	(83.5)	(80.8)	(75.4)	(82.2)
PRECIP cm	0	0	0	0.5	3.6	6.6	15.5	25.7	10.4	2.3	0	0	64.5
(in.)	(0)	(0)	(0)	(0.2)	(1.4)	(2.6)	(6.1)	(10.1)	(4.1)	(0.9)	(0)	(0)	(25.4)
POTET cm	8.6	10.9	17.0	18.3	19.1	17.8	15.7	13.5	14.2	15.5	13.5	9.4	173.0
(in.)	(3.4)	(4.3)	(6.7)	(7.2)	(7.5)	(7.0)	(6.2)	(5.3)	(5.6)	(6.1)	(5.3)	(3.7)	(68.1)

Cold Midlatitude Steppe (BSk)

Tehran, Iran (BSk): pop. 5,433,000, lat. 35° 40′ N, long. 51° 26′ E, elev. 1206 m (3957 ft)												
	Jan	Feb	Mar	Apr	May	Jun	Jul	Aug	Sep	Oct	Nov	Dec	Annual
Temperature°C	3.5	5.2	10.2	15.4	21.2	26.1	29.5	28.4	24.6	18.3	10.6	4.9	16.5
(°F)	(38.3)	(41.4)	(50.4)	(59.7)	(70.2)	(79.0)	(85.1)	(83.1)	(76.3)	(64.9)	(51.1)	(40.8)	(61.7)
PRECIP cm	3.8	2.3	3.6	3.0	1.3	0.3	0	0	0	0.5	2.8	2.8	20.3
(in.)	(1.5)	(0.9)	(1.4)	(1.2)	(0.5)	(0.1)	(0)	(0)	(0)	(0.2)	(1.1)	(1.1)	(8.0)
POTET cm.	0.3	0.5	2.3	5.3	10.7	16.0	19.8	17.5	12.2	6.6	2.0	0.5	93.7
(in.)	(0.1)	(0.2)	(0.9)	(2.1)	(4.2)	(6.3)	(7.8)	(6.9)	(4.8)	(2.6)	(0.8)	(0.2)	(36.9)

Lanzhou, China (BSk): pop. 2,000,000, lat. 36° 03′ N, long. 103° 51′ E, elev. 1506 m (4941 ft)												
	Jan	Feb	Mar	Apr	May	Jun	Jul	Aug	Sep	Oct	Nov	Dec	Annual
Temperature°C	−6.3	−1.4	5.5	12.0	17.3	20.1	22.8	21.4	16.3	10.3	1.6	−5.0	9.6
(°F)	(20.7)	(29.5)	(41.9)	(53.6)	(63.1)	(69.6)	(73.0)	(70.6)	(61.4)	(50.5)	(34.9)	(23.0)	(49.3)
PRECIP cm	0.3	0.3	0.8	1.3	1.8	3.8	6.6	9.1	5.6	1.5	0.3	0	30.7
(in.)	(0.1)	(0.1)	(0.3)	(0.5)	(0.7)	(1.5)	(2.6)	(3.6)	(2.2)	(0.6)	(0.1)	(0)	(12.1)
POTET cm	0	0	2.0	5.6	9.7	12.4	14.2	12.4	7.9	4.1	0.5	0	67.6
(in.)	(0)	(0)	(0.8)	(2.2)	(3.8)	(4.9)	(5.6)	(4.9)	(3.1)	(1.6)	(0.2)	(0)	(26.6)

Kemmerer, Wyoming (BSk): pop. 3300, lat. 41° 48′ N, long. 110° 32′ W, elev. 2120 m (6954 ft)												
	Jan	Feb	Mar	Apr	May	Jun	Jul	Aug	Sep	Oct	Nov	Dec	Annual
Temperature°C	−8.3	−6.6	−2.8	3.9	9.2	13.3	17.2	16.1	11.5	5.7	−1.9	−5.5	4.3
(°F)	(17.0)	(20.1)	(27.0)	(39.0)	(48.6)	(55.9)	(63.0)	(60.9)	(52.7)	(42.3)	(28.5)	(22.0)	(39.8)
PRECIP cm	1.8	1.7	1.9	2.1	2.6	2.8	1.9	2.1	1.6	2.1	1.7	1.7	24.0
(in.)	(0.7)	(0.7)	(0.7)	(0.8)	(1.0)	(1.1)	(0.7)	(0.8)	(0.6)	(0.8)	(0.7)	(0.7)	(9.5)
POTET cm	0	0	0	2.7	6.4	9.1	11.9	10.4	6.6	3.4	0	0	50.5
(in.)	(0)	(0)	(0)	(1.1)	(2.5)	(3.6)	(4.7)	(4.1)	(2.6)	(1.3)	(0)	(0)	(19.9)

Highland (H)

La Paz, Bolivia (H): pop. 788,000, lat. 16° 30′ S, long. 68° 10′ W, elev. 4103 m (13,461 ft)												
	Jan	Feb	Mar	Apr	May	Jun	Jul	Aug	Sep	Oct	Nov	Dec	Annual
Temperature°C	9.0	9.0	9.0	9.0	8.0	7.0	8.0	8.0	9.0	10.0	10.0	10.0	9.0
(°F)	(48.2)	(48.2)	(48.2)	(48.2)	(46.4)	(44.6)	(46.4)	(46.4)	(48.2)	(50.0)	(50.0)	(50.0)	(48.2)
PRECIP cm	13.9	10.8	5.6	2.2	0.9	0.5	0.4	1.7	3.4	3.7	4.8	7.6	55.5
(in.)	(5.5)	(4.2)	(2.2)	(0.9)	(0.4)	(0.2)	(0.2)	(0.7)	(1.3)	(1.5)	(1.9)	(3.0)	(21.9)

Lhasa, (Tibet) China (H): pop. 382,000, lat. 29° 40′ N, long. 91° 07′ E, elev. 3685 m (12,090 ft)												
	Jan	Feb	Mar	Apr	May	Jun	Jul	Aug	Sep	Oct	Nov	Dec	Annual
Temperature°C	−2.0	1.0	5.0	8.0	12.0	17.0	16.0	16.0	14.0	9.0	4.0	0	8.0
(°F)	(28.4)	(33.8)	(41.0)	(46.4)	(53.6)	(62.6)	(60.8)	(60.8)	(57.2)	(48.2)	(39.6)	(32.0)	(46.4)
PRECIP cm	0	1.3	0.8	0.5	2.5	6.3	12.2	8.9	6.6	1.3	0.2	0	40.6
(in.)	(0)	(0.5)	(0.3)	(0.2)	(1.0)	(2.5)	(4.8)	(3.5)	(2.6)	(0.5)	(0.1)	(0)	(16.0)

Appendix B

Information Sources, Organizations, and Agencies

This is a sampling of information sources for further geographic inquiry. Consult your campus library and instructor for related sources and research pathways.

General Sources

Canada Yearbook. Published annually by minister of Supply and Services. Available from Publication Sales and Services, Statistics Canada, Ottawa, Canada K1A 0T6. Also available from Statistics Canada: *Human Activity and the Environment — 1986 and State of the Environment, Report to Canada* (May 1986).

Climate Alert. Published by Climate Institute, 316 Pennsylvania Avenue, SE, Suite 403, Washington, DC 20003. A clearinghouse of information on world climate.

The Complete Guide to Environmental Careers. The CEIP Fund. Island Press, Covela, CA 95428.

Congressional Directory. Published after every biennial congressional election. Available from U.S. Government Printing Office, Washington, DC 20402.

Environment. Heldref Publications, 4000 Albemarle St., NW, Washington, DC 20016.

Geographical Bibliography for American Libraries. Chauncy D. Harris, ed. A joint project of the Association of American Geographers and the National Geographic Society, Washington, 1985. AAG, 1710 16th Street, NW, Washington, DC 20009.

Instant Information. 10,000 addresses, both public and private. Compiled by Joel Makower and Alan Green, published by Prentice Hall Press, New York, 1987.

Source Book on the Environment — A Guide to the Literature. Kenneth A. Hammond, George Macinko, and Wilma B. Fairchild, University of Chicago Press, 1978.

State of the Environment — A View Toward the Nineties. A report from the Conservation Foundation, 1987, sponsored by the Charles Stewart Mott Foundation. Conservation Foundation, 1250 42nd Street NW, Washington, DC 20076.

State of the World 1991 — Report on Progress Toward a Sustainable Society. Published annually by the Worldwatch Institute (Lester R. Brown, project director), 1776 Massachusetts Avenue NW, Washington, DC 20036.

Statistical Abstract of the United States. Published annually by the U.S. Bureau of the Census, Department of Commerce, Washington, DC. Also available from U.S. Government Printing Office, Washington, DC 20402.

U.S. Government Manual. Published by the Office of the Federal Register with annual updates. Available through the Superintendent of Documents, U.S. Government Printing Office, Washington, DC 20402.

U.S. Government Printing Office. North Capitol and H Streets NW, Washington, DC 20401. Superintendent of Documents provides monthly catalog of all U.S. government publications. Many local bookstores and offices.

The World Almanac. Published annually by World Almanac and Book of Facts, a Scripps Howard company, New York.

World Resources 1988–89. Published annually by the World Resources Institute, in collaboration with the United Nations Environment Programme, 1735 New York Avenue NW, Washington, DC 20006.

Geography Organizations

Alaska Geographical Society, P.O. Box 4–EEE, Anchorage, AK 99509. One of many state and provincial geographical societies operating in the United States and Canada.

American Geographical Society, 156 Fifth Avenue, Suite 600, New York, NY 10010–7002. Publishes *Focus* quarterly.

Association of American Geographers, 1710 16th Street NW, Washington, DC 20009. Principal academic and professional geography organization—offers student memberships and career information.

Canadian Association of Geographers, McGill University, 805 Sherbrooke Street West, Montréal, Québec H3A 2K6.

Geographic alliances, 60 state chapters. Information is available from college and university geography departments; generally coordinated by the National Geographic Society.

National Council for Geographic Education, Indiana University of Pennsylvania, Indiana, PA 15705.

National Geographic Society, 17th and M Streets NW, Washington, DC 20036. Publishes maps, atlases, *National Geographic* Magazine, and *Geography Education Update* through their Geographic Education Program.

Royal Geographical Society, 1 Kingsington Gore, London SW7 2AR, England.

Organizations of Interest

American Association for the Advancement of Science, 1333 H Street NW, Washington, DC 20005.

American Chemical Society, 1155 16th Street NW, Washington, DC 20036.

American Forestry Association, 1319 118th Street NW, Washington, DC 20036.

American Geophysical Union, 2000 Florida Avenue NW, Washington, DC 20009.

American Meteorological Society, 45 Beacon Street, Boston, MA 02108.

British Association for the Advancement of Science, Fortress House, 23 Savile Row, London W1X 1AB.

Center for Science in the Public Interest, 1501 16th Street NW, Washington, DC 20036.

Cousteau Society, 930 West 21st Street, Norfolk, VA 23517.

Environmental Action Foundation, 1525 New Hampshire Avenue NW, Washington DC 20036.

Environmental Defense Fund, 257 Park Avenue South, New York, NY 10010.

Friends of the Earth, Inc., 530 Seventh Avenue SE, Washington, DC 20003.

Greenpeace, U.S.A., Inc., 1436 U Street NW, Washington, DC 20009.

National Parks and Conservation Association, 1015 31st Street NW, Washington, DC 20007. (Natural history associations are organized in conjunction with most national parks.)

National Weather Association, 4400 Stamp Road, Suite 404, Temple Hills, MD 20748.

National Wildlife Federation, 1400 16th Street NW, Washington, DC 20036. Publishes an annual updated list of all agencies and organizations dealing with natural resources—the *Conservation Directory.*

Natural Resources Defense Council, 122 East 42nd Street, New York, NY 10168. Publishes *The Amicus Journal.*

The Nature Conservancy, 1800 North Lynn Street, Arlington, VA 22209.

Rocky Mountain Institute, 1739 Snowmass Creek Road, Snowmass, CO 81654. Research in energy-related issues and conservation.

Sierra Club, 730 Polk Street, San Francisco, CA 94109.

Soil Conservation Society of America, 7515 Northeast Ankeny Road, Ankeny, IA 50021.

Union of Concerned Scientists, 26 Church Street, Cambridge, MA 02238.

United Nations Environment Programme, Office of Public Information, New York Liaison Office, Room DC 2–0803, United Nations, New York, NY 10017; or, UNEP, P.O. Box 30552, Nairobi, Kenya.

The Wilderness Society, 1400 I Street NW, Washington, DC 20005.

Worldwatch Institute, 1776 Massachusetts Avenue NW, Washington, DC 20036.

Government Agencies

Council of Environmental Quality, 722 Jackson Place NW, Washington, DC 20006.

Department of Agriculture, 14th Street and Independence Avenue SW, Washington, DC 20250.
Forest Service
Soil Conservation Service

Department of Commerce, 14th Street between Constitution Avenue and E Street NW, Washington, DC 20230.
National Environmental Satellite, Data, and Information Service
National Geophysical Data Center
National Oceanic and Atmospheric Administration
  National Hurricane Center
  National Climate Program Office
National Oceanographic Data Center
National Severe Storms Forecast Center
National Weather Service
Department of Energy, 1000 Independence Avenue SW, Washington, DC 20585.
Department of the Interior, 1800 C Street NW, Washington, DC 20240.
Bureau of Land Management
Bureau of Mines
Bureau of Reclamation

Fish and Wildlife Service
National Park Service
Environment Canada, Ottawa, Ontario K1A OH3.
Environmental Protection Agency, 401 M Street SW, Washington, DC 20460
Geological Survey of Canada, Headquarters: 601 Booth Street, Ottawa, Ontario K1A OE8. Many divisions, centres, and branches performing specialized functions. Provincial survey offices nationwide.
National Academy of Sciences, Washington, DC 20550.
Office of Coastal Zone Management, 1825 Connecticut Avenue, Suite 700, Washington, DC 20235.
Solar Energy Research Institute, 6536 Cole Boulevard, Golden, CO 80401.
U.S. Geological Survey, 12201 Sunrise Valley Drive, Reston, VA 22092. Many regional offices, bookstores, and mapping centers.

Glossary

The chapter in which each term is **boldfaced** is designated in parentheses, followed by a specific definition relevant to the usage in the chapter.

Abiotic (1) Nonliving.

Ablation (17) Loss of glacial ice through melting, sublimation, wind removal by deflation, or the calving off of blocks of ice.

Abrasion (14) Mechanical wearing and erosion of bedrock accomplished by the rolling and grinding of particles and rocks carried in a stream or imbedded in glacial ice.

Absolute humidity (7) Actual amount of water vapor in the air, expressed as the weight of water vapor (grams) per unit volume of air (cubic meter)—an expression of weight per volume. Saturation absolute humidity refers to the maximum possible vapor content in grams per cubic meter.

Absorption (4) Assimilation and conversion of radiation from one form to another in a medium. In the process the temperature of the absorbing surface is raised, thereby affecting the rate and quality of radiation from that surface.

Actual evapotranspiration (9) Actual amount of evaporation and transpiration that occurs; derived in the water balance by subtracting the deficit from potential evapotranspiration.

Adiabatic (8) Pertaining to the heating and cooling of a descending or ascending parcel of air through compression and expansion, without any exchange of heat between the parcel and the surrounding environment.

Adsorption (18) The process whereby cations become attached to soil colloids, or the adhesion of gas molecules and ions to solid surfaces with which they come into contact.

Advection (4) Horizontal movement of air or water from one place to another.

Advection fog (7) Active condensation formed when warm, moist air moves laterally over cooler water or land surfaces, causing the lower layers of the overlying air to be chilled to the dew-point temperature.

Aggradation (14) The general building up of a land surface because of the deposition of material; opposite of degradation. When the sediment load of a stream exceeds the stream's capacity, the stream channel is filled through this process.

Air mass (8) A distinctive, homogeneous body of air in terms of temperature and humidity, that takes on the moisture and temperature characteristics of its source region.

Air pressure (3) Pressure produced by the motion, size, and number of gas molecules and exerted on surfaces in contact with the air. Normal sea level pressure, as measured by the height of a column of mercury (Hg), is expressed as 1013.2 millibars, 760 mm of Hg, or 29.92 inches of Hg. Air pressure can be measured with mercury or aneroid barometers.

Albedo (4) The reflective quality of a surface, expressed as the relationship of incoming to reflected insolation and stated as a percentage; a function of surface color, angle of incidence, and surface texture.

Aleutian low (6) See subpolar low-pressure cell.

Alfisols (18) Moderately weathered forest soils that are moist versions of Mollisols, with productivity dependent on specific patterns of moisture and temperature; rich in organics; most wide-ranging of the 10 soil orders in the Soil Taxonomy classification.

Alleghany orogeny (12) See orogenesis.

Alluvial fan (15) Fan-shaped fluvial landform at the mouth of a canyon, particularly noticeable in arid landscapes where streams are intermittent.

Alluvial terraces (14) Level areas that appear as topographic steps above the river, created by a stream as it scours with renewed downcutting into its floodplain; composed of unconsolidated alluvium.

Alluvium (14) General descriptive term for clay, silt, and sand transported by running water and deposited in sorted or semisorted sediment on a floodplain, delta, or in a stream bed.

Alpine glacier (17) A glacier confined in a mountain valley or walled basin, consisting of three subtypes: valley glacier (within a valley), piedmont glacier (coalesced at the base of a mountain, spreading freely over nearby lowlands), and outlet glacier (flowing outward from a continental glacier).

Altitude (4) The angular distance between the horizon (a horizontal) and the Sun (or any point).

Analemma (4) A convenient device to track the passage of the Sun, the Sun's declination, and the positive and negative equation of time throughout the year.

Anemometer (6) A device used to measure wind velocity.

Aneroid barometer (3) A device to measure air pressure using a partially emptied, sealed cell (see air pressure).

Antarctic geographic zone (1) The region poleward of the Antarctic Circle (66.5° S latitude), including Antarctica, which is covered by Earth's greatest ice sheet.

Antarctic high (6) A consistent high-pressure region centered over Antarctica; source region for an intense polar air mass that is dry and associated with the lowest temperatures on Earth.

Anthropogenic atmosphere (3) Earth's next atmosphere, so named because humans appear to be the principal causative agent.

Anticline (12) Upfolded strata in which layers slope away from the axis of the fold, or central ridge.

Anticyclone (6) A dynamically or thermally caused area of high atmospheric pressure with descending and diverging air flows.

Aphelion (2) The most distant point in Earth's elliptical orbit about the Sun; reached on July 4 at a distance of 152,083,000 km (94.5 million mi); variable over a 100,000-year cycle.

Apparent temperature (5) The temperature subjectively perceived by each individual, also known as sensible temperature.

Aquatic ecosystems (20) An association of plants and animals and their nonliving environment in a marine setting.

Aquiclude (9) A body of rock that does not conduct water in usable amounts; that is, an impermeable layer.

Aquifer (9) Rock strata permeable to groundwater flow.

Aquifer recharge area (9) The surface area where water enters an aquifer to recharge the water-bearing strata in the groundwater system.

Arctic geographic zone (1) The region poleward of the Arctic Circle (66.5° N latitude), dominated by the Arctic Ocean, which is covered by floating sea ice that thins in the summer, and glacial ice in the form of icebergs.

Arctic tundra (20) A biome in the northernmost portions of North America and the Soviet Union, featuring low ground-level herbaceous plants as well as some woody plants.

Aridisols (18) Largest single soil order in the Soil Taxonomy classification and typical of dry climates; low in organic matter and dominated by calcification and salinization.

Arroyo (15) A Spanish term for an intermittently dry streambed or wash.

Artesian water (9) Pressurized groundwater that rises in a well or a rock structure above the local water table; may flow out onto the ground (see piezometric surface).

Asperities (12) Sticking or friction points of high strain, which are the points that give way and release energy between two sides of a fault. Earthquake magnitude is related to the spatial extent to which asperities fail.

Asthenosphere (11) Region of the upper mantle, just below the crust, known as the plastic layer; the least rigid portion of Earth's interior; shatters if struck yet flows under extreme heat and pressure.

Atmosphere (1) The thin veil of gases surrounding Earth, which forms a protective boundary between outer space and the biosphere; generally considered to be essentially below 480 km (300 mi).

Atoll (16) A circular, or ring-shaped, island enclosing a central lagoon; created by coral colonies.

Auroras (2) A spectacular glowing light display in the ionosphere, stimulated by the interaction of the solar wind with oxygen and nitrogen gases; aurora borealis (northern) and aurora australis (southern).

Autotrophs (19) Organisms capable of using carbon dioxide directly as their sole source of carbon; known as producers. They fix carbon through photosynthesis.

Autumnal equinox (4) The time around September 22–23 when the Sun's declination crosses the equatorial parallel and all places on Earth experience days and nights of equal length. The Sun rises at the South Pole and sets at the North Pole.

Available water (9) The portion of capillary water that is accessible to plant roots; usable water held in soil moisture storage.

Axial parallelism (4) Earth's axis is parallel and aligned with itself throughout the year; the North Pole points to near Polaris.

Axis (4) An imaginary line, extending through Earth from the geographic North Pole to the geographic South Pole, around which Earth rotates.

Azores high (6) A subtropical high-pressure cell that forms in the Northern Hemisphere in the eastern Atlantic (see Bermuda high); associated with warm, clear water and large quantities of *sargassum,* or seaweed, characteristic of the Sargasso Sea.

Backswamp (14) A low-lying, swampy area of a floodplain; adjacent to a river, with the river's natural levees on one side and the sides of the valley on the other (see yazoo tributary).

Badland (15) Rugged topography, usually of relatively low, varied relief and barren of vegetation; associated with arid and semiarid regions and rocks with low resistance to weathering.

Bajada (15) A continuous apron of coalesced alluvial fans, formed along the base of mountains in arid climates; presents a gently rolling surface from fan to fan.

Barrier beaches (16) Narrow, long depositional features, generally of sand, which form offshore and are roughly parallel to the coast; may appear as barrier islands and long chains of barrier beaches.

Barrier islands (16) Generally, a broadened barrier beach.

Basalt (11) A common extrusive igneous rock, its mafic composition is fine-grained, comprising the bulk of the ocean-floor crust, lava flows, and volcanic forms; gabbro in its intrusive form.

Base level (13) A hypothetical level below which a stream cannot erode its valley, and thus the lowest operative level for denudation processes; in an absolute sense represented by sea level extending back under the landscape.

Basin and Range Province (15) A region of dry climates, few permanent streams, and interior drainage patterns in the western United States formed by a sequence of horsts and grabens.

Batholith (11) The largest plutonic form exposed at the surface; an irregular intrusive mass (>100 km^2; >40 mi^2) that invades crustal rocks, cooling slowly so that crystals develop.

Baymouth bar (16) An extensive sand spit that encloses a bay, cutting it off completely from the ocean and forming a lagoon; produced by littoral drift and wave action.

Beach (16) The portion of the coastline where an accumulation of sediment is in motion.

Beach drift (16) See littoral drift.

Beaufort wind scale (6) A descriptive scale for the visual estimation of wind speeds; originally conceived in 1806 by Admiral Beaufort of the British Navy.

Bed load (14) Coarse materials that are dragged along the bed of a stream by traction or by the rolling and bouncing motion of saltation; involves particles too large to remain in suspension.

Bedrock (13) The rocks of Earth's crust that are below the soil and are basically unweathered; such solid crust is sometimes exposed as an outcrop.

Bermuda high (6) A subtropical high-pressure cell that forms in the western North Atlantic (see Azores high).

Biogeochemical cycles (19) The various circuits of flowing elements and materials (carbon, oxygen, nitrogen, phosphorus, water) that combine Earth's biotic and abiotic systems; the cycling of materials is continuous and renewed through the biosphere and the life processes.

Biogeographical realms (20) Eight regions of the biosphere, each representative of evolutionary core areas of related flora (plants) and fauna (animals); a broad geographical classification scheme.

Biogeography (19) The study of the distribution of plants and animals and related ecosystems; the geographical relationships with their environments over time.

Biomass (19) The total mass of living organisms on Earth or per unit area of a landscape; also, the weight of the living organisms in an ecosystem.

Biome (20) A large terrestrial ecosystem characterized by specific plant communities and formations; usually named after the predominant vegetation in the region.

Biosphere (1) That area where the atmosphere, lithosphere, and hydrosphere function together to form the context within which life exists; an intricate web that connects all organisms with their physical environment.

Biotic (1) Living.

Blowout depressions (15) Eolian erosion whereby deflation forms basins in areas of loose sediment. Size may range up to hundreds of meters.

Bolson (15) The slope and basin area between the crests of two adjacent ridges in a dry region.

Boreal forest (20) See northern needleleaf forests.

Brackish (7) Seawater with a salinity of less than 35‰; for example, the Baltic Sea.

Breaker (16) The point where a wave's height exceeds its vertical stability and the wave breaks as it approaches the shore.

Breakwater (16) A structure built to block the movement of littoral currents and beach drift; used to protect harbor entrances and beach deposits.

Brine (7) Seawater with a salinity of more than 35‰; for example, the Persian Gulf.

Calcification (18) The illuviated accumulation of calcium carbonate or magnesium carbonate in the B and C soil horizons.

Caldera (12) An interior sunken portion of a composite volcanic crater; usually steep-sided and circular, sometimes containing a lake; also can be found in conjunction with shield volcanoes.

Caliche (18) A cemented or hardened subsurface diagnostic calcic soil horizon; found in the southwestern United States, usually in arid and semiarid climates.

Calorie (2) The amount of energy required to raise one gram of water $1C°$.

Capillary water (9) Soil moisture, most of which is accessible to plant roots; held in the soil by surface tension and cohesive forces between water and soil (see also available water, field capacity, and wilting point).

Carbon dioxide (3) A natural by-product of life processes and complete combustion; the principal radiatively active gas in the greenhouse effect; CO_2.

Carbon monoxide (3) An odorless, colorless, tasteless combination of carbon and oxygen produced by the incomplete combustion of fossil fuels or other carbon-containing substances; CO.

Carnivore (19) A secondary consumer that principally eats meat for sustenance. The top carnivore in a food chain is considered a tertiary consumer.

Cartography (1) The making of maps and charts; a specialized science and art that blends aspects of geography, engineering, mathematics, graphics, computer science, and artistic specialties.

Catastrophism (11) A philosophy that attempts to fit the vastness of Earth's age and the complexity of the rock record into a very shortened time span through a belief in short-lived and catastrophic worldwide events.

Cation-exchange capacity (CEC) (18) The ability of soil colloids to exchange cations between their surfaces and the soil solution; a measured potential.

Chaparral (20) Dominant shrub formations of Mediterranean dry summer climates; characterized by sclerophyllous scrub and short, stunted, and tough forests; derived from the Spanish *chapparo;* specific to California.

Chemical weathering (13) Decomposition and decay of the constituent minerals in rock through chemical alteration of those minerals. Water is essential, with rates keyed to temperature and precipitation values. Processes include hydrolysis, oxidation, carbonation, and solution.

Chemosynthesis (2) Production of organic compounds by living tissue without involving sunlight; the action of purple-sulfur bacteria, for example.

Chinook winds (8) North American term for a warm, dry, downslope air flow; characteristic of the rain shadow region on the leeward side of mountains; known as föhn, or foehn, winds in Europe.

Chlorophyll (19) A light-sensitive pigment that resides within the chloroplast bodies of plants in leaf cells; the basis of photosynthesis.

Cinder cone (12) A landform of tephra and scoria, usually small and cone-shaped and generally not more than 450 m (1500 ft) in height; with a truncated top.

Circle of illumination (4) The division between lightness and darkness on Earth; a day-night great circle.

Circum-Pacific belt (12) A tectonically and volcanically active region encircling the Pacific Ocean; also known as the "ring of fire."

Cirque (17) A scooped-out, amphitheater-shaped basin at the head of an alpine glacier valley; an erosional landform.

Cirroform (7) Wispy clouds or clouds in filaments, patches, or bands that usually are quite high in altitude and are composed of ice crystals.

Cirrus (7) Wispy filaments of ice-crystal clouds that occur above 6000 m (20,000 ft); appear in a variety of forms, from feathery hairlike fibers to veils of fused sheets of ice crystals, and sometimes to a dappled mackerel sky.

Classification (10) The process of ordering or grouping data or phenomena in related classes; results in a regular distribution of information; a taxonomy.

Climatic regions (10) Areas of similar climate, which contain characteristic regional weather and air mass patterns.

Climatology (10) A scientific study of climate and climatic patterns and the consistent behavior of weather and weather variability and extremes over time in one place or region; and the effects climate and climate change has on human society and culture.

Climographs (10) Graphs that plot daily, monthly, or annual temperature and precipitation values for a selected station; may also include additional weather information.

Closed system (1) A system that is shut off from the surrounding environment so that it is entirely self-contained in terms of energy and materials; Earth is a closed material system (see open system).

Cold front (8) The leading edge of a cold air mass; identified on a weather map by a line marked with a series of triangular spikes, pointing in the direction of frontal movement.

Collision-coalescence process (7) Principal process of raindrop formation predominant in clouds that form at temperatures above freezing. Larger moisture droplets sweep through the cloud, combining with smaller droplets, gradually coalescing and growing in size.

Community (19) A convenient biotic subdivision within an ecosystem; formed by interacting populations of animals and plants in an area.

Composite volcano (12) A volcano formed by a sequence of explosive volcanic eruptions; steep-sided, conical in shape; sometimes referred to as a stratovolcano, although composite is the preferred term.

Condensation nuclei (7) Necessary microscopic particles on which water vapor condenses to form moisture droplets; can be sea salts, dust, soot, or ash.

Conduction (4) The slow molecule-to-molecule transfer of heat through a medium, from warmer to cooler areas.

Cone of depression (9) The depressed shape of the water table around a well after active pumping from an aquifer. The water table adjacent to the well is drawn down during the process of water removal.

Confined aquifer (9) An aquifer that is bounded above and below by impermeable layers of rock or sediment.

Consumptive uses (9) Water losses that cause water to be removed from a water budget at one point and to remain unavailable further downstream.

Continental divides (14) A ridge or elevated area that determines the drainage patterns of drainage basins; specifically, that ridge in North America that separates drainage to the Pacific in the west from drainage to the Atlantic and Gulf in the east and to Hudson Bay and the Arctic Ocean in the north.

Continental drift (11) A proposal by Wegener in 1912 stating that Earth's landmasses have migrated over the past 200 million years from a supercontinent he called Pangaea to the present configuration; a widely accepted concept today (see plate tectonics).

Continental glacier (17) A continuous mass of unconfined ice, covering at least 50,000 km^2 (19,500 mi^2); most extensive as ice sheets covering Greenland and Antarctica.

Continentality (5) A qualitative designation applied to stations that lack the temperature-moderating effects of the sea and that exhibit a greater range between minimum and maximum temperatures, both daily and annually.

Continental platforms (11) The broadest category of landforms, including those masses of crust that reside above or near sea level and the adjoining undersea continental shelves along the coastline.

Continental shields (12) Generally old, low-elevation heartland regions of continental crust; various cratons (granitic cores) and ancient mountains exposed at the surface.

Contour lines (1) Isolines on a topographic map that connect all points at the same elevation relative to a reference elevation called the vertical datum. The spaces between contour lines portray slope and are known as contour intervals.

Convection (4) Vertical transfer of heat from one place to another through the actual physical movement of air; involves a strong vertical motion.

Coordinated Universal Time (UTC) (1) Formerly known as Greenwich Mean Time outside the United Kingdom; now measured by 80 atomic clocks provided by 24 nations and kept in Paris, France; assumes a prime meridian (0° longitude) passing through Greenwich, London, England.

Coral (16) A simple cylindrical marine animal with a saclike body that secretes calcium carbonate to form a hard external skeleton; lives symbiotically with nutrient-producing algae. Presently going through a bleaching and dying cycle worldwide, apparently due to increasing water temperatures in the ocean.

Cordilleran system (12) One of two large mountain systems on Earth; refers to the relatively young mountains along the western margins of North and South America, from the tip of Tierra del Fuego to the massive peaks of Alaska.

Core (11) The deepest inner portion of Earth, representing one-third of its entire mass; differentiated into two zones—a solid iron inner core surrounded by a dense, molten, fluid metallic-iron outer core.

Coriolis force (6) The apparent deflection of moving objects on Earth from a straight path, in relationship to the differential speed of rotation at varying latitudes. Deflection is to the right in the Northern Hemisphere and to the left in the Southern Hemisphere; it produces a maximum effect at the poles and zero effect along the equator.

Crater (12) A circular surface depression formed by volcanism (accumulation, collapse, or explosion); usually located at a volcanic vent or pipe; can be at the summit or on the flank of a volcano.

Crevasses (17) Vertical cracks that develop in a glacier as a result of friction between valley walls, or tension forces of extension on convex slopes, or compression forces on concave slopes.

Crust (11) Earth's outer shell of crystalline surface rock, ranging from 5 to 60 km (3 to 38 mi) in thickness from oceanic crust to mountain ranges. Average density of continental crust is 2.7 g per cm^3, whereas oceanic crust is 3.0 g per cm^3.

Cumuliform (7) Clouds that exhibit vertical development; forms are puffy and globular.

Cumulonimbus (7) A towering, precipitation-producing cumulus cloud that is vertically developed across altitudes associated with other clouds; frequently associated with lightning and thunder and thus designated a thunderhead.

Cumulus (7) Bright and puffy cumuliform clouds up to 2000 m in altitude (6500 ft).

Cut bank (14) A steep bank formed along the outer portion of a meandering stream; produced by lateral erosive action of a stream.

Cyclogenesis (8) An atmospheric process that describes the birth of a midlatitude wave cyclone; usually along the polar front; strengthening and development of a wave cyclone is also associated with the eastern slope of the Rockies, other north-south mountain barriers, and along the North American and Asian east coasts.

Cyclone (6) A dynamically or thermally caused low-pressure area of converging and ascending air flows (see wave cyclone and tropical cyclone).

Daylength (4) Duration of exposure to insolation, varying during the year depending on latitude; an important aspect of seasonality.

Debris avalanche (13) A mass of falling and tumbling rock, debris, and soil; can be dangerous because of the tremendous velocities achieved by the onrushing materials; differentiated from a debris slide by velocity.

Declination (4) The latitude that receives direct overhead (perpendicular) insolation on a particular day; migrates annually through the 47° of latitude between the tropics.

Decomposers (19) Microorganisms that digest and recycle the organic debris and waste in the environment: includes bacteria, fungi, insects, and worms.

Deep cold zone (7) The deepest part of the ocean, where temperatures gradually lower to near freezing; represents about 80% of the oceanic mass.

Deficit (9) In a water balance, the amount of unmet, or unsatisfied, potential evapotranspiration (DEFIC).

Deflation (15) A process of wind erosion that removes and lifts individual particles, literally blowing away unconsolidated, dry, or noncohesive sediments.

Delta (14) A depositional plain formed where a river enters a lake or an ocean; named after the triangular shape of the Greek letter delta.

Denudation (13) A general term that refers to all processes that cause degradation of the landscape: weathering, mass movement, erosion, and transportation; at the heart of exogenic processes.

Deposition (14) The process whereby weathered, wasted, and transported sediments are laid down. Specific depositional landforms result from the actions of air, water, and ice agents.

Desert biomes (20) Arid landscapes of uniquely adapted dry-climate plants and animals.

Desert pavement (15) An arid landscape surface formed when wind deflation and sheetwash remove smaller particles, leaving residual pebbles and gravels to concentrate at the surface; resembles a cobblestone street.

Desertification (20) The expansion of deserts worldwide, related principally to poor agricultural practices (i.e., overgrazing and inappropriate agricultural practices), improper soil-moisture management, erosion and salinization, deforestation, and the ongoing climatic change; an unwanted semipermanent invasion into neighboring biomes.

Dew-point temperature (7) The temperature at which a given mass of air becomes saturated, holding all the water it can hold. Any further cooling or addition of water vapor results in active condensation.

Differential weathering (13) The effect of different resistances in rock, coupled with variations in the intensity of physical and chemical weathering, produces a landscape exhibiting uneven landforms.

Diffuse radiation (4) The downward component of scattered incoming insolation from clouds and the atmosphere; casts a shadowless light on the ground.

Discharge (9) The measured volume of flow in a river that passes by a given cross section of a stream in a given unit of time; usually denoted in cubic meters per second or cubic feet per second.

Doldrums (6) A region of equatorial calm associated with the intertropical convergence zone; so named for the difficulty encountered in the era of sailing ships.

Downwelling current (6) An area of the sea where a convergence or accumulation of water thrusts excess water downward; occurs, for example, at the western end of the equatorial current or along the margins of Antarctica.

Drainage basin (14) The basic spatial geomorphic unit of a river system; distinguished from a neighboring basin by ridges and highlands that form divides delimiting the catchment area of the drainage basin, or its watershed.

Drainage density (14) A measure of the overall operational efficiency of a drainage basin; determined by the ratio of combined channel lengths to the unit area.

Drainage pattern (14) A geometric arrangement of streams in a region; determined by slope, differing rock resistance to weathering and erosion, climatic and hydrologic variability, and structural controls of the landscape.

Drawdown (9) See cone of depression.

Drumlin (17) A depositional landform related to glaciation that is composed of till and is streamlined in the direction of continental ice movement; blunt end upstream and tapered end downstream with a rounded summit.

Dry adiabatic rate (DAR) (8) The rate at which a parcel of air that is less than saturated cools (if ascending) or heats (if descending); a rate of 10C° per 1000 m (5.5F° per 1000 ft) (see also adiabatic).

Dune (15) A depositional feature of sand grains deposited in transient mounds, ridges, and hills; characteristic of the erg deserts of North Africa. Extensive areas of sand dunes are called sand seas.

Dust dome (4) A dome of airborne pollution associated with every major city; may be blown by winds into elongated plumes downwind from the city.

Dynamic equilibrium model (13) The balancing act between tectonic uplift and reduction rates of erosion, between the resistance of crust materials and the work of denudation processes. Landscapes exhibit an ongoing adaptation to rock structure, climate, local relief, and elevation.

Earthquake (12) A sharp release of energy that produces shaking in Earth's crust at the moment of rupture along a fault or in association with volcanic activity. Movement of the crust produces tectonic earthquakes, whereas the movement of magma produces volcanic earthquakes (harmonic tremors). Earthquake magnitude is estimated by the Richter scale; intensity is described by the Mercalli scale.

Ebb tides (16) Tides that are falling during the daily tidal cycle.

Ecological succession (19) The process whereby different and usually more complex assemblages of plants and animals replace older and usually simpler communities. Changes apparently move toward a more stable and mature condition.

Ecology (19) The science that studies the interrelationships among organisms and their environment and among various ecosystems.

Ecosphere (1) Another name for the biosphere.

Ecosystem (19) A self-regulating association of living plants, animals, and their nonliving physical and chemical environment.

Efflux (7) The reduction of oceanic salinity through organic removal, materials settling out, or precipitation of solids.

Effusive eruption (12) An eruption characterized by low-viscosity basaltic magma, with characteristic low-gas content readily escaping. Lava pours forth onto the surface with relatively small explosions and little tephra; tends to form shield volcanoes.

Elastic-rebound theory (12) A concept describing the faulting process, in which the two sides of a fault appear locked despite the motion of adjoining pieces of crust, but with accumulating strain they rupture suddenly, snapping to new positions relative to each other.

Electromagnetic spectrum (2) All the radiant energy produced by the Sun placed in an ordered range and divided according to wavelengths; travels to Earth at the speed of light in the form of waves.

Eluviation (18) The downward removal of finer particles and minerals from the upper horizons of soil.

Empirical classification (10) A classification based on weather statistics or other data; used to determine general climate categories.

Endogenic system (11) The system internal to Earth, driven by radioactive heat derived from sources within the planet. In response, the surface is fractured, mountain building occurs, and earthquakes and volcanoes are activated.

Entisols (18) A soil order in the Soil Taxonomy classification that specifically lacks vertical development of horizons; usually young or undeveloped; found in active slopes, alluvial-filled floodplains, poorly drained tundra.

Entrenched meanders (14) Excavated into the landscape, such as incised river meanders; thought to be evidence of stream rejuvenation.

Environmental lapse rate (3) The actual lapse rate in the lower atmosphere at any particular time under local weather conditions; may deviate above or below the average normal lapse rate of 6.4C° per 1000 m (3.5F° per 1000 ft).

Eolian (15) Caused by wind; refers to the erosion, transportation, and deposition of materials by wind; spelled aeolian in some countries. Such processes tend to operate more efficiently in arid and semiarid climates.

Epipedon (18) The diagnostic soil horizon that forms at the surface; not to be confused with the A horizon; may include all or part of the illuviated B horizon.

Equal area (1) A trait of a map projection; indicates the equivalence of all areas on the surface of the map, although shape is distorted.

Equatorial geographic zone (1) Generally referring to that area within the tropics that straddles the equator, stretching roughly 10° of latitude north and south.

Equatorial and tropical rain forest (20) A lush biome of tall broadleaf evergreen trees and diverse plants and animals. The dense canopy of leaves is usually arranged in three levels.

Equatorial countercurrent (6) A strong countercurrent, generated by the western intensification pile-up, that travels east along the full extent of the Pacific, Atlantic, and Indian oceans; usually flows alongside or just beneath the westward-flowing surface current.

Equatorial low-pressure trough (6) A thermally caused low-pressure area that almost girdles Earth, with air converging and ascending all along its extent; also called the intertropical convergence zone (ITCZ).

Equilibrium (1) The status of a balanced energy and material system; maintains the system's general structure and character; a balance between form and process.

Equilibrium line (17) The area of a glacier where accumulation (gain) and ablation (loss) are balanced.

Erg desert (15) Sandy deserts, or areas where sand is so extensive that it constitutes a sand sea.

Erosion (14) Denudation by wind, water, and ice, which dislodges, dissolves, or removes surface material.

Erratics (17) Large rocks and boulders left behind by retreating glaciers and strewn across the glaciated surface; usually different in composition and origin from the ground on which they were deposited.

Esker (17) A sinuously curving, narrow deposit of coarse gravel that forms along a meltwater stream channel, developing in a tunnel beneath the glacier. A retreating glacier leaves such ridges behind in a pattern that usually parallels the path of the glacier; may appear branching.

Estuary (14) The point at which the mouth of a river enters the sea and freshwater and seawater are mixed; a place where tides ebb and flow.

Eurasian-Himalayan system (12) One of two major mountain chains on Earth, stretching from southern Asia, China, and northern India and continuing in a belt through the upper Middle East to Europe and the European Alps; includes the world's highest peak, Mount Everest in the Himalayas, at an elevation of 8848 m (29,028 ft).

Eustatic (eustasy) (7) Refers to worldwide changes in sea level that are not related to movements of land but rather to a rise and fall in the volume of water in the oceans, principally through relocation in frozen ice sheets.

Eutrophication (19) A natural process in which lakes receive nutrients and sediment and become enriched; the gradual filling and natural aging of water bodies.

Evaporation (9) The movement of free water molecules away from a wet surface into air that is less than saturated; the phase change of water to water vapor; vaporization below the boiling point of water.

Evaporation fog (7) A fog formed when cold air flows over the warm surface of a lake, ocean, or other body of water; forms as the water molecules evaporate from the water surface into the cold, overlying air; also known as a steam fog or sea smoke.

Evaporation pan (9) A standardized pan from which evaporation occurs, with water automatically replaced and measured; an evaporimeter.

Evaporites (11) Chemical sediments formed from inorganic sources when water evaporates and leaves behind a residue of salts previously in solution.

Evapotranspiration (9) The merging of evaporation and transpiration water loss into one term (see potential and actual evapotranspiration).

Evolutionary atmosphere (2) Earth's atmosphere between 3.3 and 4.0 billion years ago; thought by scientists to have been composed principally of water vapor and lesser amounts of carbon dioxide and nitrogen; an anaerobic environment.

Exfoliation dome (13) A dome-shaped feature of weathering, produced by the response of granite to the overburden removal process, which relieves pressure from the rock. Layers of rock spall off in slabs or shells.

Exogenic system (11) The external surface system, powered by insolation, which energizes air, water, and ice and sets them in motion, under the influence of gravity. Includes all processes of landmass denudation.

Exosphere (3) An extremely rarified outer atmospheric halo beyond the thermopause, which occurs at an altitude of 480 km (300 mi); probably composed of hydrogen and helium atoms, with some oxygen atoms and nitrogen molecules present near the thermopause.

Exotic stream (9) A river that rises in a humid region and flows downstream through an arid region, with discharge decreasing toward the mouth; for example, the Nile River and the Colorado River.

Explosive eruption (12) A violent and unpredictable eruption, the result of magma that is thicker, stickier, higher in gas content and silica, and more viscous than that of an effusive eruption; tends to form blockages within a volcano; produces composite volcanic landforms.

Faulting (12) The process whereby displacement and fracturing occurs between two portions of Earth's crust; usually associated with earthquake activity.

Feedback loop (1) When a portion of the system output cycles back as an information input, causing changes that guide further system operations (see negative and positive feedback).

Field capacity (9) The soil moisture condition when all capillary water for a particular soil is present; the water that is generally available to plant roots.

Firn (17) Snow of a granular texture that is transitional in the slow transformation from snow to glacial ice; snow that has persisted through a summer season in the zone of accumulation.

Firn line (17) The snowline that is visible on the surface of a glacier, where winter snows survive the summer ablation season; analogous to a snowline on land.

Fjord (17) A drowned glaciated valley, or glacial trough, along a coast which is filled by the sea.

Flash flood (15) A sudden and short-lived torrent of water that exceeds the capacity of a stream channel; associated with desert and semiarid washes.

Flood basalts (12) An accumulation of horizontal flows formed when lava spreads out from elongated fissures onto the surface in extensive sheets; associated with effusive eruptions; also known as plateau basalts—for example, the Columbia Plateau.

Floodplain (14) A low-lying area near a stream channel, subject to recurrent flooding; alluvial deposits generally mask underlying rock.

Flood tides (16) Rising tides during the daily tidal cycle.

Fluvial (14) Stream-related processes; from the Latin *fluvius* for "river" or "running water."

Fog (7) A cloud, generally stratiform, in contact with the ground, with visibility usually restricted to less than 1 km (3300 ft).

Folding (12) A process that bends and deforms beds of various rocks subjected to compressional forces.

Foliated (11) Planes in rock aligned, or appearing as wavy striations, in a particular way following metamorphism; describes the mineral structure in metamorphic rocks.

Food chain (19) The circuit along which energy flows from producers, who manufacture their own food, to consumers; a one-directional flow of chemical energy, ending with decomposers.

Food web (19) A complex network of interconnected food chains, with consumers generally having several different food chains available to them.

Formation class (20) That portion of a biome that concerns the plant communities only, subdivided by size, shape, and structure of the dominant vegetation present. Six general classes are recognized: forest, woodland, grassland, shrubland, desert, and tundra.

Friction force (6) The effect of drag by the wind as it moves across a surface; may be operative through 500 m (1640 ft) of altitude. Surface friction slows the velocity of the wind and therefore reduces the effectiveness of the Coriolis force.

Front (8) The leading edge of an advancing air mass; a line of contrasting weather conditions.

Funnel cloud (8) The visible swirl extending from the bottom side of a cloud, which may or may not develop into a tornado. A tornado is a funnel cloud that has extended all the way to the ground.

Fusion (2) The process of forcibly joining, under extreme temperature and pressure, positively charged hydrogen and helium nuclei; occurs naturally in thermonuclear reactions within stars, such as our Sun.

Gelifluction (13) Refers to soil flow processes in periglacial environments; progressive, lateral movement. A type of *solifluction* under periglacial conditions.

Genetic classification (10) A type of classification that uses causative factors to determine climatic regions; for example, an analysis of the effect of interacting air masses on climate.

Geodesy (1) The science that determines Earth's shape and size through surveys, mathematical means, and remote sensing.

Geographic information system (GIS) (2) A computer-based data processing tool or methodology used for gathering, manipulating, and analyzing geographic information to produce a holistic, interactive analysis.

Geoid (1)   A word that describes Earth's shape; literally, the shape of Earth is Earth-shaped. A theoretical surface at sea level that extends through the continents; deviates from a perfect sphere.

Geologic cycle (11)   A general term characterizing the vast cycling at the Earth-atmosphere interface.

Geologic time scale (11)   A listing of eras, periods, and epochs that span Earth's history; reflects the relative relationship of various layers of rock strata and the absolute dates as determined by scientific methods such as radioactive isotopic dating.

Geomorphic cycle model (13)   A conceptual model proposed by William Morris Davis to characterize landscape development as a cyclic pattern from initial uplift to final erosional surface, or peneplain. He named the evolutionary phases youth, maturity, and old age.

Geomorphic threshold (13)   The threshold up to which landforms change before lurching to a new set of relationships, with rapid realignments of landscape materials and slopes.

Geomorphology (13)   The science that analyzes and describes the origin, evolution, form, classification, and spatial distribution of landforms.

Geostrophic winds (6)   Winds moving between pressure areas along paths that are parallel to the isobars. In the upper troposphere the pressure gradient force equals the Coriolis force so that the amount of deflection is proportional to air movement.

Geothermal energy (12)   The energy potential of boiling steam produced by subsurface magma in near contact with groundwater. Active examples of geothermal energy development include Iceland, New Zealand, Italy, and northern California.

Glacial ice (17)   A hardened form, very dense in comparison to normal snow or firn; under pressure within a glacier and capable of downhill movement.

Glacial surge (17)   The rapid, lurching, unexpected movement of a glacier. Causes are still debatable, but it appears that some surge is due to water pressure in the basal layers of the glacier.

Glaciation (7)   The formation of ice crystals in a rising cumulonimbus cloud as it enters a region of the atmosphere that is below freezing; not to be confused with the geomorphic term for a glacial phase.

Glacier (17)   A large mass of perennial ice resting on land or floating shelflike in the sea adjacent to the land; formed from the accumulation and recrystallization of snow, which then flows slowly under the pressure of its own weight and the pull of gravity.

Glacio-eustatic (7)   Changes in sea level in response to the amount of water stored on Earth as ice; the more water that is bound up in glaciers and ice sheets, the lower the sea level.

Graben (12)   Pairs or groups of faults that produce downward faulted blocks; characteristic of the basins of the interior western United States (see horst and Basin and Range Province).

Graded (14)   Refers to the mutual adjustment of the load carried by a stream and the related landscape through which the stream flows, forming a state of dynamic equilibrium between erosion, transported load, deposition, and the stream's capacity. In reality, an individual stream may have both graded and ungraded portions in its course.

Gradient (14)   The drop in elevation from a stream's headwaters to its mouth, ideally forming a concave slope. Characteristically the upper portion has a greater slope than the downstream channel.

Granite (11)   A coarse-grained (slow-cooling) intrusive igneous rock of 25% quartz and more than 50% potassium and sodium feldspars; characteristic of the continental crust.

Gravitational water (9)   That portion of surplus water that percolates downward from the capillary zone, pulled by gravity from the soil surface area to the groundwater zone.

Gravity (2)   The natural force exerted by the mass of an object; produced in an amount proportional to an object's mass.

Great circle (1)   Any circle of circumference drawn on a globe with its center coinciding with the center of the globe. An infinite number of great circles can be drawn, but only one parallel is a great circle—the equator.

Greenhouse effect (4)   The process whereby radiatively active gases absorb and delay the loss of heat to space, thus keeping the lower troposphere moderately warmed through the radiation and reradiation of infrared wavelengths. The approximate similarity between this process and that of a greenhouse explains the name.

Greenwich Mean Time (GMT) (1)   Former world standard time, now known as Coordinated Universal Time (UTC) (see Coordinated Universal Time).

Groundwater mining (9)   Pumping an aquifer beyond its flow and recharge capacities; an overutilization of a groundwater resource.

Gulf Stream (5)   A strong northward-moving warm current off the east coast of North America, which carries its water far into the North Atlantic.

Gutenberg discontinuity (11)   The transition zone between the outer core and the lower mantle.

Gyres (6)   The dominant circular ocean currents beneath subtropical high-pressure cells in both hemispheres; offset to the western margin of each ocean basin.

Habitat (19) That physical location in which an organism is biologically suited to live. Most species have specific habitat parameters, or limits.

Hadley cell (6) The vertical convection cell in each hemisphere that is generated by the low-pressure system of converging and ascending air along the equator, which then subsides and diverges in the subtropical regions.

Hail (8) A type of precipitation formed when a raindrop is repeatedly circulated above and below the freezing level in a cloud, with each cycle adding more ice to the hailstone until it becomes too heavy to stay aloft.

Hair hygrometer (7) An instrument for the measurement of relative humidity; based on the principle that human hair will change as much as 4% in length between 0 and 100% relative humidity.

Headlands (16) Protruding landforms that are extensions of the coast; generally composed of more resistant rocks. Incoming wave energy focuses on and bends around these headlands.

Herbivore (19) The primary consumer in a food chain, which eats plant material formed by a producer that has synthesized organic molecules.

Heterosphere (3) A zone of the atmosphere above the mesopause, 80 km (50 mi) in altitude; composed of rarified layers of oxygen atoms and nitrogen molecules; includes the ionosphere, which functions to absorb short wavelengths of radiation.

Heterotrophs (19) Organisms that depend on autotrophs for their source of carbon since they do not fix their own; also known as consumers.

Histosols (18) A soil order in the Soil Taxonomy classification that is formed from thick accumulations of organic matter, such as beds of former lakes, bogs, and layers of peat.

Homosphere (3) A zone of the atmosphere from the surface up to 80 km (50 mi), composed of an even mixture of gases including nitrogen, oxygen, argon, carbon dioxide, and trace gases.

Horse latitudes (6) The calms of Cancer and the calms of Capricorn, zones of windless, hot calms beneath the subtropical high-pressure cells in both hemispheres; windless areas of great difficulty for sailing ships throughout history.

Horst (12) Upward-faulted blocks produced by pairs or groups of faults; characterized by the mountain ranges in the interior of the western United States (see graben and Basin and Range Province).

Hot spots (11) Individual points of upwelling material originating in the asthenosphere, not necessarily associated with spreading centers, although Iceland is astride a mid-ocean ridge. They tend to remain fixed relative to migrating plates; some 50 to 100 are identified worldwide; exemplified by Yellowstone National Park and Hawaii.

Human-Earth relationships (1) One of the oldest themes of geography (the human-land tradition); includes settlement patterns, resource utilization and exploitation, hazard perception and planning, and the impact of environmental modification and artificial landscape creation.

Humidity (7) Water vapor content of the air. The capacity of the air to hold water vapor is mostly a function of temperature.

Humus (18) A mixture of organic debris in the soil, worked by consumers and decomposers in the humification process; characteristically formed from plant and animal litter laid down at the surface.

Hurricane (8) A tropical cyclone that is fully organized and intensified in inward-spiraling rainbands; ranges from 160 to 960 km (100 to 600 mi) in diameter, with wind speeds in excess of 119 kmph (65 knots, or 74 mph); a label used specifically in the Atlantic and eastern Pacific.

Hydration (13) A process involving water, although not involving any chemical change. This is actually a physical weathering process; water is added to a mineral, which initiates swelling and stress within the rock, mechanically forcing grains apart as the constituents expand.

Hydraulic action (14) The erosive work accomplished by the turbulence of water; causes a squeezing and releasing action in joints in bedrock; capable of prying and lifting rocks.

Hydrograph (14) A graph of stream discharge (in cms or cfs) over a period of time (minutes, hours, days, years) at a specific place on a stream. The relationship between stream discharge and precipitation input is illustrated on the graph.

Hydrologic cycle (7) A simplified model of the flow of water and water vapor from place to place as energy activates system operations. Water flows through the atmosphere, across the land where it is also stored as ice, and within groundwater.

Hydrolysis (13) A process whereby minerals chemically combine with water; a decomposition process that causes silicate minerals in rocks to break down and become altered.

Hydrosphere (1) An abiotic open system that includes all of Earth's waters.

Hydrothermal activity (11) Igneous activity that involves heated or superheated water, such as occurs with geothermal activity. Deposition can occur through chemical reactions between minerals and heated groundwater.

Hygroscopic nuclei (7) Condensation nuclei that are particularly attracted to moisture, such as certain salts derived from ocean spray.

Hygroscopic water (9) That portion of soil moisture that is so tightly bound to each soil particle that it is unavailable to plant roots; the water, along with some bound capillary water, that is left in the soil after the wilting point is reached.

Ice age (17) A cold episode, with accompanying alpine and continental ice accumulations, that has repeated roughly every 200 to 300 million years since the late Precambrian era (1.25 billion years ago); includes the most recent episode of cold during the Pleistocene Ice Age, which began 1.65 million years ago.

Icebergs (17) Floating ice created by calving ice or large pieces breaking off and floating adrift; a hazard to shipping because the bulk of the ice is submerged and can be irregular in form.

Ice cap (17) A dome-shaped form of a glacier, less extensive than an ice sheet ($<50,000$ km^2), although it buries mountain peaks and the local landscape. Several representative ice caps are located in Iceland.

Ice-crystal process (7) The feeding of ice crystals on supercooled cloud droplets in the process of raindrop formation. Since the saturation vapor pressure near an ice surface is lower than that near a water surface, supercooled water droplets evaporate near ice crystals, which then absorb the vapor.

Ice field (17) The least extensive form of a glacier, with mountain ridges and peaks visible above the ice.

Icelandic low (6) See subpolar low-pressure cell.

Ice sheet (17) An enormous continuous continental glacier. The bulk of glacial ice on Earth covers Antarctica and Greenland in two ice sheets.

Ice shelves (17) The most advanced margins of an ice sheet entering a bay along a coast. The farthest edge is marked by a sharp cliff up to 30 m (98 ft) in height above the sea; tabular islands are formed when sections of the shelf break off and move out to sea.

Igneous rocks (11) Rocks that solidify and crystallize from a hot molten state.

Illuviation (18) A depositional soil process as differentiated from eluviation; usually in the B horizon, where accumulations of clays, aluminum, iron, and some humus occur.

Inceptisols (18) An order in the Soil Taxonomy classification; weakly developed soils that are inherently infertile; usually young soils that are weakly developed, although they are more developed than Entisols.

Industrial smog (3) Air pollution associated with coal-burning industries; sulfur oxides, particulates, carbon dioxide, and exotics may comprise it.

Infiltration (7) Water access to subsurface regions of soil moisture storage through penetration of the soil surface.

Influx (7) The input of material from physical, chemical, and biological processes that causes the salinity of the ocean to increase.

Insolation (2) Solar radiation that is intercepted by Earth.

Interception (7) Delays the fall of precipitation toward Earth's surface; caused by vegetation or other ground-cover.

Internal drainage (9) Outflow of rivers through evaporation or subsurface gravitational flow in regions where rivers do not flow into the ocean. Portions of Africa, Asia, Australia, and the western United States have such drainage.

International Date Line (1) The 180° meridian; an important corollary to the prime meridian on the opposite side of the planet; established by the treaty of 1884 to mark the place where each day officially begins.

Intertropical convergence zone (ITCZ) (6) See equatorial low pressure trough.

Ionosphere (3) A layer in the atmosphere above 80 km (50 mi) where gamma, X-ray, and some ultraviolet radiation is absorbed and converted into infrared, or heat energy, and where the solar wind stimulates the auroras.

Isobar (6) An isoline connecting all points of equal pressure.

Isobaric surface (6) Isobars that mark the paths of upper air winds; an elevated surface on which all points have the same pressure, based on elevations above sea level.

Isostasy (11) A state of equilibrium formed by the interplay between portions of the lithosphere and the asthenosphere; the crust depresses with weight and recovers with the melting of the ice or removal of the load, in an uplift known as isostatic rebound.

Isotherm (5) An isoline connecting all points of equal temperature.

Jet stream (6) The most prominent movement in upper-level westerly wind flows; irregular, concentrated sinuous bands of wind, traveling at 300 kmph (190 mph), occur in both hemispheres.

Kame (17) A small hill of poorly sorted sand and gravel that accumulates in crevasses or in ice-caused indentations in the surface; may also be deposited by water in ice-contact deposits.

Karst topography (13) Distinctive topography formed in a region of limestone with poorly developed surface drainage. Weathering creates solution features that appear pitted and bumpy; originally named after the Krš Plateau of Yugoslavia.

Katabatic winds (6) Air drainage from elevated regions, flowing as gravity winds. Layers of air at the surface cool, become denser, and flow downslope, known worldwide by many local names.

Kinetic energy (3) The energy of motion in a body; derived from the vibration of the body's own movement and stated as temperature.

Lacustrine deposit (17) Deposit associated with lake level fluctuations; for example, benches or terraces marking former shorelines.

Lagoon (16) A portion of coastal seawater that is virtually cut off from the ocean by a baymouth bar or barrier beach; also, the water surrounded and enclosed by an atoll.

Landfall (8) The location along a coast where a storm moves onshore.

Landslide (13) A sudden rapid downslope movement of a cohesive mass of regolith and/or bedrock in a variety of mass-movement forms under the influence of gravity.

Land-water heating differences (5) The differences in the way land and water heat, as a result of contrasts in transmission, evaporation, mixing, and specific heat capacities. Land surfaces heat and cool faster than water and are characterized as having aspects of continentality, whereas water is regarded as producing a marine influence.

Langley (2) A gram calorie per square centimeter; a unit of measure of solar radiation.

Large marine ecosystem (LME) (20) A proposed designation of regions of the oceanic ecosystem that have unique organisms, floor topography, currents, areas of nutrient-rich upwelling, or significant predation, either human or natural.

Latent heat (7) Heat energy that is absorbed in the phase change of water and is stored in one of the three states—ice, water, or water vapor; includes the latent heat of melting, freezing, vaporization, evaporation, and condensation.

Laterization (18) A pedogenic process operating in well-drained soils that are found in warm and humid regions; typical of Oxisols. High precipitation values leach soluble minerals and soil constituents, leaving behind iron and aluminum oxides and usually reddish or yellowish colors.

Latitude (1) The angular distance measured north or south of the equator from a point at the center of Earth. A line connecting all points of the same latitudinal angle is called a parallel.

Lava (12) Magma that issues from volcanic activity onto the surface; the extrusive rock that results when magma solidifies.

Leeward slope (8) The slope opposite the windward slope; usually drier in the rain shadow.

Life zone (19) An altitudinal zonation of plants and animals that form distinctive communities. Each life zone possesses its own temperature and precipitation relationships.

Lightning (8) Flashes of light, caused by tens of millions of volts of electrical charge igniting the air to temperatures of 15,000° to 30,000°C.

Limestone (11) The most common chemical sedimentary rock (nonclastic); lithified calcium carbonate ($CaCO_3$), which is very susceptible to chemical weathering.

Limiting factor (19) The physical or chemical factor that most inhibits (either through lack or excess) biotic processes.

Lithification (11) The compaction, cementation, and hardening of sediments into sedimentary rock.

Lithosphere (1) Earth's crust and that portion of the upper mantle directly below the crust that extends down to 70 km (45 mi). Some use this term to refer to the entire Earth.

Littoral current (16) A current that forms parallel to a beach as waves arrive at an angle to the shore; generated in the surf zone by wave action, transporting large amounts of sand and sediment (also known as longshore current).

Littoral drift (16) Sand, gravels, and shells moved along the shore by the littoral current (also known as beach drift).

Littoral zone (16) A specific coastal environment; that region between the high water line during a storm and a depth at which storm waves are unable to move sea-floor sediments.

Living atmosphere (2) Earth's atmosphere between 0.6 and 3.3 billion years ago; thought by scientists to be principally composed of carbon dioxide, water vapor, and nitrogen gas (3.0 billion years ago); beginnings of photosynthesis.

Loam (18) A mixture of sand, silt, and clay in almost equal proportions, with no one texture dominant.

Location (1) A basic theme of geography dealing with the absolute and relative position of people, places, and things on Earth's surface; tells "where" something is.

Loess (15) Large quantities of fine-grained clays and silts left as glacial outwash deposits; subsequently blown by the wind great distances and redeposited as a generally unstratified, homogeneous blanket of material covering existing landscapes; in China loess is derived from desert lands.

Longitude (1) The angular distance measured east or west of a prime meridian from a point at the center of Earth. A line connecting all points of the same longitude is called a meridian.

Lysimeter (9) A device for measuring potential and actual evapotranspiration; isolates a portion of a field so that the moisture moving through the plot is measured.

Magma (11) Molten rock from beneath the surface of Earth; fluid, gaseous, under tremendous pressure, and either intruded into country rock or extruded onto the surface as lava.

Magnetic reversal (11) An important aspect of Earth's magnetic field. With an uneven regularity the field fades to zero, then phases back to full strength, but with the magnetic poles reversed. Reversals have occurred nine times during the past 4 million years.

Magnetosphere (2) Earth's magnetic force field, which is generated by dynamolike motions within the planet's outer core; works to deflect the solar wind.

Mangrove swamps (16) Wetland ecosystems between 30° N or S and the equator tend to form a distinctive community of mangrove plants, owing to the lack of freezing temperatures so that mangrove seedlings can survive.

Mantle (11) An area within the planet representing about 80% of Earth's total volume, with densities increasing with depth and averaging 4.5 g per cm^3; occurs above the core and below the crust; is rich in iron and magnesium oxides and silicates.

Map projection (1) The reduction of a spherical globe onto a flat surface in some orderly and systematic realignment of the latitude and longitude grid.

Marine (5) Descriptive of areas that are dominated by the moderating effects of the ocean and that exhibit a smaller minimum and maximum temperature range than continental locations (see land-water heating differences).

Mass movement (13) All unit movements of materials propelled and controlled by gravity; can range from dry to wet, slow to fast, small to large, and free-falling to gradual or intermittent; sometimes used interchangeably with mass wasting, although the latter term should be applied principally to gravitational movement of nonunified material.

Meandering (14) The sinuous, curving pattern common to graded streams, with the outer portion of each curve subjected to the greatest erosive action and the inner portion receiving sediment deposits. Former meanders remain on the landscape as meander scars.

Mean sea level (16) A calculated value based on the average of tidal levels recorded hourly at a given site over a long period of time, which must be at least a full lunar tidal cycle of 19 years.

Mediterranean shrubland (20) A major biome dominated by shrub formations; occurs in Mediterranean dry-summer climates and is characterized by sclerophyllous scrub and short, stunted, tough forests.

Mercury barometer (3) A device that measures air pressure with a column of mercury in a tube that is inserted in a vessel of mercury. The surrounding air exerts pressure on the mercury in the vessel (see air pressure).

Meridian (1) See longitude.

Mesocyclone (8) A large rotating circulation initiated within a parent cumulonimbus cloud at the mid-troposphere level; generally produces heavy rain, large hail, blustery winds, and lightning; may lead to tornado activity.

Mesosphere (3) The upper region of the homosphere from 50 to 80 km (30 to 50 mi) above the ground; designated by temperature criteria and very low pressures, ranging from 0.1 to 0.001 mb.

Metamorphic rock (11) Preexisting rock, both igneous and sedimentary, that goes through profound physical and chemical changes under increased pressure and temperature. Constituent mineral structures may exhibit foliated or nonfoliated textures.

Methane (5) A radiatively active gas that participates in the greenhouse effect; derived from the organic processes of burning, digesting, and rotting in the presence of oxygen; CH_4.

Microclimatology (4) The study of climates at or near Earth's surface.

Midlatitude geographic zone (1) A generalized geographic zone, extending roughly from 35° to 55° N and S latitudes, although the limits of this zone are somewhat arbitrary, given the distribution of land and water and differences in climate; home of a majority of the world's population.

Midlatitude broadleaf and mixed forest (20) A biome in moist continental climates in areas of warm-to-hot summers and cool-to-cold winters; includes several distinct communities. Relatively lush stands of broadleaf forests trend northward into needleleaf evergreen stands.

Midlatitude grasslands (20) The major biome most modified by human activity; so named because of the predominance of grasslike plants, although deciduous broadleafs appear along streams and other limited sites; location of the world's breadbaskets of grain and livestock production.

Mid-ocean ridge (11) Submarine mountain ranges that extend more than 65,000 km (40,000 mi) worldwide and average more than 1000 km (620 mi) in width; centered along sea-floor spreading centers as the direct result of upwelling areas of heat in the upper mantle.

Milky Way Galaxy (2) A flattened, disk-shaped mass estimated to contain up to 400 billion stars; is about 100,000 light years in diameter and includes our Solar System.

Mineral (11) An element or combination of elements that form an inorganic natural compound; described by a specific formula and qualities of a specific nature.

Mixing zone (7) The surface layer of the ocean, representing only 2% of the oceanic volume; Sun-warmed and wind-driven; mixes rapidly compared to lower depths.

Model (1) A simplified version of a system, representing an idealized part of the real world.

Modern atmosphere (2) The fourth distinct atmosphere on Earth, having evolved from the living atmosphere about 0.6 billion years ago; lower layers composed of nitrogen, oxygen, argon, carbon dioxide, and trace gases; an additive mixture of gases.

Mohorovičić discontinuity (11) The boundary between the crust and the rest of the lithospheric upper mantle; named for the Yugoslavian seismologist Mohorovičić; a zone of sharp material and density contrasts.

Moist adiabatic rate (MAR) (8) The rate at which a parcel of saturated air cools in ascent; a rate of 6C° per 1000 m (3.3F° per 1000 ft). This rate may vary, with moisture content and temperature, from 4C° to 10C° per 1000 m (2F° to 6F° per 1000 ft) (see adiabatic).

Moisture droplets (7) Initial composition of clouds. Each droplet measures approximately 0.002 cm (0.0008 in.) in diameter and is invisible to the human eye.

Mollisols (18) A soil order in the Soil Taxonomy classification that has a mollic epipedon and a humus-rich organic content high in alkalinity; some of the world's most significant agricultural soils.

Monsoon (6) From the Arabic word *mausim,* meaning "season"; refers to an annual cycle of dryness and wetness, with seasonally shifting winds produced by changing atmospheric pressure systems; affects India, Southeast Asia, Indonesia, northern Australia, and portions of Africa.

Montane forests (20) Needleleaf forests associated with mountain elevations (see northern needleleaf forest).

Moraine (17) Marginal glacial deposits of unsorted and unstratified material, producing a variety of depositional landforms.

Movement (1) A major theme in geography involving migration, communication, and the interaction of people and processes across space.

Natural levees (14) Long, low ridges that occur on either side of a river in a developed floodplain; sedimentary (coarse gravels and sand) depositional by-products of river-flooding episodes.

Neap tides (16) Unusually low tidal ranges produced during the first and third quarters of the Moon, with an offsetting pull from the Sun; as opposed to spring tides.

Nebula (2) A cloud of dust and gas in space.

Negative feedback (1) A feedback loop that tends to slow or dampen response in a system; promotes self-regulation in a system; far more common than positive feedback in living systems.

Net primary productivity (19) The net photosynthesis (photosynthesis minus respiration) for a given community; considers all growth and all reduction factors that affect the amount of useful chemical energy (biomass) fixed in an ecosystem.

Net radiation (NET R) (4) The net all-wave radiation available at Earth's surface; the final outcome of the radiation balance process between incoming and outgoing shortwave and longwave energy.

Niche (19) The basic function, or occupation, of a life-form within a given community; the way an organism obtains its food, air, and water.

Nickpoint (knickpoint) (14) The point at which the longitudinal profile of a stream is abruptly broken by a change in gradient; for example, a waterfall, rapids, or cascade.

Nimbostratus (7) A rain-producing, dark, grayish, stratiform cloud characterized by gentle drizzles.

Nitrogen dioxide (3) A reddish-brown choking gas produced in high-temperature combustion engines; can be damaging to human respiratory tracts and to plants; participates in photochemical reactions and acid deposition.

Noctilucent clouds (3) Rare and unusual shining bands of ice, which may be seen glowing at high latitudes long after sunset; formed within the mesosphere, where cosmic and meteoric dust acts as nuclei for the formation of ice crystals.

Normal lapse rate (3) The average rate of temperature decrease with increasing altitude in the lower atmosphere; an average value of 6.4C° per km, or 1000 m (3.5F° per 1000 ft); a rate that exists principally during daytime conditions.

Northern needleleaf forest (20) Forests of pine, spruce, fir, and larch, stretching from the east coast of Canada westward to Alaska and continuing from Siberia westward across the entire extent of the Soviet Union to the European Plain; called the taiga (a Russian word) or the boreal forest; principally in the D climates. Includes *montane forests*.

Nuclear winter hypothesis (5) An increase in Earth's albedo and upper atmospheric heating resulting in surface cooling; associated with the detonation of a relatively small number of nuclear warheads within the biosphere. Now encompasses a whole range of ecological, biological, and climatic impacts.

Nuée ardente (12) Incandescent ash and super-hot gases that often accompany explosive eruptions of composite volcanoes; also known as a glowing cloud.

Occluded front (8) In a cyclonic circulation, the over-running of a surface warm front by a cold front and the subsequent lifting of the warm air wedge off the ground. Initial precipitation is moderate to heavy.

Ocean basins (11) The physical containers for Earth's oceans.

Oceanic trenches (11) The deepest single features of Earth's crust; associated with subduction zones. The deepest is the Mariana Trench near Guam, which descends to 11,033 m (36,198 ft).

Omnivore (19) A consumer that feeds on both producers (plants) and consumers (meat)—a role occupied by humans, among other animals.

Open system (1) A system with inputs and outputs crossing back and forth between the system and the surrounding environment. Earth is an open system in terms of energy.

Orogenesis (12) The process of mountain building that occurs when large-scale compression leads to deformation and uplift of the crust; literally the birth of mountains.

Orographic lifting (8) The uplift of migrating air masses in response to the physical presence of a mountain, which acts as a topographic barrier. The lifted air cools adiabatically as it moves upslope; may form clouds and produce increased precipitation.

Outgassing (7) The release of trapped gases from rocks, forced out through cracks, fissures, and volcanoes from within Earth; the terrestrial source of Earth's water.

Outlet glacier (17) A type of alpine glacier that flows outward from a continental glacier and appears as a valley glacier in form.

Overland flow (9) Surplus water that flows across a surface toward stream channels. Together with precipitation and subsurface flows, it constitutes the total runoff from an area.

Oxbow lake (14) A lake that was formerly part of the channel of a meandering stream; isolated when a stream eroded its outer bank forming a cutoff through the neck of a looping meander.

Oxisols (18) In the Soil Taxonomy classification a soil order of tropical soils that are old, deeply developed, and lacking in horizons wherever well-drained; heavily weathered, low in cation exchange capacity, and low in fertility.

Ozone layer (3) See ozonosphere.

Ozonosphere (3) A layer of ozone (O_3) occupying the full extent of the stratosphere (20 to 50 km or 12 to 30 mi above the surface); the region of the atmosphere where ultraviolet wavelengths are principally absorbed and converted into heat.

Pacific high (6) Also known as the Hawaiian high; a high-pressure cell that dominates the Pacific in July, retreating southward in the Northern Hemisphere in January.

Paleoclimatology (10) The science that studies the climates of past ages.

Paleoseismology (12) The science that studies earthquake frequency or lack of frequency along specific faults; location, and magnitude of past earthquakes.

PAN (3) See peroxyacetyl nitrates.

Pangaea (11) The supercontinent formed by the collision of all continental masses approximately 225 million years ago; named by Wegener in 1912 in his continental drift theory.

Parallel (1) See latitude.

Parent material (13) The unconsolidated material, from both organic and mineral sources, that is the basis of soil development.

Patterned ground (13) Areas in the periglacial environment where freezing and thawing of the ground create polygonal forms of arranged rocks at the surface that may coalesce into stone polygon nets.

Pediment (15) The area created when a mountain front retreats through erosional action in an arid landscape; an inclined bedrock plain layered with a thin veneer, or coating, of alluvium; different from the depositional surface of a bajada.

Pedogenic regimes (18) Specific soil-forming processes keyed to climatic regimes: laterization, calcification, salinization, and podzolization, among others.

Pedon (18) A soil profile extending from the surface to the lowest extent of plant roots or to the depth where regolith or bedrock is encountered; imagined as a hexagonal column encompassing from 1 to 10 square meters in surface area forming the basic sampling unit.

Peneplain (14) A final old erosional surface in Davis's geomorphic cycle; a very controversial concept and one that is unproven.

Percolation (7)   The process by which water permeates through to the subsurface environment; vertical water movement through soil or porous rock.

Periglacial regions (13)   High-latitude and high-altitude regions that occupy about 20% of the land surface adjacent to glaciers; typically found in cold Dfc, Dfd, and E climates and influenced greatly by frost action.

Perihelion (2)   That point in Earth's elliptical orbit about the Sun where Earth is closest to the Sun; occurs on January 3 at 147,255,000 km (91,500,000 mi); variable over a 100,000-year cycle.

Permafrost (13)   In periglacial environments a condition whereby land below the surface may be permanently frozen; subjected to daily and seasonal cycles of freeze-thaw action down to a meter or more in depth.

Permeability (9)   The ability of water to flow through soil or rock; a function of the texture and structure of the medium.

Peroxyacetyl nitrates (PAN) (3)   A whole family of pollutants formed from photochemical reactions involving nitric oxide (NO) and hydrocarbons (HC). They produce no known human health effects but are particularly damaging to plants.

Phase change (7)   The change in phase, or state, between ice, water, and water vapor; involves the absorption or release of latent heat.

Photochemical smog (3)   Air pollution produced by the interaction of ultraviolet light, nitrogen dioxide, and hydrocarbons; produces ozone and PAN through a series of complex photochemical reactions. Automobiles are the major source of the contributive gases.

Photogrammetry (2)   An aspect of remote sensing that uses aerial photographs and orbital imagery to improve the accuracy of surface maps, usually at less expense and with greater ease than is possible with on-site surveys.

Photoperiod (19)   The duration of daylight experienced at a given location.

Photosynthesis (19)   The joining of carbon dioxide and water in plants, under the influence of certain wavelengths of visible light; releases oxygen and produces energy-rich organic material (sugars and starches).

Physical weathering (13)   The breaking and disintegrating of rock without any chemical alteration; sometimes referred to as mechanical or fragmentation weathering.

Piezometric surface (9)   A pressure level in an aquifer, creating an imaginary line above the surface in some instances and defined by the level to which water rises in wells; caused by the fact that the water in a confined aquifer is under the pressure of its own weight.

Pioneer community (19)   The initial community in an area; usually found on new surfaces or those that have been stripped of life—for example, surfaces created by mass movements of land, or land exposed by a retreating glacier, or cooled lava flows, or land disturbed by human activities.

Place (1)   A major theme in geography focused on the tangible and intangible characteristics that make each location unique.

Plane of the ecliptic (2)   A plane intersecting all the points of Earth's orbit.

Planetesimal hypothesis (2)   The idea that early protoplanets formed from the condensing masses of a nebular cloud of dust, gas, and icy comets; a formation process now being observed in other parts of the galaxy.

Planimetric map (1)   A basic map showing the horizontal position of boundaries, land-use activities, and political, economic, and social outlines.

Plate tectonics (11)   The conceptual model that encompasses continental drift, sea-floor spreading, and related aspects of crustal movement; widely accepted as the foundation of crustal tectonic processes.

Playa (15)   An area of salt crust left behind on a desert floor usually in the middle of a bolson or valley; intermittently wet and dry, as runoff evaporates in a desert climate.

Plinthite (18)   An ironstone hardpan structure formed in Oxisol subsurfaces or surface horizons that are exposed to repeated wetting and drying sequences; quarried in some areas for building materials.

Pluton (11)   A mass of intrusive igneous rock that has cooled slowly in the crust; forms in any size or shape. The largest partially exposed pluton is a batholith.

Pluvial (17)   Referring to a period in the past, principally during the Pleistocene, when there was greater moisture availability or efficiency; resulted in increased lake levels in arid and seasonally arid regions.

Podzolization (18)   A pedogenic process in cool, moist climates; forms a highly leached soil with strong surface acidity because of humus from acid-rich trees.

Point bar (14)   The inner portion of a meander, which receives sediment fill.

Polar easterlies (6)   Variable weak, cold, and dry winds moving away from the polar region; a result of anticyclonic circulation.

Polar front (6)   A significant zone of contrast between cold and warm air masses; roughly situated between 50° and 60° latitude.

Polar high-pressure cell (6)   A weak, anticyclonic, thermally produced pressure system positioned roughly above each pole; is the region of the lowest temperatures on Earth.

Polar jet stream (6) A strong wind current located at the tropopause along the polar front; occurs between 7600 m and 10,700 m (24,900 and 35,100 ft) in altitude and meanders between 30° and 70° N latitude (see jet stream).

Polypedon (18) The identifiable soil in an area, with distinctive characteristics differentiating it from surrounding polypedons forming the basic mapping unit; composed of many pedons.

Porosity (9) The total volume of available pore space in soil; expressed as a ratio and percentage; a result of the texture and structure of the soil.

Positive feedback (1) Feedback that amplifies or encourages responses in the system.

Potential evapotranspiration (9) The amount of moisture that would evaporate and transpire if adequate moisture were available; the amount lost under optimum moisture conditions—that is, the moisture demand.

Precipitation (9) Rain, snow, sleet, and hail—the moisture supply.

Pressure gradient force (6) Causes air to move from areas of higher barometric pressure to areas of lower barometric pressure due to pressure differences.

Prime meridian (1) An arbitrary meridian designated as 0° longitude; the point from which longitudes are measured east or west; agreed on by the nations of the world in an 1884 treaty.

Primordial atmosphere (2) Earth's first atmosphere, derived from the formation nebula. Lighter gases escaped early, with water vapor, hydrogen cyanide, ammonia, methane, and others predominant.

Process (1) A set of actions and changes that occur in some special order; central to modern geographic synthesis.

Province (15) A large region that shares several homogeneous traits, making it appropriate to assign a regional name.

Pyroclastics (12) Explosively ejected rock pieces and broken fragments launched aerially by a volcanic eruption.

Radiation fog (7) Formed by radiative cooling of the surface, especially on clear nights in areas of moist ground; occurs when the air layer directly above the surface is chilled to the dew-point temperature, thereby producing saturated conditions.

Radiatively active gases (5) Gases in the atmosphere such as carbon dioxide, methane, and water vapor, that absorb and radiate infrared wavelengths.

Rain gauge (9) A standardized device that catches and measures rainfall.

Rain shadow (8) The area on the leeward slopes of a mountain range; in the shadow of the mountains, where precipitation receipt is greatly reduced compared to windward slopes.

Reef (16) An organically derived formation of sedimentary coral rock, occurring in several distinctive shapes along coastal regions.

Reflection (4) The portion of arriving energy that returns directly back into space without being converted into heat or performing any work (see albedo).

Refraction (4) The bending effect that occurs when insolation enters the atmosphere or another medium; the same process by which a crystal, or prism, disperses the component colors of the light passing through it.

Region (1) A geographic theme that focuses on areas that display unity and internal homogeneity of traits; includes the study of how a region forms, evolves, and interrelates with other regions.

Regolith (13) Partially weathered rock overlying bedrock, whether residual or transported.

Relative humidity (7) A term that reflects the ratio of water vapor actually in the air (content) compared to the maximum water vapor the air is able to hold (capacity) at that temperature; expressed as a percentage.

Relief (11) Elevation differences in a local landscape as an expression of the unevenness, height, and slope variation.

Remote sensing (2) Information acquired from a distance, without physical contact with the subject; for example, photography or orbital imagery.

Respiration (19) The process other than photosynthesis by which plants derive energy for their operations; essentially, the reverse of the photosynthetic process; releases carbon dioxide, water, and heat into the environment.

Revolution (4) The annual orbital movement of Earth about the Sun; determines year length and the length of seasons.

Rhumb line (1) A line of constant compass direction, or constant bearing, which crosses all meridians at the same angle. A portion of a great circle.

Richter scale (12) An open-ended, logarithmic scale that estimates earthquake magnitude; designed by Charles Richter in 1935.

Ring of fire (12) See circum-Pacific belt.

Ripples (15) The smallest features that form in response to the distance traveled (eolian or water agents) by individual saltating grains.

Roche moutonnée (17) An asymmetrical hill of exposed bedrock; displays a gently sloping upstream side that has been smoothed and polished by a glacier and an abrupt, steep downstream side.

Rock cycle (11) A model representing the interrelationships among the three rock-forming processes: igneous, sedimentary, and metamorphic.

Rockfall (13) Free-falling movement of debris from a cliff or steep slope, generally falling straight down or bounding downslope.

Rossby waves (6) Undulating horizontal motions in the upper-air westerly circulation at middle and high latitudes.

Rotation (4) The turning of Earth on its axis; averages 24 hours in duration; determines day-night relationships.

Salinity (7) The concentration of natural elements and compounds dissolved in solution, as solutes; measured by weight in parts per thousand (‰) in seawater.

Salinization (18) A pedogenic regime that results from high potential evapotranspiration rates in deserts and semiarid regions. Soil water is drawn to surface horizons, and dissolved salts are deposited as the water evaporates.

Saltation (14) The transport of sand grains (usually larger than 0.2 mm, or 0.008 in.) by stream or wind, bouncing the grains along the ground in asymmetrical paths.

Salt marshes (16) Wetland ecosystems characteristic of latitudes poleward of the 30th parallel.

Sand sea (15) An extensive area of sand and dunes; characteristic of Earth's erg deserts.

Saturated (7) Air that is holding all the water vapor that it can hold at a given temperature.

Scale (1) The ratio of the distance on a map to that in the real world; expressed as a representative fraction, graphic scale, or written scale.

Scarification (13) A term that is generally applied to human-induced mass movements of Earth materials, such as large-scale open-pit mining and strip mining.

Scattering (4) Deflection and redirection of insolation by atmospheric gases, dust, ice, and water vapor. The amount of scattering is a function of the wavelengths involved and the scattering media; the shorter the wavelength, the greater the scattering.

Scientific method (2) An approach that uses applied common sense in an organized and objective manner; based on observation, generalization, formulation of a hypothesis, and ultimately the development of a theory.

Sea-floor spreading (11) As proposed by Hess and Dietz, the mechanism driving the movement of the continents; associated with upwelling flows of magma along the worldwide system of mid-ocean ridges.

Secondary air mass (8) An air mass that becomes modified in terms of temperature and moisture characteristics with distance and time away from its source region.

Sediment (13) Fine-grained mineral matter that is transported by air, water, or ice.

Sedimentary rocks (11) One of three basic rock types; formed from the compaction, cementation, and hardening of sediments derived from former rocks; principally quartz, feldspar, and clay minerals.

Seismic gaps (12) Areas along fault zones that form a gap in earthquake occurrence and may therefore possess the greatest accumulated strain (see paleoseismology).

Seismic waves (11) The shock waves sent through the planet by an earthquake or underground nuclear test. Transmission varies according to temperature and the density of various layers within the planet.

Seismograph (12) A device that measures seismic waves of energy transmitted throughout Earth's interior.

Sensible heat (3) Heat that can be measured with a thermometer; a measure of the concentration of kinetic energy from molecular motion.

Shield volcano (12) A symmetrical mountain landform built from effusive eruptions; gently sloped, gradually rising from the surrounding landscape to a summit crater; typical of the Hawaiian Islands.

Sinkholes (13) Nearly circular depressions created by the weathering of limestone karst landscapes; also known as dolines; may collapse through the roof of an underground space.

Sling psychrometer (7) A device for the measurement of relative humidity using two thermometers—a dry bulb and a wet bulb—mounted side-by-side.

Small circle (1) Circles on a globe's surface that do not share Earth's center; for example, all parallels other than the equator.

Snowline (17) A temporary line marking the elevation where winter snowfall persists throughout the summer; seasonally, the lowest elevation covered by snow during the summer.

Soil (18) A dynamic natural body comprised of fine materials covering Earth's surface in which plants grow, and is composed of both mineral and organic matter.

Soil colloids (18) Tiny clay and organic particles in soil; provide chemically active sites for mineral ion adsorption.

Soil horizon (18) The various layers exposed in a pedon; roughly parallel to the surface and identified as O, A, B, and C.

Soil moisture storage (9) The retention of moisture within soil; represents a savings account that can accept deposits (soil moisture recharge) or experiences withdrawals (soil moisture utilization) as conditions change.

Soil Taxonomy (18) A soil classification system based on observable soil properties actually seen in the field; a final version published in 1975 by the U.S. Soil Conservation Service.

Solar constant (2) The amount of insolation intercepted by Earth on a plane surface perpendicular to the Sun's rays when Earth is at its average distance from the Sun; a value of 1.968 calories per cm² per minute; averaged over the entire globe at the thermopause; or, 1370 watts per m².

Solar wind (2) Clouds of ionized (charged) gases emitted by the Sun and traveling in all directions from the Sun's surface. Effects on Earth include auroras, disturbance of radio signals, and possible influences on weather.

Solifluction (13) Gentle downslope movement of a saturated surface material (soil and regolith), in various climatic regimes, where temperatures are above freezing. Sometimes used in a generic sense to include *gelifluction* in periglacial environments.

Solum (18) A true soil profile in the pedon; ideally, a combination of A and B horizons.

Solution (13) The dissolved load of a stream; especially the chemical solution derived from limestone, dolomite, or salt deposits.

Spalling (13) A form of weathering associated with fracturing or fragmentation of rock; often related to exfoliation processes.

Spatial analysis (1) The examination of phenomena in the context of spatial interactions, patterns, and variations over area and/or space; the key integrative approach of geography.

Specific heat (5) The increase of temperature in a material when energy is absorbed. Because water requires far more heat to raise its temperature than does a comparable volume of land, water is said to have a higher specific heat.

Specific humidity (7) The mass of water vapor (in grams) per unit mass of air (in kilograms) at any specified temperature. The maximum mass of water vapor that a kilogram of air can hold at any specified temperature is termed its maximum specific humidity.

Specific yield (9) The groundwater that is available for utilization at the surface; expressed as a percentage of the total volume of a saturated aquifer.

Speed of light (2) Specifically, 299,792 kilometers per second (186,282 miles per second), or more than 9.4 trillion kilometers per year (5.9 trillion miles per year)—a distance known as a light year.

Spheroidal weathering (13) A chemical weathering process in which the sharp edges and corners of boulders and rocks are weathered in thin plates that create a rounded, spheroidal form.

Spit (16) A depositional form that develops when transported sand is deposited in long ridges that partially cross the mouth of a bay.

Spodosols (18) A soil order in the Soil Taxonomy classification that occurs in northern coniferous forests; best developed in cold, moist, forested climates of Dfb, Dfc, Dwc, Dwd; lacks humus and clay in the A horizons, with high acidity associated with podzolization processes.

Spring tides (16) The highest tidal ranges, which occur when the Moon and the Sun are in conjunction (at new Moon) or in opposition (at full Moon) stages.

Squall line (8) A zone slightly ahead of a fast-advancing cold front, where wind patterns are rapidly changing and blustery and precipitation is strong.

Stability (8) The condition of a parcel of air relative to whether it remains where it is or changes its initial position. The parcel is stable if it resists displacement upwards, unstable if it continues to rise; relates to adiabatic and lapse rate processes.

Staff gauge (14) A simple device to measure stream height; uses a pole stuck vertically into a stream, with various elevations marked on its face.

Steady-state equilibrium (1) The condition that occurs in a system when rates of input and output are equal and the amounts of energy and stored matter are nearly constant around a stable average.

Stem flow (7) Intercepted precipitation that drains across leaves and down plant stems to the ground.

Stilling well (14) A device positioned on a stream bank and used to measure stream height.

Stomata (19) Small openings on the undersides of leaves, through which water and gases pass.

Storm surges (8) Large quantities of seawater pushed inland by the strong winds associated with a tropical cyclone.

Stratiform (7) A class of clouds that are developed horizontally or appear layered.

Stratigraphy (11) An analysis of the sequence, spacing, and spatial distribution of rock strata; offers clues as to the ages and origins of rocks and rock structures.

Stratocumulus (7) A lumpy, grayish, low-level cloud, patchy with sky visible, sometimes present at the end of the day.

Stratosphere (3) That portion of the homosphere that ranges from 20 to 50 km (12.5 to 30 mi) above Earth's surface, with temperatures ranging from −57°C (−70°F) at the tropopause to 0°C (32°F) at the stratopause. The functional ozonosphere is within the stratosphere.

Stratus (7) A stratiform cloud generally below 2000 m (6500 ft).

Stream orders (14) A way of expressing the linear aspects of channels in a drainage basin; a convenient hierarchy of stream size.

Strike-slip fault (12) Horizontal movement along a fault-line, that is, movement in the same direction as the fault; also known as a transcurrent fault. Such movement is described as right-lateral or left-lateral, depending on the relative motion observed.

Subantarctic geographic zone (1) Related to the geographic region between 55° and 66.5° S latitude (Antarctic Circle).

Subarctic geographic zone (1) Related to the geographic region between 55° and 66.5° N latitude (Arctic Circle).

Subduction zone (11) An area where two plates of crust collide and the denser oceanic crust dives beneath the less dense continental plate, forming deep oceanic trenches and seismically active regions.

Sublimation (7) A process in which ice evaporates directly to water vapor or water vapor freezes to ice (deposition).

Subpolar low-pressure cell (6) A region of low pressure centered approximately at 60° latitude in the North Atlantic near Iceland and in the North Pacific near the Aleutians, as well as in the Southern Hemisphere. Air flow is cyclonic, counterclockwise circulation that weakens in summer and strengthens in winter; a source area of cyclonic activity affecting Europe and North America.

Subsolar point (2) The only point receiving perpendicular insolation at a given moment—the Sun directly overhead.

Subsurface diagnostic horizon (18) A soil horizon that originates below the epipedon at varying depths; may be part of the A and B horizons; important in soil description as part of the Soil Taxonomy.

Subtropical geographic zone (1) Pertaining to a geographic region that ranges from 23.5° to 35° latitude in both hemispheres; dominated by Earth's arid and semiarid lands.

Subtropical high-pressure cells (6) Dynamic high-pressure areas covering roughly the region from 20° to 35° N and S latitudes; responsible for the hot, dry areas of Earth's arid and semiarid deserts.

Subtropical jet stream (6) An upper tropospheric wind current that occurs in the subtropical latitudes, near the boundary between the tropics and the midlatitudes (see jet stream).

Sulfur dioxide (3) A colorless gas detected by its pungent odor; produced by the combustion of fossil fuels that contain sulfur as an impurity; can react in the atmosphere to form sulfuric acid, a component of acid deposition.

Summer solstice (4) The time when the Sun's declination is at the Tropic of Cancer, at 23.5° N latitude; June 20–21 each year.

Sunrise (4) That moment when the disk of the Sun first appears above the horizon.

Sunset (4) That moment when the disk of the Sun totally disappears.

Sunspots (2) Magnetic disturbances on the surface of the Sun; occurring in an average 11-year cycle; related flares, prominences, and outbreaks produce surges in solar wind.

Supercooled droplets (7) Found in clouds at temperatures ranging from −15° to −35°C. Such droplets form ice crystals in colder clouds and provide an essential mechanism of raindrop formation.

Surface creep (15) A form of eolian transport that involves particles too large for saltation; a process whereby individual grains are impacted by moving grains, sliding, and rolling.

Surplus (9) The amount of moisture that exceeds potential evapotranspiration; moisture oversupply when soil moisture storage is at field capacity.

Suspended load (14) Fine particles held in suspension in a stream. The finest particles are not deposited until the stream velocity nears zero.

Swells (16) Regular patterns of smooth, rounded waves in open water; can range from small ripples to very large waves.

Syncline (12) A trough in folded strata, with beds that slope toward the axis of the downfold.

System (1) Any ordered, interrelated set of materials or items existing separate from the environment, or within a boundary; energy transformations and energy and matter storage and retrieval occur within a system.

Taiga (20) See northern needleleaf forest.

Tectonic processes (11) Driven by internal energy from within Earth; refers to large-scale movement and deformation of the crust.

Temperate rain forest (20) A major biome of lush forests at middle and high latitudes; occurs along narrow margins of the Pacific Northwest in North America, among other locations; a mixture of broadleaf and needleleaf trees and thick undergrowth, including the tallest trees in the world.

Temperature inversion (3) A reversal of the normal decrease of temperature with increasing altitude; can occur anywhere from ground level up to several thousand meters.

Tephra (12) Pulverized rock and clastic materials ejected violently during a volcanic eruption—all pyroclastics from a volcano.

Terranes (12) Migrating crustal pieces, dragged about by processes of mantle convection and plate tectonics. Displaced terranes are distinct in their history, composition, and structure from the continents that accept them.

Terrestrial ecosystem (20) A self-regulating association of plants and animals and their abiotic environment characterized by specific plant formations; usually named for the predominant vegetation and known as a biome when large and stable.

Thermal equator (5) A line on an isothermal map that connects all points of highest mean temperature.

Thermocline transition zone (7) A region in the depths of an ocean or a lake where a strong vertical temperature gradient occurs; lacks the motion of the surface.

Thermopause (2) A zone approximately 480 km (300 mi) in altitude that serves conceptually as the top of the atmosphere; an altitude used for the determination of the solar constant.

Thermosphere (3) A region of the heterosphere extending from 80 to 480 km (50 to 300 mi) in altitude; contains the functional ionosphere layer.

Throughfall (7) Precipitation that falls directly to the ground; combines with precipitation that drips from vegetation to the ground.

Thunder (8) The violent expansion of suddenly heated air, created by lightning discharges, sending out shock waves as an audible sonic bang.

Tidal bore (16) An incoming tidal current that is constricted as it enters a river channel or coastal inlet forming a measurable wave.

Tides (16) Patterns of daily oscillations in sea level produced by astronomical relationships among the Sun, the Moon, and Earth; experienced in varying degrees around the world.

Till plains (17) Large, relatively flat plains composed of unsorted glacial deposits behind a terminal or end moraine. Low-rolling relief and unclear drainage patterns are characteristic.

Tilted-fault block (12) A tilted landscape produced by a normal fault on one side of a range; for example, the Sierra Nevada and the Grand Tetons.

Tombolo (16) A landform created when coastal sand deposits connect the shoreline with an offshore island outcrop or sea stack.

Topographic map (1) A map that portrays physical relief through the use of elevation contour lines that connect all points at the same elevation above or below a vertical datum, such as mean sea level.

Topography (11) The undulations and configurations that give Earth's surface its texture; the heights and depths of local relief including both natural and human-made features.

Tornado (8) An intense, destructive cyclonic rotation, developed in response to extremely low pressure; associated with mesocyclone formation.

Total runoff (9) Surplus water that flows across a surface toward stream channels; formed by overland flow, combined with precipitation and subsurface flows into those channels.

Traction (14) A type of sediment transport that drags coarser materials along the bed of a stream.

Tradewinds (6) Northeast and southeast winds that converge in the equatorial low pressure trough, forming the intertropical convergence zone.

Transform fault (11) An elongated zone along which faulting occurs between mid-ocean ridges; produces a relative horizontal motion with no new crust formed or destroyed; strike-slip motion, either left or right lateral.

Transmission (4) The passage of shortwave and longwave energy through the atmosphere or water.

Transparency (5) The quality of light able to shine through a medium. Light penetrates the oceans to an average depth of 60 m (200 ft).

Transpiration (9) The movement of water vapor out through the pores in leaves from water drawn by their roots from the soil moisture storage.

Transport (14) The actual movement of weathered and eroded materials by air, water, and ice.

Tropical geographic zone (1) The geographic region ranging between 10° and 23.5° N or S latitude.

Tropical cyclone (8) A cyclonic circulation originating in the tropics, with winds between 30 and 64 knots (39 to 73 mph); characterized by closed isobars, circular organization, and heavy rains (see hurricane and typhoon).

Tropical savanna (20) A major biome containing large expanses of grassland interrupted by trees and shrubs; a transitional area between the humid rain forests and tropical seasonal forests and the drier, semiarid tropical steppes and deserts.

Tropical seasonal forest and scrub (20) A variable biome on the margins of the rain forests, occupying regions of lower and more erratic rainfall; the site of transitional communities between the rain forests and tropical grasslands.

Tropic of Cancer (4) The northernmost point of the Sun's declination during the year; 23.5° N latitude.

Tropic of Capricorn (4) The southernmost point of the Sun's declination during the year; 23.5° S latitude.

Troposphere (3) The home of the biosphere; the lowest layer of the homosphere, containing approximately 90% of the total mass of the atmosphere; extends up to the tropopause, marked by a temperature of −57°C (−70°F); occurring at an average altitude of 18 km (11 mi) at the equator, 13 km (8 mi) in the middle latitudes, and at lower altitudes near the poles.

True shape (1) A map property showing the correct configuration of coastlines; a useful trait of conformality for navigational and aeronautical maps, although areal relationships are distorted.

Tsunami (16) A seismic sea wave, traveling at high speeds across the ocean, formed by sudden and sharp motions in the seafloor, such as sea-floor earthquakes, submarine landslides, or eruptions from undersea volcanoes.

Typhoon (8) A tropical cyclone in excess of 65 knots (74 mph) that occurs in the western Pacific; same as a hurricane except for location.

Ultisols (18) A soil order in the Soil Taxonomy classification, featuring highly weathered forest soils, principally in the Cfa climatic classification. Increased weathering and exposure can degenerate an Alfisol into the reddish color and texture of these more humid to tropical soils. Fertility is quickly exhausted when Ultisols are cultivated.

Unconfined aquifer (9) The zone of saturation in water-bearing rock strata, with no impermeable overburden and recharge generally accomplished by water percolating down from above.

Uniformitarianism (11) An assumption that physical processes active in the environment today are operating at the same pace and intensity that has characterized them throughout geologic time; proposed by Hutton and Lyell.

Unit hydrograph (14) A model hydrograph characterizing a drainage basin that receives a unit depth of effective rainfall over a specific unit of time, uniformly spread over the drainage basin.

Upslope fog (7) Forms when moist air is forced to higher elevations along a hill or mountain and is thus cooled; for example, along the Appalachians and the eastern slopes of the Rockies.

Upwelling current (6) Ocean currents that cause surface waters to be swept away from the coast by surface divergence or offshore winds. Cool, deep waters, which are generally nutrient rich, rise to replace the vacating water.

Urban heat islands (4) Urban microclimates, which are warmer on the average than areas in the surrounding countryside because of various surface characteristics; other climatic and weather anomalies are also related to urbanization.

Valley fog (7) The settling of cooler, more dense air in low-lying areas, produces saturated conditions and fog.

Valley glacier (17) A type of alpine, or mountain, glacier within the confines of a valley; can range from 100 m (328 ft) to 100 km (62 mi) in length.

Vapor pressure (7) That portion of total air pressure that results from water vapor molecules; expressed in millibars (mb). At a given temperature the maximum capacity of the air is termed its saturation vapor pressure.

Vascular plants (19) Those plants which have developed vascular tissues for conducting fluid and nutrients; almost 250,000 species on Earth. The most highly developed plants.

Ventifacts (15) The individual pieces and pebbles etched and smoothed by eolian erosion—abrasion by wind-blown particles.

Vernal equinox (4) The time when the Sun's declination crosses the equatorial parallel and all places on Earth experience days and nights of equal length; around March 20–21 each year. The Sun rises at the North Pole and sets at the South Pole.

Vertisols (18) A soil order in the Soil Taxonomy classification that features expandable clay soils; composed of more than 30% swelling clays such as montmorillonite; occurs in regions that experience highly variable soil moisture balances through the seasons.

Volcano (12) A landform at the end of a conduit or pipe which rises from below the crust and vents to the surface. Magma rises and collects in a magma chamber deep below, resulting in eruptions that are effusive or explosive forming the mountain landform.

Wadi (15) An Arabic term for an intermittently dry streambed or wash.

Warm front (8) The leading edge of an advancing warm air mass, which is unable to push cooler, passive air out of the way; tends to push the cooler, underlying air into a wedge shape; noted on weather maps with a series of rounded knobs heading in the direction of the frontal movement.

Wash (15) An intermittently dry streambed that fills with torrents of water after rare precipitation events in arid lands.

Water detention (9) The retaining of excess surplus water in surface puddles and indentations.

Watershed (14) The catchment area of a drainage basin; delimited by divides.

Waterspout (8) An elongated, funnel-shaped circulation formed when a tornado takes place over water.

Water table (9) The upper limit of groundwater; that contact point between zones of saturation and aeration in an unconfined aquifer.

Wave-cut platform (16) A flat, or gently sloping, table-like bedrock surface that develops in the tidal zone, where wave action cuts a bench that extends from the cliff base out into the sea.

Wave cyclone (8) An organized area of low pressure, with converging and ascending air flow producing an interaction of air masses; migrates along storm tracks. Such lows or depressions form the dominant weather pattern in the middle and higher latitudes of both hemispheres.

Wavelength (2) Measurement of waveform; the actual distance between the crests of successive waves. The number of waves passing a fixed point in one second is called the frequency of the wavelength.

Wave refraction (16) A process that concentrates energy on headlands and disperses it in coves and bays; a coastal straightening process.

Waves (16) Undulations of ocean water produced by the conversion of solar energy to wave energy; produced in a generating region or a stormy area of the sea.

Weathering (13) The related processes by which surface and subsurface rock disintegrate, or dissolve, or are otherwise broken down. Rocks at or near Earth's surface are exposed to physical, organic, and chemical weathering processes.

West Antarctic ice sheet (10) A vast grounded ice mass partially held back by the Ross ice shelf in Antarctica.

Westerlies (6) The predominant wind flow pattern from the subtropics to high latitudes in both hemispheres.

Western intensification (6) The piling up of ocean water along the western margin of each ocean basin, to a height of about 15 cm (6 in.); produced by the tradewinds that drive the oceans westward in a concentrated channel.

Wetlands (16) Unique, narrow, vegetated strips occupying many coastal areas and estuaries worldwide; highly productive ecosystems with an ability to trap organic matter, nutrients, and sediment.

Wilting point (9) That point in the soil moisture balance when only hygroscopic water and some bound capillary water remains. Plants wilt and eventually die after prolonged stress from a lack of available water.

Wind (6) The horizontal movement of air relative to Earth's surface; produced essentially by air pressure differences from place to place; also influenced by the Coriolis force and surface friction.

Wind chill factor (5) An indication of the enhanced rate at which body heat is lost to the air. As wind speed increases, heat loss from the skin also increases.

Wind vane (6) A device to determine wind direction.

Windward slope (8) The intercepting slope of a mountain range, relative to the prevailing winds (see orographic lifting).

Winter solstice (4) That time when the Sun's declination is at the Tropic of Capricorn, at 23.5° S latitude, December 21–22 each year.

Withdrawal (9) The removal of water by society from the supply, after which it is used for various purposes and then is returned to the water supply.

Wrangellia terranes (12) One of many terranes that became cemented together to form present day North America and the Wrangell Mountains; originating in a far locale approximately 10,000 km (6200 mi) away; a former volcanic island arc and associated marine sediments.

Yardangs (15) Streamlined rock structures, formed by deflation and abrasion; appear elongated and aligned with the most effective wind direction.

Yazoo tributary (14) A small tributary stream draining alongside a floodplain; blocked from joining the main river by its natural levees and elevated stream channel (see backswamp).

Zone of aeration (9) A groundwater zone above the water table, which may or may not hold water in pore spaces.

Zone of saturation (9) A groundwater zone below the water table, in which all pore spaces are filled with water.

Index

Producers, 561
Psychrometer, sling, 183
Ptolemy, 10, 13
Pumice, 308
Pyroclastics, 355
Pythagoras, 8

Qattara Depression, 427
Quadrangle maps, 26

Radiation
  daily curve, 97–98
    temperature lag, 98
  diffuse, 94
  electromagnetic spectrum,
    45–46, 51
  global imbalance, 97
  global net, 96
  longwave, 99
  map of, 92
  NET R, 99–104
  shortwave, 99
Radiating bodies, 46
Radiation balance
  Earth's surface, 97–106
  Earth-atmosphere, 95–96
  in the troposphere, 91–95
  urban, 104–106
Radiation fog, 190
Radiatively active gases, 128,
  284
Rain, 203, 214, 230, 257
Rain forests
  canopy, 586
  deforestation of, 587–88
  equatorial and tropical, 586
  extent of, 587
  light in, 586
Rain gauge, 231
Rain shadow, 203
Raindrop formation, 184–85
Reefs, 465
Reflection, 93
Refraction, 94
Reg desert, 427
Regolith, 370, 516
Relative humidity, 180–83, 184
Relief, orders of, 296–98
Remote sensing, 48–51
  GMS (Japan), 50, 220
  GOES-East, 51, 212
  Landsat, 50
  SLAR, 50, 219
Reservoirs, 419–20
  Lake Powell, 443
  local base level, 367
  Teton Dam break, 420
Respiration, 552

Reverse fault, 336
Revolution, Earth's, 85
Rhine River valley, 340, 431
Rhumb lines, 25, 26
Rhyolite, 308
Ria, 465
Richat dome, 336
Richter magnitude scale, 344
Ridge and valley province, 341,
  343
Rift valleys, 338
Ring of fire. See Circum-Pacific
  belt
Ripples, in sand, 429
River deltas, 412
Rivers. See also Streams
  management of, 416–19
  world's largest, 248
Robinson projection, 26–27
Roche moutonnée, 495
Rock, 305
Rock cycle, 305, 306–11
Rockfall, 383
Rocky Mountains, 203, 297, 341
Rossby waves, 151
Rotation, 85, 86
Runoff, distribution of, 246
Russell fjord, 487–88

Sagan, Carl, 608, 613
Sahara, 124, 281–82, 429–30,
  599–600
Sahel, 282, 599
Salic soil horizons, 531
Salinity, 170–72
Salinization, in soils, 531
Salt marshes, 467
Saltation
  eolian, 428
  fluvial, 401
Saltwater intrusion, 473
San Andreas fault system, 323,
  339, 347–48
Sand, 310, 517
Sand beaches, 471
Sand sea, 429
Sandy Hook, 460
San Francisco earthquakes, 347
San Juan River, Utah, 400
Santa Ana winds, 156
Saturated air, 180
Sauer, Carl, 571
Savanna, 593
Scale, of maps, 21–22, 28
Scarification, 314, 388
Scattering, 94
Scientific method, 38–39
Sclerophyllous shrub, 596

Seafloor
  age of, 317–18
  spreading, 294, 315–17
Sea level
  and global warming, 287,
    459
  changes, 458–59
  measurement of, 453
Seamounts, 323, 326
SEASAT, 139
Seasons, 84, 90–91
  contributing factors, 85–88
  key dates, 89
Sediment, 371
Sedimentary cycle, 558
Sedimentary processes, 308–
  10
Seismic gaps, 348
Seismic waves, 299
Seismograph, 344
Seismology. See Earthquakes
Selenium soil contamination,
  532–33
Selva, 586
Sensible heat, 61, 117
Seventh approximation, soil
  classification, 523
Shearing, 335
Sheetwash, 522
Shields, continental, 333
Shield volcano, 354
Ship Rock, New Mexico, 308
Shoreline, 452
Short-grass prairie, 598
SI. See International System of
  Units
Siberia, 122–24, 275, 594
Siberian high pressure, 150
Siberian lowland, 297
Sial, 303
Sierra Nevada, 305, 340–41,
  497
Sierra Nevadan orogeny, 341
Silicate mineral family, 305
Sills, 308
Silt, as soil component, 517
Sima, 303
Sinkholes, 378
Sinusoidal projection, 25
Sleet, 230, 257, 281
Slipface, 430–31
Slope dynamics, 369
Slopes, 381
Slump, 384
Small circles, 20–21
Smog, 72–74
Snow, 176, 185, 230, 257, 267
Snowline, 112, 480

Snowy Mountains Scheme,
  Australia, 239
Sodium chloride, evaporites,
  310
Soil characteristics, 514–22
Soil chemistry, 520–21
Soil classification, 522–42
  Marbut system, 523
  pedogenic regimes, 524
  seventh approximation, 523
  soil series, 523
  Soil Taxonomy, 523–24
Soil creep, 385
Soil fertility, 514, 520
Soil formation, 521–22
Soil horizons, 514–16
  albic, 538
  argillic, 536
  diagnostic, 524
  epipedon, 524
  salic, 531
  spodic, 538
  subsurface diagnostic, 524
Soil management, 521–22
Soil moisture, 234–35
  regimes, 519
  storage, 230, 233–36
Soil orders
  Alfisols, 535–36
  Aridisols, 529–34
  Entisols, 539
  Histosols, 541
  Inceptisols, 539
  Mollisols, 534–35
  of Soil Taxonomy, 524–42
  Oxisols, 524–29
  Spodosols, 537–39
  Ultisols, 536–37
  Vertisols, 540
  world map of, 528
Soil pollution, 532–33
Soil profile, 514
Soil properties, 516–20
Soil separates, 517
Soil series, 523
Soil solution, 520
Soil Taxonomy, 523–24, 526
Solar constant, 46–47, 51
Solar day, 86
Solar energy, 41, 45–46
  and photosynthesis, 5, 551
  collection, 101–103
  distribution, 48
  in ecosystems, 555
  variation, 37, 47
Solar fusion, 41
Solar radiation, 41–48
Solar Max, 47